Applied Mathematical Sciences
Volume 148

Springer
New York
Berlin
Heidelberg
Barcelona
Hong Kong
London
Milan
Paris
Singapore
Tokyo

Applied Mathematical Sciences

(continued following index)

Roger Peyret

Spectral Methods for Incompressible Viscous Flow

With 61 Illustrations

Springer

Roger Peyret
Université de Nice-Sophia Antipolis
Laboratoire J.A. Dieudonné
Parc Valrose
Nice 06000
France

Editors

S.S. Antman
Department of Mathematics
and
Institute for Physical Science
  and Technology
University of Maryland
College Park, MD 20742-4015
USA

J.E. Marsden
Control and Dynamical
  Systems, 107-81
California Institute of
  Technology
Pasadena, CA 91125
USA

L. Sirovich
Division of Applied
  Mathematics
Brown University
Providence, RI 02912
USA

Mathematics Subject Classification (2000): 76-08, 65-01, 35Pxx

---

Library of Congress Cataloging-in-Publication Data
Peyret Roger.
   Spectral methods for incompressible viscous flow / Roger Peyret.
     p. cm. — (Applied mathematical sciences ; 148)
   Includes bibliographical references and index.
   ISBN 978-1-4419-2913-6
    1. Viscous flow. 2. Navier-Stokes equations—Numerical solutions. 3. Spectral theory
(Mathematics) I. Title. II. Applied mathematical sciences (Springer-Verlag New York
Inc.) ; v. 148.
   QA1 .A647 vol. 148
   [QA929]
   510 s—dc21
[532˙.0533]   2001054909

Printed on acid-free paper.

Production managed by Allan Abrams; manufacturing supervised by Erica Bresler.
Photocomposed copy prepared from the author's LATEX files.

Printed in the United States of America.

9 8 7 6 5 4 3 2 1

Springer-Verlag New York Berlin Heidelberg
A member of BertelsmannSpringer Science+Business Media GmbH

Preface

The objective of this book is to provide a comprehensive discussion of Fourier and Chebyshev spectral methods for the computation of incompressible viscous flows, based on the Navier-Stokes equations.

For reasons of efficiency and confidence in the numerical results, the researchers and practitioners involved in computational fluid dynamics must be able to master the numerical methods they use. Therefore, in writing this book, beyond the description of the algorithms, I have also tried to provide information on the mathematical and computational, as well as implementational characteristics of the methods.

The book contains three parts. The first is intended to present the fundamentals of the Fourier and Chebyshev methods for the solution of differential problems. The second part is entirely devoted to the solution of the Navier-Stokes equations, considered in vorticity-streamfunction and velocity-pressure formulations. The third part is concerned with the solution of stiff and singular problems, and with the domain decomposition method.

In writing this book, I owe a great debt to the joint contribution of several people to whom I wish to express my deep gratitude. First, I express my friendly thanks to L. Sirovich, editor of the series "Applied Mathematical Sciences," who suggested that I write the book. Many thanks are also addressed to my colleagues and former students who contributed to the completion of the book in various ways. I am happy to thank P. Bontoux, O. Botella, J.A. Désidéri, U. Ehrenstein, M.Y. Forestier, J. Fröhlich, S. Gauthier, H. Guillard, P. Le Quéré, J.-M. Malé, C. Olivier, R. Pasquetti, J.-P. Pulicani, I. Raspo, C. Sabbah, E. Serre, B. Shizgal, and H. Viviand.

Lastly, I wish to acknowledge the courtesy of the International Center of Mechanical Sciences in Udine (Italy) for allowing me to use, in Chapter 4, a part of the material published in the Lectures Notes *Advanced Turbulent Flow Computations*.

Nice, France, May 2001   Roger Peyret

Contents

III   Special topics   295

Introduction

The prototype of spectral methods for the solution of differential problems is the well-known Fourier method which consists of representing the solution as a truncated series expansion, the unknowns being the expansion coefficients. The Fourier basis is appropriate for periodic problems. For nonperiodic problems, the Chebyshev or Legendre polynomial bases are commonly used, but other basis functions could be considered according to the problem under consideration.

Thus, the foundations of spectral methods are not recent. For many years, before the appearance of computers, the theoretical studies in mathematical physics, and especially in fluid mechanics, made an extensive use of series expansions. Thus, these studies have led to the important development of "special functions" which constituted a large part of the mathematical analysis during the nineteenth century and the first half of the twentieth century.

However, the series expansion methods have shown severe limitations due to the difficulty in calculating the truncated sums with a large number of terms or treating nonlinear problems. These limitations were still present in the earlier stage of computational fluid dynamics when computer performances were not sufficient for an effective use of series expansions. Consequently, this led to the disfavour of the series expansion techniques for the benefit of discrete numerical methods as finite-difference or finite-element methods. But, the relative low accuracy of these discretization methods is an obstacle to the accurate representation of complex flows with a very fine structure as encountered in many problems in fluid mechanics. Hence, in the 1970s, we saw a revival of the Fourier method which was applied to the direct numerical simulation of turbulence (Orszag and Patterson, 1972). The success of the Fourier method for turbulence computations was due to two facts : the increasing power of computers and the efficiency of the fast Fourier transform (FFT) algorithm for the calculation of the sums. These improvements were fundamental for a fast calculation of the nonlinear terms through the "pseudospectral" technique : the differentiations are made in the spectral space (i.e., the space of the expansion coefficients) and the products are performed in the physical space (i.e., the space of the values of the unknowns at some discrete points), the connection between both spaces being done through the FFT algorithm.

One of the main properties of the Fourier series is its fast rate of convergence, which is exponential for infinitely differentiable functions. This constitutes an obvious advantage over finite-difference or finite-element methods. On the other hand, the drawback of the Fourier method is its inability to handle nonperiodic problems, because of the presence of the Gibbs oscillations due to the nonuniform convergence of the Fourier series at the extremities of the domain of definition. Therefore, it was necessary to have recourse to other basis functions more appropriate to the solution of nonperiodic problems. The straightforward choice is constituted by the Chebyshev polynomials. Indeed, the Chebyshev polynomial series expansion possesses some properties of the Fourier series (rate of convergence and possibility of using the FFT algorithm) while being exempt from the Gibbs oscillations at the boundaries. Another possibility is constituted by the Legendre polynomials, which present some interesting mathematical properties. However, the fact that a fast summation algorithm does not exist for the Legendre series may be a drawback in the case of large resolution. In this book, only the Fourier and Chebyshev methods will be addressed, but the solution methods described for the Chebyshev approximation apply to the Legendre polynomial approximation as well.

The objective of the book is to present spectral methods for the solution of the Navier-Stokes equations governing the motions of incompressible viscous fluids. It contains three parts devoted, respectively, to the presentation of basic spectral methods, to the solution of the Navier-Stokes equations, and to the discussion of special topics.

The solution methods are based on algorithms involving a time-discretization scheme and, at each time-cycle, the solution of elementary differential problems, essentially Poisson, Helmholtz, or steady-state advection-diffusion equations. Consequently, the global efficiency of a spectral code requires, at one and the same time, an efficient time-discretization scheme and efficient elementary spectral solvers. These questions are addressed in the four chapters of the first part of the book.

The first chapter constitutes an introduction to the general principles on which the spectral methods are based. Chapter 2 is devoted to the discussion of the Fourier method. After a brief presentation of the properties of the Fourier series, the application to the solution of a differential problem is described. The case of equations with nonconstant coefficients is considered. The pseudospectral technique and the associated problem of aliasing are discussed in detail.

The Chebyshev method is discussed in Chapter 3. To begin with, the basic properties of the Chebyshev polynomials and the Chebyshev series are presented. Then we consider the solution of a general boundary value problem for second-order elliptic equations. The methods of approximation to such problems may be either of Galerkin-type (tau method) where the unknowns are the expansion coefficients, or of collocation-type where the unknowns are generally the values of the solution at selected points. Both types of methods lead to an algebraic system whose solution, by means of direct or iterative methods, is discussed in detail, especially in the multidimensional case. In connection with the treatment of general boundary conditions, we introduce the basic principles of the influence (or capacitance) matrix method which will be used in several circumstances throughout the book. This method, adapted to linear problems, is based on the superposition of elementary solutions. The arbitrary parameters involved in the linear combination of these elementary solutions are solutions of an algebraic system obtained by prescribing some conditions specific to the problem. The advantage of the influence matrix method is that the major part of the calculations can be done in a preprocessing stage performed before the start of the time-integration.

Chapter 4, which ends the first part, is devoted to the time-discretization of time-dependent equations, considering, more especially, the advection-diffusion equation. One-step and multistep methods are considered in various situations : explicit, semi-implicit, and fully implicit schemes. The stability of the time-discretization schemes is analyzed in the Fourier and Chebyshev cases.

The second part of the book is entirely devoted to the Navier-Stokes equations. The aim is twofold : first, to discuss the fundamental principles on which the solution methods are based and, second, to describe, analyze, and assess some of the methods which have been shown to be efficient for the computation of complex flows. We endeavour to illustrate most of the discussed methods by presenting some examples of application. These examples were chosen either to illuminate some characteristics of the method or to give the reader an idea about the kind of flows which can be computed with spectral methods.

In Chapter 5, which opens the second part, we briefly recall the Navier-Stokes equations for incompressible flows within the two classical formulations : velocity-pressure and vorticity-streamfunction. We also present the Boussinesq approximation which is of common use for convection problems.

Chapter 6 considers the vorticity-streamfunction equations. We successively discuss the three cases : periodic flow in both spatial directions, periodic flow in one direction only, and nonperiodic flow. The classical difficulty associated with the solution of these equations is connected to the boundary conditions at a no-slip wall : there are two conditions for the streamfunction and none for the vorticity. The influence matrix method, described in this chapter, permits a rigorous treatment of the no-slip conditions. In our opinion, this method constitutes the best way to solve the vorticity-streamfunction equations because it is the most straightforward, especially when it is associated with the collocation technique.

The Navier-Stokes equations in velocity-pressure variables are considered in Chapter 7. As in Chapter 6, three cases are examined : three-dimensional fully periodic flow, two-dimensional flow with one periodic direction, and two-dimensional flow without periodicity. Except in the fully periodic case, the usual difficulty associated with the Navier-Stokes equations in primitive variables is to calculate the pressure field ensuring the solenoidal character of the velocity field. Two types of approaches are distinguished according to the nature of the elliptic problem to be solved at each time-cycle of the integration process.

The first type of discretization scheme leads to a Stokes problem whose solution may be obtained either by the Uzawa method or by the influence matrix method. The Uzawa method is iterative except in the one-dimensional case where the Uzawa operator, acting on the pressure, can easily be inverted. The influence matrix method, which is based on the solution of a Poisson equation for the pressure, is direct and presents an interest to give an exact solenoidal polynomial velocity field. This method is analyzed carefully and its implementation is described in detail. Also, the possible existence of spurious pressure modes is discussed.

The second type of time-discretization method belongs to the class of fractional step methods and is based on a projection principle. In a first step, a provisional velocity field is calculated from a Helmholtz equation. This field is not solenoidal and must be corrected, in the second step, by means of an appropriate pressure gradient field. This projection step corresponds to a Darcy (or div-grad) problem. Various possibilities for solving the Darcy problem are discussed. Among them, two methods will be described more deeply. These methods differ in the polynomial spaces in which the velocity components and pressure are defined. The so-called $I\!\!P_N - I\!\!P_{N-2}$ approximation consists of approximating the pressure with a polynomial of degree $N-2$, whereas the velocity components are approximated with polynomials of degree N. The advantage of this approximation is twofold : it is exempt from spurious pressure modes, and the calculation of the pressure, through a Uzawa-type equation, does not require boundary conditions. In the second method ($I\!\!P_N - I\!\!P_N$ method), the velocity components and pressure are defined in the same space of polynomials of degree N. Then the pressure is defined as the solution of a Poisson equation with a Neumann boundary condition whose expression depends on the algorithm itself. The resulting pressure field is exempt from spurious modes, but a delicate point concerns the proper definition of the algorithm leading to a Neumann condition ensuring the accuracy of the solution. We describe a method in which a predicted pressure field is calculated so that the Neumann condition for the final pressure is accurate in time and space and, consequently, provides an accurate solution to the Stokes problem.

In the third part of the book we discuss two special topics : the solution of stiff and singular problems in Chapter 8 and the domain decomposition method in Chapter 9.

The solution of a stiff problem is regular but exhibits large variations in one region (or more) of small extent compared to the size of the computational domain. The accurate Chebyshev representation of such a solution requires special treatment. One possible way is to consider a domain decomposition method so that the rapid variation region is included within a subdomain in which the polynomial degree is taken large enough. Another approach consists of considering a suitable coordinate transform which may preferably be determined automatically by a self-adaptive procedure. The case of singular problems is more delicate because the rate of convergence of the Chebyshev approximation is then only algebraic and, when the singularity is strong, the polynomial representation is subject to oscillations. The domain decomposition method can also be used for singular problems, the purpose being to shift the singularity to a corner of a subdomain. Another way, which is more efficient, is constituted by the subtraction method: the most singular part of the solution is subtracted so that the solution of the resulting problem is less singular, if not completely regular. These various approaches will be discussed in Chapter 8.

Finally, the domain decomposition method is addressed in the last chapter. The reasons for using a domain decomposition technique in spectral methods are various. One of them is connected to the shape of the computational domain. Chebyshev methods are adapted to rectangular domains. For more complicated domains recourse to a multidomain method is unavoidable. Also, in the case of an elongated rectangular domain, the partition into square subdomains is of interest for several computational reasons. Diverse types of domain decomposition methods (influence matrix, iterative, and spectral-element methods) will be described for model differential problems. The application of the influence matrix method to the multidomain solution of the Navier-Stokes equations will be discussed and illustrated by some typical examples.

This book is not intended to cover the entire field of spectral methods from theoretical aspects to the extensive application in all fields of fluid mechanics. We restricted ourselves to the case of incompressible viscous flows governed by the Navier-Stokes equations with the purpose of transmiting our experience in a comprehensive way. Obviously, some of the fundamental topics addressed here have already been presented in other books. We have tried, as much as possible, to consider these topics under a new light and to discuss the more recent developments. Among the books devoted to spectral methods, we mention the pioneering book by Gottlieb and Orszag (1977) which opened the road to so many fruitful developments in the field.

Also, the book by Canuto *et al.* (1988) constitutes a comprehensive work on the subject including important theoretical results. More specialized are the books by Mercier (1989), Boyd (1989), Funaro (1992, 1997), Bernardi and Maday (1992) and by Karniadakis and Sherwin (1999). Concerning the general properties of Fourier series and Chebyshev polynomials we refer to classical books on mathematical and numerical analysis. Also, more specialized books on Fourier series (Körner, 1988 ; Brigham, 1988) and Chebyshev polynomials (Fox and Parker, 1968 ; Rivlin, 1974) may be of great help to the reader.

Part I

Basic spectral methods

1
Fundamentals of spectral methods

The aim of this introductory chapter is to present, in a general way, the spectral methods in their various formulations : Galerkin, tau, and collocation. By using the notion of residual, it will be shown how spectral approximation can be defined for the representation of a given function as well as for the solution of a differential problem. These questions will be addressed in detail in the following two chapters devoted, respectively, to Fourier and Chebyshev methods.

1.1 Generalities on the method of weighted residuals

Spectral methods belong to the general class of weighted residuals methods (Finlayson, 1972) for which approximations are defined in terms of a truncated series expansion, such that some quantity (error or residual) which should be exactly zero is forced to be zero only in an approximate sense. This is done through the scalar product

$$(u, v)_w = \int_\alpha^\beta u\, v\, w\, dx \,, \tag{1.1}$$

where $u(x)$ and $v(x)$ are two functions defined on $[\alpha, \beta]$ and $w(x)$ is some given weight function.

Let us consider the expansion of a function $u(x)$ in the truncated series

$$u_N(x) = \sum_{k=0}^{N} \hat{u}_k \, \varphi_k(x), \quad \alpha \leq x \leq \beta, \tag{1.2}$$

where the trial (or basis) functions $\varphi_k(x)$ are given and the expansion coefficients must be determined. In spectral methods the chosen trial functions are orthogonal : trigonometric functions e^{ikx} for periodic problems, Chebyshev $T_k(x)$ or Legendre $L_k(x)$ polynomials for nonperiodic problems. In a general way, the trial functions are orthogonal with respect to some weight $w(x)$, such that

$$(\varphi_k, \varphi_l)_w = c_k \, \delta_{k,l}, \tag{1.3}$$

where $c_k =$ constant and $\delta_{k,l}$ is the Kronecker delta.

Now we introduce the residual R_N which should be made equal to zero. For example, if the function $u(x)$ is given and if we look for its approximation $u_N(x)$, then the residual is

$$R_N(x) = u - u_N.$$

If $u_N(x)$ is an approximate solution to the differential equation

$$L\,u - f = 0,$$

then the residual R_N is defined by

$$R_N(x) = L\,u_N - f.$$

The weighted residuals method consists in annuling R_N in an approximate sense by setting to zero the scalar product

$$(R_N, \psi_i)_{w_\star} = \int_{\alpha}^{\beta} R_N \, \psi_i \, w_\star \, dx = 0, \quad i \in I_N, \tag{1.4}$$

where $\psi_i(x)$ are the test (or weighting) functions, and the weight $w_\star$ is associated with the method and trial functions. The dimension of the discrete set I_N depends on the problem under consideration as will be discussed in the sequel. For complex functions, ψ_i in the integral above is replaced by its complex conjugate $\overline{\psi}_i$.

The choice of the test functions and of the weight defines the method, for example:

(i) the Galerkin-type method corresponds to the case where the test functions are the trial functions themselves and where the weight $w_\star$ is the weight associated with the orthogonality of the trial functions, that is,

$$\psi_i = \varphi_i \quad \text{and} \quad w_\star = w, \tag{1.5}$$

(ii) the collocation method corresponds to the choice

$$\psi_i = \delta(x - x_i) \quad \text{and} \quad w_\star = 1, \tag{1.6}$$

where δ is the Dirac delta-function. The collocation points x_i are selected points on $[\alpha, \beta]$. As will be discussed later, the choice of such points is not completely arbitrary. From (1.4) and (1.6) we simply get

$$R_N(x_i) = 0. \tag{1.7}$$

Therefore, in the collocation method, the residual is exactly zero at certain points whereas in the Galerkin-type method the residual is zero in the mean.

1.2   Approximation of a given function

In this section we present the general way to determine the coefficients $\hat{u}_k$ in the expansion (1.2) corresponding to the approximation of a given function u. Specific application to Fourier and Chebyshev approximations will be detailed in Chapters 2 and 3, respectively. We refer to Isaacson and Keller (1966) for an introduction to the basic theoretical aspects of the theory of approximation. More specialized analyses can be found in Gottlieb and Orszag (1977), Canuto et al. (1988), and Funaro (1992).

1.2.1   Galerkin-type method

This type of method is characterized by the choice $\psi_i = \varphi_i$ and $w_\star = w$. Therefore, the residual

$$R_N(x) = u - u_N = u - \sum_{k=0}^{N} \hat{u}_k \varphi_k \tag{1.8}$$

is made zero in the mean according to

$$(R_N, \varphi_i)_w = \int_\alpha^\beta \left(u - \sum_{k=0}^{N} \hat{u}_k \varphi_k \right) \varphi_i \, w \, dx = 0, \quad i = 0, \ldots, N. \tag{1.9}$$

Here the set I_N introduced in (1.4) is $\{0, \ldots, N\}$, so that the $N + 1$ Galerkin equations (1.9) determine the $N+1$ coefficients $\hat{u}_k$. Now, by using the orthogonality property (1.3), we obtain the following expression for the coefficients $\hat{u}_k$:

$$\hat{u}_k = \frac{1}{c_k} \int_\alpha^\beta u \, \varphi_k \, w \, dx, \quad k = 0, \ldots, N. \tag{1.10}$$

At this point it is interesting to note that, in the present case, the least-squares method (see Isaacson and Keller, 1966; Finlayson, 1972) leads to an identical result. In the least-squares method, we look for the coefficients $\hat{u}_k$ which minimize the scalar product $(R_N, R_N)_w$. This is obtained by setting to zero the partial derivatives of $(R_N, R_N)_w$ with respect to the coefficients $\hat{u}_k$. Therefore, in the frame of the weighted residuals method, the test functions are

$$\psi_i = \frac{\partial R_N}{\partial \hat{u}_i} = -\varphi_i \tag{1.11}$$

from Eq.(1.8).

1.2.2   Collocation method

The residual $R_N = u - u_N$ is made equal to zero at the $N + 1$ collocation points x_i, $i = 0, \ldots, N$, so that $I_N = \{0, \ldots, N\}$ and

$$u_N(x_i) = u(x_i), \quad i = 0, \ldots, N, \tag{1.12}$$

hence,

$$\sum_{k=0}^{N} \hat{u}_k \, \varphi_k(x_i) = u(x_i), \quad i = 0, \ldots, N. \tag{1.13}$$

This gives an algebraic system to determine the $N + 1$ coefficients $\hat{u}_k$, $k = 0, \ldots, N$. The existence of a solution for this system implies that $\det\{\varphi_k(x_i)\} = 0$. This is a first condition to be satisfied by the collocation points.

As a matter of fact, the coefficients $\hat{u}_k$, solution of the system (1.13), can be obtained explicitly without solving the system itself. This is done by using a discrete orthogonality property of the trial functions φ_k associated to special sets $\{x_i\}$ of collocation points. As will be discussed later, this is equivalent to a numerical evaluation of the integral in Eq.(1.10) using the Gauss formula.

Finally, when applied to the approximation of a given function, the collocation method is nothing other than an interpolation technique based on the set $\{x_i\}$.

1.3   Approximation of the solution of a differential equation

Let us consider the approximation of the solution of the differential equation

$$Lu - f = 0, \quad \alpha < x < \beta, \tag{1.14}$$

where L is a differential operator assumed to be linear and of second order. To the equation (1.14) the following linear boundary conditions are associated

$$B_- u = g_- \quad \text{at } x = \alpha, \quad B_+ u = g_+ \quad \text{at } x = \beta, \qquad (1.15)$$

where B_- and B_+ correspond to Dirichlet, Neumann or Robin conditions.

1.3.1 The traditional Galerkin method

The traditional Galerkin method applies when the trial functions φ_k in the expansion (1.2) satisfy the homogeneous boundary conditions associated with (1.15), let

$$B_- \varphi_k = 0 \quad \text{at } x = \alpha, \quad B_+ \varphi_k = 0 \quad \text{at } x = \beta. \qquad (1.16)$$

In this case, the solution u of Eqs.(1.14)-(1.15) can be sought as the combination

$$u = \tilde{u} + v,$$

where $\tilde{u}$ is any function satisfying the boundary conditions (1.15). The resulting problem for v is then

$$L v - h = 0, \quad \alpha < x < \beta \qquad (1.17)$$

$$B_- v = 0 \quad \text{at } x = \alpha, \qquad (1.18)$$

$$B_+ v = 0 \quad \text{at } x = \beta, \qquad (1.19)$$

where $h = f - L\tilde{u}$. The fact that the trial functions φ_k satisfy the homogeneous boundary conditions guarantees that the approximation $v_N(x)$, defined by

$$v_N(x) = \sum_{k=0}^{N} \hat{v}_k \, \varphi_k(x), \qquad (1.20)$$

satisfies the boundary conditions whatever the values of the expansion coefficients $\hat{v}_k$. Now the residual $R_N(x)$ is

$$R_N(x) = L v_N - h. \qquad (1.21)$$

Then, according to the general formulation (1.4) with $I_N = \{0, \ldots, N\}$ and (1.5), the Galerkin equations are

$$(R_N, \varphi_i)_w = (L v_N - h, \varphi_i)_w = 0, \quad i = 0, \ldots, N, \qquad (1.22)$$

or else, by replacing v_N by its expansion (1.20),

$$\sum_{k=0}^{N} \hat{v}_k \, (L \varphi_k, \varphi_i)_w = (h, \varphi_i)_w, \quad i = 0, \ldots, N. \qquad (1.23)$$

The scalar product $(L\varphi_k, \varphi_i)_w$ is evaluated using the properties of the trial functions, in particular, their orthogonality. From Eq.(1.10), the scalar product $(h, \varphi_i)_w$ is equal to $c_i \hat{h}_i$ where $\hat{h}_i$, $i = 0, \ldots, N$, is the expansion coefficient of h. System (1.23) furnishes $N + 1$ equations for determining the $N + 1$ coefficients $\hat{v}_k$, $k = 0, \ldots, N$.

Note that, in the present case, the number of Galerkin equations (1.22) of the general type (1.4) is exactly $N + 1$. This is because the boundary conditions are satisfied by the trial functions. When this latter property does not hold, the method may be applied by constructing, from the basis $\{\varphi_k\}$, a new basis $\{\psi_k\}$ satisfying the boundary conditions; this can be done generally by defining ψ_k as a linear combination of some φ_k's. But the new basis may be nonorthogonal, so that this approach is not much used and generally one prefers a simpler method called the "tau method" which is described in the following section.

1.3.2 The tau method

The so-called tau method, introduced by Lanczos in 1938 (see Lanczos, 1956), is a modification of the Galerkin method allowing the use of trial functions not satisfying the homogeneous boundary conditions (1.16). The solution of Eqs.(1.14)-(1.15) is sought in the form

$$u_N(x) = \sum_{k=0}^{N} \hat{u}_k \, \varphi_k(x) \, . \tag{1.24}$$

The equations determining the $N + 1$ coefficients $\hat{u}_k$, $k = 0, \ldots, N + 1$ are obtained by considering the Galerkin equations of type (1.4) with $R_N = L\,u_N - f$, $\psi_i = \varphi_i$, $w_\star = w$ and $I_N = \{0, \ldots, N - 2\}$, therefore,

$$\sum_{k=0}^{N} \hat{u}_k \, (L\,\varphi_k, \varphi_i)_w = (f, \varphi_i)_w \, , \quad i = 0, \ldots, N - 2, \tag{1.25}$$

these equations being completed with the boundary conditions

$$B_-\, u_N(\alpha) = g_- \, , \quad B_+\, u_N(\beta) = g_+ \tag{1.26}$$

System (1.25)-(1.26) permits the calculation of the $N + 1$ coefficients $\hat{u}_k$, $k = 0, \ldots, N$. The non-consideration of the Galerkin equations for $i = N - 1$ and $i = N - 2$ introduces a supplementary error, the "tau error"" which has given its name to the method. The tau error will be discussed in Section 3.4.3 and we refer to Gottlieb and Orszag (1977) and to Canuto et al. (1988) for more details.

1.3.3 The collocation method

Following the general description made in Section 1.1, the collocation equations are obtained by considering the expansion (1.24) and by making the associated residual $R_N = L u_N - f$ equal to zero at the inner collocation points $x_i \in I_N = \{1, \ldots, N-1\}$, let

$$L u_N(x_i) = f(x_i), \quad i = 1, \ldots, N-1. \tag{1.27}$$

The system is closed with the boundary conditions at $x = x_0 = \alpha$ and $x = x_N = \beta$, i.e. :

$$B_- u_N(x_0) = g_- , \quad B_+ u_N(x_N) = g_+ . \tag{1.28}$$

Equations (1.25) and (1.26) give a system of $N+1$ equations for the $N+1$ coefficients $\hat{u}_k$, $k = 0, \ldots, N$. An equivalent formulation is, however, generally preferred. It consists of considering the values $u_N(x_i)$ at the collocation points x_i, $i = 0, \ldots, N$, as unknowns rather than the coefficients $\hat{u}_k$. This is possible since the $\hat{u}_k$, $k = 0, \ldots, N$, can be expressed in terms of the $u_N(x_i)$, $i = 0, \ldots, N$, as discussed in Section 1.2.2 [Eq. (1.13)]. Therefore, one can construct differentiation formulas expressing the derivative, of any order, at a given collocation point in terms of the values of the function itself at all collocation points, that is,

$$u_N^{(p)}(x_i) = \sum_{j=0}^{N} d_{i,j}^{(p)} u_N(x_j). \tag{1.29}$$

This construction will be made more precise in Sections 2.5 and 3.3.6 for the Fourier and Chebyshev approximations, respectively. Note that Eq.(1.29) may also be obtained from the Lagrange interpolation polynomial based on the grid values $u_N(x_i)$. This polynomial is trigonometric in the Fourier case and algebraic in the Chebyshev case. In this way, the collocation method may be considered as associated with a global polynomial approximation rather than with a truncated series expansion.

If $\mathcal{D}$ is the differentiation matrix, defined by $\mathcal{D} = [d_{i,j}^{(1)}]$, $i, j = 0, \ldots, N$, and if we denote by U the vector whose components are the values of u_N at the collocation points, and by $U^{(p)}$ the vector of the values of the pth derivative of U, we have

$$U^{(1)} = \mathcal{D} U$$

and $U^{(p)} = \mathcal{D}^p U$. Therefore, except if the analytical expressions of $d_{i,j}^{(p)}$ are needed, it will be sufficient and sometimes more accurate (e.g., for Chebyshev approximation), from a practical point of view, to calculate $d_{i,j}^{(1)}$ and then to evaluate $\mathcal{D}^p$ numerically.

2

Fourier Method

The more familiar spectral method is the Fourier method in which the basis functions are trigonometric functions. Such a basis is adapted to periodic problems. Spatial periodicity appears in a variety of flows. Beside the case of homogeneous turbulence, for which the assumption of periodicity in the three spatial directions is realistic, there also exists a number of physical situations where the flow is periodic in one or two directions. In these last situations, the spatial approximation makes use of Fourier series in the periodic directions associated with another type of approximation, e.g., Chebyshev polynomials, in the nonperiodic directions.

In the present chapter, we restrict ourselves to the presentation of the basic properties of the Fourier series and to the solution techniques associated with them. The purpose is to present the tools which will be used in the sequel to solve the Navier-Stokes equations. Moreover, by considering simple differential equations, we want to point out the special characteristics of the Fourier methods (Galerkin and collocation) and to emphasize their similarities and differences. Only the one-dimensional case is addressed in this chapter. The extension to multidimensional problems will be considered in Chapters 6 and 7 for the solution of the Navier-Stokes equations.

2.1 Truncated Fourier series

Let a function $u(x)$ be assumed to be 2π-periodic in $0 \leq x \leq 2\pi$. As a matter of fact, the assumption of periodicity is not necessary to define a

Fourier series approximation, but if the function under consideration is not periodic the convergence of the associated series is not uniform near the boundaries and the Gibbs oscillations contaminate the whole domain. The Gibbs phenomenon also appears when the function is discontinuous. This is the reason why the use of Fourier series should be restricted to the representation of smooth, periodic functions, although efficient ways to remove the Gibbs phenomenon exist as shown by Gottlieb and Shu (1997).

2.1.1 Calculation of Fourier coefficients

The function $u(x)$ is approximated by the truncated series expansion of type (1.2), written here as

$$u_K(x) = \sum_{k=-K}^{K} \hat{u}_k \, e^{i k x}, \qquad (2.1)$$

where $i^2 = -1$. This expansion contains $2K + 1$ complex coefficients which have to be determined. But the function $u(x)$ is assumed to be real so that the two Fourier coefficients, with an opposite value of k, are complex conjugates, i.e.,

$$\hat{u}_{-k} = \overline{\hat{u}_k}, \qquad (2.2)$$

the coefficient $\hat{u}_0$ being obviously real. Therefore, with (2.2), the expansion (2.1) also contains $2K + 1$ real unknowns. Generally, in practical calculations, the complex coefficients $\hat{u}_k$ are calculated for $k = 0, \ldots, K$ and then the remaining coefficients are given by (2.2). The complex form of expansion (2.1) is useful for applying fast Fourier transform.

The scalar product (1.1) is valid here with $w = 1$, that is,

$$(u, v) = \int_0^{2\pi} u \, \overline{v} \, dx \qquad (2.3)$$

so that the orthogonality property of the complex exponential functions is

$$\int_0^{2\pi} e^{i k x} e^{-i l x} \, dx = \begin{cases} 2\pi & \text{if} \quad k = l, \\ 0 & \text{if} \quad k \neq l. \end{cases} \qquad (2.4)$$

Now we can calculate the coefficients $\hat{u}_k$, $k = -K, \ldots, K$, by using the Galerkin-type technique described in Section 1.2.1. More precisely, the residual $R_K = u - u_K$ is set to zero in the average sense

$$(R_K, e^{i l x}) = 0, \quad l = -K, \ldots, K,$$

that is,

$$\int_0^{2\pi} \left[u \, e^{-i l x} - \sum_{k=-K}^{K} \hat{u}_k \, e^{i(k-l) x} \right] dx = 0.$$

Then, with the orthogonality property (2.4), only the term for which $k = l$ remains in the sum, and we get the expression of the Fourier coefficients

$$\hat{u}_k = \frac{1}{2\pi} \int_0^{2\pi} u\, e^{-ikx}\, dx, \quad k = -K, \ldots, K. \qquad (2.5)$$

To close this section, we consider the real form of the truncated Fourier series

$$u_K(x) = a_0 + \sum_{k=1}^{K} (a_k \cos kx + b_k \sin kx) \qquad (2.6)$$

with the well-known formulas

$$a_0 = \frac{1}{2\pi} \int_0^{2\pi} u\, dx, \quad a_k = \frac{1}{\pi} \int_0^{2\pi} u \cos kx\, dx, \quad b_k = \frac{1}{\pi} \int_0^{2\pi} u \sin kx\, dx,$$

that is, $a_0 = \hat{u}_0$, $a_k = 2\,\mathcal{R}e\,(\hat{u}_k)$, and $b_k = 2\,\mathcal{I}m\,(\hat{u}_k)$.

2.1.2 Some results on convergence

First, it is interesting to know the rate of decay of the Fourier coefficients. It can be easily shown (see, e.g., Gottlieb and Orszag, 1977) that, for a function $u(x)$, periodic, continuous on $[0, 2\pi]$ as well as its derivatives $u^{(p)}$ to the order $m-1$ included and with the mth derivative absolutely integrable, the Fourier coefficients behave like

$$\hat{u}_k = O\left(|k|^{-m}\right) \quad \text{for } k \to \infty. \qquad (2.7)$$

If the total variation of the mth derivative is bounded, the above result becomes

$$\hat{u}_k = O\left(|k|^{-m-1}\right) \quad \text{for } k \to \infty. \qquad (2.8)$$

These results show that the more regular the function u, the more rapid is the convergence toward zero of its Fourier coefficients when $k \to \infty$.

Among the results on the convergence of the approximation u_K using various norms (Canuto et al., 1988 ; Mercier, 1989), we select the estimate based on the $L^p(0, 2\pi)$-norm given by Canuto et $al.$ (1988). For $1 < p < \infty$, the error estimate is

$$\|u - u_K\|_{L^p(0,2\pi)} \leq C\, K^{-m} \|u^{(m)}\|_{L^p(0,2\pi)}, \qquad (2.9)$$

where C is a constant independent of K. If $p = 1$ or $p = \infty$, the inequality (2.9) holds with the constant C replaced by $C\,(1 + \ln K)$.

From this result it is deduced that, for an infinitely differentiable function, the approximation error is smaller than any power of $1/K$: the convergence is exponential. This behaviour is commonly called "spectral" or "infinite" accuracy. For a sufficiently large number of Fourier modes, the error depends only on the regularity of the function under consideration. Such behaviour has to be compared to the $O(1/N^p)$ error of a finite-difference approximation where $1/N$ is the mesh size and p, which depends on the scheme, is essentially finite and even relatively small. However, we point out that in the presence of singularity, the rate of convergence of the Fourier approximation is only algebraic.

2.2   Discrete Fourier series

In this section, we apply the general technique of collocation (or interpolation) outlined in Section 1.2.2. The collocation points associated with the Fourier series are defined by

$$x_i = \frac{2\pi i}{N}, \quad i = 0, \ldots, N, \tag{2.10}$$

such that $x_0 = 0$ and $x_N = 2\pi$. The function $u(x)$ is assumed to be periodic therefore it satisfies $u(x_0) = u(x_N)$ and similar equalities for its derivatives. The function $u(x)$ is approximated by the expansion (2.1), i.e.,

$$u_K(x) = \sum_{k=-K}^{K} \hat{u}_k \, e^{i k x} \tag{2.11}$$

but now the coefficients $\hat{u}_k$, $k = -K, \ldots, K$, are defined by imposing the residual $R_K(x) = u(x) - u_K(x)$ to be zero at the collocation points, let

$$R_K(x_i) = u(x_i) - u_K(x_i) = 0, \quad i = 1, \ldots, N, \tag{2.12}$$

or

$$\sum_{k=-K}^{K} \hat{u}_k \, e^{i k x_i} = u(x_i), \quad i = 1, \ldots, N. \tag{2.13}$$

As previously mentioned, the above system contains $2K + 1$ complex unknowns $\hat{u}_k$, therefore it is necessary that

$$N = 2K + 1. \tag{2.14}$$

Besides, the matrix $\mathcal{M}$ associated with the system (2.13) is unitary up to the factor N, that is $\mathcal{M}^* \mathcal{M} = N \mathcal{I}$ (where $\mathcal{M}^*$ is the conjugate transpose of $\mathcal{M}$ and $\mathcal{I}$ is the identity matrix) so that its determinant satisfies $|\det \mathcal{M}| = N^{N/2}$. Consequently, the matrix $\mathcal{M}$ is invertible and system (2.13) determines the Fourier coefficients. Since $u(x_i)$ is real, the equality $\hat{u}_{-k} = \overline{\hat{u}}_k$ holds and the above system contains $2K + 1$ real unknowns.

Now, the coefficients $\hat{u}_k$ are explicitly determined by application of the discrete orthogonality relation

$$\sum_{i=1}^{N} e^{i(k-l)\frac{2\pi i}{N}} = \begin{cases} N & \text{if } k - l = mN, \quad m = 0, \pm 1, \pm 2, \ldots, \\ 0 & \text{otherwise}. \end{cases}$$

$$\tag{2.15}$$

Therefore, we multiply both sides of Eq.(2.13) by e^{-ilx} and we sum from $i = 1$ to $i = N$. Then, owing to (2.15), we obtain

$$\hat{u}_k = \frac{1}{N} \sum_{i=1}^{N} u(x_i) e^{-ikx_i}, \quad k = -K, \ldots, K. \tag{2.16}$$

It must be remarked that the discrete orthogonality relation (2.15), as well as the formula (2.16), can be directly deduced, respectively, from (2.4) and (2.5) by the numerical quadrature of the integrals by using the trapezoidal method.

Concerning the orthogonality relations, it is important to notice that the integral in (2.4) is not zero for $k - l = 0$, while the sum in (2.15) is not zero for $k - l = mN$, $m = 0, \pm1, \pm2, \ldots$. This is a consequence of the discretization and is related to the question of sampling : two trigonometrical functions with different frequencies, e^{ik_1x} and e^{ik_2x}, are equal at collocation points $x_i = 2\pi i/N$ when $k_2 - k_1 = mN$, $m = 0, \pm1, \ldots$. Therefore, the same set of values at collocation points may represent e^{ik_1x} as well as $e^{i(k_1+mN)x}$. The consequences of this phenomenon, known as "aliasing," will be discussed in the following sections.

2.3   Relation between Galerkin and collocation coefficients

In the previous sections, the coefficients $\hat{u}_k$ of the truncated Fourier series (2.1) have been determined by means of two techniques : Galerkin (or projection) and collocation (or interpolation). The aim of the present section is to make clear the relationship between these coefficients. In order to avoid confusion, we denote here, by $\hat{u}_k^e$, the Galerkin-type coefficients as defined by the integral (2.5) and by $\hat{u}_k^c$ the collocation-type coefficients given by the sum (2.16). Let us consider the infinite Fourier series (assumed to be absolutely convergent) :

$$u(x) = \sum_{k=-\infty}^{\infty} \hat{u}_k^e \, e^{ikx}. \tag{2.17}$$

If the sum (2.17), evaluated at $x_i = 2i\pi/N$, $i = 1, \ldots, N$, that is, $u(x_i)$, is substituted into (2.16), that is,

$$\hat{u}_k^c = \frac{1}{N} \sum_{i=1}^{N} \left(\sum_{p=-\infty}^{\infty} \hat{u}_p^e \, e^{ipx_i} \right) e^{-ikx_i}, \quad k = -K, \ldots, K,$$

or, because the series is absolutely convergent,

$$\hat{u}_k^c = \frac{1}{N} \sum_{p=-\infty}^{\infty} \hat{u}_p^e \left(\sum_{i=1}^{N} e^{i(p-k)x_i} \right).$$

Finally, taking the discrete orthogonality relation (2.15) into account, we obtain

$$\hat{u}_k^c = \hat{u}_k^e + \sum_m \hat{u}_{k+mN}^e, \quad k = -K, \ldots, K, \qquad (2.18)$$

where the summation is taken for $m = \pm 1, \pm 2, \ldots$. The sum in (2.18), which characterizes the difference between the coefficients $\hat{u}_k^c$ and $\hat{u}_k^e$, is called "alias". Its presence is a consequence of the sampling phenomenon mentioned in the previous section. Note, however, that the modes appearing in the alias term correspond to frequencies larger than the cut-off frequency K. More precisely, it can be shown (Canuto *et al.*, 1988) that the L^2-norm of the aliasing error

$$E_A = \sum_{k=-K}^{K} \left(\sum_m \hat{u}_{k+mN}^e \right) e^{ikx} \qquad (2.19)$$

is bounded by $C K^{-m} \|u^{(m)}\|_{L^2(0,2\pi)}$. In other words, in the L^2-norm, the aliasing error is similar to the interpolation error (2.9).

To summarize, we have the following identity

$$\begin{aligned} u(x_i) = u_K(x_i) &= \sum_{k=-\infty}^{\infty} \hat{u}_k^e e^{ikx_i} = \sum_{k=-K}^{K} \hat{u}_k^c e^{ikx_i} \\ &= \sum_{k=-K}^{K} \left(\hat{u}_k^e + \sum_{\substack{m=-\infty \\ m\neq 0}}^{\infty} \hat{u}_{k+mN}^e \right) e^{ikx_i} \end{aligned} \qquad (2.20)$$

which is useful for discussing the relationship between the two types of approximation.

2.4 Odd and even collocation

We come back to the discrete Fourier expansion

$$u_K(x_i) = \sum_{k=-K}^{K} \hat{u}_k e^{ikx_i}, \quad x_i = \frac{2i\pi}{N}, \quad i = 1, \ldots, N = 2K+1, \qquad (2.21)$$

where

$$\hat{u}_k = \frac{1}{N} \sum_{i=1}^{N} u(x_i) e^{-ikx_i}, \quad k = -K, \ldots, K. \qquad (2.22)$$

From a practical point of view, the above sums are generally calculated by means of the fast Fourier transform (FFT) algorithm (see, e.g., Brigham, 1988; Van Loan, 1992). The first FFT algorithms (Cooley and Tukey, 1965) were working for $N = 2^p$, that is, for an even value of N. Because $N =$

$2K + 1$, such algorithms cannot be used and the expansion (2.1) should be modified. Modern FFT algorithms (Temperton, 1983) work with vector lengths equal to $2^p\,3^q\,5^r$, that is, allowing odd values of N. When N is even, the expansion (2.1) is replaced by

$$u_K(x) = \sum_{k=-K+1}^{K} \hat{u}_k\,e^{ikx} \tag{2.23}$$

so that the number of Fourier coefficients is equal to $2K$. Then the coefficients $\hat{u}_k$ are expressed by (2.22) where

$$N = 2K. \tag{2.24}$$

It must be noticed that a difficulty arises from the last mode in (2.23). The coefficient $\hat{u}_K$ given by (2.22) is real since $e^{-iK\,x_i} = e^{-i\pi i} = (-1)^i$ and $u(x)$ is a real function. The last term $\hat{u}_K\,e^{iK\,x}$ cannot be associated with its analog $u_{-K}e^{-iK\,x}$ (because the sum starts from $-K+1$) and, consequently, is not able to represent a real function, except if $\hat{u}_K = 0$. Numerical experiments have shown that the presence of this last term may lead to instabilities in a time-dependent problem. These instabilities are related to the fact that only the cosine appears for the frequency K: the basis is incomplete and the derivative $u'(x)$, which should involve $\sin K\,x$, cannot be represented in the basis. Therefore, it is recommended, when using the even collocation, to filter the last mode by simply setting $\hat{u}_K = 0$ at the end of the FFT process. In the following, the truncated Fourier series will generally be considered in the conventional form (2.1), it being understood that this equation must be replaced by Eq.(2.23) in the case of an even collocation.

2.5   Differentiation in the physical space

As mentioned in Section 1.3.3, the collocation method for the solution of differential equations usually considers as unknowns the values of the function at the collocation points rather than the expansion coefficients. Such an approach, however, is not to be recommended in the case of the Fourier approximation to a constant-coefficient equation, since it leads to the solution of an algebraic system, contrary to the case where the unknowns are the Fourier coefficients. On the other hand, for an equation with variable coefficients (e.g., an equation in spherical coordinates) the solution in the physical space may be of interest.

Therefore, this section is devoted to the construction of the differentiation matrices introduced in Section 1.3.3. More precisely, let us consider the expansion

$$u_K(x) = \sum_{k \in I_K} \hat{u}_k\,e^{ikx}, \tag{2.25}$$

where $I_K = \{-K, \ldots, K\}$ for odd collocation and $I_K = \{-K+1, \ldots, K\}$ for even collocation. By differentiating Eq.(2.25) we get, for the pth derivative,

$$u_K^{(p)}(x_i) = \sum_{k \in I_K} (i k)^p \, \hat{u}_k \, e^{i k x_i} , \qquad (2.26)$$

where $x_i = 2 i \pi / N$, $i = 1, \ldots, N$, with $N = 2K + 1$ for odd collocation and $N = 2K$ for even collocation. Note that, in this latter case, the last coefficient $\hat{u}_K$ must be discarded as explained before.

Now, by bringing $\hat{u}_k$ given by Eq.(2.16) with $k \in I_K$ into Eq.(2.26) evaluated at x_i, and after some algebra, we obtain

$$u_K^{(p)}(x_i) = \sum_{j=1}^{N} d_{i,j}^{(p)} \, u_K(x_i), \quad i = 1, \ldots, N, \qquad (2.27)$$

with

$$d_{i,j}^{(p)} = \frac{1}{N} \left[\frac{d^p}{d\xi^p} \left(\frac{\sin\left(N' \xi/2\right)}{\sin\left(\xi/2\right)} \right) \right]_{\xi = x_i - x_j} , \quad i \neq j , \qquad (2.28)$$

where $N' = N = 2K + 1$ for odd collocation and $N' = N - 1 = 2K - 1$ for even collocation. If $i = j$, we have

$$d_{i,i}^{(p)} = \begin{cases} 0 & \text{for odd } p, \\[2ex] (-1)^{p/2} \dfrac{2}{N} \displaystyle\sum_{k=1}^{(N'-1)/2} k^p & \text{for even } p. \end{cases} \qquad (2.29)$$

In particular, the expressions for the two first derivatives in the case of *odd collocation* are :

First order derivative

$$d_{i,j}^{(1)} = \begin{cases} \dfrac{(-1)^{i+j}}{2 \sin h_{i,j}} & \text{if } i \neq j, \\[2ex] 0 & \text{if } i = j, \end{cases} \qquad (2.30)$$

with $h_{i,j} = (x_i - x_j)/2$.

Second-order derivative

$$d_{i,j}^{(2)} = \begin{cases} (-1)^{i+j+1} \dfrac{\cos h_{i,j}}{2 \sin^2 h_{i,j}} & \text{if } i \neq j, \\[2ex] -\dfrac{N^2 - 1}{12} & \text{if } i = j. \end{cases} \qquad (2.31)$$

In the case of *even collocation*, the expressions are :

First-order derivative

$$
d_{i,j}^{(1)} = \begin{cases} \dfrac{1}{2}(-1)^{i+j} \cot h_{i,j} & \text{if } i \neq j, \\ 0 & \text{if } i = j. \end{cases}
\tag{2.32}
$$

Second-order derivative

$$
d_{i,j}^{(2)} = \begin{cases} \dfrac{1}{4}(-1)^{i+j} N + \dfrac{(-1)^{i+j+1}}{2 \sin^2 h_{i,j}} & \text{if } i \neq j, \\ -\dfrac{(N-1)(N-2)}{12} & \text{if } i = j. \end{cases}
\tag{2.33}
$$

2.6   Differential equation with constant coefficients

This section is devoted to the presentation of the Fourier method for solving linear differential equations with constant coefficients. The Galerkin and collocation methods will be successively shown and compared. It will be shown that the two methods differ from one another due to the approximation of the forcing term in the differential equation.

Let us consider the second-order differential equation

$$
L u \equiv -\nu u'' + a u' + b u = f,
\tag{2.34}
$$

where ν, a and b are constant and $f = f(x)$ is 2π-periodic. We are interested in the 2π-periodic solution of Eq. (2.34), represented by the truncated Fourier series expansion (2.1).

2.6.1   Galerkin method

With Fourier series, the basis functions are themselves 2π-periodic, that is we are in a situation equivalent to the case where the basis functions satisfy the homogeneous boundary condition (see Section 1.3.1). Therefore, the traditional Galerkin method can be applied. The residual R_K is defined by

$$
R_K(x) = L u_K - f = \sum_{k=-K}^{K} \hat{u}_k \, L \, e^{ikx} - f.
\tag{2.35}
$$

The Galerkin equations are constructed by setting to zero the scalar product (R_K, w_i) with $w_i = e^{iix}$, let

$$
\left(R_K, \, e^{iix}\right) = \sum_{k=-K}^{K} \left(L e^{ikx}, \, e^{iix}\right) - \left(f, \, e^{iix}\right) = 0, \quad i = -K, \ldots, K.
\tag{2.36}
$$

Taking into account that

$$L e^{i k x} = \left(\nu k^2 + \underline{i} a k + b \right) e^{i k x} \equiv G_k e^{i k x} ,$$

and that

$$\left(f , e^{\underline{i} i x} \right) = 2 \pi \hat{f}_i ,$$

where $\hat{f}_i$ is the Fourier coefficient of f corresponding to the frequency i, Eq.(2.36) can be written as

$$\sum_{k=-K}^{K} G_k \hat{u}_k \int_0^{2\pi} e^{\underline{i} (k-i) x} \, dx = 2 \pi \hat{f}_i , \quad i = -K, \ldots, K .$$

Finally, thanks to the orthogonality relation (2.4), we obtain the Galerkin equations

$$G_k \hat{u}_k = \hat{f}_k , \quad k = -K, \ldots, K , \tag{2.37}$$

from which $\hat{u}_k$ is explicitly calculated (assuming $G_k \neq 0$). In fact, because u is a real function, it is necessary to calculate $\hat{u}_k$ for $k = 0, \ldots, K$ and then the spectrum is completed by using the relation $\hat{u}_{-k} = \overline{\hat{u}}_k$, $k = 1, \ldots, K$.

2.6.2 Collocation method

According to the procedure outlined in Section 1.3.3, the collocation method consists of requiring the expansion (2.1) to satisfy exactly the differential equation at collocation points $x_i = 2\pi i/N$, $i = 1, \ldots, N$. Therefore, the collocation equations associated with the problem (2.34) are

$$L u_K(x_i) - f(x_i) = 0 , \quad i = 1, \ldots, N . \tag{2.38}$$

We systematically consider here the case of odd collocation $N = 2K + 1$, but the method also applies to even collocation $N = 2K$ with the modification already mentioned.

Now we have the choice between two possibilities according to whether the unknowns are the Fourier coefficients or the grid values. These two approaches are equivalent from the approximation point of view but the computational task is not the same. In the first case, the Fourier coefficients are explicitly determined, whereas in the second case the grid values are solutions of an algebraic system.

We begin with the second manner which considers, as unknowns, the grid values $u_K(x_i)$, $i = 1, \ldots, N$. Let us denote by U the vector of the grid values

$$U = (u_K(x_1), \ldots, u_K(x_N))^T$$

and by $\mathcal{D}$ the differentiation matrix, such that the pth derivative is expressed as

$$U^{(p)} = \mathcal{D}^p U , \tag{2.39}$$

where $U^{(p)} = \left(u_K^{(p)}(x_1), \ldots, u_K^{(p)}(x_N) \right)^T$ and $\mathcal{D} = [d_{i,j}^{(1)}]$, $i, j = 1, \ldots, N$, with $d_{i,j}^{(1)}$ defined in Section 2.5. Therefore, by combining Eqs.(2.39) and (2.38), we get the algebraic system

$$\left(-\nu \mathcal{D}^2 + a\mathcal{D} + b\mathcal{I} \right) U = F, \tag{2.40}$$

where $F = (f(x_1), \ldots, f(x_N))^T$. The solution U is obtained by solving this system.

Now we consider the equivalent way which consists in calculating the Fourier coefficients. This approach will allow us to make clear the similarities and differences with the Galerkin method. The derivatives of u_K at the collocation points x_i are given by the expression (2.26) and the exact value $f(x_i)$ in (2.38) can be expressed using the identity (2.20), i.e.,

$$f(x_i) = f_K(x_i) = \sum_{k=-K}^{K} \left(\hat{f}_k + \sum_{\substack{m=-\infty \\ m \neq 0}}^{\infty} \hat{f}_{k+mN} \right) e^{i k x_i},$$

where $\hat{f}_k$ are the exact Fourier coefficients (the superscript e is omitted) defined by the integral (2.5). By bringing these expressions into (2.38), we get

$$\sum_{k=-K}^{K} \left(G_k \hat{u}_k - \hat{f}_k^c \right) e^{i k x_i} = 0, \quad i = 1, \ldots, N, \tag{2.41}$$

with

$$G_k = \nu k^2 + i a k + b$$

and

$$\hat{f}_k^c = \hat{f}_k + \sum_{\substack{m=-\infty \\ m \neq 0}}^{\infty} \hat{f}_{k+mN}.$$

Equation (2.41) may be considered as homogeneous algebraic system for the $(2K+1)$ complex unknowns $G_k \hat{u}_k - \hat{f}_k^c$, $k = -K, \ldots, K$. Because the associated matrix is invertible (see Section 2.2), the unique solution of this system is the null solution, that is,

$$\hat{G}_k \hat{u}_k - \hat{f}_k = \sum_{\substack{m=-\infty \\ m \neq 0}}^{\infty} \hat{f}_{k+mN}, \tag{2.42}$$

which would be exactly the Galerkin equations (2.37) if the sum in the right-hand side were not present. As explained in Section 2.3, this sum is due to the phenomenon of aliasing. We remind ourselves that the alias terms in the right-hand side of Eq.(2.42) correspond to frequencies larger than the cut-off frequency, that is, of the order of the approximation.

So, the Galerkin and collocation methods applied to the solution of the differential equation (2.34) are not completely equivalent. However, when the Galerkin method is used, the Fourier coefficients of the forcing term f are practically never evaluated through the integral (2.5) because the analytical integration is generally impossible. These coefficients are then evaluated through discrete Fourier expansion and the FFT algorithm, so that the two methods become numerically identical.

2.7   Differential equation with variable coefficients

In the present section, we consider the case where the differential equation (2.34) has variable coefficients. For the sake of simplicity, only the coefficient a is assumed to be a function of x with periodicity 2π. The discussion will be obviously valid for any other variable coefficient of (2.34).

Now, whatever the method, Galerkin or collocation, an algebraic system has to be solved. The collocation method where the unknowns are the values of the function at the collocation points is the more versatile and must be recommended. However, it is also interesting to consider the Galerkin method in order to compare the two types of methods and, in particular, to make clear the question of aliasing.

2.7.1   Galerkin method

By following the general lines of the Galerkin method shown in Section 2.6.1, we obtain, instead of Eq.(2.37), the following Galerkin equations

$$\nu k^2 \hat{u}_k + i \sum_{\substack{p,q \\ p+q=k}} q\,\hat{a}_p\,\hat{u}_q + b\,\hat{u}_k = \hat{f}_k\,, \quad k = -K,\ldots,K\,, \qquad (2.43)$$

where the summation is taken on $p = -K,\ldots,K$ and $q = -K,\ldots,K$, and where $\hat{a}_k$ and $\hat{f}_k$ are, respectively, the Fourier coefficients of $a(x)$ and $f(x)$ such that

$$a(x) \cong a_K\,(x) = \sum_{k=-K}^{K} \hat{a}_k\,e^{ikx}\,, \quad f(x) \cong f_K\,(x) = \sum_{k=-K}^{K} \hat{f}_k\,e^{ikx}\,. \quad (2.44)$$

Contrary to the constant-coefficient case, the coefficients $\hat{u}_k$ are no longer obtained by an explicit calculation but are solutions of an algebraic system: the Fourier modes are coupled by the convolution sum $\sum q\,\hat{a}_p\,\hat{u}_q$. The application of the Galerkin method is not so attractive as it was in the constant-coefficient case.

2.7.2   Collocation method

Considering first the case where the unknowns are the values of u_K at collocation points x_i, we have to solve an algebraic system similar to (2.40) with an obvious modification due to the fact that the coefficient a is no longer constant. Therefore, Eq.(2.40) is replaced by

$$\left(-\nu \mathcal{D}^2 + \mathcal{D}' + b\mathcal{I}\right) U = F, \qquad (2.45)$$

where the matrix $\mathcal{D}'$ is defined by $\mathcal{D}' = [a(x_i) d_{i,j}^{(1)}]$, $i, j = 1, \ldots, N$. Therefore, the computational task is equivalent to the one associated with the constant-coefficient case. It is obvious that, in the variable-coefficient case, the collocation method with grid values as unknowns has to be applied.

Now we consider the collocation method with Fourier coefficients as unknowns, which is equivalent to Eq.(2.45) from the point of view of approximation. The discussion of this approach will allow us to make clear the difference between the Galerkin and collocation methods.

The collocation equations analogous to Eq.(2.38) are

$$\sum_k \left(\nu k^2 + b\right) \hat{u}_k\, e^{i k x_i} + a(x_i) \sum_k (i k)\, \hat{u}_k\, e^{i k x_i} = f(x_i), \quad i = 1, \ldots, N.$$
$$(2.46)$$

Here, and in the following if not specified, the sum is taken over $k = -K, \ldots, K$. Equations (2.46) constitute an algebraic system for the $N = 2K + 1$ Fourier coefficients $\hat{u}_k$, $k = -K, \ldots, K$. This system can be put in the form similar to (2.41) by expressing $a(x_i)$ and $f(x_i)$ by means of the relation (2.20).

First, the Fourier coefficients of $a(x)$ and $f(x)$ are calculated by means of the finite discrete sum of type (2.16). To avoid confusion with those associated with the Galerkin method, these coefficients will be denoted here by $\hat{a}_k^c$ and $\hat{f}_k^c$, respectively. The collocation approximation $D_1(x_i)$ of $a(x)\, u'(x)$ can be expressed as

$$D_1(x_i) = a(x_i) \sum_k (i k)\, \hat{u}_k\, e^{i k x_i} = \left(\sum_p \hat{a}_p^c\, e^{i p x_i}\right) \left(\sum_q i q\, \hat{u}_q\, e^{i q x_i}\right)$$

$$= i \sum_k \left(\sum_{\substack{p,q \\ p+q=k}} q\, \hat{a}_p^c\, \hat{u}_q + \sum_{\substack{p,q \\ p+q=k+N}} q\, \hat{a}_p^c\, \hat{u}_q + \sum_{\substack{p,q \\ p+q=k-N}} q\, \hat{a}_p^c\, \hat{u}_q \right) e^{i k x_i}$$

$$\equiv i \sum_k S_k\, e^{i k x_i},$$

where the sums (simple and double) are taken on $\{-K, \ldots, K\}$. The above equations have been obtained by remarking that $e^{i(p+q) x_i} = e^{i k x_i}$ if, and only if, $p + q = k + mN = k + m(2K + 1)$ with $m = -1, 0, 1$ (taking into account that $-K \le p, q \le K$). By bringing the expression of D_1 into (2.46), and by replacing $f(x_i)$ by its expansion, we get the collocation equations

$$\sum_k \left[(\nu k^2 + b)\, \hat{u}_k + i S_k - \hat{f}_k^c\right] e^{i k x_i} = 0, \quad i = 1, \ldots, N.$$

Now, as previously done for the constant-coefficient case, these equations constitute a homogeneous algebraic system whose unknowns are the quantities in square brackets. The associated determinant is different from zero, so that the unique solution is the null solution

$$\left(\nu\, k^2 + b\right) \hat{u}_k + \underline{i}\, S_k - \hat{f}_k^c = 0, \quad k = -K, \ldots, K. \qquad (2.47)$$

These equations are equivalent to (2.45) and (2.46). It is interesting to consider another set of equations, always equivalent to (2.46), but where $a(x_i)$ and $f(x_i)$ are expressed by the infinite Fourier series as shown in Eq. (2.20). We denote the "exact" Galerkin Fourier coefficients by $\hat{a}_k^e$ and $\hat{f}_k^e$. By following the same lines as above, the first-order derivative term is expressed as

$$D_1\left(x_i\right) = \underline{i} \sum_k \left(S_k^{(0)} + S_k^{(1)} + S_k^{(2)} + S_k^{(3)}\right) e^{i\, k\, x_i}$$

with

$$S_k^{(0)} = \sum_{\substack{p,q \\ p+q=k}} q\, \hat{a}_p^e\, \hat{u}_q, \quad S_k^{(1)} = \sum_{\substack{m=-\infty \\ m\neq 0}}^{\infty} \sum_{\substack{p,q \\ p+q=k}} q\, \hat{a}_{p+mN}^e\, \hat{u}_q$$

$$S_k^{(2)} = \sum_{m=-\infty}^{\infty} \sum_{\substack{p,q \\ p+q=k+N}} q\, \hat{a}_{p+mN}^e\, \hat{u}_q,$$

$$S_k^{(3)} = \sum_{m=-\infty}^{\infty} \sum_{\substack{p,q \\ p+q=k-N}} q\, \hat{a}_{p+mN}^e\, \hat{u}_q.$$

Then we get the collocation equations, equivalent to (2.47), but in a different form

$$\left(\nu\, k^2 + b\right) \hat{u}_k + \underline{i} S_k^{(0)} + \underline{i}\left(S_k^{(1)} + S_k^{(2)} + S_k^{(3)}\right)$$

$$= \hat{f}_k^e + \sum_{\substack{m=-\infty \\ m\neq 0}}^{\infty} \hat{f}_{k+mN}^e, \qquad k = -K, \ldots, K. \qquad (2.48)$$

These equations must be compared to the Galerkin equations (2.43), recalling that $\hat{a}_k^e$ and $\hat{f}_k^e$ in Eq. (2.48) must be, respectively, identified with $\hat{a}_k$ and $\hat{f}_k$ in Eq. (2.43). The term $\underline{i}\, S_k^{(0)}$ is exactly the convolution sum appearing in Eq. (2.43). But the term $\underline{i}\left(S_k^{(1)} + S_k^{(2)} + S_k^{(3)}\right)$, which reflects the phenomenon of aliasing, is not present in Eq. (2.43). It is the same for the forcing term in the right-hand side of Eq. (2.48).

2.8   Nonlinear differential equation

Let us assume that the first-order derivative term $a\,u'$ in Eq. (2.34) is replaced by the nonlinear term $u\,u'$. The application of any Fourier method (Galerkin or collocation) leads to a nonlinear algebraic system which has to be solved by iteration.

Although important progress has been made, in the last 15 years, on the construction of efficient iterative procedures for linear problems, the success of such procedures to the solution of complicated nonlinear problems, such as those encountered in fluid mechanics, is largely problem-dependent. However, when the problem of interest is unsteady the time-variable allows us to avoid iterations by evaluating explicitly the nonlinear terms. As a matter of fact, even if only the steady-state solution is of interest, the consideration of the time-dependent equations constitutes a way to define an iterative procedure.

As a model of a time-dependent equation we consider the Burgers equation

$$\partial_t u + u\,\partial_x u - \nu\,\partial_{xx}u = 0 \tag{2.49}$$

to be solved in $0 \le x \le 2\pi$ with the initial condition

$$u(x,0) = u_0(x), \tag{2.50}$$

where $u_0(x)$ is 2π-periodic. The coefficient ν is a positive constant.

For the description of the general approach, Eq. (2.49) is discretized in time with a very simple semi-implicit scheme (more accurate time-discretization schemes are shown in Chapter 4). Let us consider the Fourier approximation

$$u_K(x,t) = \sum_{k=-K}^{K} \hat{u}_k(t)\,e^{i\,k\,x} \tag{2.51}$$

and denote by u_K^n the approximation of u_K at time $t_n = n\,\Delta t$, $n = 0, 1, \ldots$. Equation (2.49) is discretized in time such that the residual R_K is

$$R_K = \frac{u_K^{n+1} - u_K^n}{\Delta t} + u_K^n\,\partial_x u_K^n - \nu\,\partial_{xx}u_K^{n+1}. \tag{2.52}$$

The application of the Galerkin technique gives the equations

$$\left(\frac{1}{\Delta t} + \nu\,k^2\right)\hat{u}_k^{n+1} = \frac{1}{\Delta t}\hat{u}_k^n - \hat{w}_k^n, \quad k = -K, \ldots, K, \tag{2.53}$$

where the nonlinear term, $u_K^n\partial_x u_K^n$, appears in

$$\hat{w}_k^n = i \sum_{\substack{p,q=-K \\ p+q=k}}^{K} q\,\hat{u}_p^n\,\hat{u}_q^n. \tag{2.54}$$

In these equations $\hat{u}_k^n$ is the approximation of $\hat{u}_k$ at time $n\,\Delta t$. Knowing the initial value $\hat{u}_k^0$, Eqs. (2.53) and (2.54) give the solution $\hat{u}_k^{n+1}$ directly without iteration.

The only delicate question concerns the evaluation of the convolution sum $\hat{w}_k^n$ which must be as efficient as possible. This is obtained through the so-called "pseudospectral technique". This technique consists of performing the differentiations in the spectral space (the space of coefficients $\hat{u}_k$, $k = -K, \ldots, K$) and the products in the physical space (the space of the values $u_K(x_i)$ at the collocation points $x_i = 2i\pi/N$, $i = 1, \ldots, N$). The link between the two spaces is made by the FFT. The algorithm is the following:

1. Calculate $u_K^n(x_i)$, $i = 1, \ldots, N$, by the FFT from the knowledge of $\hat{u}_k^n$, $k = -K, \ldots, K$.

2. Calculate $\partial_x u_K^n(x_i)$, $i = 1, \ldots, N$, by the FFT from the Fourier coefficients $\underline{i}\,k\,\hat{u}_k^n$, $k = -K, \ldots, K$.

3. Form the product $w_K^n(x_i) = u_K^n(x_i)\,\partial_x u_K^n(x_i)$, $i = 1, \ldots, N$.

4. Calculate the Fourier coefficients $\tilde{w}_k^n$, $k = -K, \ldots, K$, by the FFT from the values $w_K^n(x_i)$, $i = 1, \ldots, N$.

Observe that we have denoted by $\tilde{w}_k^n$ the expression of $\hat{w}_k^n$ calculated by the pseudospectral technique. Because of the aliasing phenomenon the quantity $\tilde{w}_k^n$ is different from $\hat{w}_k^n$ given by Eq.(2.54). This can be made precise by making an analysis similar to the one carried out in the previous section but with the main difference that $a(x_i)$ was exactly known while $u_K^n(x_i)$, the coefficient of the first-order derivative, is the result of the numerical calculation. Let us write

$$w_K^n(x_i) = \sum_{k=-K}^{K} \tilde{w}_k^n\, e^{i\,k\,x_i}, \qquad (2.55)$$

where $w_K^n(x_i) = u_K^n(x_i)\,\partial_x u_K^n(x_i)$. The inverse expansion is

$$
\begin{aligned}
\tilde{w}_k^n &= \frac{1}{N} \sum_{i=1}^{N} w_K^n(x_i)\, e^{-i\,k\,x_i} \\
&= \frac{1}{N} \sum_{i=1}^{N} \left(\sum_p \hat{u}_p^n\, e^{i\,p\,x_i} \right) \left(\sum_q \underline{i}\,q\,\hat{u}_q^n\, e^{i\,q\,x_i} \right) e^{-i\,k\,x_i} \\
&= \frac{1}{N} \sum_{i=1}^{N} \sum_{p,q} q\,\hat{u}_p^n\,\hat{u}_q^n\, e^{i\,(p+q-k)\,x_i},
\end{aligned}
$$

where the sums are taken on $-K \le p \le K$, $-K \le q \le K$. Now, from the discrete orthogonality relation (2.15), we obtain

$$\tilde{w}_k^n = \hat{w}_k^n + \underline{i} \sum_{\substack{p,q \\ p+q=k+N}} q\, \hat{u}_p^n \hat{u}_q^n + \underline{i} \sum_{\substack{p,q \\ p+q=k-N}} q\, \hat{u}_p^n \hat{u}_q^n\,, \quad k = -K, \ldots, K\,.$$

(2.56)

In this expression $\hat{w}_k^n$ is the convolution sum defined by (2.54) and the sums are alias terms.

Therefore, the pseudospectral equations replacing (2.53) are

$$\left(\frac{1}{\Delta t} + \nu\, k^2 \right) \hat{u}_k^{n+1} = \frac{1}{\Delta t} \hat{u}_k^n - \tilde{w}_k^n\,, \quad k = -K, \ldots, K\,,$$

(2.57)

making apparent the difference with the Galerkin approximation (2.53). The alias terms can be removed as explained in Section 2.9.

In order to be complete, the collocation method is now analyzed. The residual R_K, defined by Eq. (2.52), is set to zero at the collocation points x_i, $i = 1, \ldots, N$,

$$\frac{1}{\Delta t} \left[u_K^{n+1}(x_i) - u_K^n(x_i) \right] + u_K^n(x_i)\, \partial_x u_K^n(x_i) - \nu\, \partial_{xx} u_K^{n+1}(x_i) = 0\,. \quad (2.58)$$

The grid values and the collocation Fourier coefficients are connected with

$$u_K^n(x_i) = \sum_{k=-K}^{K} \hat{u}_k^n\, e^{i\,k\,x_i}\,, \quad \hat{u}_k^n = \frac{1}{N} \sum_{i=1}^{N} u_K^n(x_i)\, e^{-i\,k\,x_i}\,. \quad (2.59)$$

Then, by calculations analogous to those made in the previous section, it is easy to see that Eq.(2.58) leads, for the coefficients $\hat{u}_k^n$ defined by (2.59), to a system of equations identical to (2.57). Therefore, assuming that the initial condition $\hat{u}_k^0$ is calculated by a discrete Fourier series in the pseudospectral approximation (2.57) as well as in the collocation approximation, the pseudospectral and collocation methods are equivalent.

2.9 Aliasing removal

The aliasing removal technique described in the present section was considered by Orszag (1971). This technique is called the "3/2 rule" for the reason which will appear below.

The technique is very simple to implement and can be easily incorporated into any Fourier code. Moreover, it is not expensive in computing time and needs a small amount of memory. The principle consists of extending the spectrum (therefore the number of collocation points) of the involved quantities such that the alias terms, introduced in the pseudospectral calculation operating on the resulting quantities, are not actually present. We shall first describe the technique and then we briefly justify it.

Let us consider the truncated Fourier series of the two periodic functions $a(x)$ and $b(x)$:

$$a_K(x) = \sum_{k=-K}^{K} \hat{a}_k\, e^{i\,k\,x}, \quad b_K(x) = \sum_{k=-K}^{K} \hat{b}_k\, e^{i\,k\,x}. \qquad (2.60)$$

The purpose is to calculate without aliasing the convolution sum (of type (2.54))

$$\hat{w}_k = \sum_{\substack{p,q=-K \\ p+q=k}}^{K} \hat{a}_p\, \hat{b}_q \qquad (2.61)$$

using the pseudospectral technique described in Section 2.8. The straight-forward use of this technique would lead to an expression analogous to (2.56) which here takes the form

$$\tilde{w}_k = \hat{w}_k + \sum_{\substack{p,q=-K \\ p+q=k+N}}^{K} \hat{a}_p\, \hat{b}_q + \sum_{\substack{p,q=-K \\ p+q=k-N}}^{K} \hat{a}_p\, \hat{b}_q, \qquad (2.62)$$

where $N = 2K + 1$ is the number of collocation points x_i and $\hat{w}_k$ is the convolution sum (2.61).

The algorithm is as follows

1. Extend the spectra $\{\hat{a}_k\}$, $\{\hat{b}_k\}$ in $\{\hat{a}'_k\}$, $\{\hat{b}'_k\}$ according to

$$\hat{a}'_k = \begin{cases} \hat{a}_k & \text{if} \quad |k| \le K \\ 0 & \text{if} \quad K < |k| \le K' \end{cases} \qquad \hat{b}'_k = \begin{cases} \hat{b}_k & \text{if} \quad |k| \le K \\ 0 & \text{if} \quad K < |k| \le K'. \end{cases}$$
$$(2.63)$$

The frequency K' will be found later to be equal to $3K/2$.

2. Calculate (by FFT) a_K and b_K in the new physical grid $x'_i = 2\pi i/N'$, $i = 1, \ldots, N' = 2K' + 1$:

$$a_K(x'_i) = \sum_{k=-K'}^{K'} \hat{a}'_k\, e^{i\,k\,x'_i}, \quad b_K(x'_i) = \sum_{k=-K'}^{K'} \hat{b}'_k\, e^{i\,k\,x'_i}. \qquad (2.64)$$

3. Multiply $a_K(x'_i)$ and $b_K(x'_i)$:

$$w(x'_i) = a_K(x'_i)\, b_K(x'_i), \quad i = 1, \ldots, N'. \qquad (2.65)$$

4. Calculate (by FFT) the Fourier coefficients of $w'(x'_i)$:

$$\tilde{w}'_k = \frac{1}{N'} \sum_{i=1}^{N'} w(x'_i)\, e^{-i\,k\,x'_i}, \quad k = -K', \ldots, K'. \qquad (2.66)$$

This latter quantity, taken for $k = -K, \ldots, K$, gives an expression free of aliasing of the convolution sum (2.61).

Now it remains to prove that, for a suitable value of K', the aliasing is effectively removed. For that we begin by considering the product $w(x'_i)$, as defined by Eq. (2.65), in which $a'_K(x_i)$ and $b'_K(x_i)$ are replaced by their expansions (2.64). We obtain a relation similar to (2.62), i.e.,

$$\tilde{w}'_k = \sum_{\substack{p,q \\ p+q=k}} \hat{a}'_p \hat{b}'_q + \sum_{\substack{p,q \\ p+q=k+N'}} \hat{a}'_p \hat{b}'_q + \sum_{\substack{p,q \\ p+q=k-N'}} \hat{a}'_p \hat{b}'_q, \quad k = -K, \ldots, K,$$

(2.67)

where the sums are taken on $-K' \leq p, q \leq K'$ or, more simply, on $-K \leq p, q \leq K$ since $\hat{a}'_p$ and $\hat{b}'_q$ are, respectively, zero for $|p| > K$ and $|q| > K$. The first sum in Eq. (2.67) is nothing other than the Galerkin convolution sum (2.61). Now the goal is to determine $N' = 2K' + 1$ such that the alias terms in (2.67) are not present.

The second sum (alias) in Eq. (2.67) is zero if one of the frequencies, p or q appearing in each term of the sum, is larger than K since $\hat{a}_p = 0$ for $p > K$ and $\hat{a}_q = 0$ for $q > K$. Observing that $q = 2K' + 1 + k - p$ with $-K \leq p, q \leq K$, the smaller value of q is $2K' + 1 - 2K$. Therefore, the above condition is satisfied if

$$2K' + 1 - 2K > K$$

that yields $2K' > 3K - 1$. So, we may take $K' \geq 3K/2$.

The reasoning for the third sum in Eq. (2.67) is analogous : one of the frequencies p or q must be smaller than $-K$. Then the larger value of $q = -2K' - 1 + k - p$ is $-2K' - 1 + 2K$. Then, by asking this maximal value to be smaller than $-K$, we get the same condition that held previously for the first sum.

In conclusion, a convenient value of K', for removing the alias term in the pseudospectral technique, is $K' = 3K/2$. It seems important to recall that the supplementary memory necessary for the aliasing removal technique has to be used only for the evaluation of the nonconstant-coefficient or nonlinear terms. Therefore, its influence on the global memory requirements is very weak.

The aliasing removal technique also applies to the collocation method, with the evident modification due to the fact that the computed quantities are no longer the Fourier coefficients but the grid values. For example, let us consider the nonlinear collocation equations (2.58). Knowing $u_K^n(x_i)$, the corresponding Fourier collocation coefficients $\hat{u}_k^n$ are calculated by FFT. The algorithm described above applies directly with $\hat{a}_k \equiv \hat{u}_k^n$ and $\hat{b}_k \equiv i k \hat{u}_k^n$. Then it must be completed by the following step:

5. Calculate (by FFT) $a_K(x_i) b_K(x_i) \equiv u_K^n(x_i) \partial_x u_k^n(x_i)$, namely,

$$a_K(x_i) b_K(x_i) = \sum_{k=-K}^{K} \tilde{w}'_k e^{i k x_i}, \quad i = 1, \ldots, N.$$

(2.68)

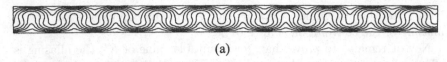

(a)

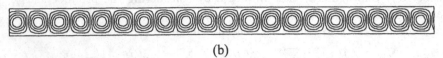

(b)

FIGURE 2.1. Rayleigh-Bénard convection at $Pr = 7$ and $Ra = 6000$: (a) isotherms, (b) streamlines.

This product is free of aliasing.

The effect of aliasing appears generally on the quality of the numerical results : oscillations and temporal instabilities. However, it is generally found that the effect of aliasing becomes negligible when the resolution is increased.

The influence of aliasing, associated to the various forms of the nonlinear terms of the incompressible Navier-Stokes equations, will be discussed in Section 7.2.2. We also refer to Canuto et al. (1988) for an extensive discussion on the phenomenon of aliasing.

Now, as an illustration of the phenomenon, we display an example where the alias frequencies appear clearly. This example (Fröhlich et al., 1991) concerns Rayleigh-Bénard convection at large aspect ratio ($A = 16.136$). We refer to Section 6.5.1 for the description of the Rayleigh-Bénard problem. The Prandtl number Pr is equal to 7 and the Rayleigh number Ra is 6000. Periodicity is assumed in the horizontal direction, so that the Fourier approximation is used in this direction. In the vertical direction the approximation makes use of a truncated expansion in Chebyshev polynomials. The solution method of the Navier-Stokes equations in Boussinesq's approximation (see Section 5.3) is described in Chapter 6.

A steady-state solution is obtained exhibiting 10 structures (Fig. 2.1) : in the horizontal direction the flow is periodic with the basic frequency $k_0 = 10$. Figure 2.2.a shows the Fourier spectrum of the third Chebyshev coefficient of the vorticity. The peaks 1 to 6 refer to the horizontal periodicity while the peaks A_1 to A_5 are aliasing errors. The phenomenon is here particularly perceptible for two reasons. First, this is due to the special shape of the spectrum, exhibiting a sequence of modes around the harmonics $(2p - 1)k_0$, $p = 1, \ldots, 6$, with nonzero amplitude separated by modes of vanishing amplitude. Second, the value of the cut-off frequency $K = N/2 = 128$ (even collocation) also has an effect on the marked appearance of alias peaks. The relationship between the computed collocation Fourier coefficients $\tilde{\omega}_k$ and the Galerkin coefficients $\hat{\omega}_k$ of the vorticity ω is

$$\tilde{\omega}_k = \hat{\omega}_k + \hat{\omega}_{k+N} + \hat{\omega}_{k-N} .$$

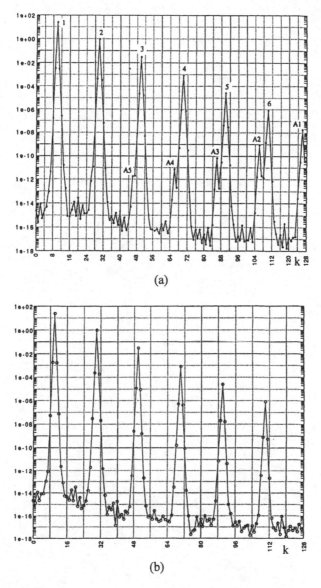

FIGURE 2.2. Fourier spectrum of the third Chebyshev coefficient : (a) with aliasing, (b) without aliasing.

The peaks A_1 to A_5 correspond to the values of k such that $k \pm N = \pm(2p-1)k_0$ with $p = 7, 8, \ldots$. As a matter of fact, the peaks A_1 to A_5 are obtained by folding the part above $k = 128$ of the entire spectrum around the cut-off frequency $K = 128$. Finally, Fig. 2.2.b show the spectrum obtained after removing the aliasing : the peaks A_1 to A_6 have completely disappeared.

3
Chebyshev method

The Fourier method is appropriate for periodic problems, but is not adapted to nonperiodic problems because of the existence of the Gibbs phenomenon at the boundaries. In the case of nonperiodic problems, it is advisable to have recourse to better-suited basis functions. Orthogonal polynomials, like Chebyshev polynomials, constitute a proper alternative to the Fourier basis. The Chebyshev series expansion may be seen as a cosine Fourier series, so that it possesses the valuable properties of the latter concerning, in particular, the convergence and the possible use of the FFT. On the other hand, the Chebyshev series expansion is exempt from the Gibbs phenomenon at the boundaries.

Another possible choice of an orthogonal polynomial basis is constituted by the Legendre polynomials. These polynomials share a number of properties with the Chebyshev polynomials. They present some advantages concerning the properties of the discrete operators and the numerical quadrature. On the other hand, no fast transform algorithm is known for Legendre polynomials. Only Chebyshev polynomials are discussed in this book, but the methods and algorithms described also apply to Legendre polynomials with only technical changes required by their specific properties. We refer to the books by Gottlieb and Orszag (1977), Canuto et al. (1988), Bernardi and Maday (1992) or Funaro (1992) for discussions on the properties and applications of the Legendre polynomials.

The present chapter is intended to give a general view of Chebyshev polynomials and their applications to the solution of boundary value problems. Two classical approaches, Galerkin-type (tau method) and collocation, will be addressed. In this latter case, it will be pointed out that the Chebyshev truncated series expansion can be seen as the Lagrange interpolation polynomial based on the collocation points. Direct and iterative methods for solving the algebraic systems, resulting from the Chebyshev approximation, will be described.

3.1   Generalities on Chebyshev polynomials

The Chebyshev polynomial of the first kind $T_k(x)$ is the polynomial of degree k defined for $x \in [-1, 1]$ by

$$T_k(x) = \cos\left(k \cos^{-1} x\right), k = 0, 1, 2, \ldots, \tag{3.1}$$

therefore, $-1 \leq T_k \leq 1$. By setting $x = \cos z$, we have

$$T_k = \cos kz, \tag{3.2}$$

from which it is easy to deduce the expressions for the first Chebyshev polynomials

$$T_0 = 1, \ \ T_1 = \cos z = x, \ \ T_2 = \cos 2z = 2\cos^2 z - 1 = 2x^2 - 1, \ldots.$$

More generally, from the Moivre formula, we get

$$\cos kz = \mathcal{R}e\left\{(\cos z + \underline{i} \sin z)^k\right\}$$

and then, by application of the binomial formula, we may express the polynomial T_k as

$$T_k = \frac{k}{2} \sum_{m=0}^{[k/2]} (-1)^m \frac{(k - m - 1)!}{m! \, (k - 2m)!} \, (2x)^{k-2m}, \tag{3.3}$$

where $[\phi]$ denotes the integer part of ϕ.

From the trigonometrical identity

$$\cos(k + 1)z + \cos(k - 1)z = 2\cos z \cos kz$$

we deduce the recurrence relationship

$$T_{k+1} - 2x\, T_k + T_{k-1} = 0, \ \ k \geq 1, \tag{3.4}$$

which allows us, in particular, to deduce the expression of the polynomials T_k, $k \geq 2$, from the knowledge of T_0 and T_1. The graph of the first polynomials is shown in Fig. 3.1.

Expression (3.3) may be useful in some special circumstances but the representation (3.2) is generally used in computational as well as theoretical studies.

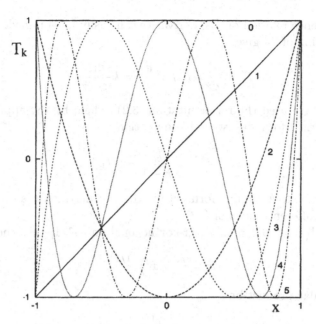

FIGURE 3.1. Graphs of the first Chebyshev polynomials, $T_k(x)$, for $k = 0, \cdots, 5$.

Now we list some properties useful for the understanding and application of Chebyshev polynomials to the solution of ordinary or partial differential equations ; other properties will be discussed in the sequel when necessary. And then, a richer set of formulas is given in Appendix A.

The values of T_k and its first-order derivative T'_k at $x = \pm 1$ are given by

$$T_k(\pm 1) = (\pm 1)^k, \quad T'_k(\pm 1) = (\pm 1)^{k+1} k^2. \tag{3.5}$$

The knowledge of these values can be of interest when prescribing boundary conditions. It is important to note that

$$T_k(-x) = (-1)^k T_k(x), \tag{3.6}$$

that is the parity of the polynomial is the same as its degree k.

The polynomial T_k vanishes at the points x_i (Gauss points) defined by

$$x_i = \cos\left(i + \frac{1}{2}\right)\frac{\pi}{k}, \quad i = 0, \ldots, k - 1 \tag{3.7}$$

and it reaches its extremal values ± 1 at the points x_i (Gauss-Lobatto points) defined by

$$x_i = \cos\frac{\pi i}{k}, i = 0, \ldots, k. \tag{3.8}$$

Note that such points are the zeros of the polynomial $\left(1 - x^2\right) T'_k(x)$.

A recurrence relation on the derivative can easily be obtained. First, the differentiation of T_k gives

$$T_k' = \frac{d}{dz}(\cos kz)\frac{dz}{dx} = k\frac{\sin kz}{\sin z},$$

where we have used the representation (3.2). Then, by the application of trigonometrical formulas, we get the relation

$$\frac{T_{k+1}'}{k+1} - \frac{T_{k-1}'}{k-1} = 2\,T_k \tag{3.9}$$

valid for $k > 1$. A similar formula for the pth derivative is obtained by successive differentiations of (3.9).

The Chebyshev polynomials are orthogonal on $[-1, 1]$ with the weight

$$w = \left(1 - x^2\right)^{-1/2}. \tag{3.10}$$

Let the scalar product be

$$(u, v)_w = \int_{-1}^{1} u\,v\,w\,dx, \tag{3.11}$$

so that the orthogonality property is

$$(T_k, T_l)_w = \int_{-1}^{1} T_k\,T_l\,w\,dx = \frac{\pi}{2}c_k\,\delta_{k,l}, \tag{3.12}$$

where $\delta_{k,l}$ is the Kronecker delta and c_k is defined by

$$c_k = \begin{cases} 2 & \text{if } k = 0, \\ 1 & \text{if } k \geq 1. \end{cases} \tag{3.13}$$

The Chebyshev approximation makes extensive use of the Gauss quadrature formulas. For the Gauss-Lobatto points $x_i = \cos \pi i/N$, $i = 0, \ldots, N$ [see Eq.(3.8)], generally used in collocation methods, the quadrature formula applied to any function $p(x)$ gives

$$\int_{-1}^{1} p\,w\,dx \cong \frac{\pi}{N}\sum_{i=0}^{N}\frac{p(x_i)}{\bar{c}_i}, \tag{3.14}$$

where

$$\bar{c}_k = \begin{cases} 2 & \text{if } \quad k = 0, \\ 1 & \text{if } \quad 1 \leq k \leq N-1, \\ 2 & \text{if } \quad k = N. \end{cases} \tag{3.15}$$

The relation (3.14) is exact if $p(x)$ is a polynomial of degree $2N - 1$ at most [see Mercier (1989) for the proof and formulas associated with other sets of points].

From Eq.(3.14) we may derive the discrete orthogonality relation based on the Gauss-Lobatto points x_i, $i = 0, \ldots, N$. For $k \neq N$ or $l \neq N$, the use of (3.14) gives an exact approximation to the integral in (3.12) since $T_k T_l$ is a polynomial of degree at most $2N - 1$:

$$\frac{\pi}{2} c_k \, \delta_{k,l} = \int_{-1}^{1} T_k \, T_l \, w \, dx = \frac{\pi}{N} \sum_{i=0}^{N} \frac{1}{\bar{c}_i} T_k(x_i) \, T_l(x_i) \, .$$

For $k = l = N$, this last formula remains exact provided c_k in the left-hand side is replaced by $\bar{c}_N$ ($= 2$). Therefore, the discrete orthogonality relation is

$$\sum_{i=0}^{N} \frac{1}{\bar{c}_i} T_k(x_i) \, T_l(x_i) = \frac{\bar{c}_k}{2} \, N \, \delta_{k,l} \qquad (3.16)$$

valid for $0 \leq k, l \leq N$.

3.2 Truncated Chebyshev series

3.2.1 Calculation of Chebyshev coefficients

Let us consider the Chebyshev approximation of the function $u(x)$ defined for $x \in [-1, 1]$:

$$u_N(x) = \sum_{k=0}^{N} \hat{u}_k \, T_k(x) \, . \qquad (3.17)$$

The expansion coefficients $\hat{u}_k$, $k = 0, \ldots, N$, are determined by following the Galerkin-type technique described in Section 1.2.1. The residual $R_N = u - u_N$ is annuled in the weak average sense

$$(R_N, T_l)_w = 0 \, , \quad l = 0, \ldots, N \, , \qquad (3.18)$$

namely,

$$\int_{-1}^{1} \left(u \, T_l \, w - \sum_{k=0}^{N} \hat{u}_k \, T_k \, T_l \, w \right) dx = 0 \, , \quad l = 0, \ldots, N \, .$$

Then, taking the orthogonality condition (3.12) into account, we obtain the expression for the Chebyshev expansion coefficients

$$\hat{u}_k = \frac{2}{\pi c_k} \int_{-1}^{1} u \, T_k \, w \, dx \, . \qquad (3.19)$$

It seems worthwhile to express (3.17) by means of the representation (3.2), that is, $T_k = \cos kz$ with $x = \cos z$. The expansion (3.17) is then written as

$$u_N = \sum_{k=0}^{N} \hat{u}_k \, \cos kz \, , \qquad (3.20)$$

showing that the Chebyshev expansion (3.17) with respect to x is equivalent to a cosine Fourier series in z. In fact, the function

$$v(z) = u(\cos z) = u(x)$$

defined in $0 \le z \le 2\pi$ is even and periodic since $v(z + 2\pi) = v(z)$. Moreover, $v(z)$ has as many bounded derivatives in $0 \le z \le \pi$ than $u(x)$ has in $-1 \le x \le 1$. Therefore, the convergence properties of the cosine Fourier series expansion (3.20) can be deduced from the results of Section 2.1.2. Moreover, since $v(z)$ is periodic, its representation (3.20) is continuous at the extremities $z = 0$ and $z = \pi$ and, consequently, is exempt from the Gibbs phenomenon at these points. More detailed results on the convergence of the Chebyshev approximation will be given in Section 3.6.

3.2.2 *Differentiation*

The expression of derivatives in the Chebyshev basis is more complicated than in the Fourier one. Indeed, the expression of the derivative of $T_k(x)$ involves all the polynomials of opposite parity and lower degree while the derivative of $e^{i k x}$ is simply $i k e^{i k x}$. This makes the computational aspects of the two approximations very different : the Chebyshev differentiation matrices in the spectral and physical spaces are full while the analogous Fourier matrices are full only in the physical space.

From the recurrence relation (3.9), one obtains

$$T_k'(x) = 2k \sum_{n=0}^{K} \frac{1}{c_{k-1-2n}} T_{k-1-2n}(x), \tag{3.21}$$

where $K = [(k-1)/2]$. Therefore, considering the first-order derivative

$$u_N'(x) = \sum_{k=0}^{N} \hat{u}_k T_k'(x) = \sum_{k=0}^{N} \hat{u}_k^{(1)} T_k(x) \tag{3.22}$$

and, taking Eq.(3.21) into account, we deduce the expression of the coefficient $\hat{u}_k^{(1)}$:

$$\hat{u}_k^{(1)} = \frac{2}{c_k} \sum_{\substack{p=k+1 \\ (p+k)\ \text{odd}}}^{N} p\,\hat{u}_p, \quad k = 0, \ldots, N-1, \tag{3.23}$$

and $\hat{u}_N^{(1)} = 0$. This can be written in matrix form as

$$\hat{U}^{(1)} = \hat{D}\,\hat{U}, \tag{3.24}$$

where $\hat{U} = (\hat{u}_0, \ldots, \hat{u}_N)^T$, $\hat{U}^{(1)} = \left(\hat{u}_0^{(1)}, \ldots, \hat{u}_N^{(1)}\right)^T$ and $\hat{D}$ is a strictly triangular upper matrix whose entries are deduced from (3.23).

The second-order derivative expansion is

$$u''_N(x) = \sum_{k=0}^{N} \hat{u}_k^{(2)} T_k(x) \tag{3.25}$$

with

$$\hat{u}_k^{(2)} = \frac{1}{c_k} \sum_{\substack{p=k+2 \\ (p+k)\ \mathrm{even}}}^{N} p\left(p^2 - k^2\right) \hat{u}_p, \quad k = 0, \ldots, N-2 \tag{3.26}$$

and $\hat{u}_{N-1}^{(2)} = \hat{u}_N^{(2)} = 0$. This is written in matrix form as

$$\hat{U}^{(2)} = \hat{D}^2\,\hat{U}, \tag{3.27}$$

where $\hat{U}^{(2)} = \left(\hat{u}_0^{(2)}, \ldots, \hat{u}_N^{(2)}\right)^T$.

The analytical expressions (3.23) and (3.25) are of interest each time the expansion coefficients of the derivatives are involved in algebraic calculations. On the other hand, if only the numerical values of the coefficients are needed, they can be calculated either from the matrix-vector products (3.24) and (3.27) or from recurrence formulas deduced from (3.9). More precisely, the expression for T_k given by (3.9) is brought into (3.21). Then, by identification of the derivative T'_k with the same index, we obtain the recurrence formula for the first-order derivative. The general recurrence formula for the coefficients $\hat{u}_k^{(p)}$ of the pth derivative is obtained by successive differentiations, let

$$c_{k-1}\,\hat{u}_{k-1}^{(p)} = \hat{u}_{k+1}^{(p)} + 2\,k\,\hat{u}_k^{(p-1)}, \quad k \geq 1, \tag{3.28}$$

be complemented with the starting values, for the first-order derivative

$$\hat{u}_N^{(1)} = 0, \quad \hat{u}_{N-1}^{(1)} = 2\,N\,\hat{u}_N, \tag{3.29}$$

and, for the second-order derivative,

$$\hat{u}_N^{(2)} = \hat{u}_{N-1}^{(2)} = 0, \quad \hat{u}_{N-2}^{(2)} = 2\,(N-1)\,\hat{u}_{N-1}^{(1)} = 2\,N\,(N-1)\,\hat{u}_N. \tag{3.30}$$

The recurrence relation (3.28) for $p = 2$ can be replaced by another connecting directly the coefficients $\hat{u}_k^{(2)}$ to $\hat{u}_k$. Such a relation is obtained by considering (3.28) with $p = 2$ and written for $k-1$ and $k+1$, such as the quantities $\hat{u}_{k-1}^{(1)}$ and $\hat{u}_{k+1}^{(1)}$ appear in the right-hand sides. Then these two equations are combined so that the quantities $\hat{u}_{k-1}^{(1)}$ and $\hat{u}_{k+1}^{(1)}$ can be eliminated thanks to (3.28) considered for $p = 1$. The resulting recurrence relation is

$$P_k\,\hat{u}_{k-2}^{(2)} + Q_k\,\hat{u}_k^{(2)} + R_k\,\hat{u}_{k+2}^{(2)} = \hat{u}_k, \quad 2 \leq k \leq N \tag{3.31}$$

with

$$P_k = \frac{c_{k-2}}{4\,k(k-1)}, \quad Q_k = \frac{-e_{k+2}}{2(k^2-1)}, \quad R_k = \frac{e_{k+4}}{4\,k(k+1)}, \qquad (3.32)$$

where

$$e_j = \begin{cases} 1 & \text{if } j \le N, \\ 0 & \text{if } j > N. \end{cases} \qquad (3.33)$$

Concerning the recurrent algorithm (3.28), Wengle and Seinfeld (1978) have remarked that it may be ill-conditioned in the sense that errors in the smallest coefficient $\hat{u}_k^{(p-1)}$ are amplified such that the accuracy of all the coefficients, even the largest ones $\hat{u}_k^{(p)}$, is destroyed. This can be avoided by simply equating to zero the coefficients smaller than a given threshold, depending on the accuracy of the computer.

3.3   Discrete Chebyshev series and collocation

This section is devoted to the Chebyshev collocation (i.e., interpolation) technique for the approximation of a given function. First, considering the discrete truncated Chebyshev series, the calculation of the expansion coefficients will be developed. Then the expression of the differentiation matrices will be established. Finally, we shall discuss an equivalent way to consider the Chebyshev expansion, namely by introducing the notion of Lagrange interpolation polynomial.

The collocation points considered here are the Gauss-Lobatto points defined by Eq.(3.8). Other sets of points, of similar nature (see Gottlieb *et al.*, 1984 ; Canuto *et al.*, 1988 ; Mercier, 1989), can be useful in some circumstances. For example, the choice of the set (3.7) may be of interest if it is not desired that the boundary points $x = \pm 1$ belong to the set of collocation points. Also, the Gauss-Radau points (see Appendix A) can be used if one wants to exclude the boundary point $x = -1$, for example in problems in cylindrical coordinates where $x = -1$ would correspond to the axis. On the other hand, for the solution of boundary value problems, the property of the set of collocation points held by the Gauss-Lobatto points to contain the boundaries is indispensable.

3.3.1   *Calculation of Chebyshev coefficients*

Considering the Chebyshev expansion (3.17), we want to calculate the coefficients $\hat{u}_k$ by means of the collocation (or interpolation) technique shown in Section 1.2.1. The technique consists of setting to zero the residual $R_N = u - u_N$ at the collocation points $x_i = \cos \pi i / N$, $i = 0, \dots, N$, let

$$u(x_i) = u_N(x_i) = \sum_{k=0}^{N} \hat{u}_k \, T_k(x_i), \quad i = 0, \ldots, N. \tag{3.34}$$

By denoting $u_i = u(x_i) = u_N(x_i)$, and using the definition (3.1), the above equation gives :

$$u_i = \sum_{k=0}^{N} \hat{u}_k \, \cos \frac{k\pi i}{N}, \quad i = 0, \ldots, N. \tag{3.35}$$

Equation (3.34) [or (3.35)] gives an algebraic system of $2N+1$ equations for determining the $2N+1$ coefficients $\hat{u}_k$. The associated matrix $\mathcal{T} = [\cos k\pi i/N]$, $k, i = 0, \ldots, N$, is invertible ; as a matter of fact, it will be found below [Eq.(3.37)] that its inverse is $\mathcal{T}^{-1} = [2\,(\cos \pi i/N)/(\bar{c}_k \bar{c}_i N)]$, $k, i = 0, \ldots, N$.

The expression for the coefficients $\hat{u}_k$ (i.e., the solution of the system (3.34)) is directly obtained by means of the discrete orthogonality relation (3.16). By multiplying each side of (3.34) by $T_l(x_i)/\bar{c}_i$, then summing from $i = 0$ to $i = N$, and using the relation (3.16), we obtain

$$\hat{u}_k = \frac{2}{\bar{c}_k N} \sum_{i=0}^{N} \frac{1}{\bar{c}_i} u_i \, T_k(x_i), \quad k = 0, \ldots, N, \tag{3.36}$$

or

$$\hat{u}_k = \frac{2}{\bar{c}_k N} \sum_{i=0}^{N} \frac{1}{\bar{c}_i} u_i \, \cos \frac{k\pi i}{N}, \quad k = 0, \ldots, N. \tag{3.37}$$

It must be noted that such expressions are nothing other than the numerical approximation (based on the Gauss-Lobatto points) of the integral appearing in Eq.(3.19).

The relations (3.35) and (3.36) show that the grid values u_i, as well as the coefficients $\hat{u}_k$, are related by truncated discrete Fourier series in cosine. Therefore, it is possible to use the FFT algorithm (in its cosine version) to connect the physical space (space of the grid values) to the spectral space (space of the coefficients). Note that it is also possible to simply make use of the matrix-vector products

$$U = \mathcal{T}\hat{U}, \quad \hat{U} = \mathcal{T}^{-1} U, \tag{3.38}$$

where U and $\hat{U}$ are, respectively, the vectors containing the grid values and the expansion coefficients. Note that the matrix-vector product for a moderate number of terms (namely 60-100) is less expensive in computing time than the FFT, depending on the computer and the routines used.

Results on the error of the collocation approximations (3.34) and (3.36) are given in Section 3.6.

3.3.2 Relation between collocation and Galerkin coefficients

The objective of this section is to make precise the relationship between the expansion coefficients defined by the integral (3.19) and those calculated from the sum (3.36). The first set of coefficients will be denoted here by $\hat{u}_k^e$ and the second set by $\hat{u}_k^c$. The general lines of the analysis made in Section 2.3 for the Fourier series apply in the present case. From Eqs.(3.19) and (3.36), we have

$$\hat{u}_k^e = \frac{2}{\pi\, c_k} \int_{-1}^{1} u(x)\, T_k(x)\, w(x)\, dx\,, \quad k = 0, \ldots, N\,, \tag{3.39}$$

$$\hat{u}_k^c = \frac{2}{\bar{c}_k\, N} \sum_{i=0}^{N} \frac{1}{\bar{c}_i} u(x_i)\, T_k(x_i)\,, \quad k = 0, \ldots, N\,. \tag{3.40}$$

Now we replace $u(x_i)$ in Eq.(3.40) by its expression in terms of the infinite series

$$u(x) = \sum_{k=0}^{\infty} \hat{u}_k^e\, T_k(x)$$

assumed to be absolutely convergent. Then, we decompose the resulting infinite sum into two partial sums according to

$$\hat{u}_k^c = \frac{2}{\bar{c}_k\, N} \sum_{l=0}^{N} \hat{u}_k^e \left[\sum_{i=0}^{N} \frac{1}{\bar{c}_i} T_k(x_i)\, T_l(x_i) \right]$$

$$+ \frac{2}{\bar{c}_k\, N} \sum_{l=N+1}^{\infty} \hat{u}_k^e \left[\sum_{i=0}^{N} \frac{1}{\bar{c}_i} T_k(x_i)\, T_l(x_i) \right]\,.$$

The expression appearing in square brackets is nothing other than the left-hand side of the discrete orthogonality relation (3.16). In the first bracket, the indices k and l vary between 0 and N, so that the relation (3.16) holds. On the other hand, in the second bracket, the index l varies between $N+1$ and infinity, so that the relation (3.16) is not applicable. The above expression for $\hat{u}_k^c$ can be written as

$$\hat{u}_k^c = \hat{u}_k^e + \frac{2}{\bar{c}_k\, N} \sum_{l=N+1}^{\infty} C_{kl}\, \hat{u}_l^e\,,$$

where

$$C_{kl} = \sum_{i=0}^{N} \frac{1}{\bar{c}_i} T_k(x_i)\, T_l(x_i) = \sum_{i=0}^{N} \frac{1}{\bar{c}_i} \cos \frac{k\, i\, \pi}{N} \cos \frac{l\, i\, \pi}{N}$$

$$= \frac{1}{2} \sum_{i=0}^{N} \frac{1}{\bar{c}_i} \left[\cos \frac{k-l}{N} i\, \pi + \cos \frac{k+l}{N} i\, \pi \right]$$

with $k = 0, \ldots, N$ and $l = N + 1, \ldots$. Then, by using the identity (valid for $p \in \mathbb{Z}$),

$$\sum_{i=0}^{N} \cos \frac{p i \pi}{N} = \begin{cases} N + 1 & \text{if } p = 2mN, \ m = 0, \pm 1, \pm 2, \ldots, \\ \frac{1}{2}[1 + (-1)^p] & \text{otherwise,} \end{cases}$$

we may calculate C_{kl} and finally get the relation connecting the collocation coefficients to the Galerkin ones

$$\hat{u}_k^c = \hat{u}_k^e + \frac{1}{\bar{c}_k} \left[\sum_{\substack{m=1 \\ 2mN > N-k}}^{\infty} \hat{u}_{k+2mN}^e + \sum_{\substack{m=1 \\ 2mN > N+k}}^{\infty} \hat{u}_{-k+2mN}^e \right]. \quad (3.41)$$

The terms in square brackets are alias terms. The reason for their presence is the same as that for discrete Fourier series (Sections 2.2 and 2.3). This is a consequence of the fact that the Chebyshev expansion in x can also be considered as a cosine Fourier in the variable $z = \cos^{-1} x$.

3.3.3 Lagrange interpolation polynomial

Let us return to the approximation

$$u_N(x) = \sum_{k=0}^{N} \hat{u}_k T_k(x), \quad (3.42)$$

where the coefficients $\hat{u}_k$, $k = 0, \ldots, N$, are determined by asking $u_N(x)$ to coincide with $u(x)$ at the collocation points $x_i = \cos \pi i / N$, $i = 0, \ldots, N$. Therefore, the polynomial of degree N defined by Eq.(3.42) is nothing other than the Lagrange interpolation polynomial based on the set $\{x_i\}$. Hence, it can also be written in the form

$$u_N(x) = \sum_{j=0}^{N} h_j(x) u(x_j) \quad (3.43)$$

with $u_N(x_j) = u(x_j)$, and $h_j(x)$ is the polynomial of degree N defined by

$$h_j(x) = \frac{(-1)^{j+1} (1 - x^2) T_N'(x)}{\bar{c}_j N^2 (x - x_j)}. \quad (3.44)$$

This expression for h_j is easily constructed by recalling that the collocation points x_j are the zeros of the polynomial $(1 - x^2) T_N'(x)$ (see Section 3.1.1) and by observing that $(1 - x^2) T_N'(x) / (x - x_j) \to (-1)^{j+1} \bar{c}_j N^2$ when $x \to x_j$, $j = 0, \ldots, N$.

Therefore, the representation (3.43) is equivalent to (3.42) and is useful in several circumstances because it does not involve the spectral coefficients.

3.3.4 Differentiation in the physical space

As mentioned in Section 1.1.3, and discussed in Section 2.7.2 for the Fourier case, the application of collocation methods to the solution of differential equations may consider as unknowns the expansion coefficients as well as the grid values. The first approach is seldom employed in the Chebyshev case (see, e.g., Marion and Gay, 1986). On the other hand, the second approach (also known as "orthogonal collocation," see Finlayson, 1972) is of current use in computational fluid mechanics since the works by Wengle (1979) for the advection-diffusion equation and by Orszag and Patera (1983) or Ouazzani and Peyret (1984) for the Navier-Stokes equations.

Therefore, in the context of the collocation method where the unknowns are the grid values, it is necessary to express the derivatives at any collocation point in terms of the grid values of the function, that is, for the pth derivative $u_N^{(p)}$:

$$u_N^{(p)}(x_i) = \sum_{j=0}^{N} d_{i,j}^{(p)}\, u_N(x_j)\,, \quad i = 0,\ldots,N\,. \tag{3.45}$$

The coefficients $d_{i,j}^{(p)}$ can be calculated according to either of the following two ways :

(i) Eliminate $\hat{u}_k$ from the derivative

$$u_N^{(p)}(x_i) = \sum_{k=0}^{N} \hat{u}_k\, T_k^{(p)}(x_i)$$

by using expression (3.36). Then, express $T_k(x_i)$ and the pth derivative $T_k^{(p)}(x_i)$ in terms of trigonometrical functions according to $T_k = \cos kz$. Finally, apply the classical trigonometrical identities to evaluate the sums.

(ii) Differentiate p times directly the interpolation polynomial (3.43) :

$$u_N^{(p)}(x) = \sum_{j=0}^{N} h_j^{(p)}(x_i)\, u_N(x_j)\,.$$

Therefore, $d_{i,j}^{(p)} = h_j^{(p)}(x_i)$ which has to be evaluated from expression (3.44).

The expression of the coefficients $d_{i,j}^{(p)}$ for the first two derivatives are :

First-order derivative

$$d_{i,j}^{(1)} = \frac{\bar{c}_i}{\bar{c}_j}\frac{(-1)^{i+j}}{(x_i - x_j)}\,, \qquad 0 \leq i,j \leq N,\ i \neq j$$

$$d_{i,i}^{(1)} = -\frac{x_i}{2(1 - x_i^2)}\,, \qquad 1 \leq i \leq N-1 \tag{3.46}$$

$$d_{0,0}^{(1)} = -d_{N,N}^{(1)} = \frac{2\,N^2 + 1}{6}\,,$$

where $x_i = \cos(\pi i/N)$, $\bar{c}_0 = \bar{c}_N = 2$, $\bar{c}_j = 1$ for $1 \le j \le N-1$.

Second-order derivative

$$d_{i,j}^{(2)} = \frac{(-1)^{i+j}}{\bar{c}_j} \frac{x_i^2 + x_i x_j - 2}{(1 - x_i^2)(x_i - x_j)^2}, \qquad \begin{array}{l} 1 \le i \le N-1, \\ 0 \le j \le N, \ i \neq j \end{array}$$

$$d_{i,i}^{(2)} = -\frac{(N^2 - 1)(1 - x_i^2) + 3}{3(1 - x_i^2)^2}, \qquad 1 \le i \le N-1$$

$$d_{0,j}^{(2)} = \frac{2}{3} \frac{(-1)^j}{\bar{c}_j} \frac{(2N^2 + 1)(1 - x_j) - 6}{(1 - x_j)^2}, \qquad 1 \le j \le N \qquad (3.47)$$

$$d_{N,j}^{(2)} = \frac{2}{3} \frac{(-1)^{j+N}}{\bar{c}_j} \frac{(2N^2 + 1)(1 + x_j) - 6}{(1 + x_j)^2}, \qquad 0 \le j \le N-1$$

$$d_{0,0}^{(2)} = d_{N,N}^{(2)} = \frac{N^4 - 1}{15}.$$

It may be useful to recall that

$$d_{i,j}^{(2)} = \sum_{k=0}^{N} d_{i,k}^{(1)} d_{k,i}^{(1)}. \qquad (3.48)$$

In vector form, the derivatives may be expressed as

$$U^{(1)} = \mathcal{D} U, \quad U^{(2)} = \mathcal{D}^2 U \qquad (3.49)$$

where

$$U = (u_N(x_0), \dots, u_N(x_N))^T, \quad U^{(p)} = \left(u_N^{(p)}(x_0), \dots, u_N^{(p)}(x_N) \right)^T$$

$$(3.50)$$

with $p = 1, 2$. The differentiation matrix $\mathcal{D}$ is defined by

$$\mathcal{D} = \left[d_{i,j}^{(1)} \right], \quad i, j = 0, \dots, N. \qquad (3.51)$$

3.3.5 Round-off errors

An important question, when dealing with Chebyshev approximations of high degree N, concerns the effect of round-off errors. A possible cause of error associated with the use of the recurrence formulas for calculating the coefficients of the Chebyshev expansion of derivatives, has been mentioned in Section 3.2.2. Another way to calculate the derivatives is to make use of the differentiation matrices whose entries have been given in the previous section. This technique is very often employed because it is not necessary to have recourse to the FFT algorithm, although this latter is much more economical for high resolution.

The possible sources of the magnification of the round-off errors, associated with the calculation of derivatives through matrix-vector products, have been analyzed in several works (Rothman, 1991 ; Breuer and Everson, 1992 ; Bayliss *et al.*, 1994).

One cause is the disparity in magnitude between the entries of the matrices. For example, in the case of the first-order differentiation matrix $\mathcal{D} = [d_{i,j}^{(1)}]$, the smallest elements are $O(1)$ while the largest ones are $O(N^2)$. Another cause can be found in the computation of the matrix elements which involve the subtraction of nearly equal numbers, as observed by Rothman (1991) who proposes a simple way to minimize these round-off errors. The technique consists of using trigonometrical identities to express the quantity $(x_i - x_j)$ in $d_{i,j}^{(1)}$ and $(1 - x_i)^2$ in $d_{i,i}^{(1)}$, let

$$x_i - x_j = 2 \sin \frac{(j+i)\pi}{2N} \sin \frac{(j-i)\pi}{2N}, \quad 1 - x_i^2 = \sin^2 \frac{i\pi}{N}. \tag{3.52}$$

Numerical experiments have shown the efficiency of this simple procedure. On the other hand, it is inefficient and even worse when applied to the second-order differentiation matrix $\mathcal{D}^2$. For such a matrix, Rothman recommends not using the analytical expressions $d_{i,j}^{(2)}$ given in the previous section but rather to calculate numerically $\mathcal{D}^2$ as the square of the matrix $\mathcal{D}$ whose entries $d_{i,j}^{(1)}$ are computed using the trigonometrical relations (3.52).

Another cause of round-off errors, identified by Bayliss *et al.* (1994), is the failure of the computed differentiation matrix to represent exactly the derivative of a constant, that is, to numerically satisfy the identity

$$\sum_{j=0}^{N} d_{i,j}^{(1)} = 0, \quad i = 0, \dots, N. \tag{3.53}$$

The remedy suggested by Bayliss *et al.* (1994) is to calculate the off-diagonal entries $d_{i,j}^{(1)}$, $j \neq i$, by means of the formulas (3.46), and the diagonal entries $d_{i,i}^{(1)}$ by

$$d_{i,i}^{(1)} = - \sum_{\substack{j=0 \\ j \neq i}}^{N} d_{i,j}^{(1)}, \quad i = 0, \dots, N. \tag{3.54}$$

Such a procedure gives a substantial improvement concerning the effect of round-off errors. Compared to the above technique consisting of the use of trigonometrical expressions, the present technique (Bayliss *et al.*, 1994) gives much better results. However, the use of the trigonometrical relations (3.52) in the correction technique of Bayliss *et al.* (1994) does not bring any improvement as shown by numerical experiments performed on various functions.

In conclusion, for the calculation of the first-order derivative, we recommend using the correction technique of Bayliss *et al.* (1994) with the entries $d_{i,j}^{(1)}$ (for $j \neq i$) calculated with the analytic expressions given in (3.46) (without the use of trigonometrical expressions) and the diagonal entries $d_{i,i}^{(1)}$ with Eq.(3.54).

For the calculation of the second-order derivative, a large number of possibilities can be envisaged by combining the various techniques invoked above. From the numerical experiments performed, any technique has been found better than the other ones for all the tested functions. However, the technique based on the Bayliss *et al.* correction often gives the best results and, when this is not the case, the results are not far from the best ones.

In conclusion, for the calculation of the second-order derivative we recommend the following technique :

1. Calculate a provisional second-order differentiation matrix $\tilde{\mathcal{D}}^2$ as the square product of the matrix $\mathcal{D}$ whose entries $d_{i,j}^{(1)}$ are calculated following the correction technique of Bayliss *et al.*, based on Eq.(3.46) and (3.54).

2. From the entries $\tilde{d}_{i,j}^{(2)}$ of $\tilde{\mathcal{D}}^2$, calculate the entries $d_{i,j}^{(2)}$ of the final second-order matrix $\mathcal{D}^2$ by a repeated application of the Bayliss *et al.* correction technique, let

$$d_{i,j}^{(2)} = \tilde{d}_{i,j}^{(2)} \text{ for } j \neq i,$$

$$d_{i,i}^{(2)} = - \sum_{\substack{j=0 \\ j \neq i}}^{N} \tilde{d}_{i,j}^{(2)}, \quad i = 0, \dots, N. \tag{3.55}$$

It is important to note that, in a general way, the effect of round-off errors on the calculation of the derivatives is significant when the polynomial degree N is larger than the value N_0 needed to represent the function itself within the machine accuracy. In other words, the value of N, for which the computation of the derivatives is strongly subject to the amplification of round-off errors, depends on the function under consideration, so that, in some cases, it can be rather large. In conclusion, it is recommended to avoid, as much as possible, over-resolution.

Moreover, it has been observed that the errors on the second-order (or higher) derivative are always larger than those associated with the first-order derivative.

The solution of differential equations makes use of the differentiation matrices to construct the approximate discrete operators. As pointed out by Breuer and Everson (1992), and generally observed in the calculations (see Section 3.4.3), the effect of round-off errors is much less marked when solving differential equations than when computing the derivatives of a function.

3.3.6   Relationship with finite-difference and similar approximations

Formula (3.45) makes clear the relationship between finite-difference and collocation Chebyshev approximations. In a finite-difference method, the approximation of a derivative at a grid point involves only very few neighbouring grid values of the function, while the Chebyshev approximation involves all the grid values. The finite-difference formula approximating a derivative can be formally obtained by representing the function under consideration through a local Lagrange interpolation polynomial of low degree (Ferziger, 1981), this polynomial changing from one discretization point to another. In the Chebyshev method, the interpolation polynomial is the same in the whole domain, it involves the values of the function at all collocation points and, consequently, the formula expressing the derivative, like (3.45), also involves all the grid values.

Higher-order finite-difference-type approximations can be constructed through Hermitian methods, which amount to considering a local Hermite interpolation polynomial (Peyret, 2000), and are closely connected to spline-function approximations (Rubin and Khosla, 1977). These approximations, like the classical finite-difference methods, the finite-volume methods and the finite-element methods are local, contrary to the Chebyshev (as well as Fourier) approximation which is of a global nature.

The global character of spectral methods is beneficial for accuracy. On the other hand, the solution of a differential problem at a given point is strongly dependent on the solution at all other points and, in particular, at the boundaries. As a consequence, the influence of the way in which the boundary conditions are handled is not localized, as is often the case with local-type methods, but extends to the whole computational domain. Hence, great care must be taken when prescribing the boundary conditions. For the same reason, the presence of local singularities may contaminate the solution everywhere even at large distances through Gibbs-type oscillations. Therefore, it is recommended employing spectral methods for computing sufficiently smooth functions. However, sophisticated filtering methods (Gottlieb and Shu, 1997), domain decomposition methods, and techniques of subtraction of the singularity constitute interesting approaches to the solution of singular problems. The latter two approaches are discussed in Chapter 8.

Moreover, the matrices associated with the Chebyshev approximation of a differential problem are full, contrary to the local-type approximations. These matrices are not symmetric nor skew-symmetric, and they are generally ill-conditioned. Therefore, great care must also be taken when solving the corresponding algebraic systems. This point will be discussed in Section 3.4.

In spite of these apparent drawbacks ("apparent" because remedies exist for curing them), use of the spectral method is recommended for the representation of smooth solutions when high accuracy is required. The error associated with the Chebyshev (as well as Fourier) approximation is $O(1/N^m)$ where N refers to the truncation and m is connected to the number of continuous derivatives of the function under consideration (see Section 3.6 for Chebyshev methods and Section 2.1.2 for Fourier methods). In particular, for infinitely differentiable functions m is larger than any integer and then exponential accuracy is obtained. Such behaviour has to be compared to the $O(1/N^p)$ error of a local-type approximation, like the finite-difference method, where $1/N$ is the mesh size and p, which depends on the method, is essentially finite and even generally small.

Another advantage of the spectral approximation is that it is defined everywhere in the computational domain. Therefore, it is easy to get an accurate value of the function under consideration at any point of the domain, beside the collocation points. This property is often exploited, in particular to get a significant graphic representation of the solution, making apparent the possible oscillations due to a wrong approximation of the derivative.

Finally, an additional property of the spectral methods is the easiness with which the accuracy of the computed solution can be estimated. This can be done by simply checking the decrease of the spectral coefficients. There is no need to perform several calculations by modifying the resolution, as is usually done in finite-difference and similar methods for estimating the "grid-convergence".

3.4   Differential equation with constant coefficients

The application of the Chebyshev method to the solution of a one-dimensional boundary-value problem with constant coefficients is the objective of the present section. The tau method and the collocation method are successively described. The properties of each of these methods are discussed and compared by considering the numerical solution of the one-dimensional Helmholtz equation.

Let us consider the second-order differential problem

$$-\nu\, u'' + a\, u' + b\, u = f\,, \quad -1 < x < 1 \tag{3.56}$$

$$\alpha_-\, u(-1) + \beta_-\, u'(-1) = g_-\,, \tag{3.57}$$

$$\alpha_+\, u(1) + \beta_+\, u'(1) = g_+\,, \tag{3.58}$$

where f is a given function of x ; ν, a, b, $\alpha_\pm$, $\beta_\pm$, and $g_\pm$ are constant. In Section 1.3.1, we have introduced the traditional Galerkin method which applies when the basis functions $\varphi_k(x)$ satisfy the homogeneous boundary conditions, namely, Eqs.(3.57) and (3.58) with $g_- = g_+ = 0$. The Chebyshev polynomials do not meet this requirement. However, in some situations, it is possible to construct a basis verifying the homogeneous boundary conditions. For example, in the case of the Dirichlet conditions $u(\pm 1) = 0$, the suitable basis functions are

$$\varphi_k(x) = \begin{cases} T_k(x) - T_0 = T_k(x) - 1 & \text{if } k \text{ is even}, \\ T_k(x) - T_1(x) & \text{if } k \text{ is odd}. \end{cases}$$

However, the basis $\{\varphi_k\}$ is not orthogonal and such a Galerkin method is seldomly used. It is generally replaced by the simpler method known under the name of tau method, although this latter itself is now superseded by the more versatile collocation method.

3.4.1   Tau method

The solution $u(x)$ of Eqs.(3.56)-(3.58) is approximated with $u_N(x)$ defined by

$$u_N(x) = \sum_{k=0}^{N} \hat{u}_k \, T_k(x) \tag{3.59}$$

and the residual associated to the differential equation (3.56) is

$$R_N = -\nu \, u_N'' + a \, u_N' + b \, u_N - f. \tag{3.60}$$

The tau equations are obtained by setting to zero the first $N - 2$ scalar products

$$(R_N, T_i)_w = 0, \quad i = 0, \ldots, N - 2, \tag{3.61}$$

and by adding the boundary conditions (3.57) and (3.58). The derivatives u_N' and u_N'' are expressed respectively by Eqs.(3.22) and (3.25) so that Eq.(3.61) gives

$$(R_N, T_i)_w = \sum_{k=0}^{N} \left(-\nu \, \hat{u}_k^{(2)} + a \, \hat{u}_k^{(1)} + b \, \hat{u}_k \right) \int_{-1}^{1} T_k \, T_i \, w \, dx - \int_{-1}^{1} f \, T_i \, w \, dx = 0$$

for $i = 0, \ldots, N - 2$. Then, thanks to the orthogonality relation (3.12), we get

$$-\nu \, \hat{u}_k^{(2)} + a \, \hat{u}_k^{(1)} + b \, \hat{u}_k = \hat{f}_k, \quad k = 0, \ldots, N - 2, \tag{3.62}$$

where $\hat{f}_k = \int_{-1}^{1} f \, T_k \, w \, dx$ is the coefficient of the Chebyshev expansion of f. These equations are supplemented with the boundary conditions (3.57) and (3.58) which yield, taking (3.5) into account,

$$\sum_{k=0}^{N} (-1)^k \left(\alpha_- - \beta_- \, k^2 \right) \hat{u}_k = g_- \tag{3.63}$$

$$\sum_{k=0}^{N} \left(\alpha_+ + \beta_+ \, k^2\right) \hat{u}_k = g_+ \,. \tag{3.64}$$

Finally, by replacing $\hat{u}_k^{(1)}$ and $\hat{u}_k^{(2)}$ in Eq.(3.62) by their respective expressions (3.23) and (3.26) in terms of the coefficients $\hat{u}_k$, $k = 0, \ldots, N$, the system (3.62)-(3.64) can be written in the form

$$\mathcal{A}\hat{U} = F \,, \tag{3.65}$$

where $\hat{U} = (\hat{u}_0, \ldots, \hat{u}_N)^T$, $F = \left(\hat{f}_0, \ldots, \hat{f}_{N-2}, g_-, g_+\right)^T$, and $\mathcal{A} = [a_{ij}]$, $i, j = 0, \ldots, N$, is the $(N + 1) \times (N + 1)$ matrix constructed from the matrix

$$\mathcal{Q} = -\nu \hat{D}^2 + a\hat{D} + b\mathcal{I} \,, \tag{3.66}$$

[where $\hat{D}$ is defined by Eqs.(3.23), (3.24), and $\mathcal{I}$ is the identity matrix] in which the two last lines are, respectively, replaced by

$$\begin{aligned} a_{N-1,j} &= \ (-1)^j \left(\alpha_- - \beta_- \, j^2\right) , \\ a_{N,j} &= \ \alpha_+ + \beta_+ \, j^2 \,, \qquad j = 0, \ldots, N. \end{aligned}$$

The matrix $\mathcal{A}$ is ill-conditioned so that the solution of the system (3.65) requires the use of adapted methods. However, we do not go further in that direction for two reasons. First, the tau method is less and less employed in favour of the collocation method. Second, when considering the Navier-Stokes equations (which is the main subject of the present book) the equations to be solved are of Helmholtz type [i.e., Eq.(3.56) with $a = 0$] which can be solved very efficiently as will be discussed below. The absence of the first-order derivative comes from the fact that the first-order derivative terms in the Navier-Stokes equations, being nonlinear, are generally handled in an explicit way and, hence, are not involved in the algebraic system to be solved.

In the case of the Helmholtz equation with Dirichlet or Neumann boundary conditions, the solution of the resulting system (3.65) reduces to the solution of two quasi-tridiagonal systems (Elliott, 1961 ; Haidvogel, 1977; Gottlieb and Orszag, 1977) whose construction is now described in the Dirichlet case.

The Dirichlet problem for the Helmholtz equation is

$$-\nu \, u'' + b \, u = f \,, \quad -1 < x < 1 \tag{3.67}$$

$$u(-1) = g_- \,, \quad u(1) = g_+ \,, \tag{3.68}$$

and the tau equations, deduced from (3.62) and (3.64), are

$$-\nu \, \hat{u}_k^{(2)} + b \, \hat{u}_k = \hat{f}_k \,, \quad k = 0, \ldots, N - 2, \tag{3.69}$$

$$\sum_{k=0}^{N} (-1)^k \hat{u}_k = g_-, \tag{3.70}$$

$$\sum_{k=0}^{N} \hat{u}_k = g_+. \tag{3.71}$$

Now the goal is to eliminate the coefficients $\hat{u}_k^{(2)}$ thanks to the recurrence relation (3.31). More precisely, denoting Eq.(3.69) by E_k, we construct the linear combination $P_k E_{k-2} + Q_k E_k + R_k E_{k+2}$, with P_k, Q_k, and R_k as defined in (3.32). Then, using Eq.(3.31), we get the equations

$$P_k' \hat{u}_{k-2} + Q_k' \hat{u}_k + R_k' \hat{u}_{k+2} = \varphi_k, \quad k = 2, \dots, N, \tag{3.72}$$

where

$$P_k' = b\, P_k, \quad Q_k' = b\, Q_k - \nu, \quad R_k' = b\, R_k,$$

and

$$\varphi_k = P_k \hat{f}_{k-2} + Q_k \hat{f}_k + R_k \hat{f}_{k+2}.$$

It is easy to see that Eqs (3.72) lead to two uncoupled systems : one for the even coefficients and the other for the odd ones. These systems are closed by adding supplementary equations obtained from the boundary conditions. By addition of Eqs (3.70) and (3.71) we get an equation involving the even coefficients only, and by subtraction a similar equation is obtained for the odd modes, that is

$$\hat{u}_0 + \hat{u}_2 + \dots = \frac{1}{2}(g_+ + g_-), \tag{3.73}$$

$$\hat{u}_1 + \hat{u}_3 + \dots = \frac{1}{2}(g_+ - g_-). \tag{3.74}$$

Therefore, the sets of even and odd coefficients are, respectively, the solutions of uncoupled algebraic systems. The associated matrices have a quasi-tridiagonal structure (three nonzero diagonals and one row) which allows a direct solution to be obtained from an extension of the usual LU decomposition algorithm. The reader will find in Appendix B the description of the algorithm as developed by Thual (1986). It is interesting to note that the operation count of the algorithm is $O(N)$ as for the pure tridiagonal system. Also, the algorithm is stable, that is, the solution is bounded, if the considered matrices satisfy the diagonally dominance-type conditions (see Appendix B). For $\nu > 0$, $b \geq 0$ (the usual case in the application to unsteady problems where b is essentially the reciprocal of the time-step), these conditions are satisfied. On the other hand, when $b < 0$, the diagonally dominant conditions are satisfied if $\nu > -b/3$.

The method described above for the Dirichlet problem applies to the case of the Neumann conditions but not to the general Robin-type boundary conditions (3.57)-(3.58). In the case where the first-order derivative term is present in the differential equation, the application of the method leads to a quasi-pentadiagonal algebraic system which can be solved with an analogous algorithm (Dennis and Quartapelle, 1985).

3.4.2 Collocation method

The fundamentals of the collocation method have been given in Section 1.3.2. It has been observed that the unknowns can be either the coefficients $\hat{u}_k$, $k = 0, \ldots, N$, of the expansion or the values $u_N(x_i)$, $i = 0, \ldots, N$, of the approximate solution at the collocation points x_i. Although both techniques are equivalent from the mathematical point of view, the discrete equations, however, are obviously different. The first technique has received little interest in the field of fluid mechanics. We refer to Marion and Gay (1986) for an application to the solution of the Navier-Stokes equations. On the other hand, the second technique, consisting of considering the grid values $u_N(x_i)$ as unknowns, is of current application. As a matter of fact, as mentioned in Section 3.3.3, this technique amounts to approximating the solution $u(x)$ with a polynomial $u_N(x)$ of degree at most equal to N, namely, $u_N \in \mathbb{P}_N$. Then, the differential equation is forced to be satisfied exactly by this polynomial at the inner collocation points.

Let us consider the differential problem (3.56)-(3.58) and the Gauss-Lobatto collocation points :

$$x_i = \cos \frac{\pi i}{N}, \quad i = 0, \ldots, N. \tag{3.75}$$

By setting to zero the residual $R_N(x)$ [defined by Eq.(3.60)] at the inner collocation points x_i, $i = 1, \ldots, N-1$, and by adding the boundary conditions, we obtain the collocation equations

$$-\nu\, u_N''(x_i) + a\, u_N'(x_i) + b\, u_N(x_i) = f(x_i), \quad i = 1, \ldots, N-1$$

$$\alpha_-\, u_N(x_N) + \beta_-\, u_N'(x_N) = g_- \tag{3.76}$$

$$\alpha_+\, u_N(x_0) + \beta_+\, u_N'(x_0) = g_+.$$

Now, from (3.45), these equations give

$$\sum_{j=0}^{N} \left(-\nu\, d_{i,j}^{(2)} + a\, d_{i,j}^{(1)} \right) u_N(x_j) + b\, u_N(x_i) = f(x_i), \quad i = 1, \ldots, N-1$$

$$\alpha_-\, u_N(x_N) + \beta_- \sum_{j=0}^{N} d_{N,j}^{(1)}\, u_N(x_j) = g_-$$

$$\alpha_+\, u_N(x_0) + \beta_+ \sum_{j=0}^{N} d_{0,j}^{(1)}\, u_N(x_j) = g_+, \tag{3.77}$$

or, in matrix form,

$$\mathcal{A}U = F, \tag{3.78}$$

where $U = (u_N(x_0), \ldots, u_N(x_N))^T$, $F = (g_+, f_1, \ldots, f_{N-1}, g_-)^T$, and $\mathcal{A}$ is the $(N+1) \times (N+1)$ matrix constructed from the matrix

$$\mathcal{Q} = -\nu\mathcal{D}^2 + a\mathcal{D} + b\mathcal{I}$$

[$\mathcal{D}$ is the differentiation matrix defined by Eq.(3.51) and $\mathcal{I}$ is the identity matrix] in which the first and last lines are replaced by quantities coming from the boundary conditions. More precisely, the entries $a_{i,j}$ of the matrix $\mathcal{A}$ have the following expression

$$a_{0,j} = \alpha_+ + \beta_+ \sum_{j=0}^{N} d_{0,j}^{(1)}, \quad a_{N,j} = \alpha_- + \beta_- \sum_{j=0}^{N} d_{N,j}^{(1)}, \quad j = 0, \ldots, N,$$

$$a_{i,j} = -\nu d_{i,j}^{(2)} + a d_{i,j}^{(1)} + b \delta_{i,j}, \quad i = 1, \ldots, N-1, \quad j = 0, \ldots, N.$$

Various methods are available for the solution of the system (3.78). However, it must be noticed that problems like (3.56)-(3.58) have generally to be solved at each time-cycle of an unsteady process. That is, they have to be solved a very large number of times, so the solution method must take this peculiarity into account. Therefore, an efficient method is the one which necessitates the least calculations possible at each time-cycle, leaving in a preprocessing stage, performed before the start of the time-integration, the calculations which can be done only once. Moreover, it is important that the calculations done at each time-cycle be adapted to modern computers taking advantage of vectorization and parallelization. In these respects, the method based on the LU decomposition, which constitutes an accurate way to solve the system (3.78), is not necessarily the most economical from the point of view of vectorization.

Moreover, the properties of the matrix $\mathcal{A}$ have also to be taken into account. The matrix $\mathcal{A}$ has no good properties : it is full, not symmetric nor skew-symmetric, and it is ill-conditioned. To be more precise, Tables 3.1 and 3.2 give information on the behaviour of the spectral radius ρ and the condition number κ of the operators associated with the first- and second-order derivatives supplied with various types of boundary conditions. We recall that, for an $n \times n$ matrix $\mathcal{M}$ with eigenvalues $\lambda_i(\mathcal{M})$, $i = 1, \ldots, n$, such that $\lambda_{max}(\mathcal{M}) = \max_i |\lambda_i(\mathcal{M})|$ and $\lambda_{min}(\mathcal{M}) = \min_i |\lambda_i(\mathcal{M})|$, the spectral radius $\rho(\mathcal{M})$ is defined by

$$\rho(\mathcal{M}) = \lambda_{max}(\mathcal{M}) \tag{3.79}$$

and the condition number $\kappa(\mathcal{M})$ is defined by

$$\kappa(\mathcal{M}) = \left[\frac{\lambda_{max}(\mathcal{M}^T\mathcal{M})}{\lambda_{min}(\mathcal{M}^T\mathcal{M})}\right]^{1/2}. \tag{3.80}$$

Operator	Boundary conditions	ρ	κ
u'	$u(-1) = 0$	$0.089N^2$ (40)	$0.049N^2$ (40)
u''	$u(-1) = 0$ $u(1) = 0$	$0.047N^4$ (50)	$0.020N^4$ (50)
u''	$u(-1) = 0$ $u'(1) = 0$	$0.047N^4$ (50)	$0.086N^4$ (50)
u''	$u'(-1) = 0$ $u'(1) = 0$	$0.014N^4$ (60)	$(\star)\ 0.0065N^4$ (30)

Table 3.1. Spectral radius ρ and condition number κ of the Chebyshev matrices approximating first- and second-order derivatives with various boundary conditions (the boundary values have been eliminated). The numbers in parentheses refer to values of N for which the given figures are correct. In $(\star)$ the null eigenvalue has been discarded.

Table 3.1 displays the behaviour, with respect to N of the spectral radius ρ and the condition number κ of the discrete operators, in the case where the boundary values have been eliminated thanks to the boundary conditions. Table 3.2 gives the same information for the full $(N + 1) \times (N + 1)$ matrices without elimination of the boundary values.

A good way to deal with the system (3.78) is to invert $\mathcal{A}$ in the pre-processing stage and to store its inverse $\mathcal{A}^{-1}$. So that, at each time-cycle, only one matrix-vector product has to be performed, this being efficiently done on vector computers. However, taking the ill-conditioning of $\mathcal{A}$ into account, the inversion has to be done by a routine whose accuracy is very reliable, for example, LINRG from IMSL library or F04AEF from NAG. Both routines are based on the LU decomposition. The first one gives directly the inverse $\mathcal{A}^{-1}$. The second one solves a set of equations

$$\mathcal{A} U^k = F^k$$

with $F^k = [\delta_{k,j}]$, $j = 0, \ldots, N$, then $\mathcal{A}^{-1}$ is the matrix constructed such that each column is made with the elements of one vector U^k.

Besides the use of an efficient inversion routine, it is recommended applying a suitable preconditionning before the inversion. A good preconditioner is the finite-difference centered operator associated with the discretization of the problem (3.56), (3.57) based on the Gauss-Lobatto points (3.75). Let $\mathcal{A}_0$ be the resulting tridiagonal matrix and let $\mathcal{A}_0^{-1}$ be its inverse, Eq.(3.78) is multiplied at the left by $\mathcal{A}_0^{-1}$, i.e.,

$$\mathcal{A}_0^{-1} \mathcal{A} U = \mathcal{A}_0^{-1} F.$$

Operator	Boundary conditions	ρ	κ
u'	$u(-1) = 0$	$0.089N^2$ (40)	$0.56N^{5/2}$ (50)
u''	$u(-1) = 0$ $u(1) = 0$	$0.047N^4$ (50)	$0.041N^{9/2}$ (30)
u''	$u(-1) = 0$ $u'(1) = 0$	$0.047N^4$ (50)	$0.088N^{9/2}$ (60)
u''	$u'(-1) = 0$ $u'(1) = 0$	$0.047N^4$ (80)	$(\star)\ 0.029N^{9/2}$ (60)

Table 3.2. Spectral radius ρ and condition number κ of the Chebyshev matrices approximating first- and second-order derivatives with various boundary conditions (the boundary values are not eliminated). The numbers in parentheses refer to values of N for which the given figures are correct. In $(\star)$ the null eigenvalue has been discarded.

The matrix $\mathcal{B} = \mathcal{A}_0^{-1}\mathcal{A}$ is found to be well-conditioned so that it can be inverted with negligible round-off errors. The solution of (3.73) is then calculated as $U = \mathcal{B}^{-1}\mathcal{A}_0^{-1}F$.

We refer to Section 3.6 for some results concerning the approximation error associated with the collocation and tau methods for solving the one-dimensional Poisson equation with Dirichlet conditions.

3.4.3 Error equation

The polynomial $u_N(x)$, obtained from the application of the tau method (Section 3.4.1) or the collocation method (Section 3.4.2), is an approximate solution of the problem (3.56)-(3.58), that is

$$Lu = f, \quad -1 < x < 1 \tag{3.81}$$

$$B_-u(-1) = g_-, \quad B_+u(1) = g_+. \tag{3.82}$$

It may be interesting to know what the equation is that is exactly solved by $u_N(x)$ (with the same boundary conditions). This equation, called the "error equation" (Canuto et al., 1988), has been considered by Gottlieb and Orszag (1977) and, subsequently, by several authors for theoretical studies on convergence and the stability of various spectral approximations. In Chapter 7, the error equation will be used to analyze the error on the divergence of the velocity in the solution of the Navier-Stokes equations when using the influence matrix technique. The construction of the error equation is described now by successively considering the tau and the collocation methods.

(a) Tau method

For the problem (3.81)-(3.82) the error equation is constructed from the equation

$$L\, u = f + p, \qquad (3.83)$$

where p is a function of x which is assumed to be represented in the Chebyshev basis by

$$p(x) = \sum_{k=0}^{\infty} \tau_k \, T_k(x). \qquad (3.84)$$

The coefficients τ_k are determined such that the polynomial u_N, the approximate solution of Eqs.(3.82)-(3.83) according to the tau method, satisfies the same problem but for all values of k, so that u_N will be the exact solution. The Galerkin equations associated with Eq.(3.83) are

$$(L\, u_N - f - p,\, T_i)_w = 0, \quad i = 0,\ldots,\infty. \qquad (3.85)$$

For $i = 0,\ldots,N-2$, these equations give

$$\frac{\pi}{2}\, c_i\, \tau_i = (L\, u_N - f_N,\, T_i)_w,$$

where $f_N = \sum_{k=0}^{N} \hat{f}_k\, T_k$ is the Chebyshev approximation to $f(x)$. Then, Eq.(3.85) yields $\tau_i = 0$, since u_N satisfies the tau equations

$$(L\, u_N - f_N,\, T_i)_w = 0, \quad i = 0,\ldots,N-2.$$

Then, for $i = N-1, N$, Eq.(3.85) yields

$$\frac{\pi}{2}\tau_{N-1} = (L\, u_N - f_N,\, T_{N-1})_w = (L\, u_N,\, T_{N-1})_w - \frac{\pi}{2}\hat{f}_{N-1},$$

$$\frac{\pi}{2}\tau_N = (L\, u_N - f_N,\, T_N)_w = (L\, u_N,\, T_N)_w - \frac{\pi}{2}\hat{f}_N.$$

Finally, for $i = N+1,\ldots,\infty$, Eq.(3.85) simply gives

$$\tau_i = -\hat{f}_i.$$

By collecting these results, we obtain, from Eq.(3.83), the error equation

$$L\, u = f_{N-2} + \tau'_{N-1}\, T_{N-1} + \tau'_N\, T_N, \qquad (3.86)$$

where f_{N-2} is the $N-2$ truncated Chebyshev expansion of f and

$$\tau'_{N-1} = \frac{2}{\pi}\, (L\, u_N,\, T_{N-1})_w, \quad \tau'_N = \frac{2}{\pi}\, (L\, u_N,\, T_N)_w. \qquad (3.87)$$

It may be observed that the error term $\tau_{N-1}T_{N-1} + \tau_N T_N$ is coming from the Galerkin equations corresponding to the last two test functions which have been discarded and replaced by the boundary conditions.

Obviously, the right-hand side of the error equation (3.86) is a polynomial of degree at most N, and the exact solution of this equation that satisfies the boundary conditions (3.82) is the polynomial u_N, the solution of the tau equations of Section 3.4.1.

For the differential operator L defined by Eq.(3.56) the error equation is

$$L\,u = f_{N-2} + (2\,a\,N\,\hat{u}_N + b\,\hat{u}_{N-1})\,T_{N-1} + b\,\hat{u}_N\,T_N,$$

where the relation $\hat{u}^{(1)}_{N-1} = 2\,N\,\hat{u}_N$ has been taken into account.

(b) Collocation method

In the collocation method the polynomial $u_N(x)$ satisfies

$$L\,u_N(x_i) = f(x_i), \quad i = 1, \ldots, N-1 \tag{3.88}$$

$$B_-u_N(x_N) = g_-\,, \quad B_+u_N(x_0) = g_+\,. \tag{3.89}$$

The error equation is the equation exactly satisfied by u_N for every x. It may be written in the form

$$L\,u = f_N + p_N\,, \tag{3.90}$$

where f_N is the polynomial interpolating f at the collocation points x_i, $i = 0, \ldots, N$, and p_N is a polynomial. This polynomial p_N is of degree N at most, since $L\,u_N - f_N$ is a polynomial of degree N. From Eq.(3.88) it follows that $p_N(x_i)$ must be zero at the inner collocation points x_i, $i = 1, \ldots, N-1$, therefore it is of the form

$$p_N = (\lambda\,x + \mu)\,T'_N(x) \tag{3.91}$$

since the inner collocation points x_i are the zeros of $T'_N(x)$. The constants λ and μ in Eq.(3.91) are uniquely determined from the identity

$$L\,u_N \equiv f_N + p_N \tag{3.92}$$

since $L\,u_N - f_N$ is a known polynomial. The expressions for λ and μ may be obtained by evaluating Eq.(3.92) at two points of the continuous interval $[-1, 1]$ distinct from the collocation points. In particular, we can use the boundary points $x = \pm 1$ and we get

$$\lambda = \frac{1}{2\,N^2}\left[R_N(1) - (-1)^{N+1}R_N(-1)\right] \tag{3.93}$$

$$\mu = \frac{1}{2\,N^2}\left[R_N(1) + (-1)^{N+1}R_N(-1)\right] \tag{3.94}$$

with $R_N = L\,u_N - f_N$.

Note that the notion of error equation also applies to unsteady problems.

In a sense, the meaning of the error equation is close to the concept of the "equivalent equation" associated with finite-difference methods (Peyret and Taylor, 1983 ; Hirsch, 1988). For a finite-difference method of order p, the equivalent equation at order q, with $q > p$, is the differential equation that is satisfied by the numerical solution with an error of order q. Therefore, the approximate solution given by the scheme must reflect the properties of the equivalent equation better than those of the original differential equation. Then the study of the properties of the equivalent equation gives valuable information on the behaviour of the approximate solution. The same idea applies to the error equation since the polynomial u_N is the exact solution of an equation which is perfectly identified.

3.5 Differential equation with nonconstant coefficients

The difficulties associated with the application of the Chebyshev method to equations with nonconstant coefficients are the same as those encountered in the Fourier case (Sections 2.7 and 2.8). The arguments developed for the Fourier method also apply to the present situation.

3.5.1 Linear equation with variable coefficients

When some coefficients of Eq.(3.56) are dependent on x, the use of the Chebyshev tau method leads to difficulties analogous to those experienced with the Fourier Galerkin method. More precisely, let us assume, for example, the coefficient a in Eq.(3.56) to be nonconstant. The scalar product $(a_N u'_N, T_i)_w$, where a_N and u'_N are replaced by their respective Chebyshev expansions, involves a convolution sum leading to a complicated algebraic system for determining the coefficients $\hat{u}_k$ of the Chebyshev expansion of the unknown u.

Consequently, it is preferable to solve nonconstant coefficient problems by means of the collocation method described in Section 3.4.2. We obtain an algebraic system of the form (3.78) whose matrix $\mathcal{A}$ is now constructed from the matrix

$$Q = -\nu \mathcal{D}^2 + \mathcal{D}' + b\mathcal{I} \tag{3.95}$$

with $\mathcal{D}' = [a(x_i)\, d_{i,j}^{(1)}]$, $i,j = 0,\ldots,N$.

3.5.2 Nonlinear equation

We consider the Burgers equation which constitutes a simple but significant model of quasi-linear unsteady problems, such as those encountered in fluid mechanics. The Burgers equation,

$$\partial_t u + u \, \partial_x u - \nu \, \partial_{xx} u = 0 \,, \tag{3.96}$$

where ν is a positive constant, has to be solved in $-1 < x < 1$ with the boundary conditions

$$u(-1, t) = g_- \,, \quad u(1, t) = g_+ \,, \tag{3.97}$$

and the initial condition

$$u(x, 0) = u_0(x) \,. \tag{3.98}$$

Let us denote by $u_N(x, t)$ the polynomial representation (with respect to x) of the solution to Eqs.(3.96)-(3.98) and by $u_N^n(x)$ the approximation of $u_N(x, t)$ at time $t_n = n \, \Delta t$, $n = 0, 1, \ldots$.

For the sake of numerical stability the time-discretization is implicit for the diffusive term $\partial_{xx} u$. As a matter of fact, an explicit evaluation of this term leads to a severe constraint on the time-step Δt which should satisfy a condition of the form $\Delta t < C / (\nu N^4)$, as proven in Section 4.3.2 for the heat equation. The nonlinear convective term $u \, \partial_x u$ is generally evaluated in an explicit way in order to avoid the use of an iterative procedure for the solution of the resulting algebraic system. Concerning the stability restriction associated with the explicit treatment of the convective term, it should be $\Delta t < C / (|u| N^2)$ in the inviscid case $\nu = 0$. More precise results are given in Chapter 4, devoted to the solution of time-dependent equations. Such a restriction on the time-step is less stringent than the one associated with the explicit evaluation of the diffusive term. Most of the spectral codes are based on the explicit treatment of the convective terms. However, in some situations, this could lead to very small time-steps and, under these conditions, it may be valuable to go further in the implicitness, even if the introduction of an iterative procedure cannot be avoided. These various possibilities are now addressed. The time-discretization is based on the simple two-level scheme already considered in the Fourier case. More accurate time-discretization schemes are discussed in Chapter 4.

(a) Explicit treatment of the convective term

Following the lines of the weighted residuals method, the residual R_N is defined by

$$R_N = \frac{u_N^{n+1} - u_N^n}{\Delta t} + u_N^n \, \partial_x u_N^n - \nu \, \partial_{xx} u_N^{n+1} \,. \tag{3.99}$$

This residual is set to zero at the inner collocation points x_i, $i = 1, \ldots, N-1$, and the system is closed by adding the boundary conditions, that is,

$$\frac{1}{\Delta t} \left[u_N^{n+1}(x_i) - u_N^n(x_i) \right] + u_N^n(x_i) \, \partial_x u_N^n(x_i) - \nu \, \partial_{xx} u_N^{n+1}(x_i) = 0 \,,$$
$$i = 1, \ldots, N-1 \tag{3.100}$$

$$u_N^{n+1}(x_N) = g_- , \quad u_N^{n+1}(x_0) = g_+ . \tag{3.101}$$

Then the spatial derivatives $\partial_x u_N^n(x_i)$ and $\partial_{xx} u_N^{n+1}(x_i)$ are expressed, respectively, in terms of the grid values $u_N^n(x_j)$ and $u_N^{n+1}(x_j)$, $j = 0, \ldots, N$, thanks to Eq.(3.45), so that these grid values are determined by the linear algebraic system

$$\sum_{j=0}^{N} d_{i,j}^{(2)} u_N^{n+1}(x_j) - \sigma u_N^{n+1}(x_i) = \frac{1}{\nu} u_N^n(x_i)\, \partial_x u_N^n(x_i) - \sigma u_N^n(x_i) , \tag{3.102}$$

$$i = 1, \ldots, N - 1$$

$$u_N^{n+1}(x_N) = g_- , \quad u_N^{n+1}(x_0) = g_+ , \tag{3.103}$$

where $\sigma = 1/(\nu \Delta t)$. Therefore, we have to solve at each time-step a system analogous to (3.78), namely,

$$\mathcal{A} U^{n+1} = F^n , \tag{3.104}$$

where $U^{n+1} = \left(u_N^{n+1}(x_0), \ldots, u_N^{n+1}(x_N) \right)^T$. The entries of the matrix $\mathcal{A}$ are constant so that it can be inverted in a precalculation stage performed before the start of the time-integration, and its inverse is stored (see Section 3.4.2). On the other hand, the forcing term F^n is changing at each time-cycle so that the efficiency of the algorithm depends partly on the way in which the nonlinear term $u_N^n(x_i)\, \partial_x u_N^n(x_i)$ is calculated. This can be done according to two different techniques.

The first technique consists of calculating the derivative $\partial_x u_N^n(x_i)$ in terms of the values $u_N^n(x_j)$, $j = 0, \ldots, N$, thanks to the differentiation matrix $\mathcal{D}$ defined by (3.51).

The second technique is based on the pseudospectral algorithm already given in the Fourier case (Section 2.8) :

1. Knowing $u_N^n(x_i)$, $i = 0, \ldots, N$, calculate by the FFT the Chebyshev coefficients $\hat{u}_k^n$, $k = 0, \ldots, N$ [from Eq.(3.37)].

2. Calculate the coefficients $\hat{u}_k^{(1)\,n}$, $k = 0, \ldots, N$, of the Chebyshev expansion of the derivative, thanks to the recurrence formula (3.28) with $p = 1$.

3. From the knowledge of the set $\hat{u}_k^{(1)\,n}$, $k = 0, \ldots, N$, calculate by the FFT the values $\partial_x u_N^n(x_i)$, $i = 0, \ldots, N$ [from Eq.(3.35)].

The efficiency of the first technique very much depends on the computer and the available matrix-vector product routines. For example, on the vector computer Cray C98, the first technique using the matrix product MXM routine, is more rapid than the second one (using the FFT) for N roughly smaller than 100. For larger values of N, the second technique becomes more rapid because of the efficiency of the FFT algorithm for large N. Finally, note that the recurrence formulas used in the second technique may be replaced by the matrix-vector product (3.24) which can be more accurate and faster.

(b) Semi-implicit treatment of the convective term
The above-mentioned constraint on the time-step, required for the stability of the explicit evaluation of the convective term, can be diminished and even avoided if this term is considered in a more implicit way. In some situations, where $u(x,t)$ can be decomposed according to $u(x,t) = \tilde{u}(x) + \overline{u}(x,t)$ with $|\tilde{u}| \geq |\overline{u}|$, we may consider the following discretization scheme

$$\frac{u_N^{n+1}(x_i) - u_N^n(x_i)}{\Delta t} + \tilde{u}_N(x_i)\,\partial_x u_N^{n+1}(x_i) - \nu\,\partial_{xx} u_N^{n+1}(x_i) \tag{3.105}$$

$$= -\overline{u}_N^n(x_i)\,\partial_x u_N^n(x_i)\,, \quad i = 1, \ldots, N-1$$

$$u_N^{n+1}(x_N) = g_-\,, \quad u_N^{n+1}(x_0) = g_+\,, \tag{3.106}$$

which is (linearly) unconditionally stable. Moreover, because $\tilde{u}_N$ is time-independent, the resulting matrix $\mathcal{A}$ of the algebraic system is also time-independent and can be inverted once and for all in the precalculation stage. If $u(x,t)$ is positive the choice $\tilde{u} = \frac{1}{2}\max_{x,t}\{u(x,t)\}$ satisfies the condition $|\tilde{u}| \geq |\overline{u}|$.

(c) Implicit treatment of the convective term
In many problems, the decomposition proposed in the previous section cannot be done. The last remedy for curing the stability problem is to treat the convective term $u\,\partial_x u$ in an implicit manner. This can be done by considering either an approximation of the type $u_N^n\,\partial_x u_N^{n+1}$ or $u_N^{n+1}\,\partial_x u_N^{n+1}$. In the first case, the resulting algebraic system is linear but its coefficients are variable in time as well as in space. So, an iterative procedure has to be devised in order to avoid the inversion of the associated matrix at each time-cycle. In the second case, we are confronted with a nonlinear system which must be solved by a similar iterative procedure.

Let us now describe the algorithm associated with the first discretization mentioned above. The discrete system is

$$\frac{u_N^{n+1}(x_i) - u_N^n(x_i)}{\Delta t} + u_N^n(x_i)\,\partial_x u_N^{n+1}(x_i) - \nu\,\partial_{xx} u_N^{n+1}(x_i) = 0\,, \tag{3.107}$$

$$i = 1, \ldots, N-1$$

$$u_N^{n+1}(x_N) = g_-\,, \quad u_N^{n+1}(x_0) = g_+\,, \tag{3.108}$$

in which the spatial derivatives are expressed by means of Eq.(3.35). The system written in compact form, is

$$\mathcal{A}\left(U^n\right)U^{n+1} = F^n\,, \tag{3.109}$$

where $\mathcal{A}\left(U^n\right)$ is a nonlinear discrete operator. The system (3.109) is iteratively solved according to :

$$\mathcal{A}_0\,\overline{U}^{m+1} = F^n - \mathcal{A}\left(U^n\right)U^{n+1,m} \tag{3.110}$$

$$U^{n+1,m+1} = U^{n+1,m} + \alpha \overline{U}^{m+1}, \tag{3.111}$$

where m refers to the iteration. The process is initiated with $U^{n+1,0} = U_0^n$. In the above equations A is the preconditioning operator and α is the relaxation parameter.

The choice of the preconditioner A_0 is important for the efficiency of the algorithm. It must be done according to the following requirements (Canuto et al., 1988 ; Funaro, 1992 ; Quarteroni and Valli, 1994) :
(i) the matrix A_0 must be easy to invert or, equivalently, the system of the form $A_0 X = S$ must be easy to solve, and
(ii) the spectral condition number κ_{sp} of the matrix $A_0^{-1}A$ must be small, that is, close to 1.
For an $n \times n$ matrix $\mathcal{M}$ with eigenvalues $\lambda_i (\mathcal{M})$, $i = 1, \ldots, n$, the spectral condition number $\kappa_{sp} (\mathcal{M})$ is defined by

$$\kappa_{sp} (\mathcal{M}) = \frac{\lambda_{max} (\mathcal{M})}{\lambda_{min} (\mathcal{M})} \tag{3.112}$$

where $\lambda_{max} (\mathcal{M})$ and $\lambda_{min} (\mathcal{M})$ are, respectively, the maximal and minimal values of $|\lambda_i (\mathcal{M})|$. Note that $1 \le \kappa_{sp} (\mathcal{M}) \le \kappa (\mathcal{M})$ where $\kappa (\mathcal{M})$ is the usual condition number defined by Eq.(3.80). In the case where $\mathcal{M}$ is symmetric, we have $\kappa_{sp} = \kappa$.

A possible choice (Orszag, 1980 ; Canuto and Quarteroni, 1985) is to use as preconditioner A_0 the finite-difference operator A_{df}, analog of A, based on the same collocation points. Therefore, the matrix $A_0 = A_{df}$ has also variable (in time and space) entries but is tridiagonal (for three-point formulas), so that the solution of the system (3.110) can be calculated at each time-cycle in a very efficient way, using the classical algorithm for the tridiagonal system (see, e.g., Isaacson and Keller, 1966 ; or Peyret and Taylor, 1983).

Another possible choice is the finite-element preconditioning (Canuto and Quarteroni, 1985 ; Deville and Mund, 1985, 1990 ; Quarteroni and Zampieri, 1992) based, for example, on linear elements. In this case, Eq.(3.109) is replaced by

$$A_{fe}\overline{U}^{m+1} = \mathcal{M} \left[F^n - A (U^n) U^{n+1,m} \right],$$

where A_{fe} is the finite-element analog to A and $\mathcal{M}$ is the mass matrix. For model elliptic problems (Deville and Mund, 1990 ; Quarteroni and Valli, 1994), the spectral condition number $\kappa_{sp} \left(A_{fe}^{-1} \mathcal{M} A \right)$ is found to be smaller than $\kappa_{sp} \left(A_{df}^{-1} A \right)$, thus leading to a better convergence of the iterative procedure.

Another possibility consists of using, as a preconditioner, a Chebyshev collocation operator A_0 defined as an approximation to the operator A (Gauthier, 1988 ; Guillard and Désidéri, 1990 ; Fröhlich et al., 1991). In the present case, A_0 may be constructed by simply neglecting the nonconstant first-order derivative term in A, that is, $A_0 = A(0)$. The resulting matrix $\mathcal{A}_0$ is then constant and can be inverted once and for all in the precalculation stage performed previously to start the time-integration as explained in Section 3.4.2.

The choice of the relaxation parameter α is also of great importance. The use of a constant α is delicate because its optimal value may depend on U^n appearing in $\mathcal{A}$, so that it has to be changed during the time-integration. Therefore, it is preferable to use a relaxation parameter $\alpha = \alpha_m$ redefined at each iteration. This dynamic calculation of α can be done in several ways (Canuto *et al.*, 1988). Among them, the preconditioned minimal residual (PMR) method gives a satisfactory convergence at a reasonable computing time. However, in several cases (Fröhlich and Peyret, 1990 ; Sabbah *et al.*, 2001), much better convergence is obtained using the preconditioned conjugate residual (PCR) method which is a little more complicated. We refer to Section 4.3.8 for the description of the PMR and PCR methods.

3.5.3 Aliasing

The existence of aliasing in the Chebyshev approximation has been shown in Section 3.3.2. Its effect on the solution of a differential equation with nonconstant coefficients is similar to the one encountered for the Fourier method (Sections 2.7 and 2.8) due to the close relationship between Fourier and Chebyshev series. As a result, the removal of aliasing can be accomplished by again using the "3/2 rule" (Section 2.9), which is applied to the Chebyshev series expansion with obvious changes. The influence of the aliasing on the solution of the Navier-Stokes equations is discussed in Section 7.8.

3.6   Some results of convergence

The similarity between Fourier and Chebyshev series leads to similar results concerning the convergence. So, it can be shown that the rate of decay of the Chebyshev coefficient $\hat{u}_k$ defined by Eq.(3.19) is again given by Eqs.(2.7) and (2.8) (see Gottlieb and Orszag, 1977).

Concerning theoretical results on the approximation error we refer to the books by Canuto *et al.* (1988), Mercier (1989), and by Bernardi and Maday (1992) in which the following estimates are given with reference to the original papers.

Error estimates are, most of the time, obtained for functional spaces weighted with the Chebyshev weight. For example, the error of the Galerkin approximation, defined by Eqs.(3.17) and (3.19) (also called the "projection error") in the $H_w^p(-1,1)$-norm, is found to satisfy

$$\|u - u_N\|_{H_w^p(-1,1)} \leq C \, N^{-1/2+2p-m} \, \|u\|_{H_w^m(-1,1)} \tag{3.113}$$

for $1 \le p \le m$, if $u \in H_w^m(-1,1)$ for some $m \ge 1$. The constant C is independent of N. The space $H_w^p(-1,1)$ is the weighted Sobolev space of order p whose norm is defined by

$$\|u\|_{H_w^p(-1,1)} = \left(\sum_{k=0}^{p} \int_{-1}^{1} |u^{(k)}(x)|^2 \, w(x) \, dx \right)^{1/2}. \tag{3.114}$$

The error estimate (3.113) is nonoptimal because the power of N appearing in (3.113) is not the difference between the order of the Sobolev spaces appearing in the left- and right-hand sides of the inequality.

A norm of current use in numerical analysis is that corresponding to $p = 0$, namely, the L_w^2-norm, associated with the space of square integrable functions. For this norm, the error satisfies the inequality

$$\|u - u_N\|_{L_w^2(-1,1)} \le C \, N^{-m} \|u\|_{H_w^m(-1,1)}. \tag{3.115}$$

This estimate is optimal. Because the error in the L_w^2-norm is of global nature and does not involve derivatives, it can be misleading in some situations where the error is not uniformly distributed and is locally large. Therefore, for practical applications, the error in the maximum norm may be of interest because it gives a finer picture of the accuracy of the approximation. Thus, for approximation (3.17), (3.19), the error estimate in the L^∞-norm is :

$$\|u - u_N\|_{L^\infty(-1,1)} \le C \, (1 + \ln N) \, N^{-m} \sum_{k=0}^{m} \|u^{(k)}\|_{L^\infty(-1,1)}. \tag{3.116}$$

Now let us consider the collocation (i.e., "interpolation") approximation u_N defined by Eq.(3.17) with coefficients $\hat{u}_k$ given by Eq.(3.36) or, equivalently, defined by Eqs.(3.43) and (3.44). The error in the H_w^p-norm satisfies the inequality

$$\|u - u_N\|_{H_w^p(-1,1)} \le C \, N^{2p-m} \|u\|_{H_w^m(-1,1)} \tag{3.117}$$

for $0 \le p \le m$. This error estimate is optimal. In the maximum norm the error is found to satisfy

$$\|u - u_N\|_{L^\infty(-1,1)} \le C \, N^{\frac{1}{2}-m} \|u\|_{H_w^m(-1,1)}. \tag{3.118}$$

Therefore, the general conclusions referred to at the end of Section 2.1.2 about the accuracy of the Fourier approximation also apply to the Chebyshev approximation : exponential accuracy is obtained for infinitely differentiable functions, but this accuracy is lost for nonsmooth functions. The case where the function is discontinuous is the worst. The Gibbs oscillations make the approximation useless unless some special treatment (e.g., filtering) is done. For continuous functions which have only a finite number of bounded derivatives, the decay of the error is only algebraic. Possible ways to solve singular problems with Chebyshev polynomial approximation are discussed in Chapter 8.

Concerning the polynomial approximation of singular functions, it is important to notice that the error decays more rapidly when the singularity is located at a boundary. The rate of convergence of the coefficients of the Chebyshev series approximating such functions has been discussed by Boyd (1986). Moreover, some theoretical results showing a doubling of the decay rate of the error have been established by Bernardi and Maday (1991) for approximations by Jacobi polynomials to functions exhibiting a singularity at an extremity of the interval.

To illustrate these properties, we now consider two examples of singular functions (compared by Botella, 1998), whose singularity is located either at an inner point of the interval or at an extremity. These functions are approximated with the collocation Chebyshev method. The first one is defined by

$$u_\alpha(x) = \begin{cases} 0 & -1 \le x < 0, \\ x^\alpha & 0 \le x \le 1, \end{cases} \tag{3.119}$$

where $\alpha > 0$. This function exhibits a singular behaviour at the center of the interval $(-1, 1)$ and belongs to $H_w^m(-1, 1)$ with $m < \alpha + 1/2$. Let $e_N(x)$ be the difference between the function $u_\alpha(x)$ and its interpolation polynomial $u_N(x)$, namely,

$$e_N(x) = u_\alpha(x) - u_N(x).$$

Equations (3.117) and (3.118) give the following estimates

$$\|e_N\|_{L_w^2(-1,1)} \le c_1 \, N^{-1/2-\alpha},$$

$$\|e_N\|_{L^\infty(-1,1)} \le c_2 \, N^{-\alpha}, \tag{3.120}$$

$$\|e_N\|_{H_w^1(-1,1)} \le c_3 \, N^{1/2-\alpha},$$

where c_1, c_2 and c_3 are positive constants independent of N. To check these estimates numerically it is necessary to evaluate the various norms with the best possible accuracy. The continuous $L_w^2(-1, 1)$- and $H_w^1(-1, 1)$- norms are calculated by evaluating the value of the interpolating polynomial $u_N(x)$ and its derivative $u_N'(x)$ on the $M + 1$ Gauss-Lobatto points $\xi_j = \cos \pi j / M, j = 0, \ldots, M$, with M much larger than N. Here, $M = 1000$ for N belonging to the range $[8, 24]$. Then the integrals are evaluated by means of the Gauss-Lobatto quadrature formula (3.14) based on the points $\xi_j, j = 0, \ldots, M$. In an analogous way, the $L^\infty(-1, 1)$-norm is calculated by taking the maximum of $e_N(x)$ on the above $M + 1$ Gauss-Lobatto points. Table 3.3 shows the numerical estimates of the order of the error, namely, $e_N = O(N^{-q})$. The calculated value of q is in agreement with the theoretical results (3.120).

	$\alpha = 2$	$\alpha = 4$	$\alpha = 6$
$L^2_w(-1,1)$	2.5	4.5	6.5
$L^\infty(-1,1)$	2.0	4.0	6.1
$H^1_w(-1,1)$	1.5	3.5	5.6

Table 3.3. Order q of the interpolation error for the function u_α defined by Eq.(3.119).

	$\alpha = 0.9$	$\alpha = 1.9$	$\alpha = 2.9$
$L^2_w(-1,1)$	2.2	4.2	6.2
$L^\infty(-1,1)$	1.8	3.8	5.9
$H^1_w(-1,1)$	0.3	2.3	4.4

Table 3.4. Order q of the interpolation error for the function u_α defined by Eq.(3.121).

The second example concerns the function $u_\alpha(x)$ defined by

$$u_\alpha(x) = \left(1 - x^2\right)^\alpha , \quad -1 \le x \le 1 \tag{3.121}$$

where the positive constant α is not an integer. This function belongs to $H^m_w(-1,1)$ with $m < \alpha + 1/4$. The singularity is no longer located in the interior of the interval but at its extremities. Table 3.4 shows the convergence rate q of the interpolation error e_N, estimated numerically as above. From these results, the following rules can be stated

$$\|e_N\|_{L^2_w(-1,1)} = O\left(N^{-1/2-2\alpha}\right) ,$$

$$\|e_N\|_{L^\infty(-1,1)} = O\left(N^{-2\alpha}\right) , \tag{3.122}$$

$$\|e_N\|_{H^1_w(-1,1)} = O\left(N^{3/2-2\alpha}\right) .$$

It is interesting to observe that for the L^2_w- and H^1_w-norms, the order of accuracy is exactly twice that given by the estimates (3.117) which is no longer optimal in this case where the singularity is located at an extremity. This behaviour is in agreement with the theoretical results obtained by Bernardi and Maday (1991).

The rate of decay of the coefficients of the Chebyshev series depends on the degree of smoothness of the function and on the location of the singularity. In these respects, it is instructive to compare the spectrum $|\hat{u}_k|$, $k = 0, \ldots, N$, calculated from formula (3.36), corresponding to each singular functions (3.119) and (3.121). The comparison is made on two bases : (i) the functions have the same kind of singularity characterized by $\alpha = 2.9$, and (ii) the functions belong to the same functional space H^m_w, namely $\alpha = 2.65$ for (3.119) and $\alpha = 2.9$ for (3.121). The difference in the decay rates is clearly seen in Fig.3.2.

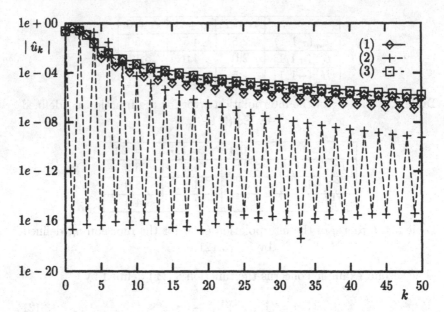

FIGURE 3.2. Chebyshev spectrum of singular functions : (1) function (3.119) with $\alpha = 2.65$, (2) function (3.119) with $\alpha = 2.9$, (3) function (3.121) with $\alpha = 2.9$.

Up to now, we have only discussed the error associated with the Chebyshev representation of a given function, either by Galerkin (projection) or collocation (interpolation) methods. When applying Chebyshev methods to calculate the solution of differential problems, the error of the approximate solution with respect to the exact one is obviously of crucial interest. Theoretical results concerning one- and two-dimensional problems are given in the books already mentioned. The nature of the error is similar to that associated with projection or interpolation errors. It depends essentially on the smoothness of the solution, the general rule being : the more the exact solution is smooth, the more the approximate solution is accurate.

However, we have to be aware that in multidimensional physically realistic problems, the solution is seldom infinitely differentiable. A very simple example is constituted by the Laplace equation in a square domain : at a corner, the intersection of two sides where homogeneous Dirichlet conditions are prescribed, the solution behaves like (Grisvard, 1985) :

$$u \sim C\, r^2 \ln r \sin 2\theta\,,$$

where r and θ are polar coordinates such that r is the distance to the corner. Therefore the third-order derivatives of u are infinite at the corner and the "infinite" spectral accuracy is lost.

Another typical case is the Stokes flow in a corner with a no-slip condition. The streamfunction ψ, the solution of the biharmonic equation, behaves like (Moffatt, 1964) :

$$\psi \sim C\, r^{3.74} \left[\cos\left(1.13 \ln r\right) g(\theta) + \sin\left(1.13 \ln r\right) h(\theta)\right] \qquad (3.123)$$

so that the second-order derivatives of the velocity become infinite at the corner. We refer to Chapter 8 for a more general discussion on the treatment of singularities.

For now, numerical examples (Botella, 1998) are displayed in order to illustrate the accuracy of spectral methods compared to finite-difference methods in the case of regular and (weakly) singular solution. More precisely, we consider the solution of the one-dimensional Helmholtz equation

$$-\nu\, u'' + u = f, \quad -1 < x < 1, \quad \nu = 10^{-2} \qquad (3.124)$$

with

$$u(-1) = g_-\,, \quad u(1) = g_+\,, \qquad (3.125)$$

using various methods :

(1) Second-order finite-difference method with uniform mesh ($\Delta x = 2/N$).

(2) Second-order finite-difference method with Gauss-Lobatto mesh (3.75).

(3) Sixth-order Hermitian method with uniform mesh ($\Delta x = 2/N$).

(4) Sixth-order Hermitian method with Gauss-Lobatto mesh (3.75).

(5) Chebyshev collocation method [Gauss-Lobatto mesh (3.75].

(6) Chebyshev tau method (polynomial degree $= N$).

The Hermitian method is based on the formula using three points for the second-order derivative and five points for the function (Collatz, 1966 ; Botella, 1988 ; Peyret, 2000). The collocation method has been described in Section 3.4.2 and the resulting algebraic system is solved by the diagonalization technique (Section 3.7.1). The tau method applied to the Helmholtz equation (Section 3.4.1) leads to the solution of quasi-tridiagonal systems obtained through the algorithm described in Appendix B.

The error is measured with the discrete L^2-norm defined by

$$\overline{E} = \left[\frac{1}{N-1} \sum_{j=1}^{N-1} |u_i - u(x_i)|^2\right]^{1/2}, \qquad (3.126)$$

where u_i refers to the approximate solution and $u(x_i)$ to the exact one.

Figure 3.3.a shows the error $\overline{E}$ found for the smooth solution

$$u(x) = 1 - \frac{\sinh\left[(x+1)/\sqrt{\nu}\right]}{\sinh\left(2/\sqrt{\nu}\right)} \qquad (3.127)$$

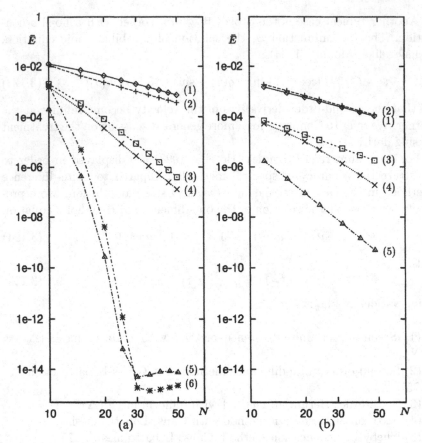

FIGURE 3.3. Error $\bar{E}$ versus the number N of degrees of freedom : (a) smooth solution (3.127), (b) singular solution (3.121) (see the text for the meaning of the labels).

obtained for $f = 1$, $g_- = 1$, and $g_+ = 0$. The exponential accuracy of the spectral methods is clearly seen in this figure. For $N < 30$ the error of the collocation method is smaller than the error of the tau method. For larger values of N, in the range of round-off errors, the accuracy of the tau method is better. This is a consequence of the ill-conditioning of the collocation matrix, although its effect is largely reduced by the use of the matrix-diagonalization method.

The second example concerns the calculation of a solution exhibiting a singularity at the boundaries $x = \pm 1$. Therefore, the computed solution is $u(x) = u_\alpha(x)$ given by Eq.(3.121), which defines f in (3.124) and $g_- = g_+ = 0$. Figure 3.3.b shows the error $\bar{E}$ in the case $\alpha = 2.9$. Results concerning the order q of the error $\bar{E} = O(N^{-q})$ for various values of α are given in Table 3.5. Some interesting conclusions can be drawn from these numerical results.

α	Sixth-order Hermitian		Chebyshev collocation
	Uniform mesh	Gauss-Lobatto mesh	
2.1	2.1	4.2	4.7
2.9	2.9	5.6	5.9
3.1	3.1	5.9	6.3
4.9	5.2	6.0	10.5

Table 3.5. Order q of the error $\overline{E}$ (discrete L^2-norm) associated with the solution (3.121).

First, it is observed that, in agreement with the general theory, the decay of the error of the Chebyshev collocation method (curve (5)) is no longer exponential but algebraic. From an evaluation of the error of the numerical solution in various continuous norms as done previously, it is found that the behaviour of this error again follows the rules stated in (3.122), namely, the same behaviour as the interpolation error.

Second, it is seen that the error of the Hermitian method is, at least in the considered range of N, $O(N^{-2.9})$ for the uniform mesh (curve (3)) and $O(N^{-5.6})$ for the variable Gauss-Lobatto mesh (curve (4)). This shows how the singular nature of the solution may have an effect on the rate of convergence of a high-order finite-difference method when the resolution near the singular point is not sufficient. Moreover, although the respective rates of convergence of the Hermitian and Chebyshev methods are close (curves (4) and (5)), the magnitude of the error of the Chebyshev method is much smaller. Therefore, in problems with weak singularities like those considered here, the Chebyshev collocation method is more efficient than the sixth-order Hermitian method, as much as the implementation of high-order Hermitian methods in the variable mesh may be complicated.

3.7   Multidimensional elliptic equation

The Chebyshev methods (tau or collocation) which have been described in the previous sections apply straightforwardly to multidimensional equations, provided the computation domain is a square or a cube. However, the resulting algebraic systems have to be solved by a more efficient technique than the direct inversion method proposed for the one-dimensional case.

The solution method usually employed, which is found to be very efficient in unsteady problems where the same equation has to be solved a large number of times, is the matrix-diagonalization procedure. This method was introduced by Lynch et al. (1964) for finite-difference approximation. It was applied to the Chebyshev tau method by Haidvogel and Zang (1979) in the two-dimensional case and by Haldenwang et al. (1984) to the three-dimensional case. The application to the collocation method was considered by Ehrenstein and Peyret (1989).

In the present section, we consider the solution of the Helmholtz equation, but the method applies to more general problems provided they are separable and the associated one-dimensional matrices are diagonalizable. This last point requires that the eigenvector matrices are well-conditioned. This requirement may not be met for the steady advection-diffusion equation and, in this case, the Schur-decomposition method constitutes an efficient way to solve the associated algebraic system.

First, in order to give an idea about its basic principles, the matrix-diagonalization method is presented for the solution of the one-dimensional Helmholtz equation. However, more general operators, for which some eigenvalues may be complex, will also be discussed. The case of the steady-state advection-diffusion equation will be considered and the Schur-decomposition method will be discussed and compared to the matrix-diagonalization method. Then the solution of the two-dimensional Helmholtz equation will be considered in detail ; in particular, the case where the coefficients, that occur in the general boundary conditions of Robin type, vary on the same side of the boundary will be addressed. Finally, the algorithm for the three-dimensional equation will be described.

Only the collocation approximation will be considered here, since the application to the tau method is similar (see the above references).

3.7.1   One-dimensional equation

Let us consider the one-dimensional Helmholtz equation with general Robin boundary conditions

$$u'' - \sigma u = f , \quad -1 < x < 1 \tag{3.128}$$

$$\alpha_- u(-1) + \beta_- u'(-1) = g_- \tag{3.129}$$

$$\alpha_+ u(1) + \beta_+ u'(1) = g_+ , \tag{3.130}$$

where σ is a positive constant. It is assumed that $\alpha_- \alpha_+ \geq 0$ without loss of generality.

The application of the collocation method (Section 3.4.2) leads to an algebraic system for the unknowns $u_N(x_i)$, $i = 0, \ldots, N$. For the diagonalization procedure, it is convenient to eliminate the boundary values, thanks to the boundary conditions (3.129) and (3.130) so that the unknown vector is

$$V = (u_N(x_1), \ldots, u_N(x_{N-1}))^T . \tag{3.131}$$

The boundary values $u_N(x_0)$ and $u_N(x_N)$ are expressed from Eqs.(3.129) and (3.130) by

$$u_N(x_0) = \frac{1}{e} \sum_{j=1}^{N-1} b_{0,j} u_N(x_j) + \frac{1}{e} (c_{0,-}g_- + c_{0,+}g_+) , \tag{3.132}$$

$$u_N(x_N) = \frac{1}{e} \sum_{j=1}^{N-1} b_{N,j} u_N(x_j) + \frac{1}{e} \left(c_{N,-}g_- + c_{N,+}g_+ \right), \qquad (3.133)$$

with

$$e = c_{0,+}c_{N,-} - c_{0,-}c_{N,+},$$

$$c_{0,-} = -\beta_+ d_{0,N}^{(1)}, \quad c_{0,+} = \alpha_- + \beta_- d_{N,N}^{(1)},$$

$$c_{N,+} = -\beta_- d_{N,0}^{(1)}, \quad c_{N,-} = \alpha_+ + \beta_+ d_{0,0}^{(1)}, \qquad (3.134)$$

$$b_{0,j} = -c_{0,+}\beta_+ d_{0,j}^{(1)} - c_{0,-}\beta_- d_{N,j}^{(1)}, \qquad j = 1, \ldots, N-1,$$

$$b_{N,j} = -c_{N,-}\beta_- d_{N,j}^{(1)} - c_{N,+}\beta_+ d_{0,j}^{(1)}, \qquad j = 1, \ldots, N-1.$$

It is recalled that $d_{i,j}^{(1)}$ and $d_{i,j}^{(2)}$ are the elements of the differentiation matrices [see Eq.(3.46) and Eq.(3.47)].

The discrete system approximating (3.128)-(3.130) can be written as

$$(\mathcal{D}_x - \sigma \mathcal{I}) V = H, \qquad (3.135)$$

where $\mathcal{I}$ is the $(N-1) \times (N-1)$ identity matrix and the matrix $\mathcal{D}_x = [d_{i,j}]$, $i, j = 1, \ldots, N-1$, is defined by

$$d_{i,j} = d_{i,j}^{(2)} + \frac{1}{e} \left(b_{0,j} d_{i,0}^{(2)} + b_{N,j} d_{i,N}^{(2)} \right)$$

and $H = [h_i]$, $i = 1, \ldots, N-1$, is the vector such that

$$h_i = f(x_i) - \frac{1}{e} \left(c_{0,-}g_- + c_{0,+}g_+ \right) d_{i,0}^{(2)} - \frac{1}{e} \left(c_{N,-}g_- + c_{N,+}g_+ \right) d_{i,N}^{(2)}.$$

The algorithm is based on the diagonalization of the matrix $\mathcal{D}_x$. It has been proven by Gottlieb and Lutsman (1983) that the eigenvalues λ_i, $i = 1, \ldots, N-1$, of $\mathcal{D}_x$ are real, negative and distinct if the following conditions are satisfied

$$\alpha_- > 0, \quad \beta_- < 0, \quad \alpha_+ > 0, \quad \beta_+ > 0. \qquad (3.136)$$

These conditions can be relaxed to include the cases $\beta_- = \beta_+ = 0$ (Dirichlet-Dirichlet), $\beta_- = \alpha_+ = 0$ (Dirichlet-Neumann), and $\alpha_- = \alpha_+ = 0$ (Neumann-Neumann). In this latter case, one eigenvalue is zero.

For the situations specified above the matrix $\mathcal{D}_x$ has $N-1$ linearly independent eigenvectors such that it can be diagonalized and

$$\mathcal{D}_x = \mathcal{P} \Lambda \mathcal{P}^{-1}, \qquad (3.137)$$

where Λ is the diagonal matrix whose entries are the eigenvalues λ_i, $i = 1, \ldots, N-1$, the transformation matrix $\mathcal{P}$ is the matrix whose columns are the eigenvectors and $\mathcal{P}^{-1}$ is its inverse. Therefore, Eq.(3.135) may be written as

$$\left(\mathcal{P}\,\Lambda\,\mathcal{P}^{-1} - \sigma\mathcal{I}\right) V = H$$

and, after left multiplication by $\mathcal{P}^{-1}$, we get

$$(\Lambda - \sigma\mathcal{I})\,\mathcal{P}^{-1} V = \mathcal{P}^{-1} H\,.$$

Then, with $\tilde{V} = \mathcal{P}^{-1}V$ and $\tilde{H} = \mathcal{P}^{-1}H$, this equation becomes

$$(\Lambda - \sigma\mathcal{I})\,\tilde{V} = \tilde{H}$$

and

$$\tilde{V} = (\Lambda - \sigma\mathcal{I})^{-1}\,\tilde{H}\,. \tag{3.138}$$

The matrix $\Lambda - \sigma\mathcal{I}$ is diagonal, such that it is easily inverted. Its inverse is simply the diagonal matrix with entries equal to $1/(\lambda_i - \sigma)$. Because $\lambda_i \leq 0$, these quantities are not zero provided $\sigma > 0$.

The general algorithm is as follows :

1. Calculate $\tilde{H} = \mathcal{P}^{-1}H$.
2. Calculate $\tilde{V} = (\Lambda - \sigma\mathcal{I})^{-1}\,\tilde{H}$.
3. Calculate $V = \mathcal{P}\tilde{V}$.
4. Calculate the boundary values $u_N(x_0)$ and $u_N(x_N)$.

In the special case of the Neumann problem for the Poisson equation ($\sigma = 0$), the matrix $\mathcal{D}_x$ has one eigenvalue equal to zero. In the calculation of the vector $\tilde{V}$, the peculiar component of $\tilde{V}$ corresponding to this null eigenvalue is simply taken arbitrarily equal to zero. This arbitrariness is consistent with the fact that the solution of the Neumann problem for the Poisson equation is defined up to a constant (obviously when the compatibility condition is satisfied).

The computational effort with this algorithm is obviously much heavier than those required by the direct inversion of $\mathcal{A} = \mathcal{D}_x - \sigma\mathcal{I}$. Even in the case of unsteady problems, for which the calculation of the λ_i, $\mathcal{P}$, and $\mathcal{P}^{-1}$ may be done once and for all in the preprocessing stage performed before the start of the time-integration, the algorithm remains more costly. However, it becomes really efficient for multidimensional problems as will be seen below.

The use of the matrix-diagonalization method necessitates the calculation of the eigenvalues of the matrix $\mathcal{D}_x$ of the associated eigenvectors and the inversion of the eigenvector matrix $\mathcal{P}$. It is known (e.g., Canuto *et al.* 1988) that the largest eigenvalues of $\mathcal{D}_x$ vary like N^4 as $N \to \infty$, so that these eigenvalues are not necessarily good approximations to the eigenvalues of the continuous operator (Weidman and Trefethen, 1988). On the other hand, the effect of round-off errors on the calculation of the eigenvalues of the second-order differentiation operator is negligible (contrary to the case of the first-order derivative). This is due to the fact that the sensitivity of the eigenvalues is measured by the condition number of the eigenvector matrix $\mathcal{P}$ (and not $\mathcal{D}_x$), as stated by the Bauer-Fike theorem (Wilkinson, 1965). In the present case, the matrix $\mathcal{P}$ is well-conditioned. For example, the condition number $\kappa(\mathcal{D}_x)$ in the Dirichlet case behaves like $0.80N^{1/4}$ (Ehrenstein, 1986). This property also ensures that no numerical difficulties are encountered when inverting the matrix $\mathcal{P}$.

Remarks

1. *Case of variable σ*
In the Helmholtz equation (3.128) the coefficient σ is assumed to be a constant. However the solution technique described above also applies if σ is a function time t, when the Helmholtz equation to be solved is the result of a time-discretization process. If σ depends on x, we diagonalize the matrix associated with u''/σ or $u'' - \sigma u$.

2. *Case of complex eigenvalues*
For elliptic equations different from the Helmholtz equation (3.128), such as for equations resulting from the application of a coordinate transformation to (3.128) or else for the advection-diffusion equation, the corresponding matrix $\mathcal{D}_x$ may have (conjugate) complex eigenvalues. In such a case, it is possible to avoid complex number computations, owing to the quasi-diagonalization procedure considered by Pasquetti and Bwemba (1994) for the solution of the equation $u'' + (2m + 1)u'/x = f$ (m = integer) coming from the Poisson equation in a cylindrical coordinate system.

For the presentation of the technique, one assumes that $\mathcal{D}_x$ has two conjugate complex eigenvalues λ_1 and $\lambda_2 = \overline{\lambda}_1$, the associated eigenvectors being W_1 and $\overline{W}_1$. Now let us introduce the following partition of the matrices Λ and $\mathcal{P}$:

$$\Lambda = \begin{bmatrix} \Lambda_1 & 0 \\ 0 & \Lambda' \end{bmatrix}, \quad \mathcal{P} = [\mathcal{P}_1 \ \mathcal{P}'],$$

where Λ_1 is the 2×2 matrix

$$\Lambda_1 = \begin{bmatrix} \lambda_1 & 0 \\ 0 & \overline{\lambda}_1 \end{bmatrix}$$

and $\mathcal{P}_1$ is the $(N-1) \times 2$ matrix constructed with the eigenvectors W_1 and $\overline{W}_1$:

$$\mathcal{P}_1 = [W_1 \ \overline{W}_1]$$

and Λ', $\mathcal{P}'$ being the complementary parts. It is easy to show that the real matrices $\mathcal{K}$ and $\mathcal{T}$ respectively defined by

$$\mathcal{K} = \begin{bmatrix} \mathcal{K}_1 & 0 \\ 0 & \Lambda' \end{bmatrix}, \quad \mathcal{T} = [\mathcal{T}_1 \ \mathcal{P}'],$$

with

$$\mathcal{K}_1 = \left[\begin{array}{cc} \mathcal{R}e(\lambda_1) & \mathcal{I}m(\lambda_1) \\ -\mathcal{I}m(\lambda_1) & \mathcal{R}e(\lambda_1) \end{array} \right], \quad \mathcal{T}_1 = [\mathcal{R}e(W_1) \quad \mathcal{I}m(W_1)],$$

are such that

$$\mathcal{D}_x = \mathcal{T} \mathcal{K} \mathcal{T}^{-1}. \tag{3.139}$$

Therefore, this equation replaces Eq.(3.137) and the algorithm described above can be applied in the same way except that the determination of $\dot{V}$, as the solution of

$$(\mathcal{K} - \sigma \mathcal{I}) \dot{V} = \ddot{H}, \tag{3.140}$$

no longer leads to uncoupled equations : two of them (corresponding to $\mathcal{K}_1$) are coupled. This technique obviously applies to more than one couple of complex eigenvalues : each couple has to be replaced by the real 2×2 matrix like $\mathcal{K}_1$ to constitute the matrix $\mathcal{K}$, as well as the associated eigenvectors are replaced by their real and imaginary parts as in $\mathcal{T}_1$ to constitute the matrix $\mathcal{T}$. The resulting system (3.140) yields a system of uncoupled equations (for the real eigenvalues) and at most coupled two-by-two (for each couple of conjugate complex eigenvalues).

3. Steady-state advection-diffusion equation

In this remark, we want to point out the difficulty associated with the application of the matrix-diagonalization method to the solution of the algebraic system resulting from the discretization of equation (3.56) with $a \neq 0$, and its multidimensional analog. Then a better adapted solution method, namely, the Schur-decomposition method, will be described and applied to two typical examples.

When $a \neq 0$ equation (3.56) is of advection-diffusion type. Such an equation arises in the course of the solution of the time-dependent advection-diffusion equation discretized with an implicit scheme [see Eq.(4.49)]. It may also appear in the solution of nonlinear equations like the Burgers equation or the Navier-Stokes equations when the nonlinear first-order derivative term is approximated in a semi-implicit way as described in Section 3.5.2 [see also Eq.(4.53)]. Such a semi-implicit technique has been used by Forestier et $al.$ (2000a,b) for the calculation of wake flows.

We consider the steady-state advection-diffusion-type equation

$$u'' - a u' - \sigma u = f, \quad -1 < x < 1 \tag{3.141}$$

where the constant coefficients a and σ are positive. Note that a could be negative without any influence on the conclusions. The Dirichlet boundary conditions

$$u(-1) = g_-, \quad u(1) = g_+, \tag{3.142}$$

are associated with Eq.(3.141), but more general conditions could be considered as well.

The problem (3.141)-(3.142) is approximated with the Chebyshev collocation method as described above in this section. So, the algebraic system determining the unknowns $u_N(x_i)$, $i = 1, \cdots, N-1$, has the form

$$(\mathcal{A} - \sigma \mathcal{I}) V = H, \qquad (3.143)$$

where V is the vector of inner unknowns [Eq.(3.131)] and $\mathcal{A}$ is a $(N-1) \times (N-1)$ matrix constructed from the Chebyshev differentiation matrix.

The eigenvalues of the matrix are generally complex (see Section 4.2.3 for some properties of these eigenvalues ; see also Nana Kouamen, 1992 ; Reddy and Trefethen, 1994). Therefore, the modification of the matrix-diagonalization method given above in Remark 2 has to be employed. Unfortunately, the transformation matrix $\mathcal{T}$ [see Eq.(3.139)] may be ill-conditioned for some values of a and N (see Fig.3.3). In such a situation, the inversion of $\mathcal{T}$ is inaccurate or even impossible.

This behaviour may be explained by examining the eigenvalue problem associated to the advection-diffusion operator, namely,

$$\varphi'' - a\,\varphi' = \lambda\varphi, \quad -1 < x < 1 \qquad (3.144)$$

$$\varphi(-1) = 0, \quad \varphi(1) = 0. \qquad (3.145)$$

The eigenvalues λ_n and the associated eigenvectors φ_n are, respectively,

$$\lambda_n = -\frac{a^2}{4} - \frac{\pi^2 n^2}{4}, \quad n = 1,2,3,\cdots, \qquad (3.146)$$

and

$$\varphi_n = \begin{cases} e^{a\,x/2} \cos\dfrac{\pi n}{2}x & \text{if } n \text{ is odd}, \\[2mm] e^{a\,x/2} \sin\dfrac{\pi n}{2}x & \text{if } n \text{ is even}. \end{cases} \qquad (3.147)$$

As stated by Reddy and Trefethen (1994), the condition number κ_C of the basis of eigenfunctions, normalized such that $\|\varphi_n\| = 1$, is given by

$$\kappa_C = e^a. \qquad (3.148)$$

Therefore, κ_C is exponentially increasing with a. We may expect the condition number κ_N of the transformation matrix $\mathcal{T}$ to have similar behaviour, at least if N is sufficiently large to ensure an accurate representation of the eigenfunctions. Figure 3.4 compares the variations of κ_C and κ_N (more precisely, their inverses) in terms of a. For a fixed value of N, κ_N follows κ_C when a is small ; then, for larger values of a, κ_N deviates from κ_C and reaches much smaller values. Thus, for any value of N, there exists a range of a where the ill-conditioning of $\mathcal{T}$ makes inaccurate, and even impossible, the use of the matrix-diagonalization method.

A simple remedy may be to set $v = \exp(-ax/2)\,u$ in Eq.(3.141) so as to get a Helmholtz equation [as done by Shizgal (2002) for another purpose],

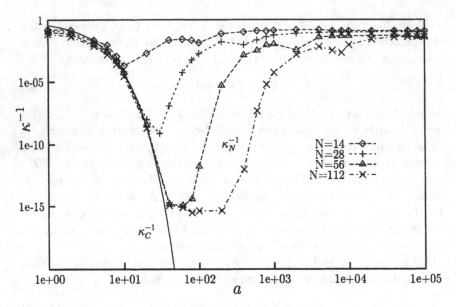

FIGURE 3.4. Dependence on a of κ_C^{-1} and κ_N^{-1} for various values of N.

accurately solved by the matrix-diagonalization technique. However, the numerical error associated with the solution of the transformed problem, which becomes stiff when a is large, restricts the application of this change of variables to relatively small values of a.

An efficient solution method of more general application is constituted by the Schur-decomposition method. The method, developed by Barthels and Stewart (1971), is based on the Schur reduction of the matrix $\mathcal{A}$ to triangular real form by orthogonal transformation. More precisely, the matrix $\mathcal{A}$ is transformed according to

$$\mathcal{A}' = \mathcal{P}^T \mathcal{A} \mathcal{P}, \tag{3.149}$$

where $\mathcal{P}$ is an orthogonal matrix. The real matrix $\mathcal{A}'$ is block-upper-triangular

$$\mathcal{A}' = \left[\mathcal{A}'_{i,j} \right], \quad i,j = 1, \cdots, p,$$

with the matrix $\mathcal{A}'_{i,j} = 0$ for $j < i$ and $p = N - 2q$, where $2q$ is the number of complex eigenvalues. Each matrix $\mathcal{A}'_{i,j}$ is of order at most 2 because of the presence of these complex eigenvalues. To be more precise, let us assume that λ_1, $\lambda_2 = \overline{\lambda}_1$ are two conjugate complex eigenvalues. Then, the 2×2 matrix $\mathcal{A}'_{1,1}$ has the form

$$\mathcal{A}'_{1,1} = \begin{pmatrix} \mathcal{R}e\,(\lambda_1) & a'_{1,2} \\ a'_{2,1} & \mathcal{R}e\,(\lambda_1) \end{pmatrix},$$

where $a'_{1,2}\,a'_{2,1} < 0$ and such that $\sqrt{|a'_{1,2}\,a'_{2,1}|} = |\mathcal{I}m\,(\lambda_1)|$.

The matrices $\mathcal{A}'$ and $\mathcal{P}$ are calculated by means of library routines (e.g., SGEES from LAPACK ; see also Barthels and Stewart, 1971). Then the solution algorithm follows that given above for the matrix-diagonalization method.

First, from Eq.(3.149), we have $\mathcal{A} = \mathcal{P}\mathcal{A}'\mathcal{P}^T$ that we bring into Eq.(3.143) which becomes

$$\left(\mathcal{P}\mathcal{A}'\mathcal{P}^T - \sigma\mathcal{I}\right) V = H.$$

Then, after left multiplication by $\mathcal{P}^T$, we obtain

$$(\mathcal{A}' - \sigma\mathcal{I})\tilde{V} = \tilde{H}, \tag{3.150}$$

where $\tilde{V} = \mathcal{P}^T V = [\tilde{u}_N(x_i)]$, $i = 1, \cdots, N-1$, and $\tilde{H} = \mathcal{P}^T H$. The matrix

$$\mathcal{A}' - \sigma\mathcal{I} = \left[a'_{i,j} - \sigma\,\delta_{i,j}\right], \quad i,j = 1, \cdots, N-1,$$

is quasi-upper-triangular. The system (3.150) is solved by back-substitution. The unknowns $\tilde{u}_N(x_i)$ are calculated explicitly for real eigenvalues but their determination necessitates the solution of a 2×2 system for each couple of conjugate complex eigenvalues.

To resume, the algorithm is as follows :

1. Calculate $\tilde{H} = \mathcal{P}^T H$.
2. Solve $(\mathcal{A}' - \sigma\mathcal{I})\tilde{V} = \tilde{H}$.
3. Calculate $V = \mathcal{P}\tilde{V}$.

Like the matrix-diagonalization method, the Schur-decomposition method is of interest for the solution of multidimensional problems arising at each time-cycle of a time-dependent process. The calculation of $\mathcal{A}'$ and $\mathcal{P}$ is done once and for all before the start of the time-integration. Then the above algorithm is applied at each time-cycle. Note that point 2 is more expensive than point 2 of the matrix-diagonalization algorithm.

The extension to multidimensional problems is analogous to that described in the following sections for the matrix-diagonalization method. Note that both methods can be applied in combination, for example, to solve the equation

$$(\partial_{xx} - a\,\partial_x)\,u + \partial_{yy}u - \sigma\,u = f, \tag{3.151}$$

Schur-decomposition is associated with the operator $\partial_{xx} - a\,\partial_x$ and matrix-diagonalization with ∂_{yy}.

Now we present two examples of application (Forestier, 2000). The first solution is very smooth and is well-represented by a polynomial of relatively low degree whatever a. The second solution is stiff and exhibits a boundary layer for large values of a. These examples allow us to compare the Schur-decomposition and matrix-diagonalization methods. The discussion is based on the level of satisfaction of the two usual requirements for a numerical solution :

1. The algebraic system accurately represents the continuous problem.

2. The solution algorithm accurately solves this system.

Both solutions are calculated with $N = 28$ and by making $\sigma = a$ in Eq.(3.141). The numerical error E is measured by the discrete L^∞-error defined by

$$E = \max_i |u_i - u(x_i)|, \tag{3.152}$$

where u_i refers to the approximate solution and $u(x_i)$ to the exact one.

The first example concerns the solution

$$u = \sin 2\pi (x + 1) \tag{3.153}$$

that defines f in Eq.(3.141) and $g_\pm = 0$ in Eq.(3.142). The function (3.153) is independent of a and is perfectly represented with $N = 28$. Figure 3.5 displays the errors in terms of a given by the Schur-decomposition method [curve (1)] and by the matrix-diagonalization [curve (2)] method. The inverse κ_N^{-1} of the condition number of the transformation matrix T involved in the diagonalization procedure is also shown in Fig.3.5. The error of the Schur-decomposition method is 10^{-14} (i.e. zero at the machine precision) whatever a. This proves, by the way, that requirement (1) is well satisfied with $N = 28$ for any value of a. Also, requirement (2) is satisfied by the Schur-decomposition method. On the other hand, requirement (2) is not satisfied for every value of a by the matrix-diagonalization method. As announced, it may be observed that the accuracy of the method is directly connected to the conditioning of the transformation matrix T.

The second example of solution is

$$u = \frac{1}{e^{2(r_2 - r_1)} - 1} \left[e^{2r_2 + r_1(x-1)} - e^{r_2(x+1)} \right] \tag{3.154}$$

with

$$r_1 = \frac{a - \sqrt{\Delta}}{2}, \quad r_2 = \frac{a + \sqrt{\Delta}}{2}, \quad \Delta = a(a + 4).$$

This function is the solution to problem (3.141)-(3.142) with $\sigma = a$, $f = 0$, $g_- = 1$, and $g_+ = 0$. For large values of a, it exhibits a boundary layer near $x = 1$ whose thickness is $O(a^{-1})$. In this case, requirement (1) is satisfied only if N is sufficiently large. Thus, for large a, we may expect the solution to be inaccurate for $N = 28$ whatever the algorithm used to solve the algebraic system. This is seen in Fig.3.6. As previously, the matrix-diagonalization method gives an error much larger than the Schur-decomposition method when κ_N is large [requirement (2) is not satisfied].

Lastly, it must be noticed that increasing N improves the error of the Schur-decomposition method. On the other hand, the error associated with the matrix-diagonalization method increases because the condition number κ_N deteriorates since it tends toward the continuous one κ_C when N increases (Fig.3.4).

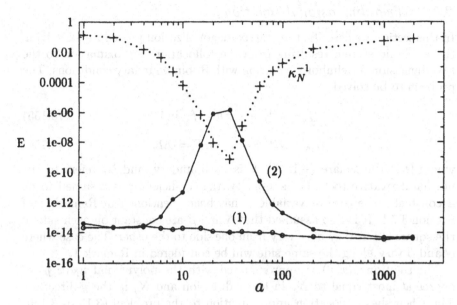

FIGURE 3.5. Error E versus a for the solution (3.153) calculated with $N = 28$: (1) Schur-decomposition method, (2) Matrix-diagonalisation method. The figure also shows the variation of κ_N^{-1}.

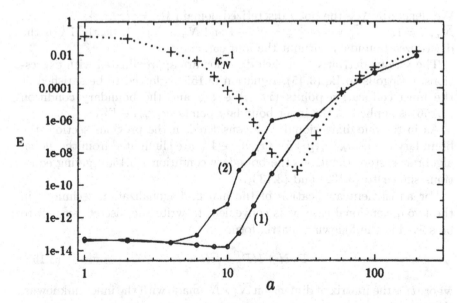

FIGURE 3.6. Error E versus a for the solution (3.154) calculated with $N = 28$: (1) Schur-decomposition method, (2) Matrix-diagonalization method. The figure also shows the variation of κ_N^{-1}.

3.7.2 Two-dimensional equation

In this section we describe the matrix-diagonalization procedure for solving the algebraic system resulting from the collocation approximation to the two-dimensional Helmholtz equation with Robin boundary conditions. The problem to be solved is

$$\partial_{xx} u + \partial_{yy} u - \sigma \, u = f \quad \text{in } \Omega \tag{3.155}$$

$$\alpha \, u + \beta \, \partial_n u = g \quad \text{on } \Gamma = \partial \Omega, \tag{3.156}$$

where Ω is the square $(-1,1)^2$, Γ is its boundary, and ∂_n refers to the normal derivative to Γ. For simplicity, the coefficient σ is assumed to be a constant ; the case of variable σ has been mentioned in Remark 1 of Section 3.7.1. It is also assumed that α and β are constant on each side of the square Ω but they may vary from one side to the other. The case where α and β vary along the same side will be considered in Remark 1.

The solution $u(x,y)$ is approximated with the polynomial $u_N(x,y)$ of degree at most equal to N_x in the x-direction and N_y in the y-direction. The Chebyshev collocation approximation to the problem (3.155)-(3.156) makes use of the Gauss-Lobatto mesh $\overline{\Omega}_N$ defined by

$$x_i = \cos \frac{\pi i}{N_x}, \quad i = 0, \dots, N_x, \qquad y_j = \cos \frac{\pi j}{N_y}, \quad j = 0, \dots, N_y.$$

We denote by Ω_N the open discretized domain $\Omega_N = \{x_i, y_j\}$, $i = 1, \dots,$ $\overline{N}_x$, $j = 1, \dots, \overline{N}_y$ where $\overline{N}_x = N_x - 1$ and $\overline{N}_y = N_y - 1$. Lastly, Γ_N^I is the discretized boundary without the four corners.

The various derivatives in each direction are approximated with expressions analogous to Eq.(3.45). Equation (3.155) is forced to be satisfied at the inner collocation points $(x_i, y_j) \in \Omega_N$ and the boundary condition (3.156) is applied at the inner boundary points $(x_i, y_j) \in \Gamma_N^I$.

As in the one-dimensional case considered in the previous section, the boundary values $u_N(x_i, y_j)$ for $(x_i, y_j) \in \Gamma_N^I$ are eliminated from the global algebraic system thanks to the boundary conditions (3.156), giving equations similar to (3.132) and (3.133).

For an efficient application of the matrix-diagonalization technique in the two-dimensional case, it is convenient to write the discrete system to be solved in the following matrix form

$$\mathcal{D}_x \mathcal{U} + \mathcal{U} \mathcal{D}_y^T - \sigma \mathcal{U} = \mathcal{H}, \tag{3.157}$$

where $\mathcal{U}$ is the matrix of dimension $\overline{N}_x \times \overline{N}_y$ made with the inner unknowns, that is,

$$\mathcal{U} = [u_N(x_i, y_j)], \quad i = 1, \dots, \overline{N}_x, \quad j = 1, \dots, \overline{N}_y.$$

In Eq.(3.157), $\mathcal{D}_x$ and $\mathcal{D}_y$ are the matrices of dimension $\overline{N}_x \times \overline{N}_x$ and $\overline{N}_y \times \overline{N}_y$, respectively, analogous to the $(N-1) \times (N-1)$ matrix $\mathcal{D}_x$

defined in the one-dimensional case considered in Section 3.7.1. Lastly, $\mathcal{H}$ is the $\overline{N}_x \times \overline{N}_y$ matrix containing the inner values of f and the values of g.

Let us denote by Λ_x and Λ_y the diagonal matrices whose entries are the eigenvalues $\lambda_{x,i}$, $i = 1, \ldots, \overline{N}_x$, and $\lambda_{y,j}$, $j = 1, \ldots, \overline{N}_y$, of the matrices $\mathcal{D}_x$ and $\mathcal{D}_y$, respectively, so that

$$\mathcal{D}_x = \mathcal{P}\Lambda_x \mathcal{P}^{-1}, \quad \mathcal{D}_y = \mathcal{Q}\Lambda_y \mathcal{Q}^{-1}, \tag{3.158}$$

where $\mathcal{P}$ and $\mathcal{Q}$ are the matrices with their columns containing the eigenvectors.

Now the aim is to use the expressions (3.158) in order to reduce the system (3.157) to a sequence of uncoupled equations. First, Eq.(3.157) is left multiplied by $\mathcal{P}^{-1}$ so that it becomes

$$\mathcal{P}^{-1}\mathcal{D}_x \mathcal{P}\tilde{\mathcal{U}} + \tilde{\mathcal{U}}\mathcal{D}_y^T - \sigma\tilde{\mathcal{U}} = \tilde{\mathcal{H}}, \tag{3.159}$$

where $\tilde{\mathcal{U}} = \mathcal{P}^{-1}\mathcal{U}$ and $\tilde{\mathcal{H}} = \mathcal{P}^{-1}\mathcal{H}$. Taking the first equation (3.158) into account, the above equation gives

$$\Lambda_x \tilde{\mathcal{U}} + \tilde{\mathcal{U}}\mathcal{D}_y^T - \sigma\tilde{\mathcal{U}} = \tilde{\mathcal{H}}. \tag{3.160}$$

Now we multiply at right this last equation by $(\mathcal{Q}^T)^{-1}$, that is,

$$\Lambda_x \hat{\mathcal{U}} + \hat{\mathcal{U}}\mathcal{Q}^T \mathcal{D}_y^T (\mathcal{Q}^T)^{-1} - \sigma\hat{\mathcal{U}} = \hat{\mathcal{H}}, \tag{3.161}$$

where we have introduced $\hat{\mathcal{U}} = \tilde{\mathcal{U}}(\mathcal{Q}^T)^{-1}$ and $\hat{\mathcal{H}} = \tilde{\mathcal{H}}(\mathcal{Q}^T)^{-1}$. Finally, taking into account that

$$\mathcal{Q}^T \mathcal{D}_y^T (\mathcal{Q}^T)^{-1} = \Lambda_y,$$

Equation (3.161) is of the form

$$\Lambda_x \hat{\mathcal{U}} + \hat{\mathcal{U}}\Lambda_y - \sigma\hat{\mathcal{U}} = \hat{\mathcal{H}}. \tag{3.162}$$

Therefore, if $\hat{\mathcal{U}} = [\hat{u}_{i,j}]$ and $\hat{\mathcal{H}} = \left[\hat{h}_{i,j}\right]$, $i = 1, \ldots, \overline{N}_x$, $j = 1, \ldots, \overline{N}_y$, Eq.(3.162) simply gives

$$\hat{u}_{i,j} = \frac{\hat{h}_{i,j}}{\lambda_{x,i} + \lambda_{y,j} - \sigma}, \quad i = 1, \ldots, \overline{N}_x, \quad j = 1, \ldots, \overline{N}_y. \tag{3.163}$$

Knowing $\hat{\mathcal{U}}$, it is easy to calculate $\tilde{\mathcal{U}}$ and $\mathcal{U}$.

To sum up, the algorithm is :

1. Calculate $\tilde{\mathcal{H}} = \mathcal{P}^{-1}\mathcal{H}$.
2. Calculate $\hat{\mathcal{H}} = \tilde{\mathcal{H}}(\mathcal{Q}^T)^{-1}$.
3. Calculate $\hat{\mathcal{U}}$ from (3.163).

4. Calculate $\tilde{\mathcal{U}} = \hat{\mathcal{U}} Q^T$.
5. Calculate $\mathcal{U} = \mathcal{P}\tilde{\mathcal{U}}$.
6. Calculate the boundary values $u_N(x_i, y_j)$ for $(x_i, y_j) \in \Gamma_N^I$.

It must be noticed that the values of $u_N(x, y)$ at the corners of the square Ω are not at all involved in the solution. If necessary, they can be computed when the solution has been calculated. More precisely, let us consider, for example, the two sides $x = -1$ and $x = 1$ of Ω. The boundary condition (3.156) relative to these sides is extended up to the corners $(\pm 1, 1)$ to give the two equations

$$\alpha\, u_N\, (\pm 1, 1) + \beta\, \partial_x u_N\, (\pm 1, 1) = g\, (\pm 1, 1)\,.$$

Then, by expressing the derivatives $\partial_x u_N\, (\pm 1, 1)$ thanks to the formula (3.45) with $p = 1$, we get two equations that determine the two corner values $u_N\, (\pm 1, 1)$ since the inner values $u_N\, (x_i, 1)$, $i = 1, \ldots, \overline{N}_x$, are known.

In the case of the Poisson equation ($\sigma = 0$) with pure Neumann conditions ($\alpha = 0$), there exists a couple (i, j) for which $\lambda_{x,i} = \lambda_{y,j} = 0$ so that Eq.(3.163) has no meaning. As done in the one-dimensional case, the associated value $\hat{u}_{i,j}$ may be taken arbitrarily, say zero.

The computational effort associated with the matrix-diagonalization procedure is made of two parts. The first part consists of the calculation of the eigenvalues and eigenvectors, as well as the inversion of the eigenvector matrices. In the second part, described in the above algorithm, essentially four matrix-matrix products have to be performed.

The overall computational effort may seem to be very large and not really competitive with other solution methods (iterative methods, for example) for algebraic systems. As a matter of fact, in the case of the solution of unsteady problems (the usual situation considered here), Helmholtz problems similar to the one studied in the present section have to be solved at each time-cycle, that is a very large number of times. The interest of the method is that the first part of the calculations, which is also the most expensive, can be done once and for all in the preprocessing stage performed before the start of the time-integration. Therefore, at each time cycle, only the second part has to be performed. That is, the solution reduces to four matrix-matrix products which are efficiently done on vector computers.

Remarks

1. Change in boundary conditions : Influence matrix method
The matrix-diagonalization procedure described above works when the coefficients α and β in the boundary conditions (3.156) are constant on a whole side of the boundary Γ, although they may have different values on each side ; for example, the boundary conditions may be of Dirichlet type on three sides and of Neumann type on the fourth side. When this is not the case, that is, when α and β vary along one side, the algebraic system cannot be put in the form (3.157) and the procedure cannot be applied.

Such a situation appears, for example, when a Dirichlet condition is applied to one part of a side of Γ and a Neumann condition to the complementary part as considered in Section 8.4.1. In such a case, an efficient solution technique (Streett and Hussaini, 1986 ; Pulicani, 1988) based on the influence matrix method can be used.

The essence of this method, which can be applied to a large variety of problems as shown in the present book, is to take advantage of the linearity of the problem to construct the solution from a linear combination of elementary solutions. Then, the coefficients of this linear combination are determined so that the constructed solution satisfies the boundary conditions. One of the advantages of the method is that only Dirichlet problems have to be solved whatever the type of boundary conditions. The method is now described.

Let us consider the Helmholtz problem (3.155)-(3.156) in which α and β vary on Γ. The solution is sought in the form

$$u = \tilde{u} + \overline{u}, \tag{3.164}$$

where the part $\tilde{u}$ satisfies Eq.(3.155) with homogeneous Dirichlet boundary conditions, namely the $\tilde{\mathcal{P}}$-Problem :

$$\partial_{xx}\tilde{u} + \partial_{yy}\tilde{u} - \sigma\tilde{u} = f \text{ in } \Omega \tag{3.165}$$

$$\tilde{u} = 0 \text{ on } \Gamma. \tag{3.166}$$

The complementary part $\overline{u}$ is defined as the solution of the $\overline{\mathcal{P}}$-Problem :

$$\partial_{xx}\overline{u} + \partial_{yy}\overline{u} - \sigma\overline{u} = 0 \text{ in } \Omega \tag{3.167}$$

$$\overline{u} = \xi \text{ on } \Gamma, \tag{3.168}$$

where the unknown function ξ has to be determined such that $u = \tilde{u} + \overline{u}$ satisfies the boundary condition (3.156).

Now it is necessary to discuss the discretization of the problems with the Chebyshev collocation method. Let

$$u_N = \tilde{u}_N + \overline{u}_N \tag{3.169}$$

be the polynomial approximation associated with (3.164). The $\tilde{\mathcal{P}}$-problem is easily solved using the matrix-diagonalization procedure. The $\overline{\mathcal{P}}$-problem is solved by introducing the linear combination

$$\overline{u}_N(x, y) = \sum_{l=1}^{L} \xi_l \overline{u}_{N,l}(x, y), \tag{3.170}$$

where L is the number of collocation points η_l on the boundary Γ_N^I, namely the discretized boundary without the corners. The elementary solution $\overline{u}_{N,l}$ satisfies the $\overline{\mathcal{P}}_l$-Problem :

$$\partial_{xx}\overline{u}_{N,l}(x_i, y_j) + \partial_{yy}\overline{u}_{N,l}(x_i, y_j) - \sigma\overline{u}_{N,l}(x_i, y_j) = f(x_i, y_j) \text{ for } (x_i, y_j) \in \Omega_N, \tag{3.171}$$

$$\overline{u}_{N,l}|_{\eta_m} = \delta_{l,m} \text{ for } \eta_m \in \Gamma_N^I, \tag{3.172}$$

where $\delta_{l,m}$ is the Kronecker delta. Taking the boundary conditions (3.172) into account in (3.170), it is easily seen that the coefficients ξ_l, $l = 1, \ldots, L$, are nothing other than the values of the function ξ at the collocation points η_l belonging to Γ_N^I. Each $\overline{\mathcal{P}}_l$-problem is again solved by means of the diagonalization procedure. Then the constants ξ_l, $l = 1, \ldots, L$, are determined such that the boundary condition (3.156) is satisfied on Γ_N^I, that is,

$$\alpha|_{\eta_m} \left(\tilde{u}_N|_{\eta_m} + \sum_{l=1}^L \xi_l \overline{u}_{N,l}|_{\eta_m} \right)$$

$$+ \beta|_{\eta_m} \left(\partial_n \tilde{u}_N|_{\eta_m} + \sum_{l=1}^L \xi_l \partial_n \overline{u}_{N,l}|_{\eta_m} \right) = f|_{\eta_m}, \quad m = 1, \ldots, L,$$

which can be written as

$$\mathcal{M}\Xi = \tilde{E}, \tag{3.173}$$

where $\Xi = (\xi_1, \ldots, \xi_L)^T$, $\mathcal{M} = [m_{i,j}]$, $i, j = 1, \ldots, N$, is the "influence matrix" defined by

$$m_{i,j} = \alpha|_{\eta_i} \overline{u}_{N,j}|_{\eta_i} + \beta|_{\eta_i} \partial_n \overline{u}_{N,j}|_{\eta_i}$$

and $\tilde{E}$ is the vector $\tilde{E} = [\tilde{e}_i]$, $i = 1, \ldots, L$, such that

$$\tilde{e}_i = f|_{\eta_i} - \alpha|_{\eta_i} \tilde{u}_N|_{\eta_i} - \beta|_{\eta_i} \partial_n \tilde{u}_N|_{\eta_i}.$$

Finally, the solution of (3.173) gives the values ξ_l, $l = 1, \ldots, L$, which enter into the linear combination (3.170), and the problem is completely solved.

The main part of the computational effort consists of the solution of $L + 1$ Dirichlet problems, the inversion of the matrix $\mathcal{M}$, and the linear combination of the elementary solutions. Again, the price may seemed to be very high if the problem has to be solved once only. However, as for the matrix-diagonalization procedure, the influence matrix method becomes very efficient when the problem has to be solved a large number of times in the course of an unsteady process. The interesting point is that, in such a case, the elementary solutions $\overline{u}_{N,l}$, $l = 1, \ldots, L$, do not depend on time. Therefore they can be calculated once and for all in the preprocessing stage. Also, the matrix $\mathcal{M}$ is inverted during this stage and its inverse $\mathcal{M}^{-1}$ is stored. Taking these precalculations into account, only one Dirichlet problem (for the determination of $\tilde{u}_N$) has to be solved, which amounts to four matrix-matrix products as seen above. Then the constants ξ_l, $l = 1, \ldots, L$, are determined from the product $\mathcal{M}^{-1}\tilde{H}$. Now it remains to evaluate the linear combination (3.170) at the collocation points $(x_i, y_j) \in \Omega_N \cup \Gamma_N^I$. The simple loop on the index l is costly. Computing time can be saved by reorganizing the evaluation as a matrix-vector product. Note that if the storage of the L elementary solutions $\bar{u}_{N,l}(x_i, y_j)$ is not possible, the determination of the solution u_N can be done by solving the system

$$\partial_{xx} u_N(x_i, y_j) + \partial_{yy} u_N(x_i, y_j) - \sigma u_N(x_i, y_j) = f(x_i, y_j) \text{ for } (x_i, y_j) \in \Omega_N,$$
$$\tag{3.174}$$
$$u_N|_{\eta_l} = \xi_l \text{ for } \eta_l \in \Gamma_N^I, \tag{3.175}$$

again with matrix diagonalization. It must be noticed that this last solution is a little more costly than the second one consisting of evaluating the linear combination by a matrix-vector product, but is much less expensive than the simple loop on l.

Finally, it is important to notice that the influence matrix technique is nothing other than a special algorithm for solving the full algebraic system obtained from the discretization of the equation (3.155) and the boundary conditions (3.156). This will be shown for the other applications of the influence matrix method in Sections 6.3.2 and 7.3.2.c.

2. Partial diagonalization procedure
In the diagonalization method described above, the operators in both directions are diagonalized (this method is sometimes called "full diagonalization") so that the system reduces to a set of uncoupled equations. However, it may be valuable, and sometimes necessary, to diagonalize in one direction only and solve the remaining system in the other direction.

For example, when solving the Helmholtz equation with the tau method, after diagonalization in one direction, the solution of the equation in the other direction is efficiently obtained by means of the technique leading to the solution of two uncoupled quasi-tridiagonal systems (see Section 3.4.1). We recall that such systems can be solved in a number of operations of the order of N, therefore, their solution is quite possible at each time-cycle of an unsteady process.

Another example is the case where one of the operators cannot be diagonalized. In such a case, the system corresponding to that direction can be solved by the direct inversion method, as explained in Section 3.4.2.

3.7.3   Three-dimensional equation

Now we consider the general three-dimensional problem for the Helmholtz equation :

$$\partial_{xx} u + \partial_{yy} u + \partial_{zz} u - \sigma u = f \text{ in } \Omega \tag{3.176}$$

$$\alpha u + \beta \partial_n u = g \text{ on } \Gamma = \partial\Omega, \tag{3.177}$$

where $\Omega = (-1, 1)^3$ and ∂_n represents the normal derivative to the boundary Γ. It is assumed that σ is a positive constant and that α and β keep a constant value on each side of the cube. The case where σ is variable has been mentioned in Remark 1 of Section 3.7.1. When α and β are variable,

nothing forbids, in principle, the use of the influence matrix method described in Remark 1 of Section 3.7.2. However, the method is not practical
because of the size of the influence matrix. Therefore, for arbitrary functions α and β, the solution of the Helmholtz problem (3.176)-(3.177) must
be obtained by techniques other than the matrix-diagonalization procedure,
for example iterative methods. This is very much problem-dependent. For
example, in the case where the boundary condition type change on the same
side (Dirichlet on one part of the side and Neumann on the complementary
part), the domain decomposition technique may be applied as mentioned
in Section 8.4.1 in the two-dimensional case.

Problem (3.176)-(3.177) is solved by means of the collocation method
based on the Gauss-Lobatto points

$$x_i = \cos \frac{\pi i}{N_x}, \qquad i = 0, \ldots, N_x,$$

$$y_j = \cos \frac{\pi j}{N_y}, \qquad j = 0, \ldots, N_y, \qquad (3.178)$$

$$z_k = \cos \frac{\pi k}{N_z}, \qquad k = 0, \ldots, N_z.$$

As previously, we denote by Ω_N the open discretized domain, that is,
$\Omega_N = \{x_i, y_j, z_k\}$, $i = 1, \ldots, \overline{N}_x = N_x - 1$, $j = 1, \ldots, \overline{N}_y = N_y - 1$,
$k = 1, \ldots, \overline{N}_z = N_z - 1$, and by Γ_N^I the discretized boundary without the
12 edges of the cube.

The solution of the problem (3.176)-(3.177) is approximated with the
polynomial $u_N(x, y, z)$ of degree at most equal to N_x, N_y, and N_z in x-, y-
and z-directions, respectively.

The derivatives are approximated with the usual expressions [see
Eq.(3.45)]. First, the Helmholtz equation (3.176) is forced to be satisfied
by the polynomial $u_N(x, y, z)$ at every inner collocation point $(x_i, y_j, z_k) \in$
Ω_N. The boundary condition (3.177) is prescribed at every collocation point
$(x_i, y_j, z_k) \in \Gamma_N^I$. Then, as previously, the boundary values $u_N(x_i, y_j, z_k)$
for $(x_i, y_j, z_k) \in \Gamma_N^I$ are eliminated thanks the boundary condition (3.177).
We may write the resulting algebraic system under the tensor product form

$$(\mathcal{I}_z \otimes \mathcal{I}_y \otimes \mathcal{D}_x + \mathcal{I}_z \otimes \mathcal{D}_y \otimes \mathcal{I}_x + \mathcal{D}_z \otimes \mathcal{I}_y \otimes \mathcal{I}_x - \sigma \mathcal{I}_z \otimes \mathcal{I}_y \otimes \mathcal{I}_x) V = H,$$
$$(3.179)$$

where V is the column vector of (inner) unknowns ordered by row and horizontal section. More precisely, let us introduce the $\overline{N}_x$-component vector
$V_{j,k}$:

$$V_{j,k} = \left(u_N(x_1, y_j, z_k), \ldots, u_N(x_{\overline{N}_x}, y_j, z_k) \right)^T, \qquad (3.180)$$

for $j = 1, \ldots, \overline{N}_y$, $k = 1, \ldots, \overline{N}_z$. Then we define the $\overline{N}_x \overline{N}_y$-component vector V_k by

$$V_k = \left(V_{1,k}, \ldots, V_{\overline{N}_y, k} \right)^T, \quad k = 1, \ldots, \overline{N}_z, \tag{3.181}$$

and, finally, V is the $\overline{N}_x \overline{N}_y \overline{N}_z$-component vector defined by

$$V = \left(V_1, \ldots, V_{\overline{N}_z} \right)^T. \tag{3.182}$$

In Eq.(3.179), $\mathcal{D}_x$, $\mathcal{D}_y$, and $\mathcal{D}_z$ are square matrices of dimensions $\overline{N}_x \times \overline{N}_x$, $\overline{N}_y \times \overline{N}_y$ and $\overline{N}_z \times \overline{N}_z$, respectively, analogous to the matrix $\mathcal{D}_x$ defined in Section 3.7.1 in the one-dimensional case ; $\mathcal{I}_x$, $\mathcal{I}_y$, and $\mathcal{I}_z$ are identity matrices in the x-, y- and z-directions having, respectively, the same dimensions as $\mathcal{D}_x$, $\mathcal{D}_y$, and $\mathcal{D}_z$. Lastly, H is the $\overline{N}_x \overline{N}_y \overline{N}_z$-component vector constructed in the same way as V, but with components $h_{i,j,k}$ calculated from the inner values of f and the values of g.

In Eq.(3.179), and in the following, the tensor product theory is used (see Halmos, 1958 ; Lynch *et al.*, 1964 ; Van Loan, 1992). Before discussing the details of the solution of (3.179) by the diagonalization process, we recall some definitions and properties of the tensor product (also called the "Kronecker product" or "direct product").

Let $\mathcal{A} = [a_{i,j}]$ and $\mathcal{B} = [b_{i,j}]$ be two (rectangular) matrices of dimensions $K \times L$ and $M \times N$, respectively. The tensor product $\mathcal{A} \otimes \mathcal{B}$ is the matrix of dimensions $KM \times LN$ defined by

$$\mathcal{A} \otimes \mathcal{B} = \begin{bmatrix} a_{1,1}\mathcal{B} & a_{1,2}\mathcal{B} & \ldots & a_{1,L}\mathcal{B} \\ a_{2,1}\mathcal{B} & a_{2,2}\mathcal{B} & \ldots & a_{2,L}\mathcal{B} \\ \ldots & \ldots & \ldots & \ldots \\ a_{K,1}\mathcal{B} & a_{K,2}\mathcal{B} & \ldots & a_{K,L}\mathcal{B} \end{bmatrix}. \tag{3.183}$$

The following relations are satisfied by tensor products

$$(\mathcal{A} + \mathcal{C}) \otimes \mathcal{B} = \mathcal{A} \otimes \mathcal{B} + \mathcal{C} \otimes \mathcal{B}, \tag{3.184}$$

$$(\mathcal{A} \otimes \mathcal{B})(\mathcal{C} \otimes \mathcal{D}) = \mathcal{A}\mathcal{C} \otimes \mathcal{B}\mathcal{D}, \tag{3.185}$$

$$(\mathcal{A} \otimes \mathcal{B})^T = \mathcal{A}^T \otimes \mathcal{B}^T. \tag{3.186}$$

and in the case of invertible square matrices

$$(\mathcal{A} \otimes \mathcal{B})^{-1} = \mathcal{A}^{-1} \otimes \mathcal{B}^{-1}. \tag{3.187}$$

The first step in the solution of system (3.179) consists of exploiting the diagonalization process to put the solution V in a form involving only the inversion of "one-dimensional" matrices. First, system (3.179) is written as

$$\mathcal{A}V = H, \tag{3.188}$$

with

$$\mathcal{A} = \mathcal{I}_z \otimes \mathcal{I}_y \otimes \mathcal{D}_x^\sigma + \mathcal{I}_y \otimes \mathcal{D}_y \otimes \mathcal{I}_x + \mathcal{D}_z \otimes \mathcal{I}_y \otimes \mathcal{I}_x \,, \qquad (3.189)$$

where $\mathcal{D}_x^\sigma = \mathcal{D}_x - \sigma \mathcal{I}_x$. The matrices $\mathcal{D}_x$, $\mathcal{D}_y$, and $\mathcal{D}_z$ are diagonalized according to

$$\mathcal{D}_x = \mathcal{P}\Lambda_x\mathcal{P}^{-1} \,, \quad \mathcal{D}_y = \mathcal{Q}\Lambda_y\mathcal{Q}^{-1} \,, \quad \mathcal{D}_z = \mathcal{R}\Lambda_z\mathcal{R}^{-1} \,, \qquad (3.190)$$

where Λ_x, Λ_y, and Λ_z are the diagonal matrices whose entries are the eigenvalues of $\mathcal{D}_x$, $\mathcal{D}_y$ and $\mathcal{D}_z$, respectively. The matrices $\mathcal{P}$, $\mathcal{Q}$, and $\mathcal{R}$ are the associated eigenvector matrices.

Now, using (3.190) into (3.189), we have

$$\mathcal{R}^{-1} \otimes \mathcal{Q}^{-1} \otimes \mathcal{P}^{-1}\mathcal{A}\mathcal{R} \otimes \mathcal{Q} \otimes \mathcal{P} = \Lambda \,, \qquad (3.191)$$

where

$$\Lambda = \mathcal{I}_z \otimes \mathcal{I}_y \otimes \Lambda_x^\sigma + \mathcal{I}_z \otimes \Lambda_y \otimes \mathcal{I}_x + \Lambda_x \otimes \mathcal{I}_y \otimes \mathcal{I}_x \,,$$

with $\Lambda_x^\sigma = \Lambda_x - \sigma \mathcal{I}_x$. The square matrix Λ of order $\overline{N}_x\overline{N}_y\overline{N}_z$ is diagonal, so that its inverse Λ^{-1} is easily calculated. Finally, from (3.191), we get

$$\mathcal{A}^{-1} = \mathcal{R} \otimes \mathcal{Q} \otimes \mathcal{P}\Lambda^{-1}\mathcal{R}^{-1} \otimes \mathcal{Q}^{-1} \otimes \mathcal{P}^{-1} \qquad (3.192)$$

and

$$V = \mathcal{A}^{-1}H \,. \qquad (3.193)$$

The second step of the algorithm is to calculate $\mathcal{A}^{-1}$ in the most efficient way. This amounts to calculating the vector

$$W = \mathcal{R}^{-1} \otimes \mathcal{Q}^{-1} \otimes \mathcal{P}^{-1}H \,, \qquad (3.194)$$

then

$$\tilde{W} = \Lambda^{-1}W \,, \qquad (3.195)$$

and, finally,

$$V = \mathcal{R} \otimes \mathcal{Q} \otimes \mathcal{P}\tilde{W} \,, \qquad (3.196)$$

by using the same technique as for W.

Therefore, it remains to evaluate a product like W given by Eq.(3.194). Let us define the vectors X and Y :

$$X = \mathcal{I}_z \otimes \mathcal{I}_y \otimes \mathcal{P}^{-1}H \qquad (3.197)$$

$$Y = \mathcal{I}_z \otimes \mathcal{Q}^{-1} \otimes \mathcal{I}_x X \,, \qquad (3.198)$$

then

$$W = \mathcal{R}^{-1} \otimes \mathcal{I}_y \otimes \mathcal{I}_x Y \,. \qquad (3.199)$$

Calculation of X. From the vector H we construct the rectangular matrix $\mathcal{H}$, of dimension $\overline{N}_x \times \overline{N}_y \overline{N}_z$, made of the juxtaposition of $\overline{N}_z$ rectangular matrices $\mathcal{H}_k$, $k = 1, \ldots, \overline{N}_z$, of dimension $\overline{N}_x \times \overline{N}_y$, let

$$\mathcal{H} = \begin{bmatrix} \mathcal{H}_1 & \mathcal{H}_2 & \cdots & \mathcal{H}_{\overline{N}_z} \end{bmatrix}.$$

The columns of each matrix $\mathcal{H}_k$ are made with the vector $H_{j,k}$, $j = 1, \ldots, \overline{N}_y$, defined similarly to Eq.(3.180), that is,

$$\mathcal{H}_k = \begin{bmatrix} H_{1,k} & H_{2,k} & \cdots & H_{\overline{N}_y,k} \end{bmatrix}, \quad k = 1, \ldots, \overline{N}_z.$$

Then the matrix product

$$\mathcal{X} = \mathcal{P}^{-1} \mathcal{H}$$

defines the matrix $\mathcal{X}$ of dimension $\overline{N}_x \times \overline{N}_y \overline{N}_z$ whose columns are the $\overline{N}_y \overline{N}_z$ vectors $X_{1,1}, \ldots, X_{\overline{N}_y,1}, \ldots, X_{1,\overline{N}_z}, \ldots, X_{\overline{N}_y,\overline{N}_z}$ ordered in $\mathcal{X}$ in the same way that $H_{1,1}, \ldots, H_{\overline{N}_y,\overline{N}_z}$ were ordered in $\mathcal{H}$, more precisely,

$$\mathcal{X} = \begin{bmatrix} \mathcal{X}_1 & \mathcal{X}_2 & \cdots & \mathcal{X}_{\overline{N}_z} \end{bmatrix}$$

where the matrix $\mathcal{X}_k$ is

$$\mathcal{X}_k = \begin{bmatrix} X_{1,k} & X_{2,k} & \cdots & X_{\overline{N}_y,k} \end{bmatrix}, \quad k = 1, \ldots, \overline{N}_z. \tag{3.200}$$

The sets of vectors $X_{1,k}, \ldots, X_{\overline{N}_y,k}$, $k = 1, \ldots, \overline{N}_z$, define the vector X by expressions similar to Eqs.(3.181) and (3.182), that is,

$$X_k = \left(X_{1,k}, \ldots, X_{\overline{N}_y,k} \right)^T, \quad k = 1, \ldots, \overline{N}_z,$$

and

$$X = \left(X_1, \ldots, X_{\overline{N}_z} \right)^T. \tag{3.201}$$

Calculation of Y. This calculation makes use of the $\overline{N}_z$ matrices $\mathcal{X}_k$ previously calculated. Then the matrix products

$$\mathcal{Y}_k = \mathcal{X}_k \left(\mathcal{Q}^{-1} \right)^T, \quad k = 1, \ldots, \overline{N}_z,$$

define the matrices

$$\mathcal{Y}_k = \begin{bmatrix} Y_{1,k} & Y_{2,k} & \cdots & Y_{\overline{N}_y,k} \end{bmatrix}, \quad k = 1, \ldots, \overline{N}_z, \tag{3.202}$$

of dimension $\overline{N}_x \times \overline{N}_y$ from which the vector Y is deduced in the same way as X.

Calculation of W. We construct the matrix $\mathcal{Z}$ of dimension $\overline{N}_x\overline{N}_y \times \overline{N}_z$:

$$\mathcal{Z} = \begin{bmatrix} \mathcal{Y}_1^T & \mathcal{Y}_2^T & \cdots & \mathcal{Y}_{\overline{N}_z}^T \end{bmatrix},$$

where $\mathcal{Y}_k^T$ is the transpose of the matrix $\mathcal{Y}_k$ calculated above. Then the product

$$\mathcal{W} = \mathcal{Z}\left(\mathcal{R}^{-1}\right)^T \tag{3.203}$$

defines the matrix $\mathcal{W}$ of dimension $\overline{N}_x\overline{N}_y \times \overline{N}_z$ written as

$$\mathcal{W} = \begin{bmatrix} \mathcal{W}_1^T & \mathcal{W}_2^T & \cdots & \mathcal{W}_{\overline{N}_z}^T \end{bmatrix}$$

with

$$\mathcal{W}_k = \begin{bmatrix} W_{1,k} & W_{2,k} & \cdots & W_{\overline{N}_y,k} \end{bmatrix}. \tag{3.204}$$

Finally, from the vectors $W_{j,k}$, we can construct the vector W, as done previously for X and Y, that is

$$W = \left(W_1, \ldots, W_{\overline{N}_z}\right)^T \tag{3.205}$$

where the vectors W_k, $k = 1, \ldots, \overline{N}_z$, are

$$W_k = \left(W_{1,k}, W_{2,k}, \ldots, W_{\overline{N}_y,k}\right)^T. \tag{3.206}$$

This ends the calculation of W. Then, as described above, the vector $\tilde{W}$ is easily calculated from Eq.(3.195) and, finally, the solution vector V, defined by (3.196), is calculated in the same way as W.

3.8   Iterative solution methods

Our opinion is that direct solution methods have to be recommended whenever possible, namely, when they are not too costly, i.e., concerning computing time or memory requirements. However, there exists a number of situations where the use of direct methods is not possible. The necessity to have recourse to iterative solution procedures appears for nonlinear or nonseparable linear equations. Also, an iterative solution is needed when the equation to be solved at each time-cycle of a time-marching procedure, although linear and separable, has time-dependent coefficients. Lastly, iterative procedures constitute an alternative to direct methods, especially in three-dimensional problems.

In this section, we briefly describe some usual iterative methods (see Eisenstat *et al.*, 1983 ; Canuto *et al.*, 1988 ; Quarteroni and Valli, 1994) for the solution of linear algebraic systems arising from Chebyshev tau or collocation approximation methods, namely

$$\mathcal{A} V = F, \tag{3.207}$$

where V is the vector made with the unknowns. Efficiency requires preconditioning techniques, in which the preconditioning operator $\mathcal{A}_0$ has to be chosen according to the requirements given in Section 3.5.1.c.

The general algorithm is as follows :

1. Initialization

$$
\begin{aligned}
V^0 \quad & \text{is given} \\
R^0 \quad & = \mathcal{A} V^0 - F \\
\mathcal{A}_0 \, Y^0 \quad & = R^0 \\
Z^0 \quad & = Y^0,
\end{aligned}
\tag{3.208}
$$

2. Current iteration

$$
\begin{aligned}
V^{m+1} \quad & = V^m - \alpha_m \, Z^m \\
R^{m+1} \quad & = R^m - \alpha_m \, A \, Z^m \\
\mathcal{A}_0 \, Y^{m+1} \quad & = R^{m+1} \\
Z^{m+1} \quad & = Y^{m+1} + \beta_m \, Z^m .
\end{aligned}
\tag{3.209}
$$

The choice of the relaxation parameters α_m and β_m characterizes the type of procedure.

Preconditioned Richardson (PR)
This method is defined by

$$\alpha_m = \alpha = \text{ constant}, \quad \beta_m = 0. \tag{3.210}$$

The constant α has to be chosen to ensure convergence, that is, the spectral radius $\rho(\mathcal{E})$ of the iterative matrix $\mathcal{E} = \mathcal{I} - \alpha \, \mathcal{A}_0^{-1} \, \mathcal{A}$ is smaller than one. Assuming the eigenvalues λ of $\mathcal{A}_0^{-1} \, \mathcal{A}$ to be real positive, such that $\lambda_{min} \leq \lambda \leq \lambda_{max}$, the condition $\rho(\mathcal{E}) < 1$ is satisfied if

$$0 < \alpha < 2/\lambda_{max}. \tag{3.211}$$

The optimal value of α is $\alpha_{opt} = 2/(\lambda_{max} + \lambda_{min})$ (see Isaacson and Keller, 1966).

Preconditioned Minimal Residual (PMR)
The technique is characterized by

$$\alpha_m = \frac{(R^m, A\,Z^m)}{(A\,Z^m, A\,Z^m)}, \quad \beta_m = 0, \tag{3.212}$$

where $(\cdot, \cdot)$ is the Euclidian scalar product.

Preconditioned Conjugate Residual (PCR)
The parameters α_m and β_m are

$$\alpha_m = \frac{(R^m, A Z^m)}{(A Z^m, A Z^m)},$$

$$\beta_m = -\frac{(A Y^{m+1}, A Z^m)}{(A Z^m, A Z^m)}. \tag{3.213}$$

Preconditioned Conjugate Gradient (PCG)
The parameters α_m and β_m are

$$\alpha_m = \frac{(R^m, Y^m)}{(Z^m, A Z^m)},$$

$$\beta_m = \frac{(R^{m+1}, Y^{m+1})}{(R^m, Y^m)}. \tag{3.214}$$

In general, PCR and PCG methods are designed for symmetric problems, but they have been found to work well for Chebyshev approximations, even if the associated matrices are not symmetric. Fröhlich and Peyret (1990) have compared the four iterative procedures described above in the case of the Stokes problem with variable density. The best convergence has been obtained with the PCR method. A similar conclusion was obtained for the solution of a steady advection-diffusion equation (Sabbah, 2000).

Among the iterative methods adapted to the solution of nonsymmetric systems, the Bi-CGSTAB (Van der Vorst, 1992) or its improved variant Bi-CGSTAB-QMR (Chan *et al.*, 1994) could be of interest. The Bi-CGSTAB(l) proposed by Sleijpen and Fokkema (1993) has been applied by Dimitropoulos and Beris (1997) to the spectral solution of the Helmholtz equation (3.155) with σ depending on x and y. We refer to Barret *et al.* (1994), Quarteroni and Valli (1994), or to Saad (1998) for a review of modern iterative techniques, efficient but which could be expensive when applied to systems derived from Chebyshev approximations.

4

Time-dependent equations

In this chapter, we discuss the time-discretization of time-dependent equations. Although the methods apply to general nonlinear time-dependent equations, their analysis is developed in the linear case and, more especially, for the advection-diffusion equation. First, we address the stability of the spectral approximation, namely, the existence of a bounded solution of the differential equations in time resulting from the spectral approximation. Then the major part of the chapter is devoted to the presentation and discussion of the accuracy and stability of the time-discretization schemes. Second- and higher-order methods are considered in the following two cases : two-step methods (essentially based on Backward-Differentiation and Adams-Bashforth schemes) and one-step methods (Runge-Kutta schemes). The chapter ends with a comparison between the different kinds of time-discretization schemes.

4.1 Introduction

This chapter is devoted to the application of spectral methods to the solution of time-dependent partial differential equations of the form

$$\partial_t u = H(u)\,, \tag{4.1}$$

where H is a spatial second-order differential operator. With a view to the application to viscous incompressible fluid mechanics, H is the sum of a nonlinear first-order term $N(u)$ and a linear second-order term $L(u)$:

$$H(u) = N(u) + L(u) \,. \tag{4.2}$$

In the one-dimensional case, these operators are explicitly

$$N(u) = -\partial_x F(u) \,, \quad L(u) = \nu \, \partial_{xx} u \,, \tag{4.3}$$

where ν is a nonnegative constant. The Burgers equation, already considered in Sections 2.8 and 3.5.2 is obtained with $F(u) = u^2/2$. When $F(u) = a\,u$ with $a = $ constant, the resulting equation is the advection-diffusion equation which will be addressed in this chapter. However, the two-dimensional advection-diffusion equation defined by

$$N(u) = -\mathbf{V}.\nabla u \,, \quad L(u) = \nu \, \nabla^2 u \,, \tag{4.4}$$

where the vector $\mathbf{V}$ is given, will also be addressed to introduce some special issues.

The advection-diffusion equation constitutes a convenient model for discussing the numerical methods devised for the solution of the equations governing fluid motion and related transport-diffusion phenomena. For example, considering the motion of an incompressible, viscous, and heat-conducting fluid, the transport-diffusion of momentum $\rho \mathbf{V}$, vorticity ω, and temperature T is governed by an advection-diffusion equation with the fluid velocity $\mathbf{V}$ as an advective velocity. Obviously, the equation for momentum also involves the pressure gradient.

However, the coupling between the various equations of motion may induce some special properties which are not apparent when the advection-diffusion is considered alone. Such a phenomenon occurs, for example, in the stability of some time-discretization schemes applied to the vorticity-streamfunction equations as will be discussed in Section 6.3.2.e.

The time-discretization is based on finite-difference approximations. The disparity between the finite accuracy of finite-differences used for time-stepping and the "infinite" accuracy reached by the spatial spectral approximation is a question which must be considered. From the conceptual point of view, it would be more satisfactory to have the same kind of accuracy for space and time discretizations. This is the approach chosen by Morchoisne (1979, 1981) who derived a Chebyshev approximation for space and time (see Peyret and Taylor, 1983). However, the modest improvement in accuracy does not justify the cost in computing time and memory of this approach.

Presently, the current spectral codes make use of finite-difference methods for time-discretization. In order to preserve the high accuracy of the spectral method, the truncation error associated with the finite-difference approximation must be sufficiently small. Second-order methods are commonly used, but the relative low order of the scheme is generally counterbalanced by the small value of the time-step required for stability. However, the use of higher-order methods (at least, third-order) is recommended, in particular, for the calculation of highly unsteady flows. Various types of time-stepping can be used, as already discussed in Sections 2.8 and 3.5.2. The delicate point is that the coefficients of the spatial operators which can be implicitly treated must be constant for Fourier methods and space-dependent but time-independent for Chebyshev methods. This last property permits the inversion, diagonalization, or triangularization of the Chebyshev derivative operators once and for all before the start of the time-integration since the inversion of these full matrices at each time-cycle is too costly. An alternate way is to solve the algebraic systems by iteration but we think that, as far as possible, an efficient spectral code should avoid recourse to iterative solutions at each time-cycle. Thus, the explicit or implicit character of the scheme is largely dependent on the nature of the spatial operators. We distinguish :

(1) The fully explicit schemes, generally used for compressible flows (e.g., Passot and Pouquet, 1987 ; Don and Gottlieb, 1990 ; Erlebacher *et al.*, 1990 ; Guillard *et al.*, 1992 ; Kopriva, 1994) or incompressible flows with variable viscosity (Malik *et al.*, 1985), since in both cases the nonlinearity is not confined to the convective terms which contain first-order spatial derivatives only, but also appears in the second-order diffusive terms. Fully explicit schemes may also be used for the calculation of high Reynolds number incompressible flows with a spectral-element approach (Gavrilakis *et al.*, 1996), such that the polynomial degree is sufficiently small so as not to induce too stringent stability conditions. The same argument is obviously valid for more general multidomain methods.

(2) The semi-implicit schemes where the linear part $L(u)$ is implicit and the nonlinear part $N(u)$ is explicit. These schemes are commonly used for the solution of the Navier-Stokes equations for incompressible fluids (Chapters 6 and 7).

(3) The semi-implicit schemes where the linear part $L(u)$ is implicit and the nonlinear part is decomposed according to $N(u) = N_1(u) + N_2(u)$ where $N_1(u)$ is linear and treated implicitly while the nonlinear term $N_2(u)$ is explicit. Such a procedure (used, e.g., by Forestier *et al.*, 2000a,b) has been mentioned in Section 3.5.2 and will be discussed again in Section 4.4.2.

(4) The fully implicit schemes whose solution generally requires an iterative procedure. Such schemes are recommended for the solution of stiff problems in which numerical stability is a crucial issue. Examples of implicit methods for incompressible flows (Fröhlich *et al.*, 1991 ; Pagès *et al.*, 1995) will be discussed in Chapters 6 and 7.

The usual time-discretization schemes can also be distinguished according to the number of involved time-levels. According to the terminology of ordinary differential equations, the one-step methods (Euler scheme, weighted two-level scheme, Runge-Kutta schemes) involve two time-levels. For this class of scheme, high-order accuracy may be obtained through the calculations of intermediate values at several intermediate stages. Runge-

Kutta schemes are based on this procedure. Multistep methods are characterized by the fact that several time-levels are simultaneously involved. The accuracy is higher as the number of time-levels is large. High-order semi-implicit schemes are currently used with spectral methods. The reasons for this are various and will be discussed in due time.

4.2   The advection-diffusion equation

As already mentioned, the advection-diffusion equation is a valuable model for the equations governing the motion of a viscous fluid, because of the simultaneous presence of the first-order spatial derivative (characteristic of the transport) and the second-order derivative (characteristic of the viscous diffusion). Before describing and analyzing the time-discretization schemes, we think it necessary to discuss briefly, in the present section, the nature of the solution in both periodic and nonperiodic cases. In this latter case, it is of interest to know under what conditions the Chebyshev approximation is stable, namely the differential equations in time, resulting from the spectral approximation, have a bounded solution.

4.2.1   The exact initial-boundary value problem

In this chapter we are mainly interested in the solution of the one-dimensional advection-diffusion equation

$$\partial_t u + a\,\partial_x u - \nu\,\partial_{xx} u = f\,, \tag{4.5}$$

where the coefficient of diffusion (simply called "viscosity" in the sequel) ν is a nonnegative constant and the advective velocity a is constant with arbitrary sign. The case where a may depend on x will occasionally be considered. The forcing term f is a function of x and t.

The initial condition associated with (4.5) is

$$u(x,0) = u_0(x)\,. \tag{4.6}$$

The periodic (Fourier approximation) and nonperiodic (Chebyshev approximation) cases of Eq.(4.5) will be considered. In the nonperiodic case, the solution is sought in the domain $-1 < x < 1$ with the general boundary conditions

$$\alpha_-\,u(-1,t) + \beta_-\,\partial_x u(-1,t) = g_-\,(t) \tag{4.7}$$

$$\alpha_+\,u(1,t) + \beta_+\,\partial_x u(1,t) = g_+\,(t) \tag{4.8}$$

where $\alpha_\pm$ and $\beta_\pm$ are constant. Sufficient conditions, ensuring the well-posedness of the problem (4.5)-(4.8), can be obtained (Hesthaven and Gottlieb, 1996) by classical energy estimates assuming $f = g_+ = g_- = 0$. More precisely, the problem is well-posed if one of the following conditions hold :

(i) $\beta_- = 0$, $\beta_+ = 0$ (pure Dirichlet problem),

(ii) $\beta_- \neq 0$, $\beta_+ = 0$, and

$$\nu\left(1 - 2\frac{\alpha_-}{\beta_-}\right) - a \geq 0, \tag{4.9}$$

(iii) $\beta_- = 0$, $\beta_+ \neq 0$, and

$$\nu\left(1 + 2\frac{\alpha_+}{\beta_+}\right) + a \geq 0, \tag{4.10}$$

(iv) $\beta_- \neq 0$, $\beta_+ \neq 0$,

$$2\nu^2\left(\frac{\alpha_+}{\beta_+} - \frac{\alpha_-}{\beta_-} - 2\frac{\alpha_-\,\alpha_+}{\beta_-\,\beta_+}\right) - 2\nu\, a\left(\frac{\alpha_+}{\beta_+} + \frac{\alpha_-}{\beta_-}\right) - a^2 \geq 0 \tag{4.11}$$

and (4.9) [or, equivalently, (4.10)].

Note that similar conditions are obtained by Rigal (1988) who considers the case where $\beta_- = -1$ and $\beta_+ = 1$. By studying the associated eigenvalue problem, Rigal also obtains necessary and sufficient conditions which guarantee the uniform boundedness of the solution.

4.2.2 Fourier approximation

Let us assume the forcing term f and the initial condition u_0 to be 2π-periodic, then the 2π-periodic solution of (4.5) is sought in the form

$$u_K(x,t) = \sum_{k=-K}^{K} \hat{u}_k(t)\, e^{ikx} . \tag{4.12}$$

Application of the Galerkin method (Section 2.6.1) leads to the differential equation

$$d_t\hat{u}_k + \left(iak + \nu k^2\right)\hat{u}_k = \hat{f}_k , \quad k = -K, \ldots, K , \tag{4.13}$$

to be solved with the initial condition

$$\hat{u}_k(0) = \hat{u}_{0k} , \tag{4.14}$$

where $\hat{u}_{0k}$ is the Fourier coefficient of $u_0(x)$ and $\hat{f}_k(t)$ in Eq.(4.13) is the Fourier coefficient of $f(x,t)$.

In the present case, assuming $f = 0$, problem (4.13)-(4.14) can be solved exactly. The solution is

$$\hat{u}_k(t) = \hat{u}_{0k}\, e^{-(iak+\nu k^2)t}$$

so that

$$u_K(x,t) = \sum_{k=-K}^{K} \hat{u}_{0k}\, e^{ik(x-a\,t)}\, e^{-\nu k^2 t} . \tag{4.15}$$

This expression makes clear the two processes, characteristic of the equation: advection with velocity a and diffusion which has the effect of damping the amplitude of all modes, except $k = 0$. The high frequencies are damped faster. Ultimately, the solution tends toward $\hat{u}_{00}$. Such a constant steady-state solution is typical of periodicity and, in the case of a boundary-value problem, the steady-state solution essentially depends on the boundary conditions.

In general applications, equations like (4.13) are solved numerically using a convenient method as will be discussed later.

4.2.3   Chebyshev approximation

The collocation Chebyshev method, described in Section 3.4.2, is used to solve problem (4.5)-(4.8). We denote by $u_N(x,t)$ the polynomial approximation of degree N, whose values $u_N(x_i,t)$, $i = 0, \ldots, N$, are obtained by imposing the equation (4.5) to be satisfied at the inner collocation points x_i, $i = 1, \ldots, N-1$ [see Eq.(3.75)], and adding the boundary conditions (4.7)-(4.8), and let

$$d_t u_N(x_i,t) + \sum_{j=0}^{N} \left(a\, d_{i,j}^{(1)} - \nu\, d_{i,j}^{(2)} \right) u_N(x_j,t) = f(x_i,t), \quad i = 1, \ldots, N-1$$

(4.16)

$$\alpha_+ u_N(x_0,t) + \beta_+ \sum_{j=0}^{N} d_{0,j}^{(1)} u_N(x_0,t) = g_+(t) \qquad (4.17)$$

$$\alpha_- u_N(x_N,t) + \beta_- \sum_{j=0}^{N} d_{N,j}^{(1)} u_N(x_N,t) = g_-(t). \qquad (4.18)$$

Then the discretization, with respect to time (as described below), replaces the above system by an algebraic system for the unknowns

$$u_N^{n+1}(x_i) \cong u_N(x_i,t_{n+1}), \quad i = 0, \ldots, N, \quad n = 0, 1, \ldots,$$

where $t_{n+1} = (n+1)\Delta t$. Equations (4.16)-(4.18) are complemented by the initial conditions

$$u_N^0(x_i) = u_0(x_i), \quad i = 0, \ldots, N. \qquad (4.19)$$

The algebraic system obtained after time-discretization of (4.16)-(4.18) has to be solved at each time-cycle, using the solution techniques described in Chapter 3.

However, before discussing the time-integration schemes, it is advisable to analyze the behaviour of the exact solution of the differential equations (4.16)-(4.19). The system may be written in the vector form

$$d_t V = \mathcal{L}_N V + G, \qquad (4.20)$$

$$V(0) = V_0 \,, \tag{4.21}$$

where $V(t)$ is the vector of the unknowns at the inner points, that is,

$$V(t) = (u_N(x_1, t), \ldots, u_N(x_i, t), \ldots, u_N(x_{N-1}, t))^T \,. \tag{4.22}$$

Therefore, in Eqs.(4.20) and (4.21), the boundary values $u_N(x_0, t)$ and $u_N(x_N, t)$ have been eliminated thanks to the boundary conditions (4.17)-(4.18). Consequently, in the $(N - 1) \times (N - 1)$ matrix

$$\mathcal{L}_N = -a \mathcal{D}_x^{(1)} + \nu \mathcal{D}_x^{(2)} \,, \tag{4.23}$$

the matrices $\mathcal{D}_x^{(1)}$ and $\mathcal{D}_x^{(2)}$ take this elimination into account, as well as the vector G (see Section 3.7.1). More precisely, the entries $D_{i,j}^{(m)}$ of the matrix $\mathcal{D}_x^{(m)}$, $m = 1, 2$, are

$$D_{i,j}^{(m)} = \frac{1}{e} \left(b_{0,j} \, d_{i,0}^{(m)} + b_{N,j} \, d_{i,N}^{(m)} \right) + d_{i,j}^{(m)} \,, \quad i, j = 1, \ldots, N - 1, \tag{4.24}$$

with e, $b_{0,j}$, and $b_{N,j}$ defined in Eq.(3.134).

The well-posedness of the semidiscrete problem (4.20)-(4.21), with $f = g_+ = g_- = 0$, is characterized by the nature of the eigenvalues λ_i, $i = 1, \ldots, N$, of the operator $\mathcal{L}_N$ and, more precisely, by the sign of their real parts $\chi_i = \mathcal{R}e(\lambda_i)$. The quantities χ_i must be negative. If there exists at least one eigenvalue with a positive real part, the solution of (4.20)-(4.21) will not be bounded in time, and any consistent time-discretization scheme would be found unstable.

Therefore, the precise analysis of the eigenvalues of $\mathcal{L}_N$ is of importance from the theoretical point of view as well as for practical purposes. It is assumed, henceforth in the present section, that $a \geq 0$. The eigenvalue problem to be considered is

$$\mathcal{L}_N V - \lambda V = 0 \,, \tag{4.25}$$

and the following special cases will be discussed :

$\beta_- = 0$, $\beta_+ = 0$: Dirichlet-Dirichlet (D-D),
$\beta_- = 0$, $\alpha_+ = 0$: Dirichlet-Neumann (D-N),
$\alpha_- = 0$, $\alpha_+ = 0$: Neumann-Neumann (N-N).

Note that these values of $\alpha_\pm$ and $\beta_\pm$ satisfy the sufficient conditions mentioned in Section 4.2.1 for the well-posedness of the continuous problem (4.5)-(4.8). The Neumann-Dirichlet (N-D) case ($\alpha_- = \beta_+ = 0$) is well-posed if $\nu \geq a$ [see Eq.(4.9)]. For $\nu < a$, pathological behaviours, as experienced by Mofid and Peyret (1993), may be attributed to the nonexistence of an exact bounded solution. This case is not addressed here.

Before discussing the advection-diffusion problem, it is advisable to state the spectral properties of the pure diffusion problem ($a = 0$). For the above special cases of boundary conditions (D-D, D-N, and N-N), the eigenvalues of the discrete second-order derivative operator are real, distinct, and negative (except, of course, in the N-N case where one eigenvalue is zero). These results have been mentioned in Section 3.7.1.

Now we shall successively consider the three sets of boundary conditions associated to the advection-diffusion discrete operator.

(a) Dirichlet-Dirichlet conditions (D-D case)

Some theoretical results are given by Canuto *et al.* (1988). In particular, the eigenvalues λ_i, which are generally complex because of the presence of the first-order derivative term, satisfy :

$$|\lambda_i| \leq \nu\, O(N^4) + a\, O(N^2)\,, \quad i = 1, \ldots, N\,. \tag{4.26}$$

Numerical studies of these eigenvalues have been done by Gottlieb and Orszag, 1977 ; Haidvogel, 1979 ; Nana Kouamen, 1992 ; Mofid and Peyret, 1993. They show that, for small values of

$$\delta = \frac{\nu}{a}\,, \tag{4.27}$$

the real part of some eigenvalues may be positive if the degree N of the polynomial approximation is smaller than a critical value N_c. Therefore, the solution of (4.20)-(4.21) is not bounded.

More precisely, it is found that for $\delta > 8.5 \times 10^{-3}$ all eigenvalues have negative real parts. On the other hand, for $\delta \leq 8.5 \times 10^{-3}$, there exist some eigenvalues (their number depends on δ and N) with a real positive part if $7 < N < N_c$ where N_c satisfies the approximate law

$$N_c \cong (0.52/\delta)^{4/7}\,. \tag{4.28}$$

A similar behaviour was found for the tau Chebyshev method (Gottlieb and Orszag, 1977), except that the critical value N_c follows the approximate law

$$N_c \cong (1.30/\delta)^{4/7}\,. \tag{4.29}$$

The reason for such an instability, which has to be seen as an "approximation instability" and not an instability associated with a time-discretization scheme, may be connected to the existence of a boundary layer which develops at $x = 1$. More precisely, the eigenfunctions of the exact continuous problem exhibit a boundary layer behaviour near $x = 1$ when $\delta \ll 1$. If N is too small, that is, if the boundary layer is not correctly described, the discrete problem is not an approximation to the advection-diffusion problem but rather to the pure advection problem with Dirichlet conditions at both boundaries. It is known that such a problem is not well-posed : no boundary condition has to be prescribed at $x = 1$, the solution at this point should be calculated from the equation itself. Although the ill-posedness of this problem is a reason for the instability, it is necessary that the associated semidiscrete problem of type (4.20)-(4.21) exhibits the same behaviour. As a matter of fact, it has been found numerically that the eigenvalues of the discrete first-order derivative operator with two boundary conditions $u(\pm 1) = 0$ are purely imaginary for $N \le 7$ while some of them have a positive real part for $N > 7$.

The above results have been obtained for the case where the advective velocity a is constant. As a matter of fact, in realistic situations, this advective velocity is generally variable and is often zero at the boundary (e.g., in problems of free convection in an enclosure). In the case where $a = (1 - x^2)^\sigma$, with $\sigma > 0$, it is found that the critical value N_c decreases when σ increases. For $\sigma = 0.1$ the critical number N_c varies according to the approximate law $N_c \cong (0.35/\delta)^{4/7}$ which has to be compared with the law (4.28). For $\sigma = 0.5$, the phenomenon of instability does not appear for the tested values of δ ($\delta \ge 5 \times 10^{-4}$).

A way to avoid the phenomenon of the approximation instability is to employ a penalty technique which consists of replacing the boundary condition by a combination of the boundary condition and the equation itself prescribed at the boundary. This penalty method has been proposed by Funaro (1988) and by Funaro and Gottlieb (1988) with the main purpose of increasing the accuracy of some differential problems. More recently, the technique has been employed to derive stable open boundary conditions (Carpenter et al., 1994, for the high-order Hermitian method ; Hesthaven and Gottlieb, 1996, for the Legendre collocation method).

We recall that the existence of some eigenvalues with positive real part can be connected to the ill-posedness of the pure advection problem when a Dirichlet condition is prescribed at $x = 1$ (when $a > 0$, as assumed in this section), besides the Dirichlet condition at $x = -1$. If the Dirichlet condition at $x = 1$ is disregarded and replaced by the equation itself, the problem becomes well-posed. This observation leads naturally to a way of avoiding the instability of the approximation. This consists of imposing, at the considered boundary $x = 1$, no longer the Dirichlet condition $u(1, t) = 0$ or, more generally, $u(1, t) = g_+(t)$, but a combination of it and of the equation itself, namely,

$$\partial_t u(1, t) + a \, \partial_x u(1, t) - \nu \, \partial_{xx} u(1, t) = f(1, t) - \gamma \, [u(1, t) - g_+(t)] \, , \quad (4.30)$$

where γ is the (constant) penalty parameter. The expression (4.30) shows that the influence of the equation at the boundary is more and more important as γ decreases. In this way, the modified problem remains well-posed, even at the limit $\nu \to 0$.

The delicate question is the choice of the parameter γ. The conditions to be simultaneously satisfied are : (1) the eigenvalues must have nonpositive real part, (2) the solution must be accurate, (3) the stability of the time-discretization scheme used to solve the modified problem must be ensured. The optimal choice obviously depends on the value $\delta = \nu/a$ and on the degree N of the polynomial approximation. An extensive numerical study (Mofid and Peyret, 1993) shows, in particular, that the influence of γ on the accuracy is generally relatively weak. From this study, we obtain an estimate of the order of magnitude of suitable values of γ. For example, for $\delta = 5 \times 10^{-4}$ and $N = 32$, good results are obtained with $\gamma \approx 100$.

(b) Dirichlet-Neumann conditions (D-N case)
In the case where the boundary conditions in (4.20) are $u(-1,t) = 0$ and $\partial_x u(1,t) = 0$, the phenomenon of instability does not appear because the eigenvalues of the collocation advection-diffusion operator $\mathcal{L}_N$ always has negative parts. Note, however, that this property does not hold for the tau approximation for which there exists, as previously for the D-D case, a critical number N_c following the approximate law (4.28).

(c) Neumann-Neumann conditions (N-N case)
When Neumann conditions are prescribed at both boundaries $\partial_x(-1,t) = 0$ and $\partial_x(1,t) = 0$, the behaviour of the eigenvalues (except now the presence of the null eigenvalue) is found to be similar to that encountered with Dirichlet-Dirichlet conditions. More precisely, the real part of all eigenvalues are negative for $\delta > 5.4 \times 10^{-3}$. For smaller values of δ, these real parts are all negative for $N \leq 10$, whereas some of them are positive for $10 < N < N_c$, where N_c follows the approximate law

$$N_c \cong (0.62/\delta)^{4/7} \,. \tag{4.31}$$

In the N-N case, the eigenfunctions (see Nana Kouamen, 1992) do not exhibit a boundary layer behaviour and the instability is most likely connected with the Neumann condition prescribed at $x = -1$. The penalty technique applied to remove the phenomenon of instability in the D-D case has no effect in the present situation.

4.3   One-step method: the weighted two-level scheme

An important question to be discussed is the numerical stability of the time-discretization schemes used to solve the time-dependent equations like the advection-diffusion equation (4.5). In this section, we are interested in the weighted two-level scheme in order to introduce the general technique of stability analysis and to point out the constraint on the time-step associated with explicit schemes.

The time-discretization of Eq.(4.5), with $f = 0$, is

$$\frac{u^{n+1} - u^n}{\Delta t} + a \, \partial_x \left[\theta \, u^{n+1} + (1 - \theta) \, u^n \right] - \nu \, \partial_{xx} \left[\theta \, u^{n+1} + (1 - \theta) \, u^n \right] = 0 \,,$$
$$\tag{4.32}$$

where θ is a parameter which generally satisfies $0 \leq \theta \leq 1$. The explicit Euler scheme corresponds to $\theta = 0$ and the implicit Euler scheme is obtained for $\theta = 1$.

In the following, we shall successively consider the Fourier and Chebyshev cases.

4.3.1 Fourier approximation

Introducing the expansion (4.12) into Eq.(4.32), we obtain, for each Fourier mode $\hat{u}_k$, the following equation

$$\left[1 + \theta \left(\nu k^2 + \underline{i}\, a\, k\right) \Delta t\right] \hat{u}_k^{n+1} = \left[1 - (1 - \theta) \left(\nu k^2 + \underline{i}\, a\, k\right) \Delta t\right] \hat{u}_k^n ,$$

then

$$\hat{u}_k^{n+1} = g\, \hat{u}_k^n , \tag{4.33}$$

where

$$g = \frac{1 - (1 - \theta) \left(\nu k^2 + \underline{i}\, a\, k\right) \Delta t}{1 + \theta \left(\nu k^2 + \underline{i}\, a\, k\right) \Delta t}$$

is the amplification factor. The stability requires

$$|g| \leq 1 \text{ for any } k \in \{-K, \ldots, K\} \tag{4.34}$$

(see, e.g., Peyret and Taylor, 1983). The condition $|g| \leq 1$ leads to the inequality

$$(1 - 2\theta) \left(\nu^2 k^2 + a^2\right) \Delta t \leq 2\nu$$

from which we can draw the conclusions :

  (i)   if $\theta \geq 1/2$, the scheme is unconditionally stable,
  (ii)   if $\theta < 1/2$, the scheme is stable under the condition

$$\Delta t \leq \frac{2\nu}{(1 - 2\theta) \left(\nu^2 K^2 + a^2\right)} . \tag{4.35}$$

This criterion shows that : (1) the scheme is unstable for $\nu = 0$, and (2) the critical time-step behaves like $1/K^2$. This behaviour may be compared to the behaviour in Δx^2 of the centered finite-difference approximation in an uniform mesh of size Δx.

4.3.2 Chebyshev approximation

Let us now consider the solution of Eq.(4.5) with $f = 0$ satisfying the initial condition (4.6) and the Dirichlet boundary conditions

$$u(-1, t) = 0, \quad u(1, t) = 0. \tag{4.36}$$

By combining the associated Chebyshev collocation equations (4.20) and (4.21) with the time-discretization scheme (4.32), we obtain the vector equation which determines V^{n+1} :

$$[\mathcal{I} - \theta \, \Delta t \, \mathcal{L}_N] \, V^{n+1} = [\mathcal{I} + (1 - \theta) \, \Delta t \, \mathcal{L}_N] \, V^n \qquad (4.37)$$

$$V^0 = V_0, \qquad (4.38)$$

where $\mathcal{I}$ is the identity matrix. It is assumed that $\nu/|a|$ and N are such that all the eigenvalues λ_i of $\mathcal{L}_N$ [defined by Eq.(4.23)] have a nonpositive real part (see Section 4.2.3) so that the semi-discrete problem (4.20) and (4.21) has a bounded solution. Therefore, the equation (4.37) can be solved according to

$$V^{n+1} = \mathcal{E} \, V^n \qquad (4.39)$$

with

$$\mathcal{E} = [\mathcal{I} - \theta \, \Delta t \, \mathcal{L}_N]^{-1} \, [\mathcal{I} + (1 - \theta) \, \Delta t \, \mathcal{L}_N] \, .$$

Then, from (4.38) and (4.39) we get

$$V^n = \mathcal{E}^n \, V^0$$

where $\mathcal{E}^n$ means the nth power of the matrix of passage $\mathcal{E}$. The property of stability (see, e.g., Mitchell and Griffiths, 1980 ; Hirsch, 1988) is that the solution V^n remains bounded, namely, the infinite set of operator $\mathcal{E}^n$ has to be uniformly bounded. A necessary condition for this is,

$$\rho(\mathcal{E}) \le 1 + O(\Delta t) \, ,$$

where $\rho(\mathcal{E})$ is the spectral radius of $\mathcal{E}$ [see Eq.(3.79)]. Note that such a condition would be sufficient if the matrix $\mathcal{E}$ were symmetric or normal, which is not the case here. When the exact solution is known to be bounded, as it is here, the above condition is replaced by

$$\rho(\mathcal{E}) \le 1 \, . \qquad (4.40)$$

The eigenvalues μ_i, $i = 1, \ldots, N - 1$, of $\mathcal{E}$ are expressed by

$$\mu_i = \frac{1 + (1 - \theta) \, \Delta t \, \lambda_i}{1 - \theta \, \Delta t \, \lambda_i} \, , \quad i = 1, \ldots, N - 1,$$

where the λ_i's are the eigenvalues of $\mathcal{L}_N$. The stability requires $|\mu_i| \le 1$, $i = 1, \ldots, N - 1$, namely,

$$(1 - 2\theta) \, \Delta t \, |\lambda_i|^2 + 2 \, \mathcal{R}e \, (\lambda_i) \le 0 \, .$$

Because $\mathcal{R}e \, (\lambda_i) \le 0$, the scheme is unconditionally stable if $\theta \ge 1/2$, as found for the Fourier approximation. If $\theta < 1/2$, the stability condition is

$$\Delta t \le \frac{-2 \, \mathcal{R}e \, (\lambda_i)}{(1 - 2\theta) \, |\lambda_i|^2} \le \frac{2}{(1 - 2\theta) \, |\lambda_i|} \, ,$$

that is, since $\rho(\mathcal{L}_N) = \max_i |\lambda_i|$:

$$\Delta t \le \frac{2}{(1 - 2\theta)\,\rho(\mathcal{L}_N)}\,. \tag{4.41}$$

For the pure explicit scheme ($\theta = 0$), we get

$$\Delta t \le \frac{2}{\rho(\mathcal{L}_N)}\,. \tag{4.42}$$

From the estimate (4.26) it is deduced that the allowable time-step is $O(N^{-4})$ when viscosity is preponderant ($\nu \gtrsim |a|$). This condition is very stringent and, generally, necessitates the use of very small time-steps. For this reason, one prefers to consider implicit schemes or, at least, semi-implicit schemes, in which the diffusive term is implicit and the convective term (nonlinear in Navier-Stokes equations) is treated in an explicit way. The stability of such time-schemes will be discussed later. However, from (4.26), one can infer that the critical time-step will be $O(N^{-2})$. Under these conditions, when the viscosity is low (and N is relatively not very large), we may have $|a|O(N^2) \approx \nu O(N^4)$ so that the implicit treatment of the diffusive term is no longer necessary. Therefore, fully explicit schemes can be used in this situation as was done by Guillard *et al.* (1992) for the solution of the Navier-Stokes equations for compressible fluids, and by Gavrilakis *et al.* (1996) in the incompressible case using a spectral-element method.

Let us now consider successively the case of pure diffusion ($a = 0$) and pure advection ($\nu = 0$). The spectral radii associated with these two operators are given in Table 3.1. Therefore, for the pure diffusion equation with Dirichlet conditions, the stability criterion (4.42) is

$$\Delta t \le \frac{2}{0.047\,\nu\,N^4} \simeq \frac{42.55}{\nu\,N^4} \tag{4.43}$$

and for Neumann conditions

$$\Delta t \le \frac{2}{0.014\,\nu\,N^4} \simeq \frac{142.85}{\nu\,N^4}\,. \tag{4.44}$$

The criterion associated with the Dirichlet-Neumann case is again (4.43). It must be observed that the stability criterion is less stringent in the case of Neumann conditions on both boundaries.

The above results refer to the Chebyshev collocation approximation. If the tau method is used for the spatial approximation the constants in (4.43) and (4.44) must be replaced, respectively, by 6.6 and 42.55. This shows that the tau method is less stable than the collocation method.

In the case of pure advection ($\nu = 0$) with $a > 0$, the Dirichlet condition is prescribed at $x = -1$ while no boundary condition is required at $x = 1$. The partial differential equation $\partial_t u + a\,\partial_x u = 0$ is prescribed to be satisfied at this point. Therefore, in (4.37), the $N \times N$ matrix $\mathcal{L}_N$ is defined by $\mathcal{L}_N = -a \left[d_{i,j}^{(1)} \right]$, $i, j = 0, \ldots, N-1$, and the vector V^n is made with the grid values $u_N^n(x_i)$, $i = 0, \ldots, N-1$. From Table 3.1, we deduce the stability criterion

$$\Delta t \leq \frac{2}{0.089 \, a \, N^2} \simeq \frac{22.47}{a \, N^2} \, . \tag{4.45}$$

4.4   Two-step methods

The general two-step (i.e. three time-level) scheme for the time-discretization of Eq.(4.5) is

$$\frac{(1+\varepsilon) \, u^{n+1} - 2\,\varepsilon \, u^n - (1-\varepsilon) \, u^{n-1}}{2\,\Delta t}$$

$$+a \, \partial_x \left[\gamma_1 u^{n+1} + \gamma_2 u^n + (1 - \gamma_1 - \gamma_2) \, u^{n-1} \right]$$

$$-\nu \, \partial_{xx} \left[\theta_1 u^{n+1} + \theta_2 u^n + (1 - \theta_1 - \theta_2) \, u^{n-1} \right]$$

$$= \theta_1 f^{n+1} + \theta_2 f^n + (1 - \theta_1 - \theta_2) \, f^{n-1} \, . \tag{4.46}$$

The parameters ε, γ_1, γ_2, θ_1, and θ_2 are arbitrary and define a large family of discretization schemes (Ouazzani *et al.*, 1986). Consistency and order of accuracy are determined from Taylor's expansion.

The above schemes are first-order accurate whatever the value of the parameters. Second-order accuracy requires

$$\frac{\varepsilon}{2} = 2\gamma_1 + \gamma_2 - 1 = 2\,\theta_1 + \theta_2 - 1 \tag{4.47}$$

and third-order is attained if we have

$$\gamma_2 = \theta_2 = \frac{2}{3} \tag{4.48}$$

in addition to (4.47).

With a view to solving the Navier-Stokes equations, the above family of schemes is generally restricted to the case $\gamma_1 = 0$, namely, the advective term is explicitly evaluated. In this case, the third-order scheme characterized by (4.47) and (4.48) has no practical interest because of its poor stability. Third-order schemes of type (4.46) will not be addressed here. Among the second-order schemes the most used in spectral methods are the following :

(i) Implicit Backward-Differentiation (BDI2) scheme ($\varepsilon = 2$, $\gamma_1 = 1$, $\gamma_2 = 0$, $\theta_1 = 1$, $\theta_2 = 0$) :

$$\frac{3u^{n+1} - 4u^n + u^{n-1}}{2\,\Delta t} + a \, \partial_x u^{n+1} - \nu \, \partial_{xx} u^{n+1} = f^{n+1} \, . \tag{4.49}$$

(ii) Semi-implicit Leap-Frog/Crank-Nicolson (LF/CN) scheme ($\varepsilon = 0$, $\gamma_1 = 0$, $\gamma_2 = 1$, $\theta_1 = 1/2$, $\theta_2 = 0$) :

$$\frac{u^{n+1} - u^{n-1}}{2\,\Delta t} + a\,\partial_x u^n - \frac{\nu}{2}\partial_{xx}\left(u^{n+1} + u^{n-1}\right) = \frac{1}{2}\left(f^{n+1} + f^{n-1}\right) \quad (4.50)$$

essentially used with Fourier approximation for the computation of homogeneous turbulence.

(iii) Semi-implicit Adams-Bashforth/Crank-Nicolson (AB/CN) scheme ($\varepsilon = 1$, $\gamma_1 = 0$, $\gamma_2 = 3/2$, $\theta_1 = \theta_2 = 1/2$) :

$$\frac{u^{n+1} - u^n}{\Delta t} + \frac{a}{2}\partial_x\left(3\,u^n - u^{n-1}\right) - \frac{\nu}{2}\partial_{xx}\left(u^{n+1} + u^n\right) = \frac{1}{2}\left(f^{n+1} + f^n\right) .$$
$$(4.51)$$

(iv) Semi-implicit Adams-Bashforth/Backward-Differentiation (AB/BDI2) scheme ($\varepsilon = 2$, $\gamma_1 = 0$, $\gamma_2 = 2$, $\theta_1 = 1$, $\theta_2 = 0$) :

$$\frac{3\,u^{n+1} - 4\,u^n + u^{n-1}}{2\,\Delta t} + a\,\partial_x\left(2\,u^n - u^{n-1}\right) - \nu\,\partial_{xx}u^{n+1} = f^{n+1}. \quad (4.52)$$

(v) Semi-implicit Weighted Adams-Bashforth/Backward-Differentiation (WAB/BDI2) scheme ($\varepsilon = 2$, $\gamma_1 = \alpha$, $\gamma_2 = 2\,(1-\alpha)$, $\theta_1 = 1$, $\theta_2 = 0$) :

$$\frac{3\,u^{n+1} - 4\,u^n + u^{n-1}}{2\,\Delta t} + \alpha\,a\,\partial_x u^{n+1}$$
$$+ (1-\alpha)\,a\,\partial_x\left(2\,u^n - u^{n-1}\right) - \nu\,\partial_{xx}u^{n+1} = f^{n+1} . \tag{4.53}$$

The parameter α lies between 0 and 1. For $\alpha = 0$, we obtain the AB/BDI2 scheme (4.52) and for $\alpha = 1$ the BDI2 scheme (4.49). Scheme (4.53) (used by Forestier *et al.* 2000a,b) belongs to the class of semi-implicit schemes (2) described in Section 4.1 and already mentioned in Section 3.5.2. Thus, the factor $\alpha\,a$ in Eq.(4.53) plays the part of $\tilde{u}(x)$ in Eq.(3.105) and the factor $(1-\alpha)\,a$ plays the part of $\bar{u}(x,t)$.

The classic difficulty associated with these three-level schemes is their inability to determine the solution u^1 at the first time-cycle $t_1 = \Delta t$, given suitable conditions at $t = 0$. This question is addressed in Section 4.5.1.d.

The stability of schemes (i) to (v) is discussed according to the type of spatial approximation : Fourier or Chebyshev.

4.4.1 Fourier approximation

Introducing the expansion (4.12) into the above schemes (with $f = 0$) we obtain, for each Fourier component, an equation of the form

$$a_1\,\hat{u}_k^{n+1} + a_0\,\hat{u}_k^n + a_{-1}\,\hat{u}_k^{n-1} = 0, \tag{4.54}$$

where a_{-1}, a_0 and a_1 are given in Table 4.1.

Scheme	a_{-1}	a_0	a_1
BDI2	1	-4	$3 + 2\left(k'_1 - k'_2\right)$
LF/CN	$-1 - k'_2$	$2\,k'_1$	$1 - k'_2$
AB/CN	$-\dfrac{k'_1}{2}$	$-1 + \dfrac{1}{2}\left(3\,k'_1 - k'_2\right)$	$1 - \dfrac{k'_2}{2}$
AB/BDI2	$1 - 2\,k'_1$	$-4\left(1 - k'_1\right)$	$3 - 2\,k'_2$
WAB/BDI2	$1 - 2\left(1 - \alpha\right)k'_1$	$-4\left[1 - \left(1 - \alpha\right)k'_1\right]$	$3 + 2\left(\alpha\,k'_1 - k'_2\right)$

Table 4.1. Coefficients of the equation (4.54) with $k'_1 = \underline{i}\,a\,k\,\Delta t$ and $k'_2 = -\nu\,k^2\,\Delta t$.

For the study of the stability of the general three-level scheme (4.54), it is suitable to replace it by a two-level scheme. This is done by introducing the auxiliary variable $\hat{v}^{n+1}_k = \hat{u}^n_k$, so that Eq.(4.54) is equivalent to the system

$$\mathcal{G}_1\,\hat{\Phi}^{n+1}_k = \mathcal{G}_0\,\hat{\Phi}^n_k \tag{4.55}$$

where $\hat{\Phi}^n_k = (\hat{u}^n_k, \hat{v}^n_k)^T$ and $\mathcal{G}_0$ and $\mathcal{G}_1$ are the 2×2 matrices

$$\mathcal{G}_0 = \begin{pmatrix} -a_0 & -a_1 \\ 1 & 0 \end{pmatrix}, \quad \mathcal{G}_1 = \begin{pmatrix} a_1 & 0 \\ 0 & 1 \end{pmatrix}.$$

Since $a_1 \neq 0$, $\mathcal{G}_1$ is invertible and Eq.(4.54) yields

$$\hat{\Phi}^{n+1}_k = \mathcal{G}\,\hat{\Phi}^n_k \tag{4.56}$$

with $\mathcal{G} = \mathcal{G}_1^{-1}\mathcal{G}_0$. Stability requires that the modulus of the eigenvalues μ_i, $i = 1, 2$, of the amplification matrix $\mathcal{G}$ is not larger than one. These eigenvalues are the zeros of the characteristic polynomial

$$F\left(\mu\right) = a_1\,\mu^2 + a_0\,\mu + a_{-1}. \tag{4.57}$$

Incidentally, it may be noticed that the stability condition can be obtained directly from (4.54). As a matter of fact, the general solution of the recurrence equation (4.54) is of the form

$$\hat{u}^n_k = c_1\,\mu^n_1 + c_2\,\mu^n_2$$

where μ^n_1 and μ^n_2 are the nth power of the roots of $F\left(\mu\right) = 0$ and c_1 and c_2 are constants to be determined by the initial condition and the "starting scheme" (see Section 4.5.1.d). The stability requires that $\hat{u}^n_k$ remains bounded. Therefore, we get, as above, the condition

$$|\mu_i| \leq 1, \quad i = 1, 2 \quad \text{for any } k \in \{-K, \ldots, K\}. \tag{4.58}$$

Among the two roots μ_1 and μ_2, one is "principal" (also called "physical") and the other is "spurious". The principal root tends toward one when Δt tends to zero, contrary to the spurious root which tends to zero. The principal root μ_1 is the one that corresponds to the exact solution of Eq.(4.5) in Fourier space, that is, Eq.(4.13), since

$$\mu_1^n = e^{-\left(i\,a\,k + \nu\,k^2\right)n\,\Delta t}\left[1 + O\left(\Delta t^2\right)\right].$$

The reason for the existence of the spurious root is that Eq.(4.46) contains too many time-levels since two time-levels are sufficient to approximate the first-order time-derivative $\partial_t u$. Note that it may happen in some very special cases (e.g., the leap-frog scheme, $\varepsilon = 0$, $\gamma_1 = \theta_1 = 0$, $\gamma_2 = \theta_2 = 1$) that the spurious root generates an unconditional instability.

The study of the zeros of $F(\mu)$ can be rather complicated due to the fact that the coefficients are complex. A possible way is to apply Miller's theorem on the zeros of a polynomial (Miller, 1971) given in Appendix C. An alternate way consists of numerically determining the region of stability in a suitable plane, for example, the (X, Y) plane with

$$X = -\nu\,k^2\,\Delta t, \quad Y = -a\,k\,\Delta t, \tag{4.59}$$

is currently used (see Section 4.5.1.e). The domain of stability is defined by $F(\mu; X, Y) = 0$ with $|\mu| = 1$, namely, $\mu = e^{i\theta}$ with $0 \le \theta \le 2\pi$. Therefore, by separately considering the real and imaginary part of the equation $F\left(e^{i\theta}; X, Y\right) = 0$, we get the parametric equations

$$X = \varphi(\theta), \quad Y = \psi(\theta), \quad 0 \le \theta \le 2\pi, \tag{4.60}$$

which define the boundary Γ of the domain of stability S. A drawback of the choice of this (X, Y) plane is that Δt appears in both coordinates. Therefore, it may be convenient to consider also the (ξ, η) plane such that

$$\xi = \nu^2 k^2 / a^2 = X^2 / Y^2, \quad \eta = \Delta t / \nu = -Y^2 / X. \tag{4.61}$$

In this plane, the stability curve Γ gives directly the critical time-step Δt.

The application of Miller's theorem to the polynomial (4.57) leads to the following conditions ensuring that the zeros of $F(\mu)$ are not larger than one :

(i) either $|a_1| > |a_{-1}|$ and

$$|a_0\,\bar{a}_1 - a_{-1}\,\bar{a}_0| \le |a_1|^2 - |a_{-1}|^2, \tag{4.62}$$

(ii) or $|a_1| = |a_{-1}|$ and

$$|a_0| \le 2\,|a_1|. \tag{4.63}$$

From these conditions, it is easy to show that the fully implicit scheme BDI2 is unconditionally stable. The semi-implicit LF/CN is found to be

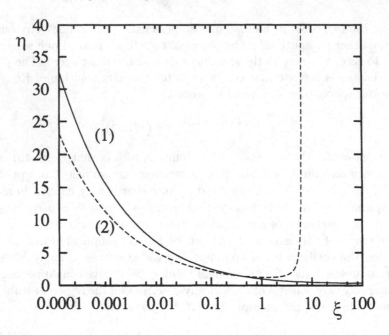

FIGURE 4.1. Domain of stability in the (ξ, η) plane : (1) scheme AB/CN, (2) scheme AB/BDI2.

stable under the Courant-type condition $|T| = |a| k \Delta t \le 1$ with $-K \le k \le K$, therefore,

$$\Delta t \le \frac{1}{|a| K}, \qquad (4.64)$$

showing that the allowable time-step is inversely proportional to the largest frequency.

For the semi-implicit schemes, the application of Miller's theorem, through the above conditions, does not yield a stability criterion in a simple form. So, we can turn to either a graphical representation of these conditions or, more simply, construct the stability curve Γ from (4.60) and (4.61). The stability curve Γ in the plane (X, Y) is shown below in Section 4.5.1.e (Fig.4.14) devoted to general multistep methods. Figure 4.1 shows the domain of stability in the (ξ, η) plane defined by (4.61). We may observe that the AB/BDI2 scheme is unconditionally stable for $\xi \ge 3 + 2\sqrt{3}$. Therefore, discarding the nonsignificant value $k = 0$ (always stable), the above condition is equivalent to $\nu/|a| \ge \sqrt{3 + 2\sqrt{3}} \simeq 2.54$. Since $k \ge 1$, the AB/CN scheme is always subjected to a constraint even for large ξ. On the other hand, for $\xi \lesssim 1$, the AB/CN scheme admits larger time-steps than AB/BDI2.

Let us now consider the semi-implicit WAB/BDI2 scheme. By application of Miller's theorem it is easy to show that the scheme is unconditionally

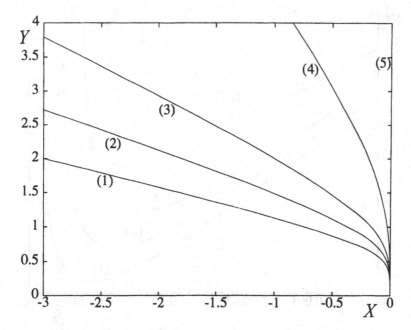

FIGURE 4.2. Domain of stability of scheme WAB/BDI2 in the (X, Y) plane for various values of α : (1) $\alpha = 0$, (2) $\alpha = 0.3$, (3) $\alpha = 0.5$, (4) $\alpha = 0.7$, (5) $\alpha \geq 3/4$.

stable for $\alpha \geq 3/4$. For $\alpha < 3/4$, the critical time-step is the larger as α is large as seen in Figs. 4.2 and 4.3. Moreover, for a given value of α the scheme is unconditionally stable provided $\nu/|a| \geq \delta_\infty (\alpha)$ where $\delta_\infty (\alpha)$ is a decreasing function of α (see Fig. 4.3).

Finally, for pure advection ($\nu = 0$) it is possible to show, using Miller's theorem, that both AB/CN and AB/BDI2 schemes are unconditionally unstable. On the other hand, the WAB/BDI2 scheme is unconditionally stable for $\alpha \geq 3/4$ and unconditionally unstable for $\alpha < 3/4$. These results can also be deduced from the analysis of Figs.4.12 to 4.14. This point will again be discussed in Section 4.5.1.c.

4.4.2   Chebyshev approximation

This section is devoted to the stability of the implicit BDI2 scheme and the semi-implicit AB/CN, AB/BDI2 and WAB/BDI2 schemes applied to the solution of the homogeneous Dirichlet problem for the advection-diffusion equation ($f = g_\pm = 0$), spatially approximated with the Chebyshev collocation method. The discrete equations in vector form are :

$$\mathcal{A}_1 V^{n+1} + \mathcal{A}_0 V^n + \mathcal{A}_{-1} V^{n-1} = 0 \qquad (4.65)$$

where V^n is the vector of the unknowns $u_N^n(x_i)$, $i = 1, \ldots, N-1$, $\mathcal{A}_{-1}$, $\mathcal{A}_0$, and $\mathcal{A}_1$ are $(N-1) \times (N-1)$ matrices whose expressions in terms of the

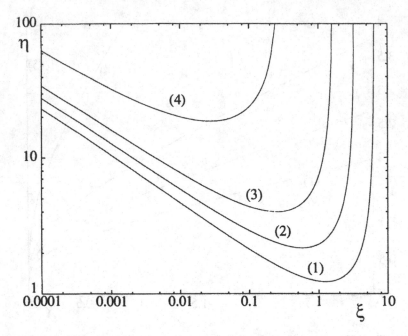

FIGURE 4.3. Domain of stability of scheme WAB/BDI2 in the (ξ, η) plane for various values of α : (1) $\alpha = 0$, (2) $\alpha = 0.3$, (3) $\alpha = 0.5$, (4) $\alpha = 0.7$.

matrices

$$\mathcal{D}_1 = \left[d_{i,j}^{(1)} \right], \quad i, j = 1, \ldots, N - 1, \tag{4.66}$$

and

$$\mathcal{D}_2 = \left[d_{i,j}^{(2)} \right], \quad i, j = 1, \ldots, N - 1, \tag{4.67}$$

are given in Table 4.2 where $\mathcal{I}$ is the $(N - 1) \times (N - 1)$ identity matrix.

Let us consider first the fully implicit BDI2 scheme defined by

$$\mathcal{A}_{-1} = \mathcal{I}, \quad \mathcal{A}_0 = -4\mathcal{I}, \quad \mathcal{A}_1 = 3\mathcal{I} + 2\Delta t \, (a\,\mathcal{D}_1 - \nu\,\mathcal{D}_2) \, .$$

We recall that it must be assumed that ν/a and N are such that the eigenvalues λ_i, $i = 1, \ldots, N - 1$, of the matrix $\mathcal{L}_N = -a\,\mathcal{D}_1 + \nu\,\mathcal{D}_2$ have a nonpositive real part. Moreover, it can be numerically checked that these eigenvalues are distinct so that the matrix $\mathcal{A}_1$ is diagonalizable according to

$$\mathcal{A}_1 = \mathcal{P} \Lambda \mathcal{P}^{-1} \, ,$$

where Λ is the diagonal matrix whose elements are equal to $3 - 2\,\Delta t\,\lambda_i$ and $\mathcal{P}$ is the $(N - 1) \times (N - 1)$ matrix whose columns are the associated eigenvectors. Let $\tilde{V}^n = \mathcal{P}^{-1} V^n$, Eq.(4.65) yields

$$(3 - 2\,\Delta t\,\lambda_i)\,\tilde{u}_N^{n+1}(x_i) - 4\,\tilde{u}_N^n(x_i) + \tilde{u}_N^{n-1}(x_i) = 0, \quad i = 1, \ldots, N - 1, \tag{4.68}$$

Scheme	$\mathcal{A}_{-1}$	$\mathcal{A}_0$	$\mathcal{A}_1$
BDI2	$\mathcal{I}$	$-4\mathcal{I}$	$3\mathcal{I} + 2\left(\mathcal{D}_1' - \mathcal{D}_2'\right)$
AB/CN	$-\dfrac{1}{2}\mathcal{D}_1'$	$-\mathcal{I} + \dfrac{1}{2}\left(3\mathcal{D}_1' - \mathcal{D}_2'\right)$	$\mathcal{I} - \dfrac{1}{2}\mathcal{D}_2'$
AB/BDI2	$\mathcal{I} - 2\mathcal{D}_1'$	$-4\left(\mathcal{I} - \mathcal{D}_1'\right)$	$3\mathcal{I} - 2\mathcal{D}_2'$
WAB/BDI2	$\mathcal{I} - 2\left(1 - \alpha\right)\mathcal{D}_1'$	$-4\left[\mathcal{I} - \left(1 - \alpha\right)\mathcal{D}_1'\right]$	$3\mathcal{I} + 2\left(\alpha\mathcal{D}_1' - \mathcal{D}_2'\right)$

Table 4.2. Coefficients of the vector equation (4.65) with $\mathcal{D}_1' = a\,\Delta t\,\mathcal{D}_1$ and $\mathcal{D}_2' = \nu\,\Delta t\,\mathcal{D}_2$.

where $\tilde{u}_N^n(x_i)$, $i = 1, \cdots, N - 1$, are the components of the vector $\tilde{V}^n$. Equation (4.68) is of the form (4.54), so that the stability can be analyzed as done in the Fourier case. More precisely, the application of Miller's theorem [Eqs.(4.62) and (4.63)] shows that scheme (4.68) is unconditionally stable [provided that $\mathcal{Re}(\lambda_i) \leq 0$, as already assumed].

Let us now consider the semi-implicit schemes AB/CN, AB/BDI2, and WAB/BDI2. The analysis developed above can no longer be done since the matrices $\mathcal{D}_1$ and $\mathcal{D}_2$ and, consequently, $\mathcal{A}_{-1}$, $\mathcal{A}_0$, and $\mathcal{A}_1$, are not diagonalizable with the same eigenvectors. In such a case, the stability analysis requires the numerical study of the eigenvalues of the matrix of passage. Therefore, it is necessary to reduce the three-level equation (4.65) into a two-level one. As usual, this is done by introducing the auxiliary unknown vector $W^{n+1} = V^n$, so that Eq.(4.65) becomes

$$\mathcal{B}_1\,\Phi^{n+1} = \mathcal{B}_0\,\Phi^n, \tag{4.69}$$

where

$$\Phi^n = \begin{pmatrix} V^n \\ W^n \end{pmatrix}, \quad \mathcal{B}_0 = \begin{pmatrix} -\mathcal{A}_0 & \mathcal{A}_1 \\ \mathcal{I} & 0 \end{pmatrix}, \quad \mathcal{B}_1 = \begin{pmatrix} \mathcal{A}_1 & 0 \\ 0 & \mathcal{I} \end{pmatrix}. \tag{4.70}$$

Therefore,

$$\Phi^{n+1} = \mathcal{E}\,\Phi^n, \tag{4.71}$$

where

$$\mathcal{E} = \mathcal{B}_1^{-1}\,\mathcal{B}_0 \tag{4.72}$$

is the matrix of passage. As previously discussed (Section 4.3.2) the stability requires the spectral radius of $\mathcal{E}$ to be smaller than or equal to one. The study of the eigenvalues μ_i of $\mathcal{E}$ is done through the numerical computation of the eigenvalues of the generalized eigenvalue problem $\mathcal{B}_0\,\Phi = \mu\,\mathcal{B}_1\,\Phi$ using a library routine.

(a) Semi-implicit AB/CN and AB/BDI2 schemes

The domains of stability in the plane $(\nu/|a|, |a|\Delta t)$ for various values of N are shown in Fig.4.4. Some typical values of the critical time-step are given in Table 4.3.

As observed for Fourier approximation, the AB/BDI2 scheme is unconditionally stable for $\nu/|a| \geq 1$, while the stability of the AB/CN scheme requires a condition on the time-step. Note, however, that for large values of $\nu/|a|$ the allowable time-step does not depend on N.

For moderately small values of $\nu/|a|$ (e.g. $\nu/|a| \approx 10^{-2}$, $N = 64$) the AB/BDI2 scheme is more stable than the AB/CN scheme. On the other hand, for much smaller values of $\nu/|a|$ (e.g. $\nu/|a| \leq 10^{-3}$) the AB/CN scheme is slightly more stable than the AB/BDI2 scheme, but this superiority is not as pronounced as for the Fourier method (see Fig. 4.1).

Until now, we have discussed only the stability of the time-discretization schemes associated with the collocation method. Figures 4.6 and 4.7 show the stability region associated with the tau method. The overall behaviour is similar to that observed for the collocation method, except : (1) the critical time-step of the AB/CN method is independent of $\nu/|a|$, and (2) the AB/BDI2 scheme is more stable than the AB/CN scheme even for small values of $\nu/|a|$. Besides, it may be observed that the collocation method is more stable than the tau method. This is one of the reasons for which the collocation method has to be used rather than the tau method.

It has been mentioned (Section 4.2.3) that the critical number N_c ensuring the stability of the Chebyshev approximation was decreased when the advective velocity a is a function of x which is zero at the boundaries $x = \pm 1$. Similar behaviour is observed for the stability of the time-discretization schemes (Le Quéré, 1987 ; Mofid, 1992) when $a = (1-x^2)^{0.1}$. The results given in Table 4.3 show that the gain obtained in the case where a is variable is more significant for the tau method (where the nonconstant term $a(x)\,\partial_x u$ is calculated with the pseudospectral technique presented in Section 3.5.2) than for the collocation method.

For the pure advection equation ($\nu = 0$) with Dirichlet condition at $x = -1$ ($a > 0$), and spatially approximated with the Chebyshev collocation method, the stability criteria found by Ouazzani *et al.* (1986) are of the form

$$\Delta t \leq \frac{C}{a\,N^2} \tag{4.73}$$

with $C = 9$ for the second-order AB2 scheme deduced from Eq.(4.51) and $C = 7.75$ for the AB/BDE2 scheme deduced from Eq.(4.52), these values being accurate for $N \geq 35$.

(b) Semi-implicit WAB/BDI2 scheme

The domain of stability (Chebyshev collocation approximation) in the $(\nu/|a|, |a|\Delta t)$ plane, for $N = 64$, obtained for various values of the parameter α, is shown in Fig.4.8. As for Fourier approximation (Fig.4.3),

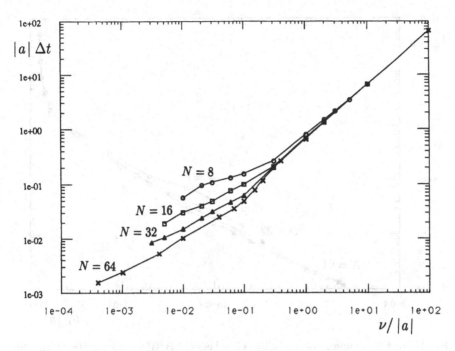

FIGURE 4.4. Domain of stability of scheme AB/CN (collocation) in the $(\nu/|a|, |a|\Delta t)$ plane, for various values of N.

ν	N	Scheme	Collocation	
			$a = 1$	$a = (1 - x^2)^{0.1}$
10^{-1}	32	AB/CN	$6. \times 10^{-2}$	6.5×10^{-2}
		AB/BDI2	1.8×10^{-1}	1.9×10^{-1}
10^{-2}	32	AB/CN	1.5×10^{-2}	2.3×10^{-2}
		AB/BDI2	1.9×10^{-2}	2.2×10^{-2}
10^{-3}	64	AB/CN	2.4×10^{-3}	4.4×10^{-3}
		AB/BDI2	2.3×10^{-3}	$5. \times 10^{-3}$

ν	N	Scheme	Tau	
			$a = 1$	$a = (1 - x^2)^{0.1}$
10^{-1}	32	AB/CN	1.7×10^{-3}	$7. \times 10^{-2}$
		AB/BDI2	1.8×10^{-1}	1.8×10^{-1}
10^{-2}	32	AB/CN	1.7×10^{-3}	1.8×10^{-2}
		AB/BDI2	1.8×10^{-2}	2.1×10^{-2}
10^{-3}	64	AB/CN	$4. \times 10^{-4}$	$4. \times 10^{-3}$
		AB/BDI2	1.3×10^{-3}	3.2×10^{-3}

Table 4.3. Critical time-step for the semi-implicit AB/CN and AB/BDI2
schemes.

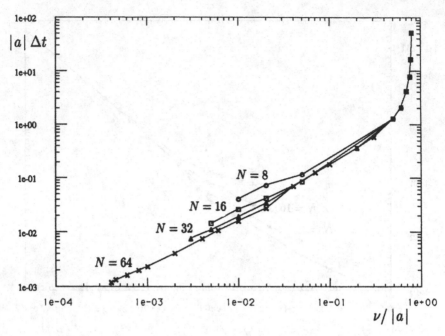

FIGURE 4.5. Domain of stability of scheme AB/BDI2 (collocation) in the $(\nu/|a|, |a|\Delta t)$ plane, for various values of N.

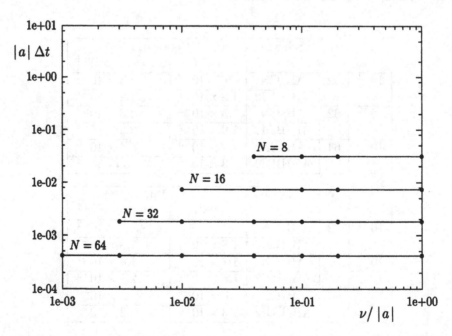

FIGURE 4.6. Domain of stability of scheme AB/CN (tau) in the $(\nu/|a|, |a|\Delta t)$ plane, for various values of N.

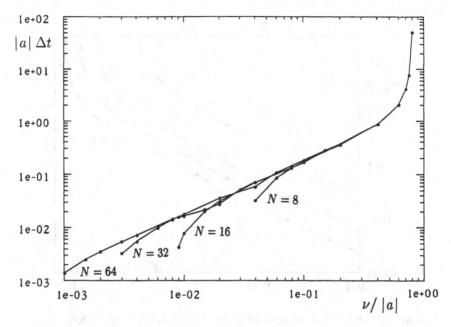

FIGURE 4.7. Domain of stability of scheme AB/BDI2 (tau) in the $(\nu/|a|, |a|\Delta t)$ plane, for various values of N.

the WAB/BDI2 scheme is unconditionally stable for $\nu/|a| \geq \delta_\infty(\alpha)$ where $\delta_\infty(\alpha)$ is a decreasing function of α. Also, as already discussed, the stability is unconditional for $\alpha \geq 3/4$. This is shown in Fig.4.9 which displays the variations of the critical time-step in terms of the parameter α, for different values of the ratio $\delta = \nu/|a|$.

Lastly, when applied to the pure advection equation ($\nu = 0$, $a > 0$), the scheme (4.53) is unconditionally stable for $\alpha \geq 3/4$, but contrary to the Fourier case, it remains conditionally stable for $\alpha < 3/4$. This is due to the fact that the eigenvalues associated with the first-order derivative operator, with Dirichlet boundary condition at $x = -1$, has a strictly positive real part, while it is zero in the Fourier case.

4.4.3 Numerical illustration

In order to illustrate the phenomenon of instability of time-discretization schemes, we consider here the advection-diffusion equation

$$\partial_t u + a\,\partial_x u - \nu\,\partial_{xx} u = f, \quad -1 < x < 1, \; t > 0 \qquad (4.74)$$

whose solution is

$$u_e(x,t) = \frac{2\pi - 1 + \sin 2\pi t}{2\pi} \sin\left[\frac{3}{4}\pi(x+1)\right]. \qquad (4.75)$$

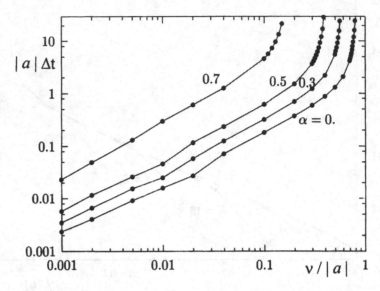

FIGURE 4.8. Domain of stability of scheme WAB/BDI2 (collocation, $N = 64$) in the $(\nu/|a|, |a|\Delta t)$ plane, for various values of α.

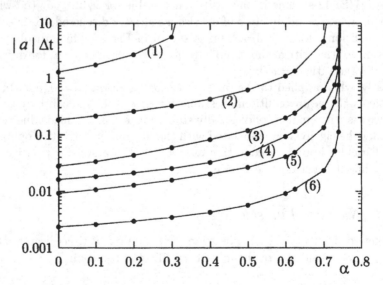

FIGURE 4.9. Domain of stability of scheme WAB/BDI2 (collocation, $N = 64$) in the $(\nu/|a|, |a|\Delta t)$ plane, for $\nu/|a| = 0.5$ (1), 0.1 (2), 0.02 (3), 0.01 (4), 0.005 (5), 0.001 (6).

This last expression defines the forcing term $f = \partial_t u_e + a \, \partial_x u_e - \nu \, \partial_{xx} u_e$, the initial condition $u(x, 0) = u_e(x, 0)$, and the Dirichlet boundary conditions $u(-1, t) = 0$ and $u(1, t) = -(2\pi - 1 + \sin 2\pi t)/2\pi$.

The problem is solved by using the AB/BDI2 scheme (4.52) associated with the Chebyshev collocation method. In order to avoid any error, due to the start of the time-integration associated with the three-level schemes, the calculation of the solution u^1 at the first-time cycle is done by defining $u^{-1} = u_e(x, -\Delta t)$. We refer to Section 4.5.1.d for a general discussion of the problem associated with the start of time-integration with a multilevel scheme. For $a = 1$, $\nu = 0.020$, and $N = 32$, the critical time-step, given by the numerical study of the eigenvalues of $\mathcal{E}$ reported in the previous section, is $\Delta t_\star = 0.003448266$ (obtained with an error of $\pm 10^{-9}$). A significant way to analyze the stability is to observe the time-behaviour of the error $\overline{E}$ defined by

$$\overline{E} = \left(\frac{1}{N-1} \sum_{i=0}^{N-1} |u_N^n(x_i) - u_e(x_i, n\Delta t)|^2 \right)^{1/2} \tag{4.76}$$

which must exhibit, if the scheme is stable, a nonincreasing amplitude as time elapses. Figure 4.10 shows the evolution of the error for four values of Δt. For $\Delta t = 0.0300$ (Fig.4.10.a), the error reaches a perfect periodic state after a transient period. For the value $\Delta t = 0.0344$, that is, close to the critical value, we observe (Fig.4.10.b) a much longer transient period during which the error becomes relatively large. This increase in the error may be attributed to the ill-conditioning of the matrix $\mathcal{P}$ made with the eigenvectors of the matrix of passage $\mathcal{E}$ defined by Eq.(4.72). In matrix form, the discrete solution of the present problem may be written

$$\Phi^{n+1} = \mathcal{E} \, \Phi^n + F^n \tag{4.77}$$

with the notation of Section 4.4.2, with F^n arising from f and the boundary conditions. Let $\hat{\Phi}^n$ be the computed solution of (4.77), that is, the solution given by the computer and spoiled by the inherent numerical errors associated with the inversion of $\mathcal{B}_1$, matrix products, etc., and essentially due to the round-off errors. The behaviour of the error $Z^n = \Phi^n - \hat{\Phi}^n$ is controlled by the equation

$$Z^n = \mathcal{E} \, Z^{n-1}$$

so that

$$\|Z^n\| \leq \|\mathcal{E}^n\| \cdot \|Z^0\|,$$

where $\mathcal{E}^n$ is the nth power of $\mathcal{E}$.

Now, assuming $\mathcal{E}$ is diagonalizable according to $\mathcal{E} = \mathcal{P}\Lambda\mathcal{P}^{-1}$, we get for the Euclidian norm

$$\|\mathcal{E}^n\|_2 \leq \|\mathcal{P}\|_2 \cdot \|\Lambda^n\|_2 \cdot \|\mathcal{P}^{-1}\|_2 = \kappa(\mathcal{P}) \, \rho^n(\mathcal{E}),$$

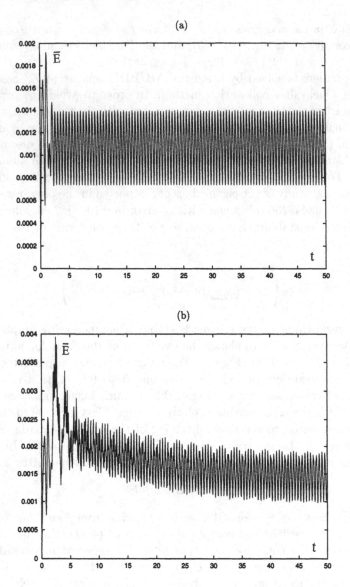

FIGURE 4.10. Time-evolution of the error $\overline{E}$ associated with the AB/BDI2 scheme, for various values of Δt : (a) $\Delta t = 0.0300$, (b) $\Delta t = 0.0344$.

where $\kappa(\mathcal{P}) = \|\mathcal{P}\|_2 \cdot \|\mathcal{P}^{-1}\|_2$ is the condition number of $\mathcal{P}$ and $\rho(\mathcal{E})$ is the spectral radius of $\mathcal{E}$. If $\rho(\mathcal{E}) < 1$ (as occurs for $0 < \Delta t < \Delta t_\star$), the error Z^n ultimately tends toward zero when $n \to \infty$, but for a finite period of time, the error Z^n is controlled by the magnitude of $\kappa(\mathcal{P})$, which can be substantially large. Such a phenomenon is the more pronounced as $\rho(\mathcal{E})$ is close to one, namely, when Δt is close to the critical value $\Delta t_\star$. For that reason, it is recommended using a time-step not too close to $\Delta t_\star$.

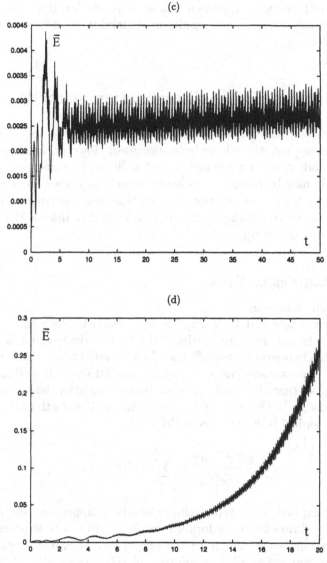

FIGURE 4.10. (continued) (c) $\Delta t = 0.03449$, (d) $\Delta t = 0.0350$

For $\Delta t = 0.03449$, namely, a little larger than the critical value, the error increases (Fig.4.10.c) continuously but remains bounded for a very long time : the instability occurs at longer times. Such behaviour is obviously dangerous in real computations, since the exact solution is not known so that there is no way to measure the growth of the error.

Finally, for $\Delta t = 0.0350$, which is clearly larger than $\Delta t_\star$, the instability appears quickly (Fig.4.10.d).

The above numerical experiments show, in particular, that the condition $\rho(\mathcal{E}) \leq 1$, which is a necessary condition of stability is, in the present case, also sufficient.

4.5   High-order time-discretization methods

In the preceding sections, second-order methods for time-discretization of time-dependent equations have been discussed. However, as already mentioned, second-order accuracy may not be sufficient for some physical problems, and it may be necessary to have recourse to higher order methods. The object of the present section is to describe the construction of multistep and one-step methods of arbitrary order and to discuss the stability properties of some of them.

4.5.1   Multistep methods

(a) Explicit schemes
The commonly used high-order explicit schemes are the well-known Adams-Bashforth schemes. In these methods the time derivative $\partial_t u$ is approximated with a two-level finite-difference formula and the spatial term H [see Eq.(4.1)] is approximated using a linear combination of H evaluated at k time-levels, so that the resulting semi-discrete equation has a truncation error of order Δt^k. Therefore, the general Adams-Bashforth (ABk) scheme of order k applied to Eq.(4.1) is of the form

$$\frac{u^{n+1} - u^n}{\Delta t} = \sum_{j=0}^{k-1} b_j H(u^{n-j}) \,. \tag{4.78}$$

Another explicit multistep scheme consists of approximating $\partial_t u$ with a high-order finite-difference formula involving several time-levels and H with an extrapolation similar to the one above. Such a discretization is particularly convenient when some part of $H(u)$, especially $L(u)$, is considered in an implicit way because it needs to be considered at level $n + 1$ only. This discretization (called "Backward-Differentiation" or "Backward-Euler") will be discussed below in detail. For the moment, only the explicit scheme (noted AB/BDEk) is considered, it is of the general form

$$\frac{1}{\Delta t} \sum_{j=0}^{k} a_j u^{n+1-j} = \sum_{j=0}^{k-1} b_j H(u^{n-j}) \,. \tag{4.79}$$

Therefore, scheme (4.78) enters the general form (4.79) with $a_0 = 1$, $a_1 = -1$, and $a_j = 0$ for $j > 1$. The coefficients a_j and b_j are determined

Scheme	Order	a_0	a_1	a_2	a_3	a_4	b_0	b_1	b_2	b_3
AB2	2	1	-1				$\dfrac{3}{2}$	$-\dfrac{1}{2}$		
AB/BDE2	2	$\dfrac{3}{2}$	-2	$\dfrac{1}{2}$			2	-1		
AB3	3	1	-1				$\dfrac{23}{12}$	$-\dfrac{16}{12}$	$\dfrac{5}{12}$	
AB/BDE3	3	$\dfrac{11}{6}$	-3	$\dfrac{3}{2}$	$-\dfrac{1}{3}$		3	-3	1	
AB4	4	1	-1				$\dfrac{55}{24}$	$-\dfrac{59}{24}$	$\dfrac{37}{24}$	$-\dfrac{9}{24}$
AB/BDE4	4	$\dfrac{25}{12}$	-4	3	$-\dfrac{4}{3}$	$\dfrac{1}{4}$	4	-6	4	-1

Table 4.4. Coefficients of the Adams-Bashforth (ABk) and
Adams-Bashforth/Backward-Differentiation (AB/BDEk) schemes.

by setting to zero the first terms of the truncation error calculated with
Taylor's expansion around $(n + 1)\,\Delta t$.

For $k = 1$, $a_0 = 1$, $a_1 = -1$, and $b_0 = 1$, we obtain the first-order forward
Euler scheme. Table 4.4 gives the values of the coefficients a_j and b_j for
ABk and AB/BDEk schemes up to fourth-order.

(b) Semi-implicit schemes

The constraint on the size of the time-step Δt, due to stability requirements
associated with a fully explicit scheme (especially restrictive for second-
order derivatives as shown in Section 4.3), leads naturally to an increase
in the degree of implicitness of the scheme. The semi-implicit methods
apply generally to nonlinear equations like (4.1)-(4.2) when the coefficients
of the linear operator $L(u)$ are time-independent. This is the case for the
Navier-Stokes equations for incompressible fluids with constant viscosity.
This linear term $L(u)$ is considered implicitly and the nonlinear term $N(u)$
is explicit, so that the resulting discrete operator is time-independent and
can be inverted or diagonalized in a preprocessing stage performed before
to start the time-integration. Second-order three-level schemes of this type
have been discussed in Section 4.4. It was mentioned that, in addition to
the other advantages, the high-order AB/BDIk schemes involve the linear
part $L(u)$ at time level $n + 1$ only. In contrast, the high-order extension
of the AB/CN scheme would necessitate a consideration of $L(u)$ at several
previous time-levels.

High-order AB/BDIk schemes, belonging to the family introduced by
Crouzeix (1980), are easily constructed from (4.79) applied to the nonlinear
part $N(u)$ with the addition of $L(u)$ considered at level $n + 1$, namely,

$$\frac{1}{\Delta t} \sum_{j=0}^{k} a_j\, u^{n+1-j} = \sum_{j=0}^{k-1} b_j\, N\left(u^{n-j}\right) + L\left(u^{n+1}\right), \qquad (4.80)$$

where the coefficients a_j and b_j are those appearing in the explicit schemes AB/BDEk (see Table 4.4). The truncation error of scheme (4.80) is $O\left(\Delta t^k\right)$.

The third-order (AB/BDI3) scheme of the family (4.80) has been applied to the solution of the Navier-Stokes equations for incompressible fluids associated with various spatial approximations: the Chebyshev tau method (Le Quéré, 1990), the Chebyshev collocation method (Botella, 1997; Botella and Peyret, 2001) or the Legendre spectral-element method (Karniadakis et al., 1991).

(c) Variable time-step

During the time-integration of an unsteady problem it may be necessary to change the value of the time-step Δt for various reasons (stability, accuracy, change in the scale of variation of the dependent variables, etc.). In such a situation the multistep schemes developed above must be modified to take the variable time-step into account.

Let us introduce $t_{n+1} - t_n = \Delta t$ and

$$t_{n+2-m} - t_{n+1-m} = \Delta t_{n+2-m} = r_{n+2-m}\,\Delta t \quad \text{for} \quad m \geq 2. \tag{4.81}$$

Then, the ABk, AB/BDEk, and AB/BDIk schemes may be written in the form (4.78), (4.79), and (4.80), respectively. The coefficients a_j and b_j are determined as previously except that the algebra is more tedious.

For the second-order AB/BDE2 (or AB/BDI2) scheme, the values of a_j and b_j are

$$a_0 = \frac{2 + r_n}{1 + r_n}, \qquad a_1 = -1 - \frac{1}{r_n}, \qquad a_2 = \frac{1}{1 + r_n},$$

$$b_0 = 1 + \frac{1}{r_n}, \qquad b_1 = -\frac{1}{r_n}, \tag{4.82}$$

where $r_n = (t_n - t_{n-1})/\Delta t$ from (4.81).

For the third-order AB/BDE3 (or AB/BDI3) scheme, these coefficients are

$$a_0 = 1 + \frac{1}{1 + r_n} + \frac{1}{1 + r_n + r_{n-1}},$$

$$a_1 = -\frac{(1 + r_n)(1 + r_n + r_{n-1})}{r_n (r_n + r_{n-1})},$$

$$a_2 = \frac{1 + r_n + r_{n-1}}{r_n r_{n-1}(1 + r_n)}, \tag{4.83}$$

$$a_3 = -\frac{1 + r_n}{r_{n-1}(r_n + r_{n-1})(1 + r_n + r_{n-1})},$$

and

$$b_0 = \frac{(1 + r_n)(1 + r_n + r_{n-1})}{r_n(r_n + r_{n-1})},$$

$$b_1 = -\frac{1 + r_n + r_{n-1}}{r_n r_{n-1}},$$

$$b_2 = \frac{1 + r_n}{r_{n-1}(r_n + r_{n-1})},$$

where $r_{n-1} = (t_{n-1} - t_{n-2})/\Delta t$.

(d) Starting scheme

With a multistep scheme, there exists a difficulty for starting the solution since the only known value is $u^0 = u(t=0)$, while the schemes (4.78)-(4.80) also need the knowledge of $u^1, \ldots, u^k$ to determine u^{k+1}. Therefore, it is necessary to use another scheme (the "starting scheme") to start the time-integration. From the practical point of view, it is interesting to note that the starting scheme may be one-order accuracy less than the general scheme. This is deduced from classical error estimates (Gear, 1971) showing that the approximation error $O(\Delta t^k)$ becomes $O(\Delta t^{k+1})$ when t is close to zero, that is, $t = m\,\Delta t$, with m sufficiently small. One can be easily convinced of this fact by considering the simple problem

$$d_t u = \lambda u$$

$$u(0) = u_0,$$

where λ is a negative constant. The exact solution is $u(t) = u_0 e^{\lambda t}$. The above differential equation is approximated by the first-order Euler scheme

$$\frac{u^{n+1} - u^n}{\Delta t} = \lambda u^{n+1}, \quad u^0 = u_0,$$

whose solution is

$$u^n = u_0(1 - \lambda \Delta t)^{-n}.$$

Now, if $t = n\,\Delta t$ is fixed ($t \leq T$) and $\Delta t \to 0$, the error $e(t)$ is found to be

$$e(t) = u^n - u(t) = \frac{u_0}{2}\lambda e^{\lambda t}\,t\,\Delta t + O\left(e^{\lambda t}t^2\,\Delta t^2\right).$$

Therefore, the error $e(t)$ is of first-order but it becomes of second-order if $t = m\,\Delta t$, with m sufficiently small, since $e^{\lambda t}t\,\Delta t = m\,\Delta t^2 + O(\Delta t^3)$.

Thus, to a second-order scheme like AB/BDI2 it is sufficient to associate a first-order starting scheme. Scheme (4.52) requires the knowledge of u^0 and u^1 to calculate u^2, or, equivalently, the knowledge of u^{-1} and u^0 to calculate u^1. Let us consider the scheme (4.52) for $n = 0$ where we pose $u^{-1} = u^0 = u_0$, namely,

$$3\frac{u^1 - u_0}{2\,\Delta t} + a\,\partial_x u_0 - \nu\,\partial_{xx}u^1 = f^1. \tag{4.84}$$

This equation defines u^1 as a first-order solution at time $2\Delta t/3$. Therefore, u^1 given by the current scheme (4.52) with $u^{-1} = u_0$, is an approximation to the solution at time $t = \Delta t$ if the initial time-step is $\Delta t_1 = 3\Delta t/2$ rather than Δt.

However, in numerical codes based on calculations done once and for all in a preprocessing stage because they are expensive [e.g., inversion of matrices (Section 3.4.2), construction and inversion of influence matrices (Sections 6.3.2.d and 7.3.2.c)], it is possible to avoid the duplication of these calculations associated with Δt_1. This is obtained by simply setting $u^{-1} = u_0$ in (4.52) while maintaining the same time-step Δt. Therefore, u^1 is the approximation to the solution at time $t = 2\Delta t/3$, provided the boundary values $g_\pm$ are themselves evaluated at $t = 2\Delta t/3$. Then the linear extrapolation $\left(3\,u^1 - u_0\right)/2$ gives the solution at time $t = \Delta t$. The error $\overline{E}$ obtained with this technique is shown in Fig.4.11 (curve (2)) corresponding to the solution of Section 4.4.3 with the parameters of Fig.4.10.a. For comparison, the error $\overline{E}$ obtained with u^{-1} equal to the exact solution at $t = -\Delta t$ (as done in Fig.4.10.a) is represented by curve (1). Note that the use of the extrapolation assumes that the initial value $u_0(x)$ satisfies the boundary conditions, that is, $u_0(\pm 1) = g_\pm(0)$. If not, it is necessary to calculate u^1 with the modified boundary conditions

$$u^1(\pm 1) = [u_0(\pm 1) + 2\,g_\pm(\Delta t)]/3\,.$$

For higher-order multistep schemes, the common way to remove the initialization difficulty is to use, at the first time-cycles, a one-step scheme like the Runge-Kutta schemes described in Section 4.5.2. An example of a second-order starting scheme is the semi-implicit two-stage Runge-Kutta/Crank-Nicolson-type scheme (4.115)-(4.116). In particular, the scheme corresponding to $\alpha = 1$ and $\beta = 0$, which is written as

$$\frac{u_1 - u^n}{\Delta t} = N(u^n) + L(u^n)$$

$$\frac{2\,u^{n+1} - u_1 - u^n}{2\,\Delta t} = \frac{1}{2}\left[N(u_1) + L(u^{n+1})\right]\,,$$

(4.85)

when applied to Eqs.(4.1)-(4.2), has been used by Botella and Peyret (2001) for the solution of the Navier-Stokes equations discretized in time with the AB/BDI3 scheme (see Section 7.4.2.b).

(e) Stability
The stability of multistep schemes up to fourth-order is analyzed on the advection-diffusion equation

$$\partial_t u + a\,\partial_x u - \nu\,\partial_{xx} u = 0\,.$$

(4.86)

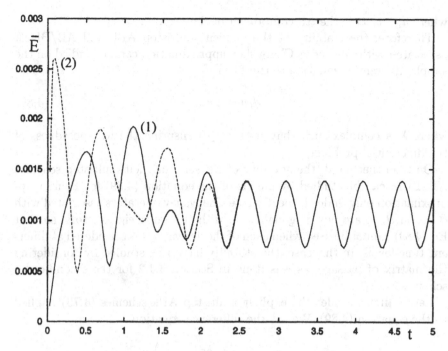

FIGURE 4.11. Influence of starting on the error $\overline{E}$: (1) u^{-1} = exact solution at $t = -\Delta t$, (2) $u^{-1} = u_0$ combined with extrapolation.

Within Fourier approximation Eq.(4.86) leads, for each Fourier coefficient, to the ordinary differential equation (see Section 4.3.2)

$$d_t \hat{u}_k = \lambda_k \, \hat{u}_k \tag{4.87}$$

with $\lambda_k = -\underline{i} \, a \, k - \nu \, k^2$.

Chebyshev approximation to the equation (4.86) with general homogeneous boundary conditions (4.7) and (4.8) with $g_\pm = 0$ leads to the system of ordinary differential equations (see Section 4.2.3)

$$d_t V = \mathcal{L}_N \, V \,, \tag{4.88}$$

where V is the vector of the inner unknowns $u_N \, (x_i, t)$, $i = 1, \ldots, N-1$, and $\mathcal{L}_N$ is a $(N-1) \times (N-1)$ matrix. If $\mathcal{L}_N$ is assumed to be diagonalizable according to

$$\mathcal{L}_N = \mathcal{P} \Lambda \mathcal{P}^{-1} \,,$$

where Λ is the diagonal matrix whose elements are the eigenvalues λ_i, $i = 1, \ldots, N-1$, of $\mathcal{L}_N$, then Eq.(4.88) is equivalent to the set of ordinary differential equations

$$d_t \, \tilde{u}_N \, (x_i, t) = \lambda_i \, \tilde{u}_N \, (x_i, t) \,,$$

where $\tilde{u}_N\,(x_i, t)$ is the ith component of the vector $\tilde{V} = \mathcal{P}^{-1}\,V$.

Therefore, the stability of the explicit multistep ABk and AB/BDEk associated with Fourier or Chebyshev approximation can be studied on the simple differential equation of the form

$$d_t v = \lambda v, \qquad (4.89)$$

where λ is complex such that $\mathcal{R}e\,(\lambda) \leq 0$ ensuring the well-posedness of the differential problem.

On the other hand, the stability of the semi-implicit multistep schemes AB/BDIk can be studied by means of the equation (4.89) for Fourier approximation only. Indeed, the discrete Chebyshev operators associated with $\partial_x u$ and $\partial_{xx} u$ are not diagonalizable with the same eigenvectors, so that Eq.(4.89) cannot be used when each of the derivatives is considered at different time-levels. In this case, the stability has to be studied by considering the matrix of passage, as was done in Section 4.4.2 for the second-order schemes.

Let us first consider the explicit multistep ABk schemes (4.79) applied to the equation (4.89). We get the difference equation

$$\sum_{j=0}^{k} a_j\, u^{n+1-j} - \lambda\,\Delta t \sum_{j=0}^{k-1} b_j\, u^{n-j} = 0 \qquad (4.90)$$

which admits solutions of the form μ^n where μ is a root of the characteristic equation

$$F\,(\mu; z) = a_0\,\mu^k + \sum_{j=0}^{k-1} (a_{j+1} - z\,b_j)\,\mu^{k-1-j} = 0 \qquad (4.91)$$

with $z = \lambda\,\Delta t$. The absolute stability requires that all roots μ_i, $i = 1, \ldots, k$, of (4.91) satisfy

$$|\mu_i| \leq 1 \quad \text{for } i = 1, \ldots, k. \qquad (4.92)$$

Therefore, the analysis of the stability leads to the study of the roots of Eq.(4.91). This is done by following the general method developed in Section 4.4.1. The domain of stability $\mathcal{S}$, in the plane

$$X = \mathcal{R}e\,(\lambda\,\Delta t), \quad Y = \mathcal{I}m\,(\lambda\,\Delta t), \qquad (4.93)$$

is determined by parametric equations like (4.60). For a given value of λ, the allowable time-step Δt is such that the point $z = \lambda\,\Delta t$ is within $\mathcal{S}$. We recall that the stability of the differential equation (4.89) requires $\mathcal{R}e\,(\lambda) \leq 0$, such that the stability domain $\mathcal{S}$ belongs to the half-plane $X \leq 0$. Moreover, due to the symmetry with respect to the axis $Y = 0$,

Scheme	Order	X_0	Y_0
AB2	2	1	0
AB/BDE2	2	$\dfrac{4}{3}$	0
AB3	3	$\dfrac{6}{11} = 0.545$	$\dfrac{12}{5\sqrt{11}} = 0.723$
AB/BDE3	3	$\dfrac{20}{21} = 0.952$	$\dfrac{59\sqrt{35}}{546} = 0.639$
AB4	4	0.3	0.43
AB/BDE4	4	0.8	2.8

Table 4.5. Coordinates of the intersections of the boundary of the stability domain with the axes, for various multistep explicit schemes.

only that part of S belonging to $Y \geq 0$ [i.e. $0 \leq \theta \leq \pi$ in Eq.(4.60)] has to be considered.

Let $\lambda_R = \mathcal{R}e\,(\lambda)$ and $\lambda_I = \mathcal{I}m\,(\lambda)$ and denote by $-X_M$ and Y_M, respectively, the coordinates of the intersection of the straight line $Y = |\lambda_I|\,X/\lambda_R$ with the boundary Γ of S, so that X_M and Y_M are nonnegative. Then the condition of stability is $|\lambda_k|\,\Delta t \leq X_M$ or, equivalently, $|\lambda_I|\,\Delta t \leq Y_M$.

When λ is real (this is the case for the pure diffusion equation), the stability requires $|\lambda|\,\Delta t \leq X_0$ where X_0 is the abscissa of the intersection of Γ with the X-axis. Also, when λ is purely imaginary (this is the case for the pure advection equation approximated with the Fourier method), the condition of stability is $|\lambda|\,\Delta t \leq Y_0$, where Y_0 characterizes the intersection of Γ with the Y-axis. The values of X_0 and Y_0 for various explicit multistep schemes are given in Table 4.5.

The stability regions S in the (X,Y) plane, defined by Eq.(4.93), are shown in Figs.4.12 and 4.13. As already mentioned, because of the symmetry with respect to the X-axis, the curves are drawn only for $Y \geq 0$. The following conclusions can be drawn :

1. For purely imaginary λ, the second-order schemes are unstable. For third- and fourth-order ABk schemes, the range of stability decreases when the order increases. An opposite behaviour is observed for the AB/BDEk schemes.

2. For negative real λ, the range of stability decreases when the order of accuracy increases. The AB/BDEk schemes are more stable than the ABk schemes.

3. For complex λ, the extent of the domain of stability generally decreases when the order increases (except in the region near the imaginary axis). The AB/BDEk schemes are more stable than the ABk schemes for $|\lambda_I/\lambda_R|$ sufficiently small.

Now we consider the semi-implicit multistep AB/BDIk schemes (4.80). As already mentioned, the use of the equation (4.89) is significant only for

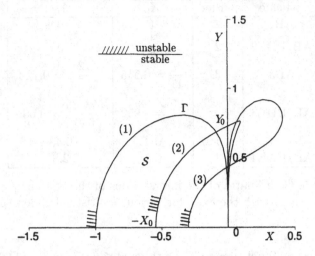

FIGURE 4.12. Domain of stability in the (X, Y) plane of explicit multistep schemes : (1) AB2, (2) AB3, (3) AB4 (the domains have to be completed by symmetry in $Y < 0$).

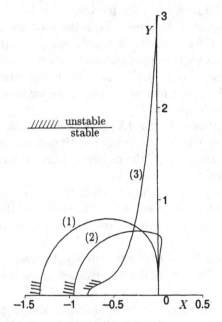

FIGURE 4.13. Domain of stability in the (X, Y) plane of explicit multistep schemes : (1) AB/BDE2, (2) AB/BDE3, (3) AB/BDE4 (the domains have to be completed by symmetry in $Y < 0$).

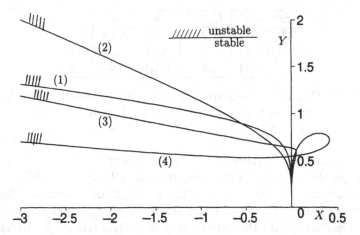

FIGURE 4.14. Domain of stability in the (X, Y) plane of semi-implicit multistep schemes : (1) AB/CN, (2) AB/BDI2, (3) AB/BDI3, (4) AB/BDI4 (the domains have to be completed by symmetry in $Y < 0$).

the Fourier method. Equation (4.89) is written

$$d_t v = (\lambda_R + \underline{i}\,\lambda_I)\, v \tag{4.94}$$

with $\lambda_R = -\nu\,k^2$ and $\lambda_I = -\underline{i}\,a\,k$. The part $\lambda_R\,v$ is treated implicitly while the part $\underline{i}\,\lambda_I\,v$ is evaluated in an explicit way by the Adams-Bashforth extrapolation. Therefore, the general semi-implicit AB/BDIk scheme (4.80) applied to Eq.(4.94) is

$$\sum_{j=0}^{k} a_j\, v^{n+1-j} - \underline{i}\,\lambda_I\,\Delta t \sum_{j=0}^{k-1} b_j\, v^{n-j} - \lambda_R\,\Delta t\, v^{n+1} = 0 \tag{4.95}$$

and the characteristic equation is

$$F(\mu; X, Y) = (a_0 - X)\,\mu^k + \sum_{j=0}^{k-1} (a_{j+1} - \underline{i}\,Y\,b_j)\,\mu^{k-1-j} = 0, \tag{4.96}$$

where

$$X = \lambda_R\,\Delta t \quad \text{and} \quad Y = \lambda_I\,\Delta t.$$

Then, the analysis is similar to the one given for the explicit methods. The regions of stability in the (X, Y) plane associated with some AB/BDIk schemes are shown in Fig. 4.14. Because of symmetry, only the part $Y \geq 0$ is represented. It is recalled that, for second-order schemes, the stability has already been discussed in Section 4.4.1, using in particular the plane $(\xi = \nu^2\,k^2/a^2,\ \eta = \Delta t/\nu)$.

Scheme	Order	a	b	c
AB/BDI2	2	2.54	0.88	1.2
AB/BDI3	3	5.61	0.69	0.6
AB/BDI4	4	11.76	0.45	1.8

Table 4.6. Coefficients entering the approximate stability criterion of semi-implicit AB/BDIk schemes.

The general following conclusions can be drawn :

1. For purely imaginary λ, the second-order AB/CN and AB/BDI2 schemes are unstable (like the explicit AB2 and AB/BDE2 schemes). The range of stability of the third-order AB/BDI3 scheme is larger ($Y_0 = 0.64$) than the one associated with the fourth-order AB/BDI4 scheme for which $Y_0 = 0.55$.

2. For negative real λ, all schemes are unconditionally stable (i.e., A-stable according to the classical terminology).

3. It is observed that the branch of the boundary Γ of the domain of stability of the AB/BDIk schemes is nearly rectilinear. From this observation, we may derive an approximate criterion for stability by replacing this branch by a straight line $Y = -X/a + b$ where a and b are determined numerically. The distance between this straight line and the curve Γ is sufficiently small if the ratio $|X/Y| = |\lambda_R/\lambda_I|$ is larger than a certain value c determined approximately. In this way, we obtain the following conditions which may be used, when $|\lambda_R/\lambda_I| > c$, as a rough guide for stability.

If $\left| \dfrac{\lambda_R}{\lambda_I} \right| > a,$ the scheme is unconditionally stable.

If $\left| \dfrac{\lambda_R}{\lambda_I} \right| < a,$ the stability condition is

$$\Delta t < \frac{a\,b}{a|\lambda_I| - |\lambda_R|}. \tag{4.97}$$

The values of a, b, and c are given in Table 4.6.

To close this section, devoted to the stability of multistep schemes, it is interesting to consider the case of fully implicit BDIk schemes. For Eq.(4.89) the schemes are

$$\sum_{j=0}^{k} a_j \, v^{n+1-j} - \lambda \Delt t \, v^{n+1} = 0. \tag{4.98}$$

It has been shown in Section 4.4.2 that the second-order BDI2 is unconditionally stable. This property is not true for higher-order schemes. More precisely, Dahlquist (1963) has proven that implicit linear multistep

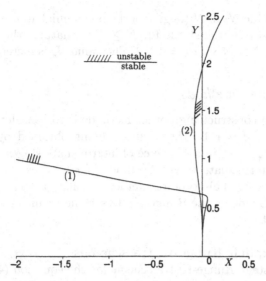

FIGURE 4.15. Domain of stability in the (X, Y) plane of third-order multistep schemes : (1) semi-implicit AB/BDI3 scheme, (2) fully implicit BDI3 scheme (the domains have to be completed by symmetry in $Y < 0$).

schemes of order higher than two cannot be A-stable (i.e., unconditionally stable, assuming $\mathcal{R}e\,(\lambda) \leq 0$). For example, Fig.4.15 shows the stablility domain of the third order fully implicit BDI3 scheme

$$\frac{11\,v^{n+1} - 18\,v^n + 9\,v^{n-1} - 2\,v^{n-2}}{6\,\Delta t} = \lambda\,v^{n+1} \qquad (4.99)$$

compared to the semi-implicit AB/BDI3 scheme. It may be seen that the implicit AB/BDI3 scheme is stable in the region of the X-negative half-plane delimited by the lines $Y/X = \pm \tan\alpha$ with $\alpha = 86.03°$ [see Hairer and Wanner (1991)]. The scheme is said to be A(86.03) stable, according to the definition of A(α)-stability introduced by Widlund (1967).

Let us consider the stability of the scheme (4.99) applied to the advection-diffusion equation.

For the Fourier method, the slope $p = |\lambda_I/\lambda_R|$ is equal to $|a|/(\nu\,k)$. Discarding the nonsignificant value $k = 0$ (constant mode), the maximum value of p is obtained for $k = 1$ and equals $|a|/\nu$. Therefore, the scheme (4.99) is unconditionally stable if $\tan^{-1}(|a|/\nu) < 86.03°$, namely, if $\delta = \nu/|a| > \delta_\star \simeq 0.0694$. On the other hand, if $\delta < \delta_\star$ the scheme (4.99) is unstable, whatever the time-step Δt and the cut-off frequency K.

For the Chebyshev collocation method, applied to the Dirichlet problem, the numerical study of the eigenvalues of the spatial discrete operator shows that the slope $p = |\lambda_I/\lambda_R| < \tan(86.03°)$ (i.e., the scheme is uncondition-ally stable) provided that $\delta = \nu/|a| > \delta_\star \simeq 0.01375$. Nevertheless, if $\delta < \delta_\star$, there exists a value $N_\star$ of N (degree of the polynomial approximation) such that the scheme (4.99) is stable for $N \geq N_\star$, whatever the time-step Δt. As an example, $N_\star = 58$ for $\delta = 10^{-3}$. The value $N_\star$ is increasing with $1/\delta$.

4.5.2   One-step methods

Another way to construct high-order methods is to consider only two levels of time (n and $n + 1$), the accuracy being obtained by dividing the time interval $t_{n+1} - t_n$ in a sequence of intermediate stages, each of which furnishes an intermediate value. Accuracy of sth order is reached with s intermediate stages. This type of method is called a "one-step method," the main example being the Runge-Kutta schemes, which are the object of the present section.

(a) Classic explicit Runge-Kutta schemes

The general s-stage Runge-Kutta scheme for the equation (4.1) is

$$K_1 = H(u^n),$$

$$K_i = H\left(u^n + \Delta t \sum_{j=1}^{i-1} a_{i,j} K_j\right), \quad i = 2, \ldots, s \tag{4.100}$$

$$u^{n+1} = u^n + \Delta t \sum_{j=1}^{s} b_j K_j.$$

The sets of coefficients $a_{i,j}$ and b_j characterize the scheme. When H depends explicitly on time, that is, $H = H(t, u)$, the quantity K_i is defined by evaluating H at time $t_n + c_i \Delta t$. These coefficients c_i which characterize the intermediate time-levels satisfy the consistency conditions

$$c_i = \sum_{j=1}^{i-1} a_{i,j}, \quad i = 2, \ldots, s, \tag{4.101}$$

with $c_1 = 0$. It is convenient to preserve the coefficients c_i even if H does not explicitly depend on t.

Another way to write the general Runge-Kutta scheme (4.100) is

$$u_0 = u^n,$$

$$u_i = u_0 + \Delta t \sum_{j=0}^{i-1} a_{i+1,j+1} H(u_j), \quad i = 1, \ldots, s - 1 \tag{4.102}$$

$$u^{n+1} = u_0 + \Delta t \sum_{j=0}^{s-1} b_{j+1} H(u_j).$$

Finally, a third form of Runge-Kutta scheme may be found in the literature, it is

$$u_0 = u^n \,,$$

$$u_i = u_{i-1} + \Delta t \sum_{j=0}^{i-1} \alpha_{i,j} H(u_j) \,, \quad i = 1,\dots,s-1 \tag{4.103}$$

$$u^{n+1} = u_{s-1} + \Delta t \sum_{j=0}^{s-1} \beta_j H(u_j) \,,$$

with

$$\alpha_{i,j} = a_{i+1,j+1} - a_{i,j+1} \,, \quad i = 2,\dots,s-1 \,, \quad j = 0,\dots,i-2 \,,$$

$$\alpha_{i,j} = a_{i+1,j+1} \,, \quad i = 1,\dots,s-1 \,, \quad j = i-1 \,,$$

$$\beta_j = b_{j+1} - a_{s,j+1} \,, \quad j = 0,\dots,s-2 \,,$$

$$\beta_{s-1} = b_s \,.$$

The coefficients $a_{i,j}$ and b_j are determined, as explained above for the multistep method, by setting to zero the first terms of the truncation error calculated from Taylor's expansion. However, this is generally not sufficient to determine completely the coefficients, and other constraints have to be added according to various criteria : minimization of the truncation error, low-storage property. . . .

The first-order one-step Runge-Kutta scheme is again the classic forward Euler scheme.

The general second-order two-stage scheme may be written

$$u_1 \quad = u^n + \alpha \, \Delta t \, H\left(u^n\right)$$

$$u^{n+1} \quad = u^n + \frac{\Delta t}{2\,\alpha} \left[(2\,\alpha - 1) \, H\left(u^n\right) + H\left(u_1\right)\right] \,, \tag{4.104}$$

where the parameter $\alpha \, (= a_{2,1})$ is arbitrary. The case $\alpha = 1/2$ corresponds to the well-known modified Euler scheme.

Among the third-order three-stage Runge-Kutta schemes, we note the "optimal TVD" scheme (Shu and Osher, 1988 ; Shu, 1998) often used in association with ENO (Essentially Non Oscillating) methods. The corresponding coefficients are

$$a_{2,1} = 1 \,, \quad a_{3,1} = a_{3,2} = \frac{1}{4} \,, \quad b_1 = b_2 = \frac{1}{6} \,, \quad b_3 = \frac{2}{3} \,.$$

Note that this scheme may be written under the form

$$
\begin{aligned}
u_1 &= u^n + \Delta t\, H\left(u^n\right) \\
u_2 &= \frac{3}{4}u^n + \frac{1}{4}u_1 + \frac{1}{4}\Delta t\, H\left(u_1\right) \\
u^{n+1} &= \frac{1}{3}u^n + \frac{2}{3}u_2 + \frac{2}{3}\Delta t\, H\left(u_2\right) ,
\end{aligned}
\tag{4.105}
$$

which is advantageous since only one evaluation of H is required at each stage. Such a scheme, however, is not "low-storage" in the sense that will be discussed below.

The classic fourth-order fourth-stage Runge-Kutta scheme is defined by the following parameters

$$
a_{2,1} = \frac{1}{2}, \quad a_{3,1} = 0, \quad a_{3,2} = \frac{1}{2}, \quad a_{4,1} = a_{4,2} = 0, \quad a_{4,3} = 1,
$$
$$
b_1 = \frac{1}{6}, \quad b_2 = b_3 = \frac{1}{3}, \quad b_4 = \frac{1}{6}.
\tag{4.106}
$$

This scheme has been used, for example, by Gavrilakis et al. (1996) for the Navier-Stokes equations spatially approximated with a Legendre spectral-element method.

(b) Low-storage explicit Runge-Kutta schemes

One important aspect of any time-discretization scheme is to minimize the storage requirement. This requirement may be crucial for complex flow calculations needing a very large number of degrees of freedom. Therefore, a significant effort continues to be devoted to reduce the storage needed by the Runge-Kutta schemes.

If the spatial discretization of $H\left(u\right)$ involves N degrees of freedom (modes or grid values), a Runge-Kutta scheme requiring $2N$ memory locations is called a $2N$-storage scheme, a $3N$-storage scheme if it requires $3N$ locations, etc. Note that $2N$ is the storage required by the usual first-order Euler scheme.

Williamson (1980) has shown that all second-order schemes can be cast in $2N$-storage form. On the other hand, only some three-stage third-order schemes can be cast in a $2N$-storage version. For four-stage fourth-order schemes, it has been shown (Fyfe, 1966) that all of them can be $3N$-storage but only very special schemes (not commonly used) can be cast in the $2N$-storage form.

In the present section, after a presentation of the general $2N$-storage scheme, we shall present the $2N$-storage version of the second-order scheme (4.104) and the three-stage third-order scheme proposed by Williamson (1980). Then we shall discuss the four-stage, third-order, $2N$-storage scheme proposed by Carpenter and Kennedy (1994), in which the additional stage is used to improve the stability properties. Finally, fourth-order $3N$- and $2N$-storage schemes will be described.

b.1. General 2N-storage scheme
The algorithm is of the form

$$u_0 = u^n$$

$$Q_j = A_j \, Q_{j-1} + \Delta t \, H\,(u_{j-1})$$

$$u_j = u_{j-1} + B_j \, Q_j \,, \qquad j = 1,\ldots,s \tag{4.107}$$

$$u^{n+1} = u_s \,,$$

with $A_1 = 0$. Williamson (1980) gives the expressions for A_j and B_j in terms of the coefficients of the general Runge-Kutta scheme (4.100). This can be done only if these coefficients have certain ratios to each other. This is the reason why all the Runge-Kutta schemes cannot be put into the 2N-storage form. The expressions are

$$B_j = a_{j+1,j} \,, \quad j \neq s$$

$$B_s = b_s$$

$$A_j = \frac{b_{j-1} - B_{j-1}}{b_j} \,, \quad j \neq 1, \quad b_j \neq 0 \tag{4.108}$$

$$A_j = \frac{a_{j+1,j-1} - c_j}{B_j} \,, \quad j \neq 1, \quad b_j = 0 \,.$$

b.2. Second-order schemes
    The second-order two-stage scheme (4.104) can be cast in the 2N-storage form (4.108) with

$$A_1 = 0, \quad A_2 = -2\,\alpha^2 + 2\,\alpha - 1, \quad B_1 = \alpha, \quad B_2 = \frac{1}{2\,\alpha},$$

and is written as

$$u_0 = u^n$$

$$Q_1 = \Delta t \, H\,(u_0)$$

$$u_1 = u_0 + \alpha \, Q_1$$

$$Q_2 = (-2\,\alpha^2 + 2\,\alpha - 1)\, Q_1 + \Delta t \, H\,(u_1) \tag{4.109}$$

$$u_2 = u_1 + \frac{1}{2\,\alpha} Q_2$$

$$u^{n+1} = u_2 \,.$$

b.3. Third-order schemes

The three-stage third-order scheme proposed by Williamson (1980) is defined by the values

$$A_1 = 0, \quad A_2 = -\frac{5}{9}, \quad A_3 = -\frac{153}{128},$$

$$B_1 = \frac{1}{3}, \quad B_2 = \frac{15}{16}, \quad B_3 = \frac{8}{15}, \qquad (4.110)$$

and can be written as :

$$u_0 = u^n$$
$$Q_1 = \Delta t H(u_0)$$

$$u_1 = u_0 + \frac{1}{3}Q_1$$

$$Q_2 = -\frac{5}{9}Q_1 + \Delta t H(u_1)$$

$$u_2 = u_1 + \frac{15}{16}Q_2 \qquad (4.111)$$

$$Q_3 = -\frac{153}{128}Q_2 + \Delta t H(u_2)$$

$$u_3 = u_2 + \frac{8}{15}Q_3$$

$$u^{n+1} = u_3.$$

This scheme is broadly used in spectral methods, for example, for the Navier-Stokes equations for compressible fluids spatially approximated with the Chebyshev collocation method (Guillard *et al.*, 1992).

It will be shown later that the allowable time-step required for (linear) stability increases with the order of accuracy. Taking advantage of this property, Carpenter and Kennedy (1994) have devised a $2N$-storage, four-stage, third-order accurate scheme possessing the same stability constraint as the four-stage, fourth-order scheme. More precisely, thanks to supplementary coefficients introduced by the additional stage, it is possible to have an amplification factor of fourth-order type. Incidentally, it is interesting to observe that the above property makes this third-order scheme of fourth-order accuracy in the linear case.

The requirement on the amplification factor is not sufficient to determine all the coefficients. Therefore, the remaining freedom is exploited by imposing the coefficients to be rational and to minimize the main part of the truncation error.

The four-stage third-order RK3/CK scheme constructed, consistent with the above-described requirements, is defined by the set

$$A_1 = 0, \quad A_2 = -\frac{205}{243}, \quad A_3 = -\frac{243}{38}, \quad A_4 = -\frac{2}{9},$$

$$B_1 = \frac{19}{36}, \quad B_2 = \frac{27}{19}, \quad B_3 = \frac{2}{9}, \quad B_4 = \frac{1}{4}. \tag{4.112}$$

b.4. Fourth-order scheme

The classic four-stage scheme defined by the coefficients (4.106) can be written (Blum, 1962) in the following $3N$-storage form

$$
\begin{aligned}
u_0 &= u^n \\
P_0 &= 0 \\
Q_0 &= u^n \\
P_j &= A_j\, P_{j-1} + B_j\, Q_{j-1} \\
Q_j &= C_j\, Q_{j-1} + H(u_{j-1}) \\
u_j &= u_{j-1} + \Delta t\, (D_j\, P_j + E_j\, Q_j), \quad j = 1,\ldots, s = 4 \\
u^{n+1} &= u_s
\end{aligned}
\tag{4.113}
$$

with coefficients given in Table 4.7.

j	A_j	B_j	C_j	D_j	E_j
1	0	1	0	0	$\frac{1}{2}$
2	0	1	0	$-\frac{1}{2}$	$\frac{1}{2}$
3	$\frac{1}{6}$	0	$-\frac{1}{2}$	0	1
4	1	-1	2	1	$\frac{1}{6}$

Table 4.7. Coefficients of the classic four-stage fourth-order $3N$-storage Runge-Kutta scheme.

As for the third-order scheme, Carpenter and Kennedy (1994) have considered fourth-order schemes with five stages. By increasing the number of stages, that is, the number of arbitrary coefficients, they constructed a scheme (RK4/CK scheme) according to the following properties: (1) $2N$-storage, (2) large domain of stability, (3) small truncation error.

The values of the coefficients A_j and B_j of equations (4.107)-(4.108) are

$$
\begin{aligned}
A_1 &= 0 & B_1 &= 0.1028639988105 \\
A_2 &= -0.4801594388478 & B_2 &= 0.7408540575767 \\
A_3 &= -1.4042471952 & B_3 &= 0.7426530946684 \\
A_4 &= -2.016477077503 & B_4 &= 0.4694937902358 \\
A_5 &= -1.056444269767 & B_5 &= 0.1881733382888.
\end{aligned}
\tag{4.114}
$$

A rational form of these coefficients is given by Carpenter and Kennedy (1994).

The above scheme has been used for the study of the interaction shock-vortex by means of the solution of the Euler equations with the ENO method (Erlebacher *et al.*, 1997). These authors report that this scheme is more efficient than Williamson's third-order 2N-storage scheme (4.111) because of the larger allowable time-step (higher by a factor of 1.9). The 2N-storage third-order and fourth-order schemes respectively defined by (4.112) and (4.114) were applied (Wilson *et al.*, 1998) to the Navier-Stokes equations for incompressible fluids in association with fourth- and sixth-order Hermitian finite-difference approximations. Presently, these 2N-storage schemes do not seem to have been applied in association with spectral approximations.

(c) Semi-implicit Runge-Kutta schemes

As already explained, the constraint on the time-step due to the explicit nature of the scheme can be reduced by considering a semi-implicit time-discretization. The linear viscous part $L(u)$ of $H(u)$ [see Eqs.(4.1)-(4.3)] which leads to a very restrictive time-step (for low or moderate viscosity) is treated in an implicit way. The nonlinear convective part $N(u)$ is evaluated explicitly.

In this section, we present semi-implicit two- and three-stage Runge-Kutta schemes, respectively, which are extensions of (4.104) and (4.111). For fully implicit Runge-Kutta schemes, we refer to Hairer *et al.* (1987) and Hairer and Wanner (1991).

c.1. Two-stage schemes (RK2/CN)

For the equations (4.1)-(4.2) the family of two-stage second-order semi-implicit schemes, extensions of (4.104), is of the form

$$u_1 = u^n + \Delta t \left[\alpha N(u^n) + \beta L(u^n) + (\alpha - \beta)L(u_1)\right]$$

(4.115)

$$u^{n+1} = u^n + \frac{\Delta t}{2\alpha} \left[(2\alpha - 1)N(u^n) + N(u_1) + \alpha L(u^n) + \alpha L(u^{n+1})\right] .$$

(4.116)

The parameters α and β are arbitrary ($\alpha > 0$ and $\alpha \geq \beta$ to ensure resolvability). The case $\alpha = \beta$ avoids the solution of an implicit equation at the first stage but the stability is reduced (see below). The case $\beta = 0$ avoids the calculation of $L(u^n)$ at the first stage. Note that for $\beta = 1/2$, the scheme can be written in the 2N-storage form considered below [Eq.(4.117)]. The above general scheme will be denoted by RK2/CN since the discretization of the viscous term is of Crank-Nicolson-type when $\alpha = 2\beta$.

c.2. Three-stage scheme (RK3/CN)

Zang and Hussaini (1985) have proposed an extension of Williamson's scheme that consists of retaining the explicit scheme for the nonlinear term while adding a Crank-Nicolson-type treatment of the viscous linear term. The scheme is

$$u_0 = u^n$$
$$Q_j = A_j Q_{j-1} + \Delta t N(u_{j-1})$$
$$u_j = u_{j-1} + B_j Q_j + B'_j \Delta t \left[L(u_{j-1}) + L(u_j) \right], \quad j = 1, \ldots, s = 3$$
$$u^{n+1} = u_s.$$

$$(4.117)$$

The coefficients A_j and B_j are given in (4.110) and the coefficients B'_j are

$$B'_1 = \frac{1}{6}, \quad B'_2 = \frac{5}{24}, \quad B'_3 = \frac{1}{8}.$$

The introduction of the implicit terms reduces the theoretical accuracy to second-order while the error associated with the explicit part is third-order. This scheme was applied by Zang and Hussaini (1985) to the Navier-Stokes equations for incompressible fluids approximated in space with a spectral method, the viscous terms being considered implicitly. The authors reported that, for low viscosity flows, the error was at most 50% greater than that associated with a true third-order scheme. Besides, they observe that the error decreased by a factor of 8 when Δt was halved. This is typical third-order behaviour. The scheme was applied to the Navier-Stokes equations for the spectral computations of various compressible flows including thermal convection and mixing-layer flows (Gauthier, 1991 ; Gauthier *et al.*, 1996).

(d) Stability

d.1. Explicit Runge-Kutta schemes

The analysis developed in Section 4.5.1.e is directly applicable to explicit Runge-Kutta schemes. Equation (4.89) can be used to study the stability of these schemes associated with Fourier as well as Chebyshev methods.

The application of the general explicit Runge-Kutta scheme (4.100) to the differential equation (4.89) yields

$$v^{n+1} = g v^n \qquad (4.118)$$

with

$$g(z) = 1 + \left(\sum_{i=1}^{s} b_i \right) z + \left(\sum_{i=1}^{s} b_i c_i \right) z^2$$
$$+ \left(\sum_{i,j=1}^{s} b_i a_{i,j} c_j \right) z^3 + \left(\sum_{i,j,k=1}^{s} b_i a_{i,j} a_{j,k} c_k \right) z^4 + \ldots,$$

$$(4.119)$$

where $z = \lambda \Delta t$. In particular, for the classical four-stage fourth-order (RK4) scheme, we get

$$g(z) = 1 + z + \frac{z^2}{2} + \frac{z^3}{6} + \frac{z^4}{24}. \tag{4.120}$$

In other words, g is the Taylor expansion of $e^{\lambda \Delta t}$ since the exact solution of (4.89) satisfies

$$v(t + \Delta t) = v_0 \, e^{\lambda(t+\Delta t)} = v_0 \, e^{\lambda t} e^{\lambda \Delta t} = v(t) \, e^{\lambda \Delta t}.$$

Therefore, the amplification factor g is the same for any Runge-Kutta scheme such that $s = k$, that is, the number of intermediate stages equals the required order of accuracy. As a result, the domain of stability will be the same for these schemes (RKk schemes).

The stability requires $|g| \leq 1$. As for the multistep methods it is convenient to determine the domain of stability S in the complex plane $X = \mathcal{R}e(z)$, $Y = \mathcal{I}m(z)$. This can be done numerically by setting

$$g(z) = e^{i\theta}, \quad 0 \leq \theta \leq 2\pi, \tag{4.121}$$

and then, after discretization in θ, by determining the roots using a computer routine. Figure 4.16 shows the domain of stability of Runge-Kutta (RKk) schemes of various order. Note that the $2N$-storage, four-stage third-order RK3/CK scheme introduced by Carpenter and Kennedy (1994) and defined by the set of coefficients (4.112), has the same domain of stability as the classical fourth-order scheme (curve (4)). Also, note that the $2N$-storage five-stage fourth-order scheme defined by (4.114), and whose amplification factor is

$$g = 1 + z + \frac{z^2}{2} + \frac{z^3}{6} + \frac{z^4}{24} + \frac{z^5}{200}, \tag{4.122}$$

is represented by curve (5) in Fig.4.16. The limiting values X_0 and Y_0 (see the previous section) are given in Table 4.8.

From the above stability analysis the following conclusions can be drawn:

1. For purely imaginary λ, the first- and second-order schemes are unstable. This applies to the case of Fourier approximation to the pure advection equation. On the other hand, in the case of Chebyshev approximation, the real parts of the eigenvalues are not zero, so that these low-order schemes can be stable provided the time-step Δt is sufficiently small. For the first-order scheme the stability criterion is given by Eq.(4.45). For the second-order scheme, the constant C in the criterion $a\,\Delta t \leq C/N^2$ has been found by Fulton and Taylor (1984) to be approximately equal to 18 (for $N \geq 32$) for the Chebyshev collocation method and to 3.5 (for $N > 3$) for the Chebyshev tau method.

2. For real λ, namely, in the case of pure diffusion, the fourth-order RK4/CK scheme allows the use of larger time-steps.

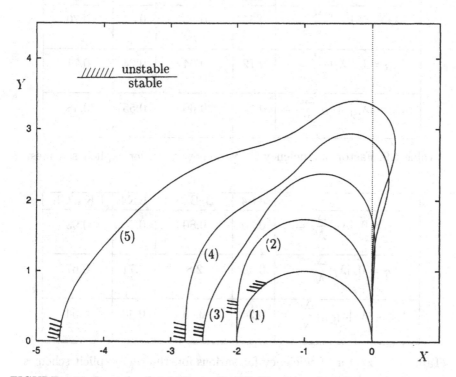

FIGURE 4.16. Domain of stability in the (X, Y) plane of explicit one-step schemes : (1) first-order Euler, (2) RK2, (3) RK3, (4) RK4, (5) RK4/CK (the domains have to be completed by symmetry in $Y < 0$).

Scheme	Order	Number of stages	X_0	Y_0
RK2	2	2	2	0
RK3	3	3	2.512	$\sqrt{3}$
RK3/CK	3	4	2.785	$2\sqrt{2}$
RK4	4	4	2.785	$2\sqrt{2}$
RK4/CK	4	5	4.65	3.34

Table 4.8. Coordinates of the intersections of the boundary of the stability domain with the axes, for various explicit Runge-Kutta schemes.

	AB3	AB/BDE3	RK3	RK3/CK
$\lambda_I = 0, \ \|\lambda_R\|\dfrac{\Delta t}{\mathcal{N}} =$	0.55	0.95	0.84	0.70
$\lambda_R = 0, \ \|\lambda_I\|\dfrac{\Delta t}{\mathcal{N}} =$	0.72	0.64	0.24	0.71
$\|\lambda_I\| = \|\lambda_R\|, \ \|\lambda_R\|\dfrac{\Delta t}{\mathcal{N}} =$	0.40	0.60	0.55	0.48

Table 4.9. Factor of efficiency for various third-order explicit schemes.

	AB4	AB/BDE4	RK4	RK4/CK
$\lambda_I = 0, \ \|\lambda_R\|\dfrac{\Delta t}{\mathcal{N}} =$	0.30	0.80	0.70	0.93
$\lambda_R = 0, \ \|\lambda_I\|\dfrac{\Delta t}{\mathcal{N}} =$	0.43	2.8	0.71	0.67
$\|\lambda_I\| = \|\lambda_R\|, \ \|\lambda_R\|\dfrac{\Delta t}{\mathcal{N}} =$	0.24	0.38	0.48	0.54

Table 4.10. Factor of efficiency for various fourth-order explicit schemes.

3. The size of the time-step Δt is not sufficient by itself to characterize the efficiency of a scheme. The volume of computations required for a given advancement in time is more significant. Therefore, we may define a factor of efficiency by the ratio $\Delta t/\mathcal{N}$ where Δt is the maximum allowable time-step and $\mathcal{N}$ is the number of evaluations of $H(u)$ within a time-cycle. Tables 4.9 and 4.10 show this factor of efficiency for various third- and fourth-order schemes in the following cases : $\lambda_I = 0$, $\lambda_R = 0$, and $|\lambda_I| = |\lambda_R|$. It may be observed that, on the whole, the most efficient schemes are the multistep AB/BDE3 scheme and the one-step RK4/CK scheme.

d.2. Semi-implicit schemes
As already explained in Section 4.5.1.e, the use of the differential equation (4.89) is restricted to the Fourier method. The semi-implicit schemes (4.115)-(4.116) and (4.117) applied to the equation (4.89) leads to (4.118) where the amplification factor g is of the form

$$g = \frac{P(X, iY)}{Q(X)}, \tag{4.123}$$

where P and Q are polynomials and $X = \lambda_R \Delta t$, $Y = \lambda_I \Delta t$. The domain of stability $|g| \leq 1$ in the (X, Y) plane is determined by numerically

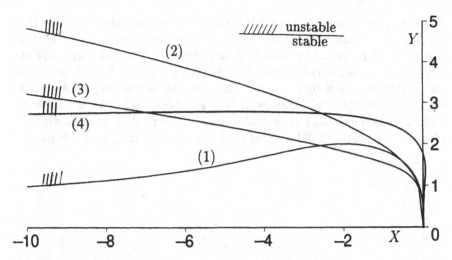

FIGURE 4.17. Domain of stability in the (X, Y) plane of semi-implicit one-step schemes: RK2/CN scheme (4.115)-(4.116) for various values of γ : (1) $\gamma = 0$, (2) $\gamma = 1/2$, (3) $\gamma = 1$; RK3/CN scheme (4.117) : curve (4) (the domains have to be completed by symmetry in $Y < 0$).

calculating the roots of

$$|P(X, \underline{i}Y)|^2 - Q^2(X) = 0 \qquad (4.124)$$

considered as a polynomial in Y, for various values of $X \in (-\infty, 0]$.

Because of the linearity of Eq.(4.89), the amplification factor g of the two-parameter family RK2/CN schemes (4.115)-(4.116) depends only on the parameter $\gamma = \alpha - \beta \geq 0$. The domain of stability in the (X, Y) plane is shown in Fig.4.17 for $\gamma = 0$, 1/2 and 1. It is observed that stability is improved as soon as $\gamma > 0$, that is, when the viscous term $L(u)$ is included in the implicit part of the first stage. Figure 4.17 also shows the domain of stability (curve (4)) of the RK3/CN scheme (4.117). For large viscosity, this latter scheme is less stable than the second-order RK2/CN scheme with $\gamma = 1/2$ or 1.

4.5.3 Conclusion

The choice between multistep and one-step methods depends on the problem under consideration and on the available computer, especially concerning its memory capacity.

In cases where explicit schemes are of use, the Runge-Kutta schemes, especially Carpenter and Kennedy's fourth-order five-stage RK4/CK scheme [(4.107), (4.108), and (4.114)] can be recommended.

On the other hand, when some degree of implicitness is required, as is generally the case with the Navier-Stokes equations for incompressible fluids, the multistep schemes AB/BDIk are recommended, because they are less time-consuming. However, a possible alternative way is the splitting method proposed by Maday *et al.* (1990). In the first step, only the convective terms are considered in a characteristic-type formulation using an explicit Runge-Kutta scheme. Then, in the second-step, the remaining part of the Navier-Stokes equations are discretized with a Backward-Differentiation formula (BDIk-type scheme). This method has recently been (Xu and Pasquetti, 2001) applied to the solution of various problems.

Part II

Navier-Stokes equations

5

Navier-Stokes equations for incompressible fluids

In this short chapter, the Navier-Stokes equations governing the motion of a viscous incompressible fluid are recalled. Two formulations are considered : the velocity-pressure formulation (also called "primitive variables" formulation) and the vorticity-streamfunction formulation. The formulation using vorticity and velocity as dependent variables will not be addressed here. This formulation, commonly used with finite-difference methods, has not yet received large attention in the field of spectral methods. However, recent works (Clercx, 1997 ; Trujillo and Karniadakis, 1999) have shown that Chebyshev solutions of the vorticity-velocity equations could be envisaged with success.

5.1 Velocity-pressure equations

The Navier-Stokes equations governing the motion of a viscous incompressible fluid are

$$\partial_t \mathbf{V} + \mathbf{A}\left(\mathbf{V}, \mathbf{V}\right) + \nabla p - \nu \nabla^2 \mathbf{V} = \mathbf{f} \tag{5.1}$$

$$\nabla . \mathbf{V} = 0 , \tag{5.2}$$

where $\mathbf{V} = (u, v, w)$ is the velocity vector, p is the pressure (divided by the constant density ρ), ν is the kinematic viscosity (ν = positive constant), and $\mathbf{f}$ is a forcing term. The term $\mathbf{f}$ is either given or dependent on other variables like temperature or concentration. In these latter cases, equations governing the evaluation of these quantities are needed (Section 5.3).

The nonlinear term $\mathbf{A}(\mathbf{V}, \mathbf{V})$ in Eq.(5.1) may be written either in the *convective* form

$$\mathbf{A}(\mathbf{V}, \mathbf{V}) = (\mathbf{V} \cdot \nabla)\mathbf{V} \qquad (5.3)$$

or in the *conservative* (or *divergence*) form

$$\mathbf{A}(\mathbf{V}, \mathbf{V}) = \nabla \cdot (\mathbf{V}\mathbf{V}). \qquad (5.4)$$

These two forms are equivalent, taking the incompressibility condition (5.2) into account in the identity

$$\nabla \cdot (\mathbf{V}\mathbf{V}) = (\mathbf{V} \cdot \nabla)\mathbf{V} + (\nabla \cdot \mathbf{V})\mathbf{V}.$$

The momentum equation (5.1) can also be written in the so-called *rotational* form where $\mathbf{A}(\mathbf{V}, \mathbf{V}) + \nabla p$ in (5.1) is expressed as

$$\mathbf{A}(\mathbf{V}, \mathbf{V}) + \nabla p = \boldsymbol{\omega} \times \mathbf{V} + \nabla P, \qquad (5.5)$$

where $\boldsymbol{\omega} = \nabla \times \mathbf{V}$ is the vorticity vector and $P = p + |\mathbf{V}|^2/2$ is the total pressure.

Lastly, the *skew-symmetric* form defined by

$$\mathbf{A}(\mathbf{V}, \mathbf{V}) = \frac{1}{2}\nabla \cdot (\mathbf{V}\mathbf{V}) + \frac{1}{2}(\mathbf{V} \cdot \nabla)\mathbf{V} \qquad (5.6)$$

may also be considered. A comparison of the properties of these various forms will be discussed in Sections 7.3.2 and 7.4.5.

Equations (5.1)-(5.2) are solved in the bounded domain Ω whose boundary is Γ. If the cases where the solution is assumed to be spatially periodic (in one or more directions) are omitted, equations (5.1)-(5.2) have to be supplemented by boundary conditions. Typical boundary conditions consist of prescribing the velocity on Γ :

$$\mathbf{V} = \mathbf{V}_\Gamma \quad \text{on } \Gamma = \partial\Omega. \qquad (5.7)$$

The incompressibility condition (5.2) imposes the total flow rate through the boundary to be zero, let

$$\int_\Gamma \mathbf{V}_\Gamma \cdot \mathbf{n} \, d\Gamma = 0, \qquad (5.8)$$

where $\mathbf{n}$ is the outward unit vector normal to Γ.

Now the initial condition is

$$\mathbf{V} = \mathbf{V}_0 \quad \text{at } t = 0. \qquad (5.9)$$

The initial velocity field $\mathbf{V}_0$ is assumed to be solenoidal

$$\nabla \cdot \mathbf{V}_0 = 0, \qquad (5.10)$$

and satisfies the compatibility condition with the boundary condition (5.7):

$$\mathbf{V}_0.\mathbf{n} = \mathbf{V}_\Gamma.\mathbf{n} \quad \text{on } \Gamma. \tag{5.11}$$

This condition ensures a minimal regularity to the solution at the initial time. In the applications to some flows, for instance, impulsive start (see Guermond and Quartapelle, 1994), the compatibility condition (5.11) cannot be satisfied. Experience shows that if the numerical method is sufficiently robust, the initial singularity is quickly damped and is not troublesome. Note also the possibility of replacing the impulsive start by a progressive one applied on a very short interval of time. Nevertheless, from the mathematical point of view, high regularity of the solution as $t \to 0$ needs some complementary compatibility conditions (Heywood and Rannacher, 1982 ; Marion and Temam, 1998).

In the problem stated above, neither boundary condition nor initial condition is required for the pressure. This is due to the fact that the pressure appears only through its gradient. As a consequence, the pressure is determined within an arbitrary function of time. We refer to Quarteroni and Valli (1994) for the case of boundary conditions involving pressure.

5.2 Vorticity-streamfunction equations

For two-dimensional flows, plane or axisymmetric, the vorticity-streamfunction equation constitutes an interesting alternative to the velocity-pressure equations.

5.2.1 Plane flow

In a two-dimensional plane geometry the vorticity vector

$$\boldsymbol{\omega} = \nabla \times \mathbf{V} \tag{5.12}$$

has only one nonzero component ω, such that

$$\boldsymbol{\omega} = \omega\,\mathbf{k}, \tag{5.13}$$

where $\mathbf{k}$ is the unit vector normal to the plane (x,y) of the flow. The vorticity ω is expressed by

$$\omega = \partial_x v - \partial_y u. \tag{5.14}$$

The equation satisfied by ω is obtained by applying the curl operator to Eq.(5.1), so that the pressure gradient term disappears. The result is

$$\partial_t \omega + B\left(\mathbf{V}, \omega\right) - \nu\,\nabla^2 \omega = F, \tag{5.15}$$

where $B(\mathbf{V}, \omega)$ is deduced from $\mathbf{A}(\mathbf{V}, \mathbf{V})$. In *convective* form, we have

$$B(\mathbf{V}, \omega) = \mathbf{V}.\nabla\omega \tag{5.16}$$

and in *conservative* (or *divergence*) form

$$B(\mathbf{V}, \omega) = \nabla.(\omega\mathbf{V}). \tag{5.17}$$

Finally, the forcing term F is

$$F = (\nabla \times \mathbf{f}).\mathbf{k}. \tag{5.18}$$

The velocity vector $\mathbf{V}$ is defined in terms of the streamfunction ψ by

$$\mathbf{V} = \nabla \times (\psi\mathbf{k}) \tag{5.19}$$

so that

$$u = \partial_y\psi, \quad v = -\partial_x\psi. \tag{5.20}$$

The equation satisfied by ψ is obtained by applying the curl operator to Eq.(5.19) and using the definition (5.12), namely,

$$\nabla^2\psi + \omega = 0. \tag{5.21}$$

Equations (5.15) and (5.21), associated with (5.19), are solved in the domain Ω assumed to be simply connected. The initial and boundary conditions are deduced from those prescribed for the velocity field.

Let us assume that $\mathbf{V}$ satisfies the boundary condition (5.7). From the definition of the streamfunction (5.19) we have, on the boundary,

$$\nabla\psi.\mathbf{t} = -\mathbf{V}_\Gamma.\mathbf{n}, \quad \nabla\psi.\mathbf{n} = \mathbf{V}_\Gamma.\mathbf{t} \quad \text{on } \Gamma, \tag{5.22}$$

where $\mathbf{t}$ is the unit vector tangent to Γ with clockwise orientation. Now, by integrating the first equation (5.22) along the boundary Γ, we obtain

$$\psi|_s - \psi|_{s_0} = \int_{s_0}^s (\nabla\psi.\mathbf{t}) \, ds = -\int_{s_0}^s \mathbf{V}_\Gamma.\mathbf{n} \, ds, \tag{5.23}$$

where s is the curvilinear abscissa along Γ and s_0 corresponds to a given arbitrary fixed point of Γ. An inspection of the equations (5.15), (5.19), (5.21), and boundary conditions (5.22) shows that the streamfunction ψ appears only through its derivatives. Therefore, the streamfunction ψ is defined up to an arbitrary function of time (a constant in the steady case), and we may assume $\psi|_{s_0} = 0$. Then, from Eq.(5.23), we deduce the Dirichlet boundary condition for ψ :

$$\psi = g \tag{5.24}$$

with

$$g = -\int_{s_0}^s \mathbf{V}_\Gamma.\mathbf{n} \, ds. \tag{5.25}$$

Now the second equation (5.22) gives another condition for ψ :

$$\partial_n \psi = h \tag{5.26}$$

with

$$h = \mathbf{V}_\Gamma.\mathbf{t}. \tag{5.27}$$

We remark that the knowledge of the velocity vector $\mathbf{V}$ on Γ leads to two boundary conditions for ψ and zero for ω. This is the classical difficulty associated with the vorticity-streamfunction equations (see, e.g., Peyret and Taylor, 1983).

The initial condition for ω is deduced from the initial condition (5.9) for the velocity, namely,

$$\omega = \omega_0 = \partial_x v_0 - \partial_y u_0 \quad \text{at } t = 0. \tag{5.28}$$

Note that the compatibility condition (5.11) implies

$$\partial_s g|_{t=0} = -\mathbf{V}_0.\mathbf{n}|_\Gamma. \tag{5.29}$$

It must be noticed that the solution of the problem does not need the knowledge of the streamfunction ψ at the initial time, a consequence of the fact that the time derivative $\partial_t \psi$ does not appear in the equations.

We refer to Guermond and Quartapelle (1994) for the theoretical proof of the equivalence between the vorticity-streamfunction and velocity-pressure formulations.

If the domain Ω is not simply connected, as assumed here, the boundary value of ψ has to be determined such that the pressure is a single-valued function. The implementation of such a condition with spectral methods is difficult even in relatively simple problems like the flow between two eccentric cylinders (Elghaoui and Pasquetti, 1999). As a matter of fact, spectral methods (except for the spectral-element method developed by Patera, 1984) are not adapted to complicated multiply-connected domains.

Lastly, it may be useful to calculate the pressure from the knowledge of the velocity field. One possible way is to consider the pressure as the solution of the Poisson equation obtained by taking the divergence of Eq.(5.1), namely,

$$\nabla^2 p = \nabla.\left[\mathbf{f} - \mathbf{A}\left(\mathbf{V}, \mathbf{V}\right)\right] \quad \text{in } \Omega \tag{5.30}$$

associated with the Neumann condition deduced from the projection of Eq.(5.1) on the normal to the boundary

$$\partial_n p = \left[\mathbf{f} - \mathbf{A}\left(\mathbf{V}, \mathbf{V}\right) + \nu\nabla^2\mathbf{V} - \partial_t\mathbf{V}\right].\mathbf{n} \quad \text{on } \Gamma. \tag{5.31}$$

In deriving Eq.(5.30) we have taken into account that $\nabla \cdot \mathbf{V} = 0$. Note that the compatibility condition associated with the Neumann problem (5.30)-(5.31) is automatically satisfied since Eqs.(5.30) and (5.31) are both deduced from the same equation.

5.2.2   Axisymmetric flow

In axisymmetric flow, the velocity vector $\mathbf{V}$ may have a component in the azimuthal direction θ, although all the flow quantities are independent of θ. Let (r, θ, z) be the cylindrical coordinate system and let $\mathbf{V} = (u, v, w)$, the vorticity vector $\boldsymbol{\omega}$ defined by Eq.(5.12) has three components ω_r, $\omega_\theta = \omega$, and ω_z, such that

$$\omega_r = -\partial_z v, \quad \omega_\theta = \omega = \partial_z u - \partial_r w, \quad \omega_z = \frac{1}{r}\partial_r (r v) . \tag{5.32}$$

Therefore, the components ω_r and ω_z are determined from the knowledge of the azimuthal velocity v. The equation satisfied by v is

$$\partial_t v + B_v (\mathbf{U}, v) - \nu \left(\nabla^2 v - \frac{v}{r^2}\right) = f_v , \tag{5.33}$$

where f_v is the forcing term, $\mathbf{U} = (u, w)$, and

$$B_v (\mathbf{U}, v) = u\partial_r v + w\partial_z v + \frac{u}{r}v \tag{5.34}$$

and

$$\nabla^2 = \partial_{rr} + \frac{1}{r}\partial_r + \partial_{zz} . \tag{5.35}$$

The azimuthal vorticity ω is determined by the equation

$$\partial_t \omega + B_\omega (\mathbf{U}, \omega) - \nu \left(\nabla^2 \omega - \frac{\omega}{r^2}\right) - 2\frac{v}{r}\partial_z v = F_\omega \tag{5.36}$$

where F_ω is the forcing term and

$$B_\omega (\mathbf{U}, \omega) = u\,\partial_r \omega + w\,\partial_z \omega - \frac{u}{r}\omega . \tag{5.37}$$

In this case, there also exists a streamfunction $\psi (r, z, t)$ such that

$$u = \frac{1}{r}\partial_z \psi, \quad w = -\frac{1}{r}\partial_r \psi , \tag{5.38}$$

where ψ satisfies the equation

$$\nabla^2 \psi - \frac{2}{r}\partial_r \psi - \omega r = 0 . \tag{5.39}$$

In summary, the equations to be solved are Eqs.(5.33), (5.36), and (5.39). As for plane flows, the associated boundary conditions are deduced from the boundary conditions prescribed on the velocity.

5.3   Boussinesq approximation

The Navier-Stokes equations within the Boussinesq approximation are derived from the Navier-Stokes equations governing the motion of compressible fluids in some simplifying circumstances [see, e.g., Joseph (1976), where the approximation is called "Oberbeck-Boussinesq"]. By assuming the Mach number of the flow and the variations in temperature sufficiently small, one can derive (Mihaljan, 1962) the Boussinesq equations as an asymptotic limit of the complete Navier-Stokes equations. The main features are :

(1) In the momentum equation the density ρ is constant except in the buoyant force $\mathbf{f}$. Therefore, the momentum equation (5.1) is valid with

$$\mathbf{f} = \frac{\rho}{\rho_0}\mathbf{g}\,, \tag{5.40}$$

where $\mathbf{g}$ is the gravitational vector field. Also, in Eq.(5.1), p refers to the pressure divided by the constant reference density ρ_0.

(2) The continuity equation is simply the equation (5.2).

(3) The energy equation is simplified by neglecting the density variations and the dissipation function. It is

$$\partial_t T + B\,(\mathbf{V},T) - \kappa_T\,\nabla^2 T = 0\,, \tag{5.41}$$

where T is the temperature and κ_T is the thermal diffusivity. The transport term $B\,(\mathbf{V},T)$ has the same expression as for vorticity ω, namely, Eq.(5.16) or (5.17).

The density ρ is connected to the temperature T through the linear equation of state

$$\rho = \rho_0\left[1 - \alpha_T\,(T - T_0)\right]\,, \tag{5.42}$$

where T_0 is the reference temperature associated with ρ_0 and the positive constant α_T is the thermal expansion coefficient. Equation (5.42) is obtained from the first term of the Taylor expansion of the equation of state $\rho = \rho\,(p,T)$ in which the pressure variations are neglected.

In summary, the system of the Navier-Stokes equations within the Boussinesq approximation is constituted by Eq.(5.1) with (5.40) and (5.42), Eq.(5.2) and Eq.(5.41).

Initial and boundary conditions for the velocity have been discussed in Section 5.1. For the temperature, the boundary conditions are of the general form

$$k_T\partial_n T + h_T\,T = \varphi_T \tag{5.43}$$

and the initial condition is

$$T = T^0 \quad \text{at } t = 0\,. \tag{5.44}$$

In case of a nonhomogeneous fluid characterized by the presence of a solute with concentration C, the density ρ is defined by the equation of state

$$\rho = \rho_0 \left[1 - \alpha_T \left(T - T_0 \right) + \alpha_C \left(C - C_0 \right) \right],\qquad(5.45)$$

where C_0 is the reference state and the positive constant α_C is the solutal density coefficient. The momentum and continuity equations are, respectively, Eq.(5.1) with (5.40) and Eq.(5.2). The evolution of concentration C is governed by the transport-diffusion equation

$$\partial_t C + B\left(\mathbf{V}, C\right) - \kappa_C \nabla^2 C = 0,\qquad(5.46)$$

where κ_C is the solute diffusivity. Associated boundary conditions are of the general form

$$k_C \partial_n C + h_C C = \varphi_C \quad \text{on } \Gamma\qquad(5.47)$$

and the initial condition is

$$C = C^0 \quad \text{at } t = 0.\qquad(5.48)$$

Finally, it is useful to note that source terms may be added to the equations governing the evolution of temperature and concentration.

5.4   Semi-infinite domain

In many situations of physical interest, the flow takes place in an unbounded domain, at least in the streamwise x-direction. Possible ways to handle the problem are :

(1) Transform the semi-infinite domain $0 \le x \le \infty$ onto a finite domain $-1 \le \xi \le 1$ by a suitable mapping $x = f(\xi)$ (see Section 8.2.2) and prescribe convenient boundary conditions at $\xi = 1$. For example, in the case of a channel flow with no-slip or stress-free walls at $y = \pm 1$ and $z = \pm 1$, possible boundary conditions (fully developed flow) are $v = 0$, $w = 0$, and $\partial_\xi u = \text{constant}$. This ensures that $\partial_x u \to 0$ for $x \to \infty$ whatever the value of the constant since $f'(\pm 1) = 0$. The case of the vorticity-streamfunction equations is considered in Section 9.5.3.b.

(2) Bound the domain with an artificial boundary $x = x_\infty$ and prescribe suitable conditions on this boundary.

(3) Introduce a "buffer-domain" in the downstream region of the flow in which the equations are modified in order to avoid the upstream influence.

The second approach is the most commonly used in computational fluid dynamics, but the difficulty is to devise the nonreflecting boundary conditions which prevent the disturbance created at the outlet to contaminate the upstream flow. This problem is enhanced in spectral methods because of their global character. As a matter of fact, there do not exist outflow boundary conditions valid at all events, their choice is highly dependent on the physical problem. We present a set of outflow boundary conditions which have been found successful by Forestier *et al.* (2000a,b) for the computation of two- and three-dimensional wake flows in a semi-infinite channel in the streamwise x-direction with stress-free walls at $y = \pm 1$ and periodicity in the spanwise z-direction. The velocity $\mathbf{V}_I = (u_I(y, z), 0, 0)$ is prescribed at the entrance $x = 0$. It is convenient to consider as unknown the disturbance velocity

$$\mathbf{V}' = \mathbf{V} - \overline{\mathbf{V}} \qquad (5.49)$$

with $\mathbf{V}' = (u', v', w')$ and $\overline{\mathbf{V}} = (\overline{u}, 0, 0)$, where the constant $\overline{u}$ is the mean value of $u_I(y, z)$. The proposed boundary conditions consist of applying, at the outlet $x = x_\infty$, an advection equation for the normal velocity component u' and for the tangential vorticity vector $\boldsymbol{\omega}'_t = \boldsymbol{\omega}_t = (\nabla \times \mathbf{V}') \times \mathbf{i}$ where $\mathbf{i}$ is the unit normal in the x-direction. These conditions are

$$\overline{D}_t u' = 0 \quad \text{at} \quad x = x_\infty, \qquad (5.50)$$

$$\overline{D}_t \boldsymbol{\omega}'_t = 0 \quad \text{at} \quad x = x_\infty, \qquad (5.51)$$

where $\overline{D}_t = \partial_t + \overline{u}\,\partial_x$. The constant advective velocity $\overline{u}$ could be replaced by a local one, as proposed by Orlanski (1976), provided the stability is ensured. Taking Eq.(5.50) and the definition of $\boldsymbol{\omega}'_t$ into account, we replace Eq.(5.51) by

$$\overline{D}_t(\partial_x u') = 0, \quad \overline{D}_t(\partial_x w') = 0 \quad \text{at} \quad x = x_\infty. \qquad (5.52)$$

Therefore Eqs.(5.50) and (5.52) provide boundary conditions for the three velocity components. It may be observed that, in the limit $\nu = 0$ and when the resulting Euler equations are linearized about the mean flow, the condition (5.51) is nothing other than the exact vorticity equation. This means that the only condition effectively imposed in this case is condition (5.50). This is in agreement with the mathematical property of the Euler equations which require only one boundary condition at the outlet.

To limit the influence of the outflow boundary condition on the upstream flow due to the global character of the Chebyshev approximation, it is recommended to approximate Eqs.(5.50) and (5.52) with a low-order method, for example, the characteristic method.

An alternative to the advection of the tangential vorticity vector is constituted by the advection of the velocity components v' and w'. Thus, Eq.(5.52) is replaced by

$$\overline{D}_t v' = 0, \quad \overline{D}_t w' = 0 \quad \text{at} \quad x = x_\infty. \qquad (5.53)$$

The use of the set of boundary conditions (5.50) and (5.53) in the computations of the wake flows mentioned above led to oscillations in the vorticity field close to the outlet. On the other hand, such conditions are applied with success for two-dimensional boundary layer flows (Marquillie and Ehrenstein, 2002). The solution is obtained by a mixed finite-difference/Chebyshev method where the streamwise derivatives are approximated with fourth-order Hermitian formulas.

A completely different way to handle the problem is to consider a buffer domain located in the downstream region in which the equations are modified in order to reduce and, if possible, destroy their elliptic character responsible of the upstream influence. The approach proposed by Streett and Macaraeg (1989/90) is closely connected to the numerical method they used, namely, the projection method (see Section 7.4.2) in which the pressure is determined by a Poisson equation with homogeneous Neumann boundary conditions. As above, the decomposition (5.49) is introduced into the equations. Thus, the essence of the method is to reduce smoothly to zero the streamwise viscous terms and the nonlinear disturbance terms resulting from the decomposition (5.49), using a hyperbolic tangent as an attenuation function. Thus, the velocity calculated in the first step of the projection method is a solution of a parabolic equation which does not require boundary conditions at the downstream boundary of the buffer domain. Additionally, the upstream influence due to the ellipticity of the pressure field is reduced by progressively setting to zero the forcing term in the pressure equation. More details on the method are given by Streett and Macaraeg (1989/90) and by Joslin et al. (1992).

6

Vorticity-Streamfunction Equations

The vorticity-streamfunction equations present some advantages over the velocity-pressure equations in the case of two-dimensional flows in simply connected domains. These advantages are well known : (1) the velocity field is automatically divergence-free, (2) the mathematical properties of the equations permit the construction of simple and robust solution methods, (3) computing time is saved because of the smaller number of equations. In the present chapter, Fourier and Chebyshev methods for the solution of the vorticity-streamfunction equations are discussed. The classical difficulty associated with this formulation is the lack of boundary conditions for the vorticity at a no-slip wall, whereas the streamfunction and its normal derivative are prescribed. This difficulty is surmounted thanks to the influence matrix method which will be described in the following. The stability of the Chebyshev approximation and the stability of the time-discretization schemes will be discussed. Finally, examples of applications (Rayleigh-Bénard convection and flow in a rotating annulus) will be presented.

6.1 Introduction

In this chapter we are interested in the solution of the Navier-Stokes equations within the vorticity-streamfunction formulation. These equations (see Chapter 5) are

$$\partial_t \omega + B\left(\mathbf{V}, \omega\right) - \nu \nabla^2 \omega = F \qquad (6.1)$$

$$\nabla^2 \psi + \omega = 0, \qquad (6.2)$$

where $\mathbf{V} = (u, v)$ is the velocity vector connected to the vorticity ω by

$$\omega = \partial_x v - \partial_y u \qquad (6.3)$$

and to the streamfunction ψ by

$$u = \partial_y \psi, \quad v = -\partial_x \psi. \qquad (6.4)$$

The initial conditions associated with Eq.(6.1)-(6.2) are

$$\mathbf{V} = \mathbf{V}_0 = (u_0, v_0) \qquad (6.5)$$

thus,

$$\omega = \omega_0 = \partial_x v_0 - \partial_y u_0. \qquad (6.6)$$

Typical boundary conditions have been discussed in Section 5.2. Various configurations will be addressed in the following : (1) fully periodic flow, (2) flow in a channel with one periodic direction, and (3) flow in a square domain with various boundary conditions.

6.2   Fourier-Fourier method

Let us assume that the forcing term F and the initial velocity field $\mathbf{V}_0$ are periodic with a period equal to 2π in both the x- and y-directions. The solution (ω, ψ) is 2π-periodic and is sought in a truncated Fourier series of the form

$$\phi_K(x, y, t) = \sum_{\mathbf{k}} \hat{\phi}_{\mathbf{k}}(t) \, e^{i(k_1 x + k_2 y)} \qquad (6.7)$$

with $\phi = \omega, \psi$. In Eq.(6.7), the sum is taken on k_1 and k_2, components of the wavenumber vector $\mathbf{k}$, such that $-K_j \leq k_j \leq K_j$, $j = 1, 2$.

Now Eq.(6.1) is discretized with respect to time with the semi-implicit scheme AB/BDI2 (see Section 4.4), but obviously other schemes can be used, in particular, the "semi-exact" scheme proposed by Basdevant (1982) and described in Section 7.3. The elliptic equation (6.2) is evaluated at time $(n + 1)\Delta t$. Let $\hat{\omega}_{\mathbf{k}}^n$ and $\hat{\psi}_{\mathbf{k}}^n$ be the, respective, approximations to $\hat{\omega}_{\mathbf{k}}$ and $\hat{\psi}_{\mathbf{k}}$ at time $n\,\Delta t$ where the notation $\hat{\phi}_{\mathbf{k}}$ means $\hat{\phi}_{k_1, k_2}$. Then the discrete version of Eqs.(6.1) and (6.2) is

$$\frac{3\hat{\omega}_{\mathbf{k}}^{n+1} - 4\hat{\omega}_{\mathbf{k}}^n + \hat{\omega}_{\mathbf{k}}^{n-1}}{2\,\Delta t} + \left(2\,\hat{B}_{\mathbf{k}}^n - \hat{B}_{\mathbf{k}}^{n-1}\right) + \nu\,k^2\,\hat{\omega}_{\mathbf{k}}^{n+1} = \hat{F}_{\mathbf{k}}^{n+1} \qquad (6.8)$$

$$-k^2\,\hat{\psi}_{\mathbf{k}}^{n+1} + \hat{\omega}_{\mathbf{k}}^{n+1} = 0, \qquad (6.9)$$

where $k^2 = |\mathbf{k}|^2 = k_1^2 + k_2^2$ and $\hat{B}_{\mathbf{k}}^n$ is the Fourier coefficient of the non-linear term $B(\mathbf{V}, \omega)$ calculated by means of the pseudospectral technique presented in Section 2.8. Note that the Fourier coefficients $\left(\hat{u}_k^{n+1}, \hat{v}_k^{n+1}\right)$ of the velocity (u, v) are determined by

$$\hat{u}_\mathbf{k}^{n+1} = \underline{i}\, k_2\, \hat{\psi}_\mathbf{k}^{n+1}\,, \quad \hat{v}_\mathbf{k}^{n+1} = -\underline{i}\, k_1\, \hat{\psi}_\mathbf{k}^{n+1}\,. \tag{6.10}$$

The initial conditions (6.5) and (6.6) are approximated according to

$$\hat{\mathbf{V}}_\mathbf{k}^0 = \hat{\mathbf{V}}_{0\mathbf{k}}\,, \quad \hat{\omega}_\mathbf{k}^0 = \hat{\omega}_{0\mathbf{k}}\,. \tag{6.11}$$

Starting procedures for the three-level scheme (6.8) are described in Section 4.5.1.d.

Equation (6.8) gives explicitly $\hat{\omega}_\mathbf{k}^{n+1}$ for every $\mathbf{k}$, in particular $\hat{\omega}_0^{n+1} = 0$ since $\hat{B}_0(t) = 0$ from its definition (5.17) [or, equivalently, (5.16) since $\nabla.\mathbf{V} = 0$] and $\hat{F}_0(t) = 0$ from its definition (5.18). Then, Eq.(6.9) gives explicitly $\hat{\psi}_\mathbf{k}^{n+1}$ for every $\mathbf{k}$ except $\mathbf{k} = \mathbf{0}$ (i.e., $k^2 = 0$). This is in accordance with the fact that the streamfunction ψ is defined up to a function of the time. We may simply set $\hat{\psi}_0^{n+1} = 0$, namely, the mean value of ψ is taken equal to zero.

6.3   Fourier-Chebyshev method

6.3.1   Flow in a plane channel

We assume that the flow takes place in an infinite channel in the x-direction and is 2π-periodic in this direction. The domain Ω in which the equations (6.1)-(6.2) are solved is defined by $0 \le x \le 2\pi$, $-1 \le y \le 1$. The forcing term F and the initial velocity field are assumed to be 2π-periodic in the x-direction.

At the boundaries $y \pm 1$, the no-slip condition $\mathbf{V} = 0$ is prescribed. As discussed in Section 5.2, this condition leads to

$$\psi(x, -1, t) = 0\,, \quad \psi(x, 1, t) = g_+(t) \tag{6.12}$$

$$\partial_y \psi(x, -1, t) = 0\,, \quad \partial_y \psi(x, 1, t) = 0\,. \tag{6.13}$$

The quantity $g_+(t)$ is the flow rate through the channel section. In some situations this flow rate is zero, for example, if the flow is assumed to take place in an elongated, but finite, tank with no inflow and outflow at the lateral boundaries. The periodic approximation is then supposed to be valid far from the lateral walls. In this case, we have $g_+(t) = 0$. The possibility of having a nonzero flow rate is discussed in Section 6.2.4.

Another type of boundary is the so-called "shear-stress-free boundary". It constitutes a model for a nondeformed free surface or simply a slippery boundary. Assuming the boundary $y = 1$ to be of this type, the associated conditions are

$$\partial_y u(x, 1, t) = 0\,, \quad v(x, 1, t) = 0\,. \tag{6.14}$$

As a result, the second condition in Eq.(6.13) has to be replaced by

$$\omega(x, 1, t) = 0.\tag{6.15}$$

6.3.2   Time-dependent Stokes equations

(a) Semi-discrete problem
Let us consider the unsteady Stokes problem

$$\partial_t \omega - \nu \nabla^2 \omega = F \quad \text{in } \Omega\tag{6.16}$$

$$\nabla^2 \psi + \omega = 0 \quad\quad \text{in } \Omega\tag{6.17}$$

$$\psi = g_-, \quad \partial_y \psi = h_- \quad \text{at } y = -1\tag{6.18}$$

$$\psi = g_+, \quad \partial_y \psi = h_+ \quad \text{at } y = 1\tag{6.19}$$

$$\omega = \omega_0 \quad \text{at } t = 0,\tag{6.20}$$

where we assume that $g_\pm$ and $h_\pm$ are given functions of x and t. These functions, as well as F and ω_0, are assumed to be 2π-periodic with respect to x.

First, we address the stability of the Fourier-Chebyshev approximation to this problem. Let us introduce the truncated Fourier expansions

$$\omega_K(x, y, t) = \sum_{k=-K}^{K} \hat{\omega}_k(y, t) e^{ikx},$$

$$\psi_K(x, y, t) = \sum_{k=-K}^{K} \hat{\psi}_k(y, t) e^{ikx}.$$

The Fourier Galerkin equations satisfied by $\left(\hat{\omega}_k, \hat{\psi}_k\right)$, $k = -K, \ldots, K$ (see Section 2.6.1) are

$$\partial_t \hat{\omega}_k - \nu \left(\partial_{yy} - k^2\right) \hat{\omega}_k = \hat{F}_k \quad \text{in } -1 < y < 1\tag{6.21}$$

$$\left(\partial_{yy} - k^2\right) \hat{\psi}_k + \hat{\omega}_k = 0 \quad \text{in } -1 < y < 1\tag{6.22}$$

$$\hat{\psi}_k = \hat{g}_{-,k}, \quad \partial_y \hat{\psi}_k = \hat{h}_{-,k} \quad \text{at } y = -1\tag{6.23}$$

$$\hat{\psi}_k = \hat{g}_{+,k}, \quad \partial_y \hat{\psi}_k = \hat{h}_{+,k} \quad \text{at } y = 1\tag{6.24}$$

$$\hat{\omega}_k = \hat{\omega}_{0,k},\tag{6.25}$$

where $\hat{F}_k$, $\hat{g}_{\pm,k}$, $\hat{h}_{\pm,k}$, and $\hat{\omega}_{0,k}$ are the Fourier coefficients, respectively, of the functions $F(x, y, t)$, $g_\pm(x, t)$, $h_\pm(x, t)$, and $\omega_0(x, y)$.

Then this one-dimensional problem in the y-direction is solved with the Chebyshev collocation method (see Section 3.4.2) based on the collocation points

$$y_j = \cos \frac{\pi j}{N}, \quad j = 0, \ldots, N. \tag{6.26}$$

Let $\hat{\omega}_{kN}$ and $\hat{\psi}_{kN}$ be the polynomial approximations, respectively, of $\hat{\omega}_k$ and $\hat{\psi}_k$. The collocation equations are

$$d_t \hat{\omega}_{kN} (y_j, t) - \nu \left[\sum_{l=0}^{N} d_{j,l}^{(2)} \hat{\omega}_{kN} (y_l, t) - k^2 \hat{\omega}_{kN} (y_j, t) \right]$$
$$= \hat{F}_k (y_j, t), \quad j = 1, \cdots, N-1$$

$$\sum_{l=0}^{N} d_{j,l}^{(2)} \hat{\psi}_{kN} (y_l, t) - k^2 \hat{\psi}_{kN} (y_j, t) + \hat{\omega}_{kN} (y_j, t) = 0,$$
$$j = 1, \cdots, N-1$$

$$\hat{\psi}_{kN} (y_0, t) = \hat{g}_{+,k} (t), \tag{6.27}$$

$$\hat{\psi}_{kN} (y_N, t) = \hat{g}_{-,k} (t),$$

$$\sum_{l=0}^{N} d_{0,l}^{(1)} \hat{\psi}_{kN} (y_l, t) = \hat{h}_{+,k} (t),$$

$$\sum_{l=0}^{N} d_{N,l}^{(1)} \hat{\psi}_{kN} (y_l, t) = \hat{h}_{-,k} (t),$$

supplemented with the initial conditions

$$\hat{\omega}_{kN} (y_j, 0) = \hat{\omega}_{0,k} (y_j, t), \quad j = 1, \ldots, N. \tag{6.28}$$

To analyze the stability of this differential problem, we pose $\hat{F}_k = \hat{g}_{\pm,k} = \hat{h}_{\pm,k} = 0$ and we look for a solution of the form

$$\hat{\omega}_{kN} (y_j, t) = e^{\lambda t} \hat{\omega}_{kN}^{\star} (y_j), \quad \hat{\psi}_{kN} (y_j, t) = e^{\lambda t} \hat{\psi}_{kN}^{\star} (y_j). \tag{6.29}$$

This leads to an eigenvalue problem whose solution $\Lambda = (\lambda_1, \ldots, \lambda_{N-1})^T$ characterizes the time-behaviour of the discrete solution $(\hat{\omega}_{kN}, \hat{\psi}_{kN})$. It has been proven (Ehrenstein, 1990) that the eigenvalues $\lambda_j, j = 1, \ldots, N-1$, are real and negative. Therefore, the solution (6.29) is damped as $t \to \infty$, that is, the Fourier-Chebyshev approximation to the Stokes problem within the vorticity-streamfunction is stable. It is interesting to compare this result with the one given by Gottlieb and Orszag (1977) which states that the Chebyshev tau approximation of the fourth-order equation for the streamfunction induces complex eigenvalues with positive real part, and therefore yields an unstable time-dependent problem.

(b) Time-discretization scheme

Among the second-order time-schemes currently used in the field of spectral methods applied to the Navier-Stokes equations are the semi-implicit Adams-Bashforth/Crank-Nicolson [AB/CN, Eq.(4.51)] and the Adams-Bashforth/Backward-Differentiation [AB/BDI2, Eq.(4.52)] scheme. In Section 4.4.2.a, the stability properties of each of these schemes have been compared for the advection-diffusion equation. Except for small values of the viscosity, the AB/CN scheme was found to require a much smaller time-step than the AB/BDI2 scheme. As a matter of fact, the AB/CN scheme was found to be unstable when applied to the Navier-Stokes equations approximated with the Chebyshev tau method, so that the AB/BDI2 scheme was preferred by Vanel *et al.* (1986). This choice can be justified by analyzing the stability of a more general class of three-level schemes applied to the unsteady Stokes equations. This class is a subset of schemes (4.46) where $\theta_1 = \theta = 1 - \theta_2$ (and $a = 0$).

To simplify the notations, we set $\omega_j^n = \hat{\omega}_{kN}^n(y_j)$ and $\psi_j^n = \hat{\psi}_{kN}^n(y_j)$ for approximation at time $n\,\Delta t$. The same type of simplification is done for the given quantities $\hat{F}_{kN}^n(y_j)$, $\hat{g}_{\pm,k}^n$, and $\hat{h}_{\pm,k}^n$. Thus, the time-discretization of Eqs.(6.16)-(6.19) yields

$$\frac{(1+\varepsilon)\,\omega_j^{n+1} - 2\,\varepsilon\,\omega_j^n - (1-\varepsilon)\,\omega_j^{n-1}}{2\,\Delta t}$$

$$-\nu\sum_{l=0}^{N} d_{j,l}^{(2)}\left[\theta\,\omega_l^{n+1} + (1-\theta)\,\omega_l^n\right] + \nu\,k^2\left[\theta\,\omega_j^{n+1} + (1-\theta)\,\omega_j^n\right]$$

$$= \theta\,F_j^{n+1} + (1-\theta)\,F_j^n\,; \quad j = 1,\ldots,N-1,$$

$$\sum_{l=0}^{N} d_{j,l}^{(2)}\,\psi_l^{n+1} - k^2\,\psi_j^{n+1} + \omega_j^{n+1} = 0, \qquad j = 1,\ldots,N-1,$$

$$\psi_0^{n+1} = g_+^{n+1}\,,$$
$$\psi_N^{n+1} = g_-^{n+1}\,,$$

$$\sum_{l=0}^{N} d_{0,l}^{(1)}\,\psi_l^{n+1} = h_+^{n+1}\,,$$

$$\sum_{l=0}^{N} d_{N,l}^{(1)}\,\psi_l^{n+1} = h_-^{n+1}\,.$$

$$(6.30)$$

It may be observed that such a system furnishes $2(N+1)$ equations for the $2(N+1)$ unknowns $(\omega_j^{n+1}, \psi_j^{n+1})$, $j = 0,\ldots,N$. The solvability of the system (6.30) is addressed in the following section. It is convenient to write the system in vector form

$$\mathcal{A}_1\,\Phi^{n+1} + \mathcal{A}_0\,\Phi^n + \mathcal{A}_{-1}\,\Phi^{n-1} = H^{n+1}\,, \qquad (6.31)$$

where Φ^n is the vector of unknowns defined by

$$\Phi^n = \begin{pmatrix} \Omega^n \\ \Psi^n \end{pmatrix} , \quad \Omega^n = \begin{pmatrix} \omega_0^n \\ \vdots \\ \omega_N^n \end{pmatrix} , \quad \Psi^n = \begin{pmatrix} \psi_0^n \\ \vdots \\ \psi_N^n \end{pmatrix} , \tag{6.32}$$

and $\mathcal{A}_1$, $\mathcal{A}_0$ and $\mathcal{A}_{-1}$ are $(N+1) \times (N+1)$ matrices whose expression is now given.

To do this, we introduce rectangular matrices of various order whose elements are zero:

$\mathcal{O}_1 = (N-1) \times (N+1)$ matrix,
$\mathcal{O}_2 = 2 \times (N+1)$ matrix,
$\mathcal{O}_3 = (N+3) \times 2(N+1)$ matrix,

the $(N-1) \times (N+1)$ matrix $\mathcal{I}_1$ is such that

$$\mathcal{I}_1 = [\delta_{i,j-1}] , \quad i = 0, \cdots, N-2, \quad j = 0, \cdots, N , \tag{6.33}$$

where $\delta_{i,j}$ is the Kronecker delta. We also introduce the $2 \times (N+1)$ matrix $\mathcal{I}_2$ whose elements are zero except the extreme elements on the main diagonal

$$\mathcal{I}_2 = \begin{pmatrix} 1 \, 0 \cdots 0 \\ 0 \cdots 0 \, 1 \end{pmatrix} . \tag{6.34}$$

Now we define the $2 \times (N+1)$ matrix $\mathcal{D}_3$ such that

$$\mathcal{D}_3 = \left[d_{i,j}^{(1)} \right] , \quad i = 0, N, \quad j = 0, \cdots, N , \tag{6.35}$$

and the $(N-1) \times (N+1)$ matrix $\mathcal{D}_4$ such that

$$\mathcal{D}_4 = \left[d_{i,j}^{(2)} \right] , i = 1, \cdots, N-1, \quad j = 0, \cdots, N , \tag{6.36}$$

where $d_{i,j}^{(1)}$ and $d_{i,j}^{(2)}$ are, respectively, the coefficients of the first- and second-order differentiation matrices. Also, we construct the following $(N-1) \times (N+1)$ matrices

$$\mathcal{B}_1 = \theta \, \mathcal{D}_4 - \sigma_1 \, \mathcal{I}_1$$
$$\mathcal{B}_2 = \mathcal{D}_4 - k^2 \, \mathcal{I}_1$$
$$\mathcal{B}_3 = (1 - \theta_0) \, \mathcal{D}_4 - \sigma_0 \, \mathcal{I}_1$$
$$\mathcal{B}_4 = -\sigma_{-1} \mathcal{I}_1 ,$$

where

$$\sigma_1 = \theta \, k^2 + \frac{1+\varepsilon}{2 \nu \, \Delta t} , \quad \sigma_0 = (1 - \theta) \, k^2 - \frac{\varepsilon}{\nu \, \Delta t} , \quad \sigma_{-1} = -\frac{1-\varepsilon}{2 \nu \, \Delta t} .$$

Now the $(N+1) \times (N+1)$ matrices $\mathcal{A}_1$, $\mathcal{A}_0$, and $\mathcal{A}_{-1}$ have, respectively, the following expression

$$\mathcal{A}_1 = \begin{pmatrix} \mathcal{B}_1 & \mathcal{O}_1 \\ \mathcal{I}_1 & \mathcal{B}_2 \\ \mathcal{O}_2 & \mathcal{I}_2 \\ \mathcal{O}_2 & \mathcal{D}_3 \end{pmatrix} , \quad \mathcal{A}_0 = \begin{pmatrix} \mathcal{B}_3 & \mathcal{O}_1 \\ \mathcal{O}_3 & \mathcal{O}_3 \end{pmatrix} , \quad \mathcal{A}_{-1} = \begin{pmatrix} \mathcal{B}_4 & \mathcal{O}_1 \\ \mathcal{O}_3 & \mathcal{O}_3 \end{pmatrix} ,$$

and the vector H^{n+1} is defined by

$$H^{n+1} = \left(\varphi_1^{n+1}, \cdots, \varphi_{N-1}^{n+1}, 0, \cdots, 0, g_+^{n+1}, g_-^{n+1}, h_+^{n+1}, h_-^{n+1}\right)^T,$$

(6.37)

where $\varphi_j^{n+1} = -\left[\theta F_j^{n+1} + (1-\theta) F_j^n\right]/\nu$.

c) Solution of the algebraic system

The system (6.31), which defines the solution Φ^{n+1}, is written in the form

$$\mathcal{A}_1 \Phi^{n+1} = S,$$

(6.38)

where

$$S = H^{n+1} - \mathcal{A}_0 \Phi^n - \mathcal{A}_{-1} \Phi^{n-1}.$$

(6.39)

The solution of this large system is obtained through an algorithm which is now described. The basic principle of the algorithm is to reduce the solution of the large system, of dimensions $2(N+1) \times 2(N+1)$, to the solution of six smaller systems whose dimensions are $(N+1) \times (N+1)$. Moreover, four of these smaller systems are time-independent and can be solved once and for all in a preprocessing stage performed previously to start the time-integration.

The algorithm is the discrete equivalent to the influence matrix technique usually presented on the continuous problem and which will be discussed in Section 6.3.2.d. The solution of (6.38) is sought in the form

$$\Phi^{n+1} = \tilde{\Phi}^{n+1} + \xi_1^{n+1} \overline{\Phi}_1 + \xi_2^{n+1} \overline{\Phi}_2,$$

(6.40)

where ξ_1^{n+1} and ξ_2^{n+1} are scalars and $\tilde{\Phi}^{n+1}$, $\overline{\Phi}_1$, and $\overline{\Phi}_2$ are vectors, respectively, solutions to

$$\hat{\mathcal{A}} \tilde{\Phi}^{n+1} = \tilde{S}$$

(6.41)

$$\hat{\mathcal{A}} \overline{\Phi}_1 = \overline{S}_1$$

(6.42)

$$\hat{\mathcal{A}} \overline{\Phi}_2 = \overline{S}_2,$$

(6.43)

where $\hat{\mathcal{A}}$ is the matrix constructed from $\mathcal{A}_1$ such that

$$\hat{\mathcal{A}} = \begin{pmatrix} \mathcal{B}_1 & \mathcal{O}_1 \\ \mathcal{I}_1 & \mathcal{B}_2 \\ \mathcal{O}_2 & \mathcal{I}_2 \\ \mathcal{I}_2 & \mathcal{O}_2 \end{pmatrix}.$$

(6.44)

Therefore, $\hat{\mathcal{A}}$ is the matrix which would result from the discrete problem (6.30) if the Neumann conditions on ψ^{n+1} (last two equations) were replaced by Dirichlet conditions on ω^{n+1}.

The vector $\tilde{S}$ is deduced from S by replacing in H^{n+1} the last two components h_+^{n+1} and h_-^{n+1} with zero, namely,

$$\tilde{S} = H_0^{n+1} - \mathcal{A}_0 \Phi^n - \mathcal{A}_{-1} \Phi^{n-1}$$

with

$$H_0^{n+1} = \left(\varphi_1^{n+1}, \; \cdots, \; \varphi_{N-1}^{n+1}, \; 0, \; \cdots, \; 0, \; g_+^{n+1}, \; g_-^{n+1}, \; 0, \; 0 \right)^T .$$

The vectors $\overline{S}_1$ and $\overline{S}_2$ are respectively defined by

$$\overline{S}_1 = (0, \; \cdots, \; 0, \; 0, \; 1)^T , \quad \overline{S}_2 = (0, \; \cdots, \; 0, \; 1, \; 0)^T .$$

Therefore, Φ^{n+1}, as defined by (6.40), is solution of

$$\hat{\mathcal{A}} \, \Phi^{n+1} = S_\star \tag{6.45}$$

where

$$S_\star = H_\star - \mathcal{A}_0 \, \Phi^n - \mathcal{A}_{-1} \, \Phi^{n-1}$$

and

$$H_\star = \left(\varphi_1^{n+1}, \; \cdots, \; \varphi_{N-1}^{n+1}, \; 0, \; \cdots, \; 0, \; g_+^{n+1}, \; g_-^{n+1}, \; \xi_2^{n+1}, \; \xi_1^{n+1} \right)^T .$$

It is easy to check that the first $2N$ equations of system (6.45) coincide with the first $2N$ equations of system (6.30). Therefore, the vector Φ^{n+1} defined by (6.45) satisfies these first $2N$ equations whatever the constants ξ_1^{n+1} and ξ_2^{n+1}. It remains to determine these constants in order to ensure that the last two equations of system (6.30) are satisfied, that is the Neumann conditions on the streamfunction. We get

$$\mathcal{C} \left(\tilde{\Phi}^{n+1} + \xi_1^{n+1} \overline{\Phi}_1 + \xi_2^{n+1} \overline{\Phi}_2 \right) = E , \tag{6.46}$$

where the $2 \times 2(N+1)$ matrix $\mathcal{C}$ and the vector E are given by

$$\mathcal{C} = [\mathcal{O}_2 \;\; \mathcal{D}_1] , \quad E = (h_+ , \; h_-)^T .$$

The vectors $\tilde{\Phi}^{n+1}$, $\overline{\Phi}_1$ and $\overline{\Phi}_2$ are defined by

$$\tilde{\Phi}^{n+1} = \begin{pmatrix} \tilde{\Omega}^{n+1} \\ \tilde{\Psi}^{n+1} \end{pmatrix} , \quad \overline{\Phi}_l = \begin{pmatrix} \overline{\Omega}_l \\ \overline{\Psi}_l \end{pmatrix} , \quad l = 1, 2 ,$$

with

$$\tilde{\Omega}^{n+1} = \begin{pmatrix} \tilde{\omega}_0^{n+1} \\ \vdots \\ \tilde{\omega}_N^{n+1} \end{pmatrix} , \quad \tilde{\Psi}^{n+1} = \begin{pmatrix} \tilde{\psi}_0^{n+1} \\ \vdots \\ \tilde{\psi}_N^{n+1} \end{pmatrix} ,$$

and

$$\overline{\Omega}_l = \begin{pmatrix} \overline{\omega}_{l,0} \\ \vdots \\ \overline{\omega}_{l,N} \end{pmatrix} , \quad \overline{\Psi}_l = \begin{pmatrix} \overline{\psi}_{l,0} \\ \vdots \\ \overline{\psi}_{l,N} \end{pmatrix} , \quad l = 1, 2 .$$

Due to the special structure of $\mathcal{C}$, the system (6.46) is written as

$$\xi_1^{n+1} \mathcal{D}_3 \overline{\Psi}_1 + \xi_2^{n+1} \mathcal{D}_3 \overline{\Psi}_2 = \tilde{E},$$

namely,

$$\mathcal{M} \, \Xi^{n+1} = \tilde{E}, \tag{6.47}$$

where

$$\tilde{E} = E - \mathcal{D}_3 \, \tilde{\Psi}^{n+1} \tag{6.48}$$

and $\Xi^{n+1} = \left(\xi_1^{n+1}, \, \xi_2^{n+1}\right)^T$. The matrix $\mathcal{M}$ is defined by

$$\mathcal{M} = \begin{bmatrix} \mathcal{D}_3 \overline{\Psi}_1 & \mathcal{D}_3 \overline{\Psi}_2 \end{bmatrix} = [m_{i,j}], \quad i = 0, N, \quad j = 1, 2,$$

such that

$$m_{i,j} = \sum_{l=0}^{N} d_{i,l}^{(1)} \, \overline{\psi}_{j,l} \,. \tag{6.49}$$

The solution of the system (6.47) furnishes the constants ξ_1^{n+1} and ξ_2^{n+1}:

$$\Xi^{n+1} = \mathcal{M}^{-1} \tilde{E} \tag{6.50}$$

and, hence, the solution Φ^{n+1} is completely determined since $\tilde{\Phi}^{n+1}$, $\overline{\Phi}_1$, and $\overline{\Phi}_2$ are known from Eqs.(6.41)-(6.43). However, these large systems can be split into smaller ones by carefully inspecting their structure. Thus, the system (6.41) is decomposed into the two following systems. The first system determines $\tilde{\Omega}^{n+1}$:

$$\hat{\mathcal{A}}_1 \, \tilde{\Omega}^{n+1} = \tilde{S}_1, \tag{6.51}$$

where the vector $\tilde{S}_1$ is defined by

$$\tilde{S}_1 = H_1^{n+1} - \hat{\mathcal{A}}_0 \, \Omega^n - \hat{\mathcal{A}}_{-1} \, \Omega^{n-1}$$

with

$$H_1^{n+1} = \left(\varphi_1^{n+1}, \, \cdots, \, \varphi_{N-1}^{n+1}, \, 0, \, 0\right)^T,$$

and the matrices $\hat{\mathcal{A}}_1$, $\hat{\mathcal{A}}_0$, and $\hat{\mathcal{A}}_{-1}$ are

$$\hat{\mathcal{A}}_1 = \begin{pmatrix} \mathcal{B}_1 \\ \mathcal{I}_2 \end{pmatrix}, \quad \hat{\mathcal{A}}_0 = \begin{pmatrix} \mathcal{B}_3 \\ \mathcal{I}_2 \end{pmatrix}, \quad \hat{\mathcal{A}}_{-1} = \begin{pmatrix} \mathcal{B}_4 \\ \mathcal{I}_2 \end{pmatrix}.$$

Then, $\tilde{\Omega}^{n+1}$ having been calculated, the second system determines $\tilde{\Psi}^{n+1}$:

$$\hat{\mathcal{A}}_2 \, \tilde{\Psi}^{n+1} = \tilde{S}_2 - \mathcal{I}_1 \, \tilde{\Omega}^{n+1}, \tag{6.52}$$

where

$$\hat{\mathcal{A}}_2 = \begin{pmatrix} \mathcal{B}_2 \\ \mathcal{I}_2 \end{pmatrix}, \quad \tilde{S}_2 = (0, \, \cdots, \, 0, \, g_+, \, g_-)^T.$$

The dimensions of both systems are $(N+1) \times (N+1)$. In the same way, the system (6.42) is split into the system determining $\overline{\Omega}_1$:

$$\hat{\mathcal{A}}_1 \overline{\Omega}_1 = \overline{S}_{1,1} , \qquad (6.53)$$

where $\overline{S}_{1,1} = (0, \cdots, 0, 1)^T$, and the system determining $\overline{\Psi}_1$:

$$\hat{\mathcal{A}}_2 \overline{\Psi}_1 = -\mathcal{I}_1 \overline{\Omega}_1 . \qquad (6.54)$$

At last, we obtain for system (6.43) the same decomposition, namely,

$$\hat{\mathcal{A}}_1 \overline{\Omega}_2 = \overline{S}_{2,1} , \qquad (6.55)$$

where $\overline{S}_{2,1} = (0, \cdots, 0, 1, 0)^T$, and

$$\hat{\mathcal{A}}_2 \overline{\Psi}_2 = -\mathcal{I}_1 \overline{\Omega}_2 . \qquad (6.56)$$

The matrices $\hat{\mathcal{A}}_l$, $l = 1, 2$, and $\mathcal{M}$ are inverted once and for all before the start of the time-integration, and their inverses $\hat{\mathcal{A}}_l^{-1}$ and $\mathcal{M}^{-1}$ are stored. In the same way, the vectors $\overline{\Omega}_l$, $\overline{\Psi}_l$, $l = 1, 2$, are calculated in the preprocessing stage and stored. So that, at each time-cycle, the algorithm reduces to :

1. Calculate $\tilde{\Omega}^{n+1}$ and $\tilde{\Psi}^{n+1}$ by

$$\tilde{\Omega}^{n+1} = \hat{\mathcal{A}}_1^{-1} \tilde{S}_1 , \quad \tilde{\Psi}^{n+1} = \hat{\mathcal{A}}_1^{-1} \left(\tilde{S}_2 - \mathcal{I}_1 \tilde{\Omega}^{n+1} \right) .$$

2. Calculate $\tilde{E}$ by Eq.(6.48).
3. Calculate Ξ^{n+1} by Eq.(6.50).
4. Calculate Ω^{n+1} and Ψ^{n+1} by Eq.(6.40).

Therefore, the computational effort to be done, at each time-cycle and for each Fourier mode, amounts to matrix-vector products of moderate size.

Now, thanks to the splitting of the large system (6.38) into a set of smaller systems solved successively, we are in a position to state the resolvability of the system (6.38). Thus, the six systems (6.51)-(6.56) represent the Chebyshev collocation approximation to Dirichlet problems for Poisson or Helmholtz equations. The eigenvalues of the associated matrices are negative (see Section 3.7.1) and, hence, the systems are solvable.

It remains to prove that the matrix $\mathcal{M}$ is invertible. For this we consider the system satisfied by

$$\overline{\Phi}^{n+1} = \xi_1^{n+1} \overline{\Phi}_1 + \xi_2^{n+1} \overline{\Phi}_2 , \qquad (6.57)$$

namely,

$$\mathcal{A}_1 \overline{\Phi}^{n+1} = S_0$$

with

$$S_0 = \left(0, \cdots, 0, h_+^{n+1}, h_-^{n+1}\right)^T.$$

Now let us assume $\mathcal{M}$ is not invertible, that is, $\det \mathcal{M} = 0$. This means that the homogeneous problem

$$\mathcal{A}_1 \overline{\Phi}^{n+1} = 0 \tag{6.58}$$

admits nonzero solutions. This system (6.58) can be formally identified with the algebraic system defining the eigenvalues λ of the Stokes equations considered in Section 6.2.2.a. The identification is obtained by setting

$$\frac{1+\varepsilon}{2\theta \nu \Delta t} = \lambda.$$

Therefore, if $\det \mathcal{M} = 0$, the quantity $\sigma = (1 + \varepsilon)/(2\theta \nu \Delta t)$ is an eigenvalue of the Stokes problem. But this is impossible since $\sigma > 0$ while the eigenvalues λ are negative. The conclusion is that $\det \mathcal{M} \neq 0$ and the matrix $\mathcal{M}$ is invertible.

To conclude, it must be remarked that this algorithm for solving the full discrete system can be obtained from the discrete version of an algorithm devised for solving the set of differential equations defining the continuous system. This algorithm, usually known as the denomination influence (or capacitance) matrix method, is now described.

(d) Influence matrix method

The solution algorithm described in the above section can also be derived from the discretization of a set of differential equations involved in the solution method for the Stokes problem. This method, usually called the "influence matrix method," was presented in Section 3.7.2 in the purpose of replacing the solution of a problem with arbitrary boundary conditions by one of pure Dirichlet problems.

In a general way, this method is of interest for solving linear differential problems in which the boundary conditions refer only to one of the dependent variables. In the present case, we have two boundary conditions for the streamfunction and none for the vorticity. Another example is given by the solution of the Navier-Stokes equations in primitive variables, based on a Poisson equation for the pressure associated with the requirement that the divergence of the velocity is zero at the boundary. Therefore, we have two boundary conditions for the velocity vector (i.e., its value and its divergence) and no condition for the pressure. The influence matrix method introduced for this problem by Kleiser and Schumann (1980) will be addressed in Chapter 7.

The influence matrix method applies to linear problems and is based on the superposition of particular solutions. Generally, the coefficients of the linear combination refer to the boundary value of the missing variable (i.e., vorticity or pressure). For the vorticity-streamfunction equations in one-dimensional problems, the method has been introduced in the same time by Tuckerman (1983) and by Dennis and Quartapelle (1983). The extension to two-dimensional problems, which will be presented in next section, has been done by Vanel *et al.* (1986) and Ehrenstein and Peyret (1986, 1989). Note that the same basic idea was proposed by Glowinski and Pironneau (1979) (see also Dean *et al.*, 1991) in the context of finite-element approximations.

Let us consider the system (6.21)-(6.25) discretized with respect to time by the same three-level scheme as in Eqs.(6.30). In this way, the couple $\left(\hat{\omega}_k^{n+1}, \hat{\psi}_k^{n+1} \right)$ is the solution of a Stokes-type problem of the form

$$\omega'' - \sigma \omega = f, \quad -1 < y < 1 \tag{6.59}$$

$$\psi'' - k^2 \psi + \omega = 0, \quad -1 < y < 1 \tag{6.60}$$

$$\psi(-1) = g_-, \quad \psi(1) = g_+, \tag{6.61}$$

$$\psi'(-1) = h_-, \quad \psi'(1) = h_+, \tag{6.62}$$

where $\omega = \hat{\omega}_k^{n+1}$, $\psi = \hat{\psi}_k^{n+1}$, and $\sigma = k^2 + (1+\varepsilon)/(2\theta\nu\Delta t)$. Note that the classical Stokes problem corresponds to $\sigma = 0$. In the following, we use the denomination "Stokes problem" for the problem (6.59)-(6.62) even if $\sigma \neq 0$.

The solution (ω, ψ) is sought according to the decomposition

$$\omega = \tilde{\omega} + \overline{\omega}, \quad \psi = \tilde{\psi} + \overline{\psi}. \tag{6.63}$$

The pair $\left(\tilde{\omega}, \tilde{\psi} \right)$ satisfies the $\tilde{\mathcal{P}}$-*Problem* defined by

$$\tilde{\omega}'' - \sigma \tilde{\omega} = f, \quad -1 < y < 1 \tag{6.64}$$

$$\tilde{\omega}(-1) = 0, \quad \tilde{\omega}(1) = 0, \tag{6.65}$$

and

$$\tilde{\psi}'' - k^2 \tilde{\psi} = -\tilde{\omega} \quad -1 < y < 1 \tag{6.66}$$

$$\tilde{\psi}(-1) = g_-, \quad \tilde{\psi}(1) = g_+. \tag{6.67}$$

The problem (6.64)-(6.65) is first solved. It completely determines $\tilde{\omega}$ which is then used in (6.66)-(6.67) to calculate $\tilde{\psi}$.

Then the pair $(\overline{\omega}, \overline{\psi})$ is the solution of the $\overline{\mathcal{P}}$-*Problem* :

$$\overline{\omega}'' - \sigma \overline{\omega} = 0, \quad -1 < y < 1 \tag{6.68}$$

$$\overline{\psi}'' - k^2 \overline{\psi} + \overline{\omega} = 0, \quad -1 < y < 1 \tag{6.69}$$

$$\overline{\psi}(-1) = 0, \quad \overline{\psi}(1) = 0, \tag{6.70}$$

$$\overline{\psi}'(-1) = h_- - \tilde{\psi}'(-1), \quad \overline{\psi}'(1) = h_+ - \tilde{\psi}'(1). \tag{6.71}$$

Now this problem is transformed into a problem for $\bar{\omega}$ and $\bar{\psi}$ where the boundary values $\bar{\omega}(\pm 1)$ are determined such that the Neumann conditions on $\bar{\psi}$ are satisfied. This defines the $\overline{\mathcal{P}}_0$-*Problem* :

$$\bar{\omega}'' - \sigma\bar{\omega} = 0, \quad -1 < y < 1$$
$$\bar{\psi}'' - k^2\bar{\psi} + \bar{\omega} = 0, \quad -1 < y < 1$$
$$\bar{\omega}(-1) = \xi_1, \quad \bar{\omega}(1) = \xi_2,$$
$$\bar{\psi}(-1) = 0, \quad \bar{\psi}(1) = 0,$$

where ξ_1 and ξ_2 have to be determined such that conditions (6.71) are satisfied.

The $\overline{\mathcal{P}}_0$-problem is solved by introducing the linear combination

$$\bar{\omega} = \xi_1\bar{\omega}_1 + \xi_2\bar{\omega}_2, \quad \bar{\psi} = \xi_1\bar{\psi}_1 + \xi_2\bar{\psi}_2, \tag{6.72}$$

where the elementary solutions $(\bar{\omega}_l, \bar{\psi}_l)$, $l = 1, 2$, satisfy the $\overline{\mathcal{P}}_l$-problems, namely,
$\overline{\mathcal{P}}_1$-*Problem* :

$$\bar{\omega}_1'' - \sigma\bar{\omega}_1 = 0, \quad -1 < y < 1$$
$$\bar{\omega}_1(-1) = 1, \quad \bar{\omega}_1(1) = 0,$$

and

$$\bar{\psi}_1'' - k^2\bar{\psi}_1 = -\bar{\omega}_1, \quad -1 < y < 1$$
$$\bar{\psi}_1(-1) = 0, \quad \bar{\psi}_1(1) = 0.$$

$\overline{\mathcal{P}}_2$-*Problem* :

$$\bar{\omega}_2'' - \sigma\bar{\omega}_2 = 0, \quad -1 < y < 1$$
$$\bar{\omega}_2(-1) = 0, \quad \bar{\omega}_2(1) = 1,$$

and

$$\bar{\psi}_2'' - k^2\bar{\psi}_2 = -\bar{\omega}_2, \quad -1 < y < 1$$
$$\bar{\psi}_2(-1) = 0, \quad \bar{\psi}_2(1) = 1.$$

Finally, the constants ξ_1 and ξ_2 are determined by Eq.(6.71), namely by the algebraic system

$$\bar{\psi}_1'(-1)\,\xi_1 + \bar{\psi}_2'(-1)\,\xi_2 = h_- - \tilde{\psi}'(-1)$$
$$\bar{\psi}_1'(1)\,\xi_1 + \bar{\psi}_2'(1)\,\xi_2 = h_+ - \tilde{\psi}'(1),$$

which is written

$$\mathcal{M}\,\Xi = \tilde{E}, \tag{6.73}$$

where $\Xi = (\xi_1, \xi_2)^T$ and the matrix $\mathcal{M}$ is the influence (or capacitance) matrix.

Now the various Dirichlet problems $(\tilde{\mathcal{P}}, \overline{\mathcal{P}}_1, \text{ and } \overline{\mathcal{P}}_2)$ are solved using a Chebyshev (tau or collocation) method. In the case of Chebyshev collocation, it is easy to check the identity between the influence method developed in this section and the algorithm for solving the full discrete system (6.38). More precisely, the discretized $\tilde{\mathcal{P}}$-problem is identified with the systems (6.51) and (6.52) while the discretized $\overline{\mathcal{P}}_1$- and $\overline{\mathcal{P}}_2$-problems are identified, respectively, with the systems (6.53), (6.54) and (6.55), (6.56). At last, the system (6.73) is identified with (6.50). As a matter of fact, the continuous formulation developed in the present section is much simpler to conceive and, for this reason, constitutes a guide for the construction of the discrete solution technique of Section 6.3.2.c, particularly for the two-dimensional case addressed in Section 6.4.

As already mentioned in Section 6.3.2.c, the $\overline{\mathcal{P}}_1$- and $\overline{\mathcal{P}}_2$-problems are time-independent and are solved in the preprocessing stage performed before the start of the time-integration. Note that the elementary solutions $(\overline{\omega}_1, \overline{\psi}_1)$ and $(\overline{\omega}_2, \overline{\psi}_2)$ satisfy the symmetry property

$$\overline{\omega}_2(y) = \overline{\omega}_1(-y), \quad \overline{\psi}_2(y) = \overline{\psi}_1(-y),$$

so that the solution of $\overline{\mathcal{P}}_2$ is unnecessary and only $(\overline{\omega}_1, \overline{\psi}_1)$ has to be stored.

Then, the influence matrix $\mathcal{M}$ is calculated and inverted in the preprocessing stage and its inverse is stored.

Therefore, at each time-cycle, the algorithm is :

1. Solve the $\tilde{\mathcal{P}}$-problem.
2. Calculate $\tilde{E}$.
3. Calculate Ξ by $\Xi = \mathcal{M}^{-1}\tilde{E}$.
4. Calculate (ω, ψ) by Eq.(6.72).

To conclude this section, we make two remarks.

Remarks
1. Due to the construction of the solution, ξ_1 and ξ_2 are the values of ω, respectively, at $y = -1$ and $y = 1$ which ensures that the final solution satisfies the Neumann conditions on ψ. Therefore, the storage of the K couples $(\overline{\omega}_1, \overline{\psi}_1)$, where K is the cut-off frequency of the Fourier series, may be avoided by replacing *step 4* of the above algorithm by the successive solution of the problems

$$\omega'' - \sigma\omega = f, \quad -1 < y < 1$$

$$\omega(-1) = \xi_1, \quad \omega(1) = \xi_2,$$

and

$$\psi'' - k^2\psi = -\omega, \quad -1 < y < 1$$

$$\psi(-1) = g_+, \quad \psi(1) = g_-,$$

which gives the final solution (ω, ψ). This technique is commonly used in the two-dimensional case.

Note that this variant has its counterpart in the algebraic algorithm described in Section 6.2 : *step 4* is replaced by the solution of the two following systems

$$\hat{A}_1 \Omega^{n+1} = \hat{G}_1,$$

where

$$\hat{G} = \hat{H}_1 - \hat{A}_0 \Omega^n - \hat{A}_{-1} \Omega^{n-1}$$

and

$$\hat{H}_1 = \left(\varphi_1^{n+1}, \cdots, \varphi_{N-1}^{n+1}, \xi_2, \xi_1\right)^T$$

and

$$\hat{A}_2 \Phi^{n+1} = -\mathcal{I}_1 \Omega^{n+1} + \hat{H}_2,$$

where

$$\hat{H}_2 = (0, \cdots, 0, g_+, g_-)^T.$$

2. Taking into account the simplicity of problems $\overline{\mathcal{P}}_1$ and $\overline{\mathcal{P}}_2$, one could be tempted to solve them analytically rather than numerically. This has to be avoided because, in such a case, the equivalence between the full discrete system (6.38) and the collocation influence matrix method is lost. This has, as a consequence, a serious loss in the accuracy of the final solution when σ and k are large. In this case, the exact solution of the problem $\overline{\mathcal{P}}_1$ (or $\overline{\mathcal{P}}_2$), which exhibits a boundary layer behaviour, is very far from the Chebyshev collocation solution. As a matter of fact, this latter solution is very poor compared to the exact one, but that is unimportant since it is only an intermediate solution and the accuracy is recovered when the combination (6.40) is carried out. Note that the error obtained with the use of the analytical solution can be strongly diminished if the degree N of the polynomial approximation is sufficiently large to allow a correct representation of the boundary layer, but that would necessitate a much larger value of N than needed to represent the final solution.

(e) Stability

The stability of the time-discretization schemes (6.30) is now examined. However, instead of attempting to analyze directly the large system (6.31) where we assume $F = g_\pm = h_\pm = 0$, it is better to take advantage of the decomposition introduced in the influence matrix method and, more precisely, in its discrete equivalent. We present here the general lines of the analysis developed by Ehrenstein (1990).

First of all, we remark that the streamfunction ψ appears only at the time-level $n+1$. Therefore, its approximation is stable if the approximation of the vorticity ω is stable. So, it is sufficient to study the stability of the time-evolution of ω, while taking into account the effect of ψ through its equation and boundary conditions.

Now, from Eq.(6.40), it is seen that the evolution in time of Φ^{n+1} is controlled by the evolution of $\tilde{\Phi}^{n+1}$ and of $\Xi^{n+1} = \left(\xi_1^{n+1}, \, \xi_2^{n+1}\right)^T$ which, itself, depends only on $\tilde{\Psi}^{n+1}$, that is on $\tilde{\Omega}^{n+1}$.

From these remarks, it is possible to construct the equation governing the evolution of the $N-1$ nonzero components of the vector $\tilde{\Omega}^{n+1}$, that is,

$$\tilde{\Omega}_0^{n+1} = \mathcal{H}_0 \, \tilde{\Omega}_0^n + \mathcal{H}_{-1} \, \tilde{\Omega}_0^{n-1} , \tag{6.74}$$

where $\tilde{\Omega}_0^{n+1} = \left(\tilde{\omega}_1^{n+1}, \, \cdots, \, \tilde{\omega}_{N-1}^{n+1}\right)^T$. Then, by introducing, as usual, the auxiliary quantity

$$\tilde{\Theta}^{n+1} = \tilde{\Omega}^n$$

we transform the three-level scheme (6.30) into the two-level one

$$\left(\begin{array}{c} \tilde{\Omega}^{n+1} \\ \tilde{\Theta}^{n+1} \end{array} \right) = \mathcal{E} \left(\begin{array}{c} \tilde{\Omega}^n \\ \tilde{\Theta}^n \end{array} \right) . \tag{6.75}$$

A necessary condition for stability is that the spectral radius $\rho\left(\mathcal{E}\right)$ is not larger than 1. The matrix $\mathcal{E}$ is found to have at least one eigenvalue equal to $(1 - \theta)/\theta$. Therefore, a necessary condition of stability is $-1 \leq (1 - \theta)/\theta \leq 1$, that is, $\theta \geq 1/2$. This important result means that if $\theta < 1/2$, the schemes (6.30) are unstable whatever the value of the time-step Δt. Such behaviour is very different from the usual stability of these schemes when applied to the heat equation for which a conditional stability is obtained when $\theta < 1/2$ (see Section 4.3.2).

It is possible to obtain more precise results in the case where $0 < \theta \leq 1$ and either $\varepsilon = 1$ (two-level θ-scheme) or $\varepsilon = 2\theta$ (second-order schemes). So, the spectral radius $\rho\left(\mathcal{E}\right)$ is such that :

(i) $\rho\left(\mathcal{E}\right) > 1$ if $0 < \theta < 1/2$,

(ii) $\rho\left(\mathcal{E}\right) = 1$ if $\theta = 1/2$,

(iii) $\rho\left(\mathcal{E}\right) < 1$ if $\theta \geq 1/2$.

These results show, in particular, that the Crank-Nicolson scheme ($\varepsilon = 1$, $\theta = 1/2$) is only marginally stable. It was found to be unstable in the nonlinear case of the Navier-Stokes equations (Vanel et al., 1986). In conclusion, among the schemes of the family (6.30) we recommend the use of the fully implicit (BDI2) scheme defined by

$$\varepsilon = 2, \quad \theta = 1.$$

Finally, we again point out that the requirement $\rho\left(\mathcal{E}\right) \leq 1$ is a necessary condition only, since the matrix $\mathcal{E}$ does not have the property (Section 4.3.2) ensuring that it is also a sufficient condition. Numerical experiments, however, have shown that schemes such that $\theta \geq 1/2$ are actually stable.

6.3.3 Navier-Stokes equations

Now we consider the Navier-Stokes problem constituted by the equations (6.1)-(6.2) solved in $\Omega = \{0 \leq x \leq 2\pi,\ -1 < y < 1\}$ with 2π-periodicity in the x-direction and the boundary conditions (6.12)-(6.13). The initial conditions are given by Eqs.(6.5)-(6.6). The only difference with the solution method developed in the previous section is that we have to handle the nonlinear term $B(\mathbf{V}, \omega)$. This term is generally treated explicitly but, for stability reasons, one can also consider an implicit treatment. These two options will be successively considered.

(a) Semi-implicit scheme

The nonlinear term is approximated with respect to time by means of an Adams-Bashforth extrapolation. For example, in the case of the AB/BDI2 scheme [see Eqs.(4.52), (6.8)], corresponding here to scheme (6.30) with $\varepsilon = 2,\ \theta = 1$, the time-approximation for $B(\mathbf{V}, \omega)$ is

$$B^{n+1} \cong 2\,B^n - B^{n-1}, \qquad (6.76)$$

where $B^n = B(\mathbf{V}^n, \omega^n)$. The point is to evaluate this nonlinear term in an efficient way within the Fourier-Chebyshev approximation. This can be done by means of the pseudospectral technique presented in Section 2.3.

Therefore, let us introduce the truncated Fourier series expansion

$$B_K^n = \sum_{k=-K}^{K} \hat{B}_k^n(y)\, e^{i k x} \qquad (6.77)$$

for approximating B^n. The aim is to evaluate $\hat{B}_k^n(y_j)$ where y_j, $j = 1, \cdots, N_y - 1$, are the collocation points defined by (6.26). More precisely, $\hat{u}_k^n(y_j)$, $\hat{v}_k^n(y_j)$, and $\hat{\omega}_k^n(y_j)$ being known, we have to calculate the Fourier coefficients of $(u\,\partial_x\omega + v\,\partial_y\omega)_k^n$ for every y_j, $j = 1, \cdots, N_y - 1$, in the case where B is expressed in the convective form (5.14). In the pseudospectral technique, differentiations are performed in spectral space and products in physical space (x_i, y_j), where x_i are defined by $x_i = 2\pi i / N_x$, $i = 0, \cdots, N_x = 2K + 1$. We denote by u_{KN}^n, v_{KN}^n and ω_{KN}^n the Fourier-Chebyshev approximation to u^n, v^n, and ω^n, respectively. The calculations are performed as follows.

Calculation of $\left(u_{KN} \widehat{\partial_x \omega_{KN}}\right)_k^n$

For every y_j, $j = 1, \cdots, N_y - 1$:
1. Calculate $u_{KN}^n(x_i, y_j)$, $i = 1, \cdots, N_x$, by Eq.(2.21).
2. Calculate the Fourier coefficients of $\partial_x \omega_{KN}^n$ by $\left(\widehat{\partial_x \omega_{KN}}\right)_k^n = i\,k\,\hat{\omega}_k^n$, $k = -K, \cdots, K$.
3. Calculate $\partial_x \omega_{KN}^n(x_i, y_j)$, $i = 1, \cdots, N_x$, by Eq.(2.21).

4. Calculate the products $u_{KN}^n(x_i, y_j)\, \partial_x \omega_{KN}^n(x_i, y_j)$, $i = 1, \cdots, N_x$.

5. Calculate the Fourier coefficients $(u_{KN}\widehat{\partial_x \omega_{KN}})_k^n(y_j)$, $k = -K, \cdots, K$, by Eq.(2.22).

Calculation of $(v_{KN}\widehat{\partial_y \omega_{KN}})_k^n$

1. For every y_j, $j = 1, \cdots, N_y - 1$, calculate $v_{KN}^n(x_i, y_j)$, $i = 1, \cdots, N_x$, by Eq.(2.21).

2. For every y_j, $j = 0, \cdots, N_y$, calculate $\omega_{KN}^n(x_i, y_j)$, $i = 1, \cdots, N_x$, by Eq.(2.21).

3. For every x_i, $i = 1, \cdots, N_x$, calculate $\partial_y \omega_{KN}^n(x_i, y_j)$, $j = 1, \cdots, N_y - 1$, by Eqs. (3.45)-(3.46) [or, alternatively by the FFT after calculation of the Chebyshev coefficients by Eq.(3.28)].

4. For every x_i, $i = 1, \cdots, N_x$, calculate $v_{KN}(x_i, y_j)\, \partial_y \omega_{KN}(x_i, y_j)$, $j = 1, \cdots, N_y - 1$.

5. For every y_j, $j = 1, \cdots, N_y - 1$, calculate the Fourier coefficients $\left(v_{KN}\widehat{\partial_y \omega_{KN}}\right)_k^n$, $k = -K, \cdots, K$, by Eq.(2.22).

Then the term $2\hat{B}_k^n(y_j) - \hat{B}_k^{n-1}(y_j)$ is included in the forcing term of the vorticity equation and the solution procedure described above for the Stokes problem applies without any change.

Because of the explicit evaluation of the nonlinear term, the time-step has to be chosen in order to ensure stability (see Section 4.4.2 for the advection-diffusion equation). If the allowed time-step is too small, then it may be advantageous to envisage a fully implicit scheme.

(b) Fully implicit scheme

An efficient fully implicit scheme is constituted by the BDI2 scheme [see Eq.(4.49)], such that the nonlinear term B is considered at level $n + 1$. Therefore, at each time-cycle, we have to solve the nonlinear problem

$$L\left(\omega^{n+1}\right) + B\left(\mathbf{V}^{n+1}, \omega^{n+1}\right) = G^{n+1} \quad \text{in } \Omega \tag{6.78}$$

$$\nabla^2 \psi^{n+1} + \omega^{n+1} = 0 \quad \text{in } \Omega \tag{6.79}$$

$$\psi^{n+1}(x, \pm 1) = g_\pm^{n+1} \tag{6.80}$$

$$\partial_y \psi^{n+1}(x, \pm 1) = h_\pm^{n+1}, \tag{6.81}$$

where L is the linear operator defined by

$$L(\omega) = -\nu \nabla^2 \omega + \sigma \omega, \quad \sigma = 3/(2\Delta t), \tag{6.82}$$

and G^{n+1} is defined by

$$G^{n+1} = F^{n+1} - \left(4\omega^n - \omega^{n-1}\right)/(2\Delta t).$$

The problem (6.78)-(6.81) has to be solved iteratively. One possible way (Fröhlich *et al.*, 1991) is to use the following preconditioned relaxation procedure characterized by the superscript m :

$$L\left(\overline{\omega}^{m+1}\right) = G^{n+1} - B\left(\mathbf{V}^{n+1,m}, \omega^{n+1,m}\right) - L\left(\omega^{n+1,m}\right) \quad \text{in } \Omega \quad (6.83)$$

$$\nabla^2 \overline{\psi}^{m+1} + \overline{\omega}^{m+1} = 0 \quad \text{in } \Omega \tag{6.84}$$

$$\overline{\psi}^{m+1}(x, \pm 1) = g_{\pm}^{n+1} - \psi^{n+1,m}(x, \pm 1) \tag{6.85}$$

$$\partial_y \overline{\psi}^{m+1}(x, \pm 1) = h_{\pm}^{n+1} - \partial_y \psi^{n+1,m}(x, \pm 1) \tag{6.86}$$

and

$$\phi^{n+1,m+1} = \phi^{n+1,m} + \alpha \overline{\phi}^{m+1} \tag{6.87}$$

for $\phi = \psi, \omega$. The velocity $\mathbf{V}^{n+1,m+1} = \left(u^{n+1,m+1}, v^{n+1,m+1}\right)$ is calculated from $\psi^{n+1,m+1}$ by Eq.(6.4). Therefore, at each iteration, we have a Stokes problem which is solved by the influence matrix method described in Section 6.3.2. The iterative procedure is started with

$$\phi^{n+1,0} = \phi^n .$$

In the simpler case, the relaxation parameter α is constant (PR method), but the convergence can be improved by the dynamic calculation of α as indicated in Section 3.8. An application of the above iterative procedure is described in Section 6.5.1.

6.3.4 Case of nonzero flow rate

When the flow rate in the channel is assumed to be null, the values of ψ at $y = \pm 1$ can be set to zero [Eq.(6.12) with $g_+ = 0$]. If this is not the case, the value at the boundary $\psi(x, 1, t) = g_+(t)$, which determines the flow rate, is unknown and has to be determined as part of the solution. Such a situation appears, for example, in the case where the flow in the infinite channel is driven by a given mean pressure gradient γ (Poiseuille flow) or by the motion of the boundary $y = 1$ with a given velocity U (Couette flow). Another example is constituted by the flow in a toroidal domain with a square section driven, for example, by the motion of a lateral wall (rotating Couette flow) or by a vertical thermal gradient (Rayleigh-Bénard flow). The flow is periodic in the azimuthal direction. Two-dimensional equations for plane motion can be used as an approximation if the curvature of the torus can be neglected, as well as the axial motion in the first case and radial motion in the second case. The above situations are quite comparable to the flow in a doubly-connected domain. The usual requirement in this latter case is the uniformity of the pressure, which allows the determination of the value $g_+(t)$. Here the use of Fourier series in the periodic direction makes this requirement (for the perturbation pressure) automatically satisfied and the value $g_+ (t)$ is determined from the flow rate condition.

If we remember that the indetermination comes from the use of vorticity-streamfunction equations rather than from the primitive variable equations, it is reasonable to try to overcome the difficulty by returning to this last formulation.

Let us consider the case where the boundary conditions are

$$u(x, -1, t) = 0, \quad v(x, -1, t) = 0, \tag{6.88}$$

$$u(x, 1, t) = U, \quad v(x, 1, t) = 0, \tag{6.89}$$

but the stress-free conditions (6.14) could also be handled in the same way. From the condition $v(x, 1, t) = 0$, we have

$$v(x, 1, t) = -\partial_x \psi(x, 1, t) = - \sum_{k=-K}^{K} i k \hat{\psi}_k (1, t) e^{i k x} = 0.$$

Therefore, the boundary condition is $\hat{\psi}_k (1, t) = 0$ for $k \neq 0$. So, the flow rate condition is satisfied through the zero-frequency mode, namely,

$$\hat{\psi}_0 (-1, t) = 0, \quad \hat{\psi}_0 (1, t) = g_+ (t) .$$

Then the total flow rate $g_+ (t)$ is determined by

$$g_+ (t) = \hat{\psi}_0 (1, t) = \int_{-1}^{1} \hat{u}_0 (y, t) \, dy,$$

where $\hat{u}_0$ is the zero-frequency component of the x-velocity u. The quantity $\hat{u}_0$ satisfies the x-momentum equation

$$\partial_t \hat{u}_0 + \hat{A}_{x,0} + \gamma - \nu \partial_{xx} \hat{u}_0 = \hat{f}_{x,0}, \tag{6.90}$$

where we have introduced the given mean pressure gradient γ. The x-component $\hat{A}_{x,0}$ of the nonlinear convective term is defined by

$$\hat{A}_{x,0} = \left[\partial_y \widehat{(u\, v)} \right]_0.$$

The time-discretization (e.g., using the AB-BDI2 scheme) of Eq.(6.90) leads to the following problem

$$\left(\hat{u}_0^{n+1} \right)'' - \sigma \, \hat{u}_0^{n+1} = \hat{c}_0^{n, n-1}, \quad -1 < y < 1$$

$$\hat{u}_0^{n+1} (-1) = 0, \quad \hat{u}_0^{n+1} (1) = U,$$

where $\sigma = 2/ (3 \Delta t \, \nu)$. This simple problem of Helmholtz type is solved by means of the Chebyshev method and, finally, we have that

$$g_+^{n+1} = \int_{-1}^{1} \hat{u}_0^{n+1} (y) \, dy = - \sum_{\substack{m=0 \\ m \text{ even}}}^{N_y} \frac{2}{m^2 - 1} \hat{u}_{0,m}^{n+1}, \tag{6.91}$$

where the quantities $\hat{u}_{0,m}^{n+1}$ are the Chebyshev coefficients of $\hat{u}_0^{n+1}$.

6.4   Chebyshev-Chebyshev method

This section is devoted to the solution of the unsteady vorticity-streamfunction equations in the domain $\Omega = \{x, y; -1 < x < 1, -1 < y < 1\}$. The solution method is a direct extension of the one developed in the previous section. This extension, however, is not completely straightforward because the influence matrix is singular ; in other words, the full algebraic system obtained from the Chebyshev approximation (tau or collocation) is not solvable. Therefore, a special treatment has to be applied in order to recover a well-posed problem. These points will be discussed in the following.

6.4.1   Time-discretization

As already discussed, the semi-implicit AB/BDIk schemes (4.80) constitute an efficient way to discretize the Navier-Stokes equations with respect to time. Fully implicit schemes BDIk (4.99) are free of stability constraints but necessitate the use of an iterative procedure (see Section 6.2.3.b).

The application of the general AB/BDIk scheme to the vorticity-streamfunction equations leads to the semi-discrete equations

$$\frac{1}{\Delta t} \sum_{j=0}^{k} a_j \, \omega^{n+1-j} + \sum_{j=0}^{k-1} b_j \, B^{n-j} - \nu \, \nabla^2 \omega^{n+1} = F^{n+1} \quad \text{in } \Omega \qquad (6.92)$$

$$\nabla^2 \psi^{n+1} + \omega^{n+1} = 0 \quad \text{in } \Omega \qquad (6.93)$$

$$\psi^{n+1} = g^{n+1} \quad \text{on } \Gamma = \partial\Omega \qquad (6.94)$$

$$\partial_n \psi^{n+1} = h^{n+1} \quad \text{on } \Gamma = \partial\Omega, \qquad (6.95)$$

with the coefficients a_j and b_j given in Table 4.4. The nonlinear term B is generally considered in the convective form (5.16), but the convective form (5.17) can be used as well. We refer to Section 7.4.5 for a short discussion on this point.

6.4.2   The influence matrix method

In this section we set $\omega^{n+1} = \omega$ and $\psi^{n+1} = \psi$ for the sake of simplification. So, from (6.92)-(6.95), we get the following Stokes problem which has to be solved at each time-cycle

$$\nabla^2 \omega - \sigma \, \omega = f \quad \text{in } \Omega \qquad (6.96)$$

$$\nabla^2 \psi + \omega = 0 \quad \text{in } \Omega \qquad (6.97)$$

$$\psi = g \quad \text{on } \Gamma \qquad (6.98)$$

$$\partial_n \psi = h \quad \text{on } \Gamma, \qquad (6.99)$$

with $\sigma = a_0 \, \Delta t / \nu$.

Let $\Gamma = \bigcup_{i=1}^{4} \Gamma_i$, such that Γ_1 and Γ_3 correspond, respectively, to the sides $y = 1$ and $y = -1$, and Γ_2 and Γ_4 correspond, respectively, to the sides $x = 1$ and $x = -1$. Then, we set $g|_{\Gamma_i} = g_i$ and $h|_{\Gamma_i} = h_i$, $i = 1, \cdots, 4$. The functions g_i and h_i are assumed to satisfy the following conditions of compatibility at the corners of Ω :

(i) continuity of g :

$$g_1\,(1) = g_2\,(1)\,, \qquad g_1\,(-1) = g_4\,(1)\,,$$
$$g_3\,(1) = g_2\,(-1)\,, \qquad g_3\,(-1) = g_4\,(-1)\,. \tag{6.100}$$

(ii) compatibility of g and h :

$$h_1\,(1) = g_2'\,(1)\,, \qquad h_1\,(-1) = g_4'\,(1)\,,$$
$$h_2\,(1) = g_1'\,(1)\,, \qquad h_2\,(-1) = g_3'\,(1)\,,$$
$$-h_3\,(1) = g_2'\,(-1)\,, \qquad -h_3\,(-1) = g_4'\,(-1)\,,$$
$$-h_4\,(1) = g_1'\,(-1)\,, \qquad -h_4\,(-1) = g_3'\,(-1)\,. \tag{6.101}$$

(iii) crossed derivatives :

$$h_1'\,(1) = h_2'\,(1)\,, \qquad h_1'\,(-1) = -h_4'\,(1)\,,$$
$$h_3'\,(1) = -h_2'\,(-1)\,, \qquad h_3'\,(-1) = h_4'\,(-1)\,. \tag{6.102}$$

These conditions ensure that the trace of ψ and of its derivatives on the boundary Γ are defined at the corners, and also that the equality $\partial_x\,(\partial_y\psi) = \partial_y\,(\partial_x\psi)$ is satisfied at the corners.

Now let us assume the spatial approximation based on the Chebyshev collocation method using the collocation points

$$x_i = \cos\frac{\pi i}{N_x}\,, \quad i = 0, \cdots, N_x\,, \quad y_j = \cos\frac{\pi j}{N_y}\,, \quad j = 0, \cdots, N_y\,. \tag{6.103}$$

We define by Ω_N the set of collocation points in the interior of Ω :

$$\Omega_N = \{x_i\,, y_j\,;\ i = 1, \cdots, N_x - 1,\ j = 1, \cdots, N_y - 1\}$$

and by Γ_N^I the set of collocation points on the boundary Γ except the corners. We denote by $I\!\!P_N$ the set of polynomials of degree at most equal to N_x in x and to N_y in y. Therefore, the solution (ω, ψ) is approximated with (ω_N, ψ_N) such that $\omega_N \in I\!\!P_N$, $\psi_N \in I\!\!P_N$.

Then the Chebyshev collocation approximation to the system (6.96)-(6.99) consists of searching the solution of the discrete $\mathcal{P}$-Problem :

$$\nabla^2 \omega_N - \sigma \omega_N = f \quad \text{in } \Omega_N \tag{6.104}$$

$$\nabla^2 \psi_N + \omega_N = 0 \quad \text{in } \Omega_N \tag{6.105}$$

$$\psi_N = g \quad \text{on } \Gamma_N^I \tag{6.106}$$

$$\partial_n \psi_N = h \quad \text{on } \Gamma_N^I. \tag{6.107}$$

The various derivatives occurring in this system are expressed by formulas of the type (3.45), so that Eqs. (6.104)-(6.107) form an algebraic system for the values of ω_N and ψ_N at the collocation points except the four corners. For the solution of this large algebraic system one could design an algorithm similar to that considered in Section 6.3.2.c. The construction of such an algorithm directly on the vector (or matrix) form of the equations is, however, rather complex and, hence, it is preferable to have recourse to the influence matrix method as given in Section 6.3.2.d.

The solution (ω_N, ψ_N) is decomposed according to

$$\omega_N = \tilde{\omega}_N + \overline{\omega}_N, \quad \psi_N = \tilde{\psi}_N + \overline{\psi}_N, \tag{6.108}$$

where $\tilde{\omega}_N$, $\tilde{\psi}_N$, $\overline{\omega}_N$, and $\overline{\psi}_N$ are polynomials belonging to $I\!\!P_N$.

The grid values of the pair $\left(\tilde{\omega}_N, \tilde{\psi}_N\right)$ are the solution of the discrete $\tilde{\mathcal{P}}$-Problem :

$$\nabla^2 \tilde{\omega}_N - \sigma \tilde{\omega}_N = f \quad \text{in } \Omega_N \tag{6.109}$$

$$\tilde{\omega}_N = 0 \quad \text{on } \Gamma_N^I, \tag{6.110}$$

and

$$\nabla^2 \tilde{\psi}_N = -\tilde{\omega}_N \quad \text{in } \Omega_N \tag{6.111}$$

$$\tilde{\psi}_N = g \quad \text{on } \Gamma_N^I. \tag{6.112}$$

The solution of this problem reduces to the successive solution of two algebraic systems coming from Dirichlet problems (see Section 3.7.2).

The grid values of the pair $\left(\overline{\omega}_N, \overline{\psi}_N\right)$ are the solution of the discrete $\overline{\mathcal{P}}$-Problem :

$$\nabla^2 \overline{\omega}_N - \sigma \overline{\omega}_N = 0 \quad \text{in } \Omega_N \tag{6.113}$$

$$\nabla^2 \overline{\psi}_N + \overline{\omega}_N = 0 \quad \text{in } \Omega_N \tag{6.114}$$

$$\overline{\psi}_N = 0 \quad \text{on } \Gamma_N^I \tag{6.115}$$

$$\partial_n \overline{\psi}_N = h - \partial_n \tilde{\psi}_N \quad \text{on } \Gamma_N^I. \tag{6.116}$$

This problem is similar to the original one except that the equation for $\overline{\omega}_N$ and the Dirichlet condition for $\overline{\psi}_N$ are homogeneous. This allows us to transform this problem into a set of time-independent Dirichlet problems.

First, the $\overline{\mathcal{P}}$-problem is replaced by Dirichlet problems for $\overline{\omega}_N$ and $\overline{\psi}_N$ where the boundary values of $\overline{\omega}_N$ have to be determined such that the Neumann condition on $\overline{\psi}_N$ is satisfied. This defines the $\overline{\mathcal{P}}_0$-Problem :

Find $\overline{\omega}_N$, $\overline{\psi}_N$, and ξ satisfying to

$$
\begin{aligned}
\nabla^2 \overline{\omega}_N - \sigma \overline{\omega}_N &= 0 && \text{in } \Omega_N \\
\nabla^2 \overline{\psi}_N + \overline{\omega}_N &= 0 && \text{in } \Omega_N \\
\overline{\omega}_N &= \xi && \text{on } \Gamma_N^I \\
\overline{\psi}_N &= 0 && \text{on } \Gamma_N^I,
\end{aligned}
\tag{6.117}
$$

such that the Neumann condition (6.116) is satisfied. Therefore, the main point is the determination of ξ. This is done through the influence matrix method, which uses the linearity of the $\overline{\mathcal{P}}_0$-problem to search for its solution under the form

$$
\overline{\omega}_N = \sum_{l=1}^{L} \xi_l \overline{\omega}_l, \quad \overline{\psi}_N = \sum_{l=1}^{L} \xi_l \overline{\psi}_l.
\tag{6.118}
$$

The grid values of each pair $(\overline{\omega}_l, \overline{\psi}_l)$, $l = 1, \cdots, L$, are solutions of the $\overline{\mathcal{P}}_l$-Problem :

$$
\nabla^2 \overline{\omega}_l - \sigma \overline{\omega}_l = 0 \quad \text{in } \Omega_N
\tag{6.119}
$$

$$
\overline{\omega}_l|_{\eta_m} = \delta_{m,l} \quad \text{for } \eta_m \in \Gamma_N^I,
\tag{6.120}
$$

$$
\nabla^2 \overline{\psi}_l = -\overline{\omega}_l \quad \text{in } \Omega_N
\tag{6.121}
$$

$$
\overline{\psi}_l|_{\eta_m} = 0 \quad \text{for } \eta_m \in \Gamma_N^I,
\tag{6.122}
$$

where η_m, $m = 1, \ldots, 2(N_x + N_y - 2)$, refer to the collocation points on Γ_N^I. From Eqs.(6.118) and (6.120), it is seen that the constants ξ_l, $l = 1, \cdots, L$, are the values of $\overline{\omega}_N$ (thus of ω_N) at the boundary Γ_N^I. These values are determined by asking the streamfunction $\overline{\psi}_N$ to satisfy the condition (6.116). We get the algebraic system

$$
\sum_{l=0}^{L} \left(\partial_n \overline{\psi}_l|_{\eta_m} \right) \xi_l = \left(h - \partial_n \tilde{\psi}_N \right)_{\eta_m} \equiv \tilde{h}_m, \quad \eta_m \in \hat{\Gamma}_N^I \subset \Gamma_N^I, \quad m = 1, \cdots, L
\tag{6.123}
$$

or

$$
\mathcal{M} \Xi = \tilde{E},
\tag{6.124}
$$

where $\Xi = (\xi_1, \cdots, \xi_L)^T$ and $\mathcal{M}$ is the influence matrix.

The remaining question is to determine the proper value of L, namely, $\hat{\Gamma}_N^I$. For the Chebyshev collocation method based on polynomials belonging to $\mathbb{P}_N$, the number of collocation points on Γ_N^I (i.e., except the corners) are $2(N_x + N_y - 2)$, so that the total number of elementary solutions $(\overline{\omega}_l, \overline{\psi}_l)$ should be $L = 2(N_x + N_y - 2)$. For this value of L, however, the matrix $\mathcal{M}$ is found to have four null eigenvalues, therefore it is not invertible. The proof of this result is given by Ehrenstein (1986) and the general lines of the proof are sketched by Ehrenstein and Peyret (1989). Here we restrict ourselves to giving a heuristic explanation of the behaviour of the influence matrix.

As already mentioned, the influence matrix method developed above is nothing but a special algorithm for solving the full discrete system (6.104)-(6.107). Thus, the singular nature of the influence matrix $\mathcal{M}$ simply reflects the singular nature of the full discrete system itself. This singular nature comes from the property of any polynomial $p_N(x, y) \in \mathbb{P}_N$ to which, as well as to its normal derivative, one requires that it be zero at collocation points belonging to Γ_N^I. To be more precise, let us consider the simpler case where the conditions $\psi = g$ and $\partial_n \psi = h$ are prescribed on the two adjacent sides Γ_1 and Γ_2. On the other sides, Γ_3 and Γ_4, stress-free conditions are assumed, that is, $\omega = 0$, $\psi = 0$. The conditions of compatibility (6.100)-(6.102) apply here with $g_3 = h_3 = g_4 = h_4 = 0$. We get

(i) continuity of g :

$$g_1(1) = g_2(1), \quad g_1(-1) = 0, \quad g_2(-1) = 0, \tag{6.125}$$

(ii) compatibility of g and h :

$$h_1(1) = g_2'(1), \quad h_1(-1) = 0, \quad h_2(1) = g_1'(1), \quad h_2(-1) = 0, \tag{6.126}$$

(iii) crossed derivatives :

$$h_1'(1) = h_2'(1). \tag{6.127}$$

Then we consider the streamfunction ψ which is approximated by the polynomial $\psi_N(x, y) \in \mathbb{P}_N$, and we write the equations expressing that ψ_N satisfies the boundary conditions at every collocation point on Γ_N^I. Because the values of ψ_N are given on Γ_N^I, the only equations to be considered are those expressing the Neumann conditions $\partial_y \psi = h_1$ on Γ_1 and $\partial_x \psi = h_2$ on Γ_2. If we assume that $N_x = N_y = N$, these equations are, respectively,

$$E_i^{(1)} \equiv \sum_{j=1}^{N-1} d_{0,j}^{(1)} \psi_N(x_i, y_j) + d_{0,0}^{(1)} g_1(x_i) - h_1(x_i) = 0, \quad i = 1, \cdots, N-1,$$

$$\tag{6.128}$$

$$E_j^{(2)} \equiv \sum_{i=1}^{N-1} d_{0,i}^{(1)} \psi_N(x_i, y_j) + d_{0,0}^{(1)} g_2(y_j) - h_2(y_j) = 0, \quad j = 1, \cdots, N-1.$$

$$\tag{6.129}$$

Now it is easy to check that the linear combination

$$\sum_{i=1}^{N-1} d_{0,i}^{(1)} E_i^{(1)} - \sum_{j=1}^{N-1} d_{0,j}^{(1)} E_j^{(2)} = 0$$

makes the coefficients of every $\psi_N(x_i, y_j)$ identically equal to zero. This is a direct consequence of the following two facts :

(1) the conditions of compatibility (6.125)-(6.127) which are satisfied (up to the accuracy of the Chebyshev approximation) by the polynomials interpolating g and h.

(2) the identity, at the corner $(1,1)$, valid for any polynomial $\psi_N(x,y)$:

$$\partial_x(\partial_y\psi_N) = \partial_y(\partial_x\psi_N)$$

which, after discretization and simplification, yields

$$\sum_{i=1}^{N-1} d_{0,i}^{(1)} \sum_{j=1}^{N-1} d_{0,j}^{(1)} \psi_N(x_i,y_j) = \sum_{j=1}^{N-1} d_{0,j}^{(1)} \sum_{i=1}^{N-1} d_{0,i}^{(1)} \psi_N(x_i,y_j) \ .$$

Thus, from the above analysis, it results that one (any one, but only one) of the equations (6.128)-(6.129) can be expressed as a linear combination of the others. The meaning of this result is : if ψ_N satisfies the Dirichlet conditions at all collocation points of Γ_N^I and the Neumann conditions at all inner collocation points of Γ_1 and Γ_2 except one, then the Neumann condition will be automatically satisfied at this removed point.

Obviously, the above results do not apply if the two sides where $\partial_n\psi$ is prescribed are no longer adjacent but opposite, for example, Γ_1 and Γ_3, the vorticity ω being prescribed on Γ_2 and Γ_4 (ψ is imposed on the four sides). In that case the equations analogous to (6.128) and (6.129) are linearly independent and the corresponding influence matrix is nonsingular. This reasoning leads to a simple rule for counting the number of null eigenvalues of the influence matrix according to the type of boundary conditions (prescription of $\partial_n\psi$ or ω), moreover, the Dirichlet condition on ψ. The rule is: considering a corner, there is a zero eigenvalue associated to this corner if and only if $\partial_n\psi$ is prescribed on both sides having this corner in common.

Returning to the case where ψ and $\partial_n\psi$ are prescribed on the four sides of Γ, then the influence matrix has four eigenvalues equal to zero. Therefore, four collocation points on the boundary have to be removed when the influence matrix is constructed so that $L = 2(N_x + N_y)$. The choice of these four points is not completely arbitrary and must follow some rules as shown by Bwemba and Pasquetti (1995). One possible choice (Ehrenstein and Peyret, 1989) is constituted by the set of points

$$P_1 = (x_1, -1) \ , \quad P_2 = (x_{N_x-1}, -1) \ , \quad P_3 = (x_1, 1) \ , \quad P_4 = (x_{N_x-1}, 1) \ .$$
$$(6.130)$$

Another possible choice is

$$Q_1 = (x_1, 1) \ , \quad Q_2 = (1, y_{N_y-1}) \ , \quad Q_3 = (x_{N_x-1}, -1) \ , \quad Q_4 = (-1, y_1) \ .$$
$$(6.131)$$

The location of the removed points can have an effect on the conditioning of the influence matrix (Raspo et al., 1996). With the set of points (6.130), the condition number of the influence matrix behaves like $a(\sigma)N^{1.96}$ with $a(0) = 0.46$ and $a(1000) = 0.03$. This behaviour, determined numerically assuming $N_x = N_y = N$, shows that the influence matrix is relatively well-conditioned and can be inverted.

As discussed above, the Neumann condition on ψ is automatically satisfied at the removed points, although it has not been prescribed when constructing the influence matrix. Consequently, the calculated approximate solution satisfies the equations at the inner collocation points and satisfies the boundary conditions. The polynomial approximation to the streamfunction is determined in an unique way, but the vorticity is uniquely determined only at the inner collocation points. The values of ω at the boundary are not given by the solution method. To get these values, it suffices to employ the definition of the vorticity, that is, to evaluate $\omega = -\nabla^2 \psi$ at the boundary.

Therefore, by using the influence matrix method, the solution of the Stokes problem is reduced to the solution of a number of Dirichlet problems for the Helmholtz (or Poisson) equation. The largest part of these Dirichlet problems (those determining the elementary solution $\overline{\omega}_l$, $\overline{\psi}_l$, $l = 1, \cdots, L$) are time-independent and, hence, are solved in the preprocessing stage performed before the start of time-integration. Moreover, because of the inherent symmetry properties of the elementary solutions, the number of Dirichlet problems to be solved, in the precalculation stage, is reduced. Thus, the influence matrix $\mathcal{M}$ is constructed and inverted during this stage, and its inverse $\mathcal{M}^{-1}$ is stored. In general, it is better, for storage reasons, not to store the elementary solutions in order to reconstruct the solution from Eqs.(6.108) and (6.118) but, rather, to solve again two Dirichlet problems to obtain the final solution.

The first problem defines ω_N :

$$\nabla^2 \omega_N - \sigma \omega_N = f \quad \text{in } \Omega_N \tag{6.132}$$

$$\omega_N|_{\eta_m} = \gamma_m \quad \text{for } \eta_m \in \Gamma_N^I, \tag{6.133}$$

where $\gamma_m = 0$ at the four removed collocation points [Eq.(6.130) or (6.131)] and $\gamma_m = \xi_m$ at all other points of Γ_N^I.

The second problem determines ψ_N :

$$\nabla^2 \psi_N = -\omega_N \quad \text{in } \Omega_N \tag{6.134}$$

$$\psi_N|_{\eta_m} = g|_{\eta_m} \quad \text{for } \eta_m \in \Gamma_N^I. \tag{6.135}$$

To summarize, the algorithm for solving the time-dependent Navier-Stokes equations is the following:

A. Preprocessing stage
A.1. Calculate $(\overline{\omega}_l, \overline{\psi}_l)$, where $l = 1, \cdots, L$, by solving the $\overline{\mathcal{P}}_l$-problems [Eqs.(6.119)–(6.122)]

A.2. Construct [Eqs.(6.123), (6.124)] and invert the influence matrix $\mathcal{M}$. Store the matrix $\mathcal{M}^{-1}$.

B. *At each time-cycle*

B.1. Calculate the right-hand side f in Eq.(6.96) by matrix products or the pseudospectral technique (FFT).

B.2. Calculate $(\tilde{\omega}_N, \tilde{\psi}_N)$ by solving the $\tilde{\mathcal{P}}$-problem [Eqs.(6.109)-(6.112)].

B.3. Calculate $\tilde{E}$.

B.4. Calculate $\Xi = (\xi_1, \cdots, \xi_L)^T = \mathcal{M}^{-1} \tilde{E}$, Eq.(6.124).

B.5. Calculate the final solution ω_N, ψ_N by solving (6.133)-(6.135).

B.6. Calculate ω_N at the boundary Γ_N^I by $\omega_N = -\nabla^2 \psi_N$.

All discrete systems are solved by means of the diagonalization technique (Section 3.7.2), so that the solution at each time-cycle reduces to matrix-matrix products.

6.4.3   Other influence matrix methods

The influence matrix method given in the preceding sections constitutes an efficient way to solve the large discrete system derived from the Chebyshev collocation approximation of the Navier-Stokes equations in the physical space. As a matter of fact, the method has to be seen as the most natural way to solve the problem in a strong sense.

Another way is constituted by Galerkin-type methods. To this class of methods belongs the tau method developed by Vanel *et al.* (1986) in which the influence matrix is constructed, as above, by a pointwise enforcement of the Neumann condition on the streamfunction. The problem of the singular behaviour of the influence matrix is surmounted by approximating the streamfunction with a truncated Chebyshev expansion containing two more polynomials (in each direction) than the approximation of the vorticity. Note that the application of the tau method requires the values of the unknowns at the corners of the computational domain. Therefore, it is necessary to define the vorticity at the corners from the equation. A Galerkin method with better theoretical properties has been developed by Auteri and Quartapelle (1999). The Galerkin approximation makes use of the special basis constructed on Legendre polynomials as introduced by Shen (1994). The Neumann condition is replaced by an integral condition for the vorticity which is enforced through an influence matrix technique. The influence matrix has regular behaviour. The method is able to handle infinite values of the vorticity at the corners as it appears in the classical nonregularized lid-driven cavity flow, although the solution is not free of Gibbs oscillations.

6.5   Examples of application

This section presents two examples of application of the influence matrix method. The first example concerns the application of the Fourier-Chebyshev method to the calculation of Rayleigh-Bénard convection in a large aspect-ratio domain. The second example deals with the calculation by the Chebyshev-Chebyshev method of the axisymmetric flow in a rotating annulus. Some results concerning the flow in a square cavity will be given in Section 7.4.4.

6.5.1 Rayleigh-Bénard convection

The Rayleigh-Bénard problem concerns the motion of a fluid heated from below. Mathematical modelling makes use of the vorticity-streamfunction formulation of the Navier-Stokes equations within the Boussinesq approximation (Section 5.3). These equations are solved in an infinite horizontal channel assuming periodicity in the x-direction, so that the Fourier-Chebyshev method can be applied. Spectral calculations of non-Boussinesq convection in the same configuration have been presented by Fröhlich and Peyret (1990, 1992), using the low Mach number approximation of the primitive-variable Navier-Stokes equations. Comparisons between Boussinesq, low Mach number approximations, and compressible flow equations are discussed by Fröhlich et al. (1992) and by Fröhlich and Gauthier (1993). A Chebyshev-Chebyshev solution of the low Mach number equations in a square enclosure is presented by Le Quéré et al. (1992).

The aim of the present section is twofold :

(1) compare the efficiency of semi-implicit and fully implicit schemes, and

(2) show the ability of the method to calculate temporal and spatial chaotic flows.

The flow takes place in an infinite two-dimensional channel of height H. It is assumed to be periodic in the horizontal x-direction with a period equal to L. This is an idealization of the flow in a long rectangular enclosure, so that the solution is periodic at a distance long enough from the end lateral walls. Moreover, as discussed in Section 6.3.1, the total flow rate through a vertical section of the channel is zero, so that the boundary conditions for the streamfunction are simply

$$\psi = 0, \quad \partial_y \psi = 0 \quad \text{at } y = \pm H/2 \,. \tag{6.136}$$

The temperature at the walls is fixed

$$T = T_1 \quad \text{at } y = -H/2, \quad T = T_2 \quad \text{at } y = H/2, \tag{6.137}$$

with $\Delta T = T_1 - T_2 > 0$.

The problem is made dimensionless by using the following characteristic quantities: H for length, H^2/κ_T for the time where κ_T is the thermal diffusivity, and H/κ_T for velocity. The dimensionless temperature is defined by $\theta = (T - T_2)/\Delta t$.

The motion equations are constituted by Eq.(5.15) where the term F arises from the buoyant force $\mathbf{f}$ defined by Eqs.(5.40), (5.21), and (5.41).

After rendering these equations dimensionless a change of variable is introduced in order to transform the computational domain into the usual domain adapted to Fourier-Chebyshev approximation, namely, $0 \leq X \leq 2\pi$, $-1 \leq Y \leq 1$. The flow quantities are rescaled for taking this coordinate transformation into account. We respectively denote by $\mathbf{W}$, Ω, and Ψ the resulting velocity vector, vorticity and streamfunction. The equations are then

$$\partial_t \theta + \mathbf{W}.\nabla\theta - \nabla^2\theta = 0 \tag{6.138}$$

$$\partial_t \Omega + \mathbf{W}.\nabla\Omega - Pr\,\nabla^2\Omega = \frac{a}{8}Pr\,Ra\,\partial_X\theta \tag{6.139}$$

$$\nabla^2\Psi + \Omega = 0, \tag{6.140}$$

where the components U and V of the velocity vector $\mathbf{W}$ are connected to the streamfunction Ψ by

$$U = \partial_Y\Psi, \quad V = -a\,\partial_X\Psi, \tag{6.141}$$

where $a = \pi H/L$. The vorticity Ω is defined by

$$\Omega = a\,\partial_X V - \partial_Y U \tag{6.142}$$

and

$$\nabla = (a\,\partial_X, \partial_Y), \quad \nabla^2 = a^2\,\partial_{XX} + \partial_{YY}. \tag{6.143}$$

In Eq.(6.139), Pr and Ra are the Prandtl number and the Rayleigh number, respectively, defined by

$$Pr = \frac{\nu}{\kappa_T}, \quad Ra = \frac{g\,\alpha_T\,\Delta T\,H^3}{\kappa_T\,\nu}, \tag{6.144}$$

where g is the gravitational constant and α_T is the coefficient of thermal expansion (see Section 5.3). The boundary conditions associated with Eqs.(6.138)-(6.140) are

$$\theta = 1, \quad \Psi = 0, \quad \partial_Y\Psi = 0 \quad \text{at } Y = -1, \tag{6.145}$$

$$\theta = 0, \quad \Psi = 0, \quad \partial_Y\Psi = 0 \quad \text{at } Y = +1. \tag{6.146}$$

Initial conditions will be specified later.

Now we are in a position to solve the problem by the Fourier-Chebyshev method given in Section 6.3. After time-discretization, the Stokes problems are solved by the influence matrix method in which the Helmholtz and Poisson equations are approximated with the Chebyshev tau method (Section 3.4.1) leading to the solution of quasi-tridiagonal systems. Obviously, the collocation method could be used with similar success.

Two time-discretization schemes are considered : the semi-implicit AB/BDI2 scheme (4.52) and the fully implicit BDI2 scheme (4.49). When the semi-implicit scheme is used, the temperature equation is solved first, so that θ^{n+1} can be included in the forcing term F^{n+1} of the vorticity equation. For the fully implicit scheme, the iterative procedure described in Section 6.3.3.b is employed to solve the nonlinear system at each time-cycle.

In the calculations reported below, the aspect ratio of the domain is rather large and is $A = L/H = 16.136$, that is, eight times the most unstable wavelength. In such a geometry, flow dynamics can evolve, according to the value of the Rayleigh number, from perfectly regular regimes to chaotic spatio-temporal behaviours generated by strong nonlinearities. Moreover, the time-scale governing the selection of convective structures is very large since it is defined by the diffusive time in the horizontal direction L^2/κ_T, namely, a dimensionless time of the order of A^2. Therefore, the calculation of such kinds of flows, necessitating a long time-integration, constitutes a significant test for the evaluation of time-discretization schemes.

(a) Comparison of time-discretization schemes

In this section, we want to compare the efficiency of the semi-implicit AB/BDI2 scheme and the fully implicit BDI2 scheme with iterative procedure. This comparison is extracted from the work by Fröhlich et al. (1991) who also considered the case of velocity-pressure equations. The comparison is done by considering the calculation of the convective flow at $Pr = 0.71$ and $Ra = 6 \times 10^3$. For these values the flow is steady. This steady state is obtained as the large time-limit of the unsteady flow calculated from the perturbed conductive state as an initial condition. The conductive solution is $\theta = (1 - Y)/2$, $\mathbf{W} = \mathbf{0}$, and $\Omega = 0$. Thermal convective instability is generated by a random disturbance of the temperature with an amplitude equal to 10^{-5}.

The spatial resolution is characterized by $K = 128$ Fourier modes (i.e., 256 collocation points, using the even collocation as described in Section 2.4) in the x-direction and $N_y = 25$ for the Chebyshev polynomial approximation in the y-direction. The convergence toward the steady-state is measured by the residual

$$R_\phi = \max_{k,j} \left| \hat{\phi}_k^{n+1}(y_j) - \hat{\phi}_k^n(y_j) \right| / \Delta t \qquad (6.147)$$

for $\phi = \theta, U, V$, where k refers to the Fourier modes and j to the collocation points y_j.

Note that, when dealing with the vorticity-streamfunction equations, the use of the residual R_ϕ with $\phi = \theta, \omega$ is more natural. The choice $\phi = \theta, U, V$ was made by Fröhlich et al. (1991) in view of a comparison with the results given by the primitive-variable equations. When the fully implicit scheme is used, the convergence of the iterative procedure (superscript m) is assumed to be reached when

$$\max_{k,j} \left| \hat{\phi}_k^{n+1,m+1}(y_j) - \hat{\phi}_k^{n+1,m}(y_j) \right| < \varepsilon_\phi, \qquad (6.148)$$

where ε_ϕ is a small number whose magnitude is adapted to the associated variable ϕ. Here we simply have $\varepsilon_\theta = \varepsilon_U = \varepsilon_V = \varepsilon$. When dealing with an iterative method to be used at each time-cycle of an unsteady process, the usual difficulty concerns the choice of the criterion ε and that of the relaxation parameter α (when the PR method is used, as here) in relation to the value of the time-step Δt. A compromise between these quantities has to be found.

For highly unsteady flows, corresponding, for example, to the case $Ra = 8 \times 10^4$, $Pr = 7.5$ considered later, the time-step Δt and the parameter ε must be small for accuracy reasons, and numerical experiments show that the fully implicit scheme is more expensive than the semi-implicit one. On the other hand, in the present situation where the ultimate flow is steady, Δt and ε can be larger by accepting a loss in time-accuracy during the transient stage.

The critical time-step Δt_* associated with the AB/BDI2 scheme is such that $10^{-2} < \Delta t_* < 2.10^{-2}$. The calculations are done with $\Delta t = 10^{-2}$ and, for the BDI2 scheme, also with $\Delta t = 10^{-1}$. There is no aliasing removal. The isotherms and the streamlines at steady state are very similar to those shown in Fig.2.1. Comparison of the results is made on the residuals R_φ, on the maximal values U_M and V_M of the velocity components, and on the Nusselt number Nu calculated at the lower wall. Table 6.1 shows the results at $t = 150$ given by :
(1) the semi-implicit AB/BDI2 scheme with $\Delta t = 0.01$, and
(2) the fully implicit BDI2 scheme with $\Delta t = 0.1$, $\alpha = 0.75$ (the optimal value) and $\varepsilon = 10^{-5}$.

In these calculations, the level of convergence of the iterative procedure is sufficient to give a consistent approximation (although not a highly accurate one due to the size of Δt) to the transient solution. The efficiency is measured by the computing (CPU) time needed to reach a given time or a given level of convergence toward the steady state. As shown in Table 6.1, at $t = 150$, both schemes give the same results but the CPU time (CRAY 2 computer) is less in the case of the BDI2 scheme showing its superiority in the present problem.

However, when only the steady-state solution is of interest as it is here, it may be advantageous to perform, at each time-cycle, the minimal number of iterations necessary to ensure the stability of the time-integration scheme. In this way, by taking $\Delta t = 0.1$, $\alpha = 0.75$ but $\varepsilon = 0.1$, the number of iterations decreases rapidly as time elapses, until it equals 1. This does not mean that the steady state is reached and more time-cycles are necessary to get it, with only one iteration at each time-cycle. As a consequence, and because of the relaxation, the scheme losses its consistency and the solution is shifted in time. Table 6.2 shows the results given by the fully implicit BDI2 scheme at three different times. It may be seen that the correct steady-state solution obtained at $t = 150$ with consistent schemes

	Semi-implicit $\Delta t = 0.01$	Fully implicit $\Delta t = 0.1$ $\alpha = 0.75,\ \varepsilon = 10^{-5}$
R_θ	1.468×10^{-5}	1.466×10^{-5}
R_U	5.255×10^{-4}	5.244×10^{-4}
R_V	5.904×10^{-4}	5.903×10^{-4}
U_M	8.717427	8.717090
V_M	11.150456	11.149789
Nu	2.253975	2.253975
CPU	867 s.	559 s.

Table 6.1. Comparison of semi-implicit AB/BDI2 and fully implicit BDI2
schemes. The results concern the solution at $t = 150$ in the case
$Pr = 0.71$, $Ra = 6 \times 10^3$.

	$t = 150$	$t = 250$	$t = 280$
R_θ	1.917×10^{-4}	1.870×10^{-5}	9.308×10^{-6}
R_U	6.837×10^{-3}	6.683×10^{-4}	3.326×10^{-4}
R_V	7.687×10^{-3}	7.515×10^{-4}	3.740×10^{-4}
U_M	8.736361	8.718469	8.717462
V_M	11.153931	11.150738	11.150287
Nu	2.253968	2.253975	2.253975
CPU	92 s.	150 s.	168 s.

Table 6.2. Convergence toward the steady state for the implicit BDI2
scheme. The results concern the solution obtained with $\Delta t = 0.1$,
$\alpha = 0.75$, and $\varepsilon = 0.1$ in the case $Pr = 0.71$, $Ra = 6 \times 10^3$.

(Table 6.1) is now obtained much later. In spite of this shift, the CPU time
to get the correct steady state is very short, about five time less.

   With this example, we showed that, under some circumstances, the use
of a fully implicit scheme may be of interest. The efficiency of such a scheme
is directly connected to the properties of the iterative method : choice of
the preconditioner and relaxation. In these respects, the use of a modern
iterative procedure (see Section 3.8) should improve the efficiency of the
method.

(b) Unsteady convection

This case illustrates the ability of the method to calculate highly unsteady
flows on a long time integration. The problem is the same as the previous
one but the physical parameters are $Pr = 7.5$ and $Ra = 8 \times 10^4$. The
semi-implicit AB/BDI2 scheme is used with $\Delta t = 5 \times 10^{-4}$, $K = 300$, and
$N_y = 25$. Moreover, the aliasing in the Fourier approximation is removed
by the "3/2 rule" technique presented in Section 2.9.

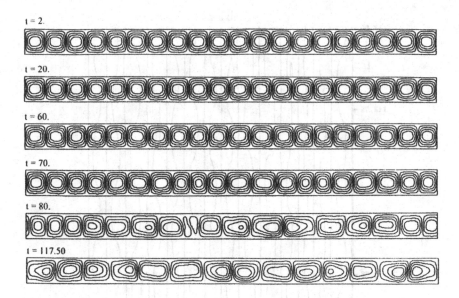

FIGURE 6.1. Instantaneous streamlines at various times in the case $Pr = 7.5$, $Ra = 8 \times 10^4$.

As usual, when dealing with high values of the physical parameters (Rayleigh number, Reynolds number...), it is worthwhile (and often absolutely necessary) to go step by step by increasing the value of the parameters. Thus, starting from the solution at $Ra = 6 \times 10^3$ as the initial condition, several calculations are done by increasing the value of Ra until $Ra = 8 \times 10^4$. For this value the established flow is highly unsteady exhibiting a chaotic behaviour in time as well as in space. As an illustration, Fig.6.1 shows the instantaneous streamlines at various times (the initial condition at $t = 0$ was the solution for $Ra = 7 \times 10^4$). One may observe the existence of a long period of time during which the flow evolves very slowly. Then the flow losses its regularity and is characterized by the vanishing of some rolls while the shape and size of the structures become irregular. An interesting tool to investigate such kinds of unsteady flows is constituted by the (x, t)-diagram of characteristic quantities. Such a diagram can make clear some time-dependent behaviours that instantaneous iso-contour lines at more or less spaced times cannot show. Thus, Fig.6.2 displays the trajectories of the zeros (curves a), minima (curves b), and maxima (curves c) of the streamfunction on the horizontal axis during a very short period of time $(112.60 \leq t \leq 112.80)$. It may be observed that the zeros remain nearly stationary. On the other hand, the extrema oscillate around a medium location, the period and amplitude being different according to the size of the rolls.

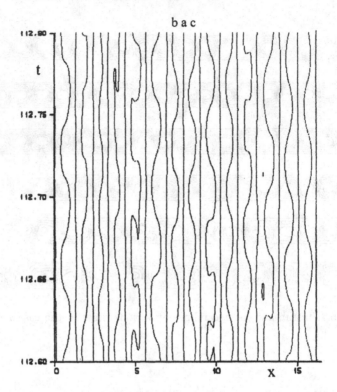

FIGURE 6.2. Trajectories of the zeros (curves a), maxima (curves b) and minima (curves c) of the streamfunction in the case $Pr = 7.5$, and $Ra = 8 \times 10^4$.

6.5.2   Axisymmetric flow in a rotating annulus

The purpose of this example is twofold : (1) to illustrate the application of the influence matrix method to the calculation of axisymmetric flows, and (2) to show the capacity of the Chebyshev collocation approximation to represent boundary-layer-type solutions.

The extension of the influence matrix method to the axisymmetric flow equations have been done by Chaouche (1990) and applied to the calculation of flows in a rotating annulus by Crespo del Arco *et al.* (1996). The following developments and results are extracted from this latter work.

As shown in Fig.6.3, the flow takes place between two cylinders of radius R_0 and R_1, and height equal to $2h$. It is driven by the constant rotation Ω of the whole system and by a forced flux entering at the inner cylindrical wall with flow rate Q and coming out at the outer cylinder. From the numerical point of view, the main difficulty is the accurate capture of the Ekman layers which develop at the plane walls $z = \pm h$. The thickness δ of these layers is such that $\delta/h = O\left(E^{1/2}\right)$ where E is the Eckman number defined by

$$E = \frac{\nu}{\Omega h^2},\tag{6.149}$$

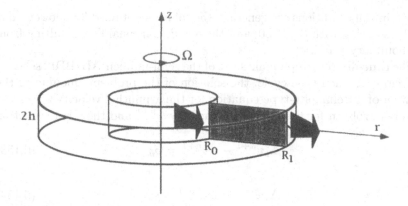

FIGURE 6.3. Geometrical configuration of the source-sink flow in the rotating cavity.

where ν is the kinematic viscosity. In the reported calculations $E = 2.24 \times 10^{-3}$ and the mesh ($N = 40$) is such that about eight collocation points belong to the Eckman layer since the space between collocation points is $O\left(N^{-2}\right)$ near the boundaries.

The equations of motion (in convective form) are given in Section 5.2.2. These equations are made dimensionless with the following characteristic quantities : h for length, h^2/ν for time, ν/h for velocity, ν/h^2 for vorticity, and ν for the streamfunction. Moreover, the equations are considered in a reference frame rotating with the angular velocity Ω. Therefore, by keeping the same notation for the dimensionless quantities, the dimensionless equations keep a form identical to Eqs.(5.33), (5.36), and (5.39), with $f_v = -2u/E$, $F_\omega = 2\partial_z v/E$, and $\nu = 1$. The radial coordinate r varies in the interval $[(R_m - 1)L, (R_m + 1)L]$, with $R_m = (R_1 + R_0)/(R_1 - R_0)$ and $L = (R_1 - R_0)/(2h)$. Therefore, the affine transformation $\rho = r/L - R_m$ is needed to transform this interval into the usual interval $[-1, 1]$. The dependent variables are unchanged except the streamfunction which is divided by LR_m.

Concerning the boundary conditions, the fluid enters at $\rho = -1$ with the dimensionless flow rate $C_w = Q/(\nu R_1)$ and leaves the domain at $\rho = 1$. The u-velocity profiles are sine-type. Therefore, in the v-ω-ψ formulation the boundary conditions are

$$v = 0, \quad \psi = \frac{C_w}{4\pi R_m}(R_m + 1)\left(1 + \sin\frac{\pi z}{2}\right), \quad \partial_\rho \psi = 0 \quad \text{at } \rho = \pm 1,$$

$$\tag{6.150}$$

and

$$v = 0, \quad \psi = 0, \quad \partial_z \psi = 0 \quad \text{at } z = -1, \tag{6.151}$$

$$v = 0, \quad \psi = \frac{C_w}{2\pi R_m}(R_m + 1), \quad \partial_z \psi = 0 \quad \text{at } z = 1. \tag{6.152}$$

The initial conditions are generally the solution obtained for a lower value of C_w, starting with $C_w = 80$, and the one-dimensional flow resulting from the boundary conditions.

The time-discretization makes use of the semi-implicit AB/BDI2 scheme. Therefore, at each time-cycle, the solution of the problem amounts to the solution of a Helmholtz-type equation for the azimuthal velocity v^{n+1}, and a Stokes problem for the azimuthal vorticity ω^{n+1} and the streamfunction ψ^{n+1}, namely,

$$\Delta_1 v^{n+1} - \sigma v^{n+1} = \varphi_v, \tag{6.153}$$

and

$$\Delta_1 \omega^{n+1} - \sigma \omega^{n+1} = \varphi_\omega \tag{6.154}$$

$$\Delta_2 \psi^{n+1} - \left(1 + \frac{r}{R_m}\right) \omega^{n+1} = 0, \tag{6.155}$$

in the square domain $\Omega = (-1,1)^2$, with boundary conditions (6.150)-(6.152). In Eqs. (6.153)-(6.155) $\sigma = 3/(2\Delta t)$ and the operators Δ_1 and Δ_2 are, respectively, $(\nabla^2 - r^{-2}I)$ and $(\nabla^2 - 2r^{-1}\partial_r)$ expressed in terms of ∇^2 defined by Eq.(5.35). Note that the eigenvalues of the discrete Chebyshev collocation form of the r-part of these operators associated with Dirichlet conditions are found to be real and distinct. The azimuthal velocity v^{n+1} is calculated first, then its value is included in the forcing term φ_ω of Eq.(6.154), and the Stokes problem is solved by means of the influence matrix method presented in Section 6.4.2. The algebraic systems are solved by partial diagonalization (Section 3.7.2, Remark 2) but the full matrix-diagonalization procedure could have been used as well.

Among the various flow regimes discussed by Crespo del Arco et al. (1996), we display here two typical cases.

The first case concerns the steady flow obtained for $C_w = 100$, $E = 2.24 \times 10^{-3}$, $L = 3.37$, and $R_m = 1.22$. Calculations are done with 41 collocation points in each direction and with Δt varying between 3×10^{-6} and 4×10^{-5}. These values of Δt may seem to be very small but the steady state is reached after a dimensionless time equal to 0.01, that is, relatively small. As a matter of fact, the rotation period $t_\Omega^* = 2\pi/\Omega$ constitutes a time scale more appropriate to the physical situation and the associated dimensionless Δt changes into the order of 10^{-4}. Figure 6.4.a shows the iso-streamfunction lines (projection of the streamlines in the meridian plane) exhibiting the presence of the Eckman layers at the horizontal walls. The profiles of the radial velocity u (Fig.6.4.b) and the azimuthal velocity v (Fig.6.4.c) at the midsection $\rho = 0$, show that the representation of the Eckman layer is quite satisfactory thanks to the well-adapted distribution of the collocation points near the boundaries.

The second case concerns the time-dependent periodic flow occurring when the flow rate is increased, namely, $C_w = 120$. The instantaneous iso-streamfunction line pattern presents a close resemblance to that shown in

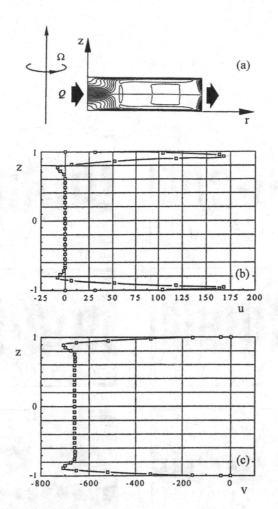

FIGURE 6.4. Steady flow for $C_w = 100$, $E = 2.24 \times 10^{-3}$, $L = 3.37$, and $R_m = 1.22$: (a) iso-streamfunction lines, (b) profile of the radial velocity at $\rho = 0$, (c) profile of the azimuthal velocity at $\rho = 0$.

Fig.6.4.a. The unsteady behaviour of the flow is illustrated (Fig.6.5) by the iso-lines of the perturbation of the axial velocity component w in the core region (i.e., excluding the inlet and outlet). The perturbations refer to a steady-state flow determined by extrapolation of the steady flows obtained for lower values of C_w.

The flow considered in this section constitutes a typical example to which a spectral method is well-adapted : a simple rectangular domain, the presence of boundary layers, and the superposition of large structures and fine structures of small relative amplitudes and small scales.

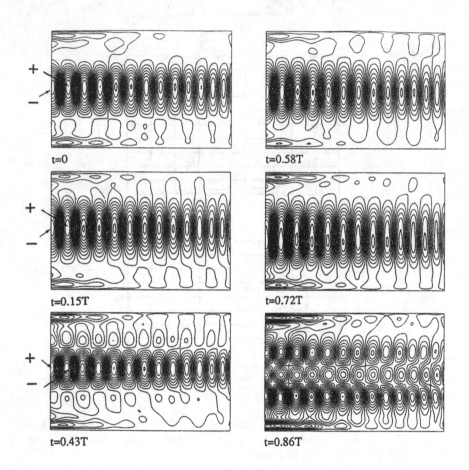

FIGURE 6.5. Iso-lines of the perturbation of the axial velocity w of the periodic flow (period T) in the core region, for $C_w = 120$, $E = 2.24 \times 10^{-3}$, $L = 3.37$, and $R_m = 1.22$.

7
Velocity-Pressure Equations

The solution of the Navier-Stokes equations in velocity-pressure variables is the subject of this chapter. This set of equations is of broader application than the vorticity-streamfunction equations which are restricted to two-dimensional flows. First, the Fourier method for computing fully periodic flows is discussed. Then the major part of the chapter is devoted to the case of one or more nonperiodic directions. In such a situation, the classical difficulty is the determination of the pressure field ensuring that the velocity field is solenoidal. This problem is discussed according to the kind of method used for the advancement in time. Time-discretization methods leading, at each time-cycle, either to a Stokes problem or to a combination of Helmholtz and Darcy problems will be considered. Various solution methods will be discussed and compared in typical examples. Finally, an application to the calculation of a three-dimensional rotating flow will be presented.

7.1 Introduction

This chapter is devoted to the solution of the Navier-Stokes equations in the primitive variable formulation. These equations (see Chapter 5) are

$$\partial_t \mathbf{V} + \mathbf{A}\left(\mathbf{V}, \mathbf{V}\right) + \nabla p - \nu \nabla^2 \mathbf{V} = \mathbf{f} \tag{7.1}$$

$$\nabla \cdot \mathbf{V} = 0\,, \tag{7.2}$$

where $\mathbf{V} = (u, v, w)$ is the velocity vector and p is the pressure. The initial condition associated with Eqs.(7.1) and (7.2) is

$$\mathbf{V} = \mathbf{V}_0 = (u_0, v_0, w_0) \tag{7.3}$$

and typical boundary conditions have been discussed in Chapter 5. Several configurations will be considered in the present chapter : (1) fully periodic three-dimensional flows, (2) flow in a two-dimensional channel with a periodic direction, (3) flow in a two-dimensional square domain.

7.2   Fourier-Fourier-Fourier method

In this section we present the Fourier method for the calculation of the three-dimensional spatially periodic solutions. This kind of solution is adapted to the study of homogeneous turbulence. The application of truncated Fourier series expansion to the direct numerical simulation (DNS) of turbulence has been initiated by the works by Orszag (1969, 1971, 1972) and Orszag and Patterson (1972). The success of these pioneering works was due to a clever combination of the efficiency of the FFT and of high-speed computing facilities.

Two points are important in the design of an efficient method : (1) the time-discretization scheme, and (2) the form and treatment of the nonlinear term. These questions are addressed now.

We assume that the forcing term $\mathbf{f}$ in Eq.(7.1) and the initial velocity field $\mathbf{V}_0$ are 2π-periodic in the three spatial directions. The velocity vector $\mathbf{V}(\mathbf{x}, t)$ and the pressure $p(\mathbf{x}, t)$ with $\mathbf{x} = (x_1, x_2, x_3)$ are approximated according to

$$\mathbf{V}(\mathbf{x}, t) \cong \mathbf{V}_K(\mathbf{x}, t) = \sum_{\mathbf{k}} \hat{\mathbf{V}}_{\mathbf{k}}(t) \, e^{i\,\mathbf{k}\cdot\mathbf{x}}, \tag{7.4}$$

$$p(\mathbf{x}, t) \cong p_K(\mathbf{x}, t) = \sum_{\mathbf{k}} \hat{p}_{\mathbf{k}}(t) \, e^{i\,\mathbf{k}\cdot\mathbf{x}}. \tag{7.5}$$

In these expressions, $\mathbf{k} = (k_1, k_2, k_3)$ is the wavenumber vector, and the sums are taken over k_1, k_2 and k_3 such that $-K_j \le k_j \le K_j$, $j = 1, 2, 3$. Therefore, the notation $\hat{\phi}_{\mathbf{k}}$ means $\hat{\phi}_{k_1, k_2, k_3}$, and we set $k^2 = |\mathbf{k}|^2 = k_1^2 + k_2^2 + k_3^2$.

Now we apply the Fourier Galerkin method to the Navier-Stokes equations (7.1) and (7.2) with the approximation (7.4)-(7.5). We get a set of differential equations for determining the Fourier coefficients $\hat{\mathbf{V}}_{\mathbf{k}}$ and $\hat{p}_{\mathbf{k}}$:

$$d_t \hat{\mathbf{V}}_{\mathbf{k}} + \hat{\mathbf{A}}_{\mathbf{k}} + i\,\mathbf{k}\,\hat{p}_{\mathbf{k}} + \nu\,k^2\,\hat{\mathbf{V}}_{\mathbf{k}} = \hat{\mathbf{f}}_{\mathbf{k}} \tag{7.6}$$

$$i\,\mathbf{k} \cdot \hat{\mathbf{V}}_{\mathbf{k}} = 0, \tag{7.7}$$

to be solved for $t > 0$, with the initial condition

$$\hat{\mathbf{V}}_{\mathbf{k}}(0) = \hat{\mathbf{V}}_{0,\mathbf{k}}.$$ (7.8)

In Eqs.(7.6) and (7.8), $\hat{\mathbf{A}}_{\mathbf{k}}$, $\hat{\mathbf{f}}_{\mathbf{k}}$, and $\hat{\mathbf{V}}_{0,\mathbf{k}}$ are the Fourier coefficients of $\mathbf{A}$, $\mathbf{f}$, and $\mathbf{V}_0$, respectively. The pressure is eliminated by applying the divergence operator in the Fourier space to Eq.(7.6), that is, by taking the scalar product of Eq.(7.6) with $i\,\mathbf{k}$, and using Eq.(7.7). We obtain

$$-k^2\,\hat{p}_{\mathbf{k}} = i\,\mathbf{k}\cdot\hat{\mathbf{f}}_{\mathbf{k}} - i\,\mathbf{k}\cdot\hat{\mathbf{A}}_{\mathbf{k}}$$

and Eq.(7.6) becomes

$$d_t\hat{\mathbf{V}}_{\mathbf{k}} + \nu\,k^2\,\hat{\mathbf{V}}_{\mathbf{k}} = \hat{\mathbf{g}}_{\mathbf{k}} - \hat{\mathbf{C}}_{\mathbf{k}}$$ (7.9)

with

$$\hat{\mathbf{g}}_{\mathbf{k}} = \hat{\mathbf{f}}_{\mathbf{k}} - \frac{\mathbf{k}\cdot\hat{\mathbf{f}}_{\mathbf{k}}}{k^2}\,\mathbf{k},$$ (7.10)

$$\hat{\mathbf{C}}_{\mathbf{k}} = \hat{\mathbf{A}}_{\mathbf{k}} - \frac{\mathbf{k}\cdot\hat{\mathbf{A}}_{\mathbf{k}}}{k^2}\,\mathbf{k}.$$ (7.11)

It is assumed that $k^2 \neq 0$. If $\mathbf{k} = \mathbf{0}$, Eq.(7.7) is automatically satisfied and Eq.(7.6) determines completely the velocity $\hat{\mathbf{V}}_{\mathbf{0}}$. The pressure $\hat{p}_{\mathbf{0}}$ is arbitrary in agreement with the property of the Navier-Stokes equations. We may simply set $\hat{p}_{\mathbf{0}} = 0$, implying that the mean value of the pressure is zero.

7.2.1   Time-discretization schemes

In the commonly used time-discretization schemes the nonlinear term $\hat{\mathbf{C}}_{\mathbf{k}}$ is treated in an explicit way. For example, Orszag and Patterson (1972) used the second-order Leap-Frog/Crank-Nicolson scheme (4.50). The advantage of this scheme is that it does not introduce numerical dissipation. On the other hand, the leap-frog time-stepping slightly uncouples even and odd time-cycles, thus producing oscillations. This is generally avoided by averaging the numerical solution periodically every few time-cycles by a procedure that conserves the second-order accuracy (see Peyret and Taylor, 1983). Moreover, the Crank-Nicolson scheme presents the drawback of not damping sufficiently the high-frequency modes. Higher-order methods can be used like the Runge-Kutta method (Rogallo, 1977). We describe now a "semi-exact" scheme proposed by Basdevant (1982) and used in some studies of homogeneous turbulence (Roy, 1982, 1986 ; Vincent and Meneguzzi, 1991).

The semi-exact scheme consists of integrating formally Eq.(7.9) between $t_\star = (n+1-m)\,\Delta t$ and $t_{n+1} = (n+1)\,\Delta t$. Then the integrals involving the right-hand side of Eq.(7.9) are approximated by means of classical integration formulas. More precisely, the integration of Eq.(7.9) gives

$$\hat{\mathbf{V}}_{\mathbf{k}}^{n+1} = \hat{\mathbf{V}}_{\mathbf{k}}^{n+1-m}\, e^{-\nu k^2 m\,\Delta t} + e^{-\nu k^2 (n+1)\,\Delta t}\, [\mathbf{I}_1\,(t_{n+1}, t_\star) - \mathbf{I}_2\,(t_{n+1}, t_\star)]\ ,$$

where

$$\mathbf{I}_{1.}(t_{n+1}, t_\star) = \int_{t_\star}^{t_{n+1}} e^{\nu k^2 \tau}\, \hat{\mathbf{g}}_{\mathbf{k}}\,(\tau)\ d\tau\ ,$$

$$\mathbf{I}_2\,(t_{n+1}, t_\star) = \int_{t_\star}^{t_{n+1}} e^{\nu k^2 \tau}\, \hat{\mathbf{C}}_{\mathbf{k}}\,(\tau)\ d\tau\ .$$

The choice of m, namely, $t_\star$, and that of the integration formula define the scheme. For $m = 2$, $\mathbf{I}_1$ and $\mathbf{I}_2$, evaluated with the midpoint rule, give the second-order scheme of leap-frog type :

$$\hat{\mathbf{V}}_{\mathbf{k}}^{n+1} = \hat{\mathbf{V}}_{\mathbf{k}}^{n-1}\, e^{-2\nu k^2 \Delta t} - 2\,\Delta t\, e^{-\nu k^2 \Delta t}\, \left(\hat{\mathbf{C}}_{\mathbf{k}}^{n} - \hat{\mathbf{g}}_{\mathbf{k}}^{n}\right)\ . \tag{7.12}$$

Another choice, considered by Vincent and Meneguzzi (1991), consists of setting $m = 1$, namely, $t_\star = t_n$. Then, $\mathbf{I}_1\,(t_{n+1}, t_n)$ is evaluated with the trapezoidal rule. The integral $\mathbf{I}_2\,(t_{n+1}, t_n)$ is written as

$$\mathbf{I}_2\,(t_{n+1}, t_n) = \mathbf{I}_2\,(t_{n+1}, t_{n-1}) - \mathbf{I}_2\,(t_n, t_{n-1})\ ,$$

then $\mathbf{I}_2\,(t_{n+1}, t_{n-1})$ is evaluated with the midpoint rule and $\mathbf{I}_2\,(t_n, t_{n-1})$ with the trapezoidal rule. Finally we get the second-order scheme of Adams-Bashforth type

$$\hat{\mathbf{V}}_{\mathbf{k}}^{n+1} = \hat{\mathbf{V}}_{\mathbf{k}}^{n}\, e^{-\nu k^2 \Delta t}\quad - \frac{\Delta t}{2}\left(3\, e^{-\nu k^2 \Delta t}\, \hat{\mathbf{C}}_{\mathbf{k}}^{n} - e^{-2\nu k^2 \Delta t}\, \hat{\mathbf{C}}_{\mathbf{k}}^{n-1}\right)$$
$$\tag{7.13}$$
$$+ \frac{\Delta t}{2}\left(\hat{\mathbf{g}}_{\mathbf{k}}^{n+1} + e^{-\nu k^2 \Delta t}\hat{\mathbf{g}}_{\mathbf{k}}^{n}\right)\ .$$

7.2.2   The nonlinear term

The form and numerical treatment of the nonlinear term $\hat{\mathbf{C}}_{\mathbf{k}}$ play a fundamental part in the stability of the scheme and in the quality of the numerical results. As mentioned in Chapter 5, the nonlinear term in the Navier-Stokes equations characterizing the convection of momentum can be considered in any equivalent form : convective, conservative, skew-symmetric, or rotational. This latter form is, however, somewhat different because the term $|\mathbf{V}|^2 /2$ is included in the pressure term. In the discrete problem, the equivalence may be lost so that different behaviours can be expected. This question has been discussed by Canuto et al. (1988), Zang (1991), Blaisdell et *al.* (1996), and by Kravchenko and Moin (1997). We resume here the main results of these investigations.

In case of zero viscosity ($\nu = 0$) and without the forcing term ($\mathbf{f} = \mathbf{0}$), the resulting Euler equations conserve momentum Q and energy E, namely,

$$Q = \int_\Omega \mathbf{V} \, d\Omega \, , \quad E = \frac{1}{2} \int_\Omega |\mathbf{V}|^2 \, d\Omega \, , \quad \Omega = [0, 2\pi]^3 \, . \tag{7.14}$$

It is highly desirable that these conservation properties be preserved in the discrete framework. The conservation of energy is especially important because it controls the stability of the numerical approximations (when $\nu = 0$). These properties are denominated "semiconservation properties" when the Navier-Stokes equations are considered.

If the nonlinear term $\hat{\mathbf{C}}_{\mathbf{k}}$ was approximated by the Galerkin method, the various form would be equivalent and the semiconservation properties would be satisfied. In practical calculations, however, the nonlinear term $\hat{\mathbf{C}}_{\mathbf{k}}$ at time-levels n or $n - 1$ is evaluated through the pseudospectral technique (Section 2.8). Due to aliasing errors, the resulting discrete term is different according to the form of the nonlinear term, thus implying different properties. More precisely, it is found that the momentum Q and energy E are semiconserved by the skew-symmetric and rotational forms. On the other hand, only Q is semiconserved by the conservative form and neither Q nor E is semiconserved by the convective form. Consequently, we may expect these last two forms to be unstable for low viscosity, while the first two forms are stable. This has effectively been observed in the numerical calculations reported in the above-referenced works.

It is interesting to compare the discrete version of the conservative and convective forms with a simple one-dimensional example (Blaisdell *et al.*, 1996 ; Kravchenko and Moin, 1997). Let us consider $u(x)$ and $v(x)$, and the equivalent two quantities

$$C^{(1)} = (u\,v)' \, , \quad C^{(2)} = u\,v' + v\,u' \, .$$

When determined with the pseudospectral technique, the Fourier coefficients $\hat{C}_k^{(1)}$ and $\hat{C}_k^{(2)}$ are, respectively, [see Eq.(2.56)] :

$$\hat{C}_k^{(1)} = \imath\,k \sum_{\substack{p,q \\ p+q=k}} \hat{u}_p\,\hat{v}_q + \imath \left[\sum_{\substack{p,q \\ p+q=k+N}} (p+q)\,\hat{u}_p\,\hat{v}_q \right.$$

$$\left. + \sum_{\substack{p,q \\ p+q=k-N}} (p+q)\,\hat{u}_p\,\hat{v}_q \right] \tag{7.15}$$

and

$$\hat{C}_k^{(2)} = \underline{i}\,k \sum_{\substack{p,q \\ p+q=k}} \hat{u}_p\,\hat{v}_q + \underline{i}\left[\sum_{\substack{p,q \\ p+q=k+N}} (p+q-N)\,\hat{u}_p\,\hat{v}_q \right.$$

$$\left. + \sum_{\substack{p,q \\ p+q=k-N}} (p+q+N)\,\hat{u}_p\,\hat{v}_q \right], \tag{7.16}$$

where the sums are taken over $p = -K, \ldots, K$, $q = -K, \ldots, K$, and $N = 2K + 1$. In each of these expressions, the first sum is exactly the Fourier coefficient which would be obtained with the Galerkin method. The other sums in square brackets are aliasing terms. Moreover, it is easy to see that the aliasing terms in $\hat{C}_k^{(1)}$ and $\hat{C}_k^{(2)}$ have opposite signs. Indeed, the first sums in the square brackets are taken for $0 \le p + q \le 2K$ so that $-(2K + 1) \le p + q - N \le -1$ and the second sums are taken for $-2K \le p + q \le 0$ so that $1 \le p + q + N \le 2K + 1$. Therefore, the aliasing error associated with the skew-symmetric form $\left(\hat{C}_k^{(1)} + \hat{C}_k^{(2)}\right)/2$ is much reduced, although not completely removed. This explains why the skew-symmetric form, even without aliasing removal, generally·gives stable and accurate results and is recommended.

Concerning the rotational form, it has been observed (Zang, 1991 ; Kravchenko and Moin, 1997) that, even with aliasing errors, the numerical integration is stable but the results show an inability to maintain the turbulence at the proper level. Nevertheless, Vincent and Meneguzzi (1991), using a high spatial resolution, have obtained satisfactory results which did not significantly differ from those obtained after removing the aliasing errors.

When aliasing is removed using the 3/2 rule, for example (or other techniques as discussed by Canuto et al., 1988), the discrete version of any form of the nonlinear term is equivalent to the Galerkin approximation and, consequently, is stable.

7.3 Fourier-Chebyshev method

The present section is devoted to the solution of the Navier-Stokes equations in a two-dimensional channel with one periodic direction. Besides its intrinsic physical interest, such a configuration allows us to discuss the fundamentals of the numerical methods in a simple case where the necessary algebra remains simple.

7.3.1 Flow in a plane channel

In this section, we consider the flow in an infinite channel in the x-direction with 2π-periodicity in this direction. Thus, the domain Ω in which Eqs. (7.1) and (7.2) are solved is defined by $0 \leq x \leq 2\pi$, $-1 \leq y \leq 1$. The forcing term $\mathbf{f} = (f_u, f_v)$ and the initial velocity field

$$\mathbf{V} = \mathbf{V}_0 \tag{7.17}$$

are assumed to be 2π-periodic with respect to x. The boundary conditions are

$$\mathbf{V}\left(x, \pm 1, t\right) = \mathbf{V}_\Gamma^\pm \left(x, t\right) = \left(u_\Gamma^\pm, v_\Gamma^\pm\right), \tag{7.18}$$

where $\mathbf{V}_\Gamma^\pm$ is also 2π-periodic. The conservation of the total flow rate demands

$$\int_0^{2\pi} \left(v_\Gamma^+ - v_\Gamma^-\right) dy = 0, \tag{7.19}$$

that is, the mean values of v_Γ^+ and v_Γ^- must be equal. Note that in the case of no-slip, impermeable walls, we have simply

$$\mathbf{V}\left(x, \pm 1, t\right) = \mathbf{0}. \tag{7.20}$$

Also, the case of a shear-stress-free wall may be considered, as defined by Eq.(6.14).

Now the solution of the problem is such that the velocity $\mathbf{V}$ and the pressure gradient ∇p are 2π-periodic. Under this circumstance, the mean pressure gradient γ is assumed to be known (see Section 6.3.4).

7.3.2 Time-dependent Stokes equations

For a presentation of the basic properties of the solution methods, we begin with the time-dependent Stokes equations

$$\partial_t \mathbf{V} - \nu \nabla^2 \mathbf{V} + \nabla p = \mathbf{f}, \quad -1 < y < 1 \tag{7.21}$$

$$\nabla \cdot \mathbf{V} = 0, \quad -1 < y < 1 \tag{7.22}$$

to be solved with the initial condition (7.17) and the boundary conditions (7.18). The solution $(\mathbf{V}, p)$ is sought in the form of truncated Fourier series with respect to x :

$$\mathbf{V}_K \left(x, y, t\right) = \sum_{k=-K}^{K} \hat{\mathbf{V}}_k \left(y, t\right) e^{ikx},$$

$$p_K \left(x, y, t\right) = \gamma x + \sum_{k=-K}^{K} \hat{p}_k \left(y, t\right) e^{ikx}.$$

The Fourier Galerkin equations (see Section 2.6.1), satisfied by $\left(\hat{\mathbf{V}}_k, \hat{p}_k\right)$, $k = -K, \ldots, K$, with $\hat{\mathbf{V}}_k = (\hat{u}_k, \hat{v}_k)$, are

$$\partial_t \hat{u}_k - \nu \left(\partial_{yy} - k^2\right) \hat{u}_k + \underline{i} k \hat{p}_k = \hat{f}_{u,k} \tag{7.23}$$

$$\partial_t \hat{v}_k - \nu \left(\partial_{yy} - k^2\right) \hat{v}_k + \partial_y \hat{p}_k = \hat{f}_{v,k} \tag{7.24}$$

$$\underline{i} k \hat{u}_k + \partial_y \hat{v}_k = 0, \tag{7.25}$$

where γ has been included in $\hat{f}_{u,k}$. These equations are solved in $-1 < y < 1$ and $t > 0$ with the boundary conditions

$$\hat{u}_k \left(\pm 1, t\right) = \hat{u}_{\Gamma,k}^{\pm} \left(t\right), \quad \hat{v}_k \left(\pm 1, t\right) = \hat{v}_{\Gamma,k}^{\pm} \left(t\right) \tag{7.26}$$

and the initial conditions

$$\hat{u}_k \left(y, 0\right) = \hat{u}_{0,k} \left(y\right), \quad \hat{v}_k \left(y, 0\right) = \hat{v}_{0,k} \left(y\right) \tag{7.27}$$

satisfying the incompressibility condition

$$\underline{i} k \hat{u}_{0,k} + \partial_y \hat{v}_{0,k} = 0. \tag{7.28}$$

The usual time-discretization schemes for the Stokes (or Navier-Stokes) equations can be classified into two categories according to the type of elliptic problem which has to be solved at each time-cycle. The first category leads to the solution of a Stokes problem and the second one to the solution of Helmholtz and Darcy problems. These two types of time-discretizations and the associated solution methods are now discussed. The discussion will concern, more precisely, the following points : (1) the formulation of the discrete problem, (2) the resolvability of the resulting algebraic system, (3) the stability of the time-scheme and (4) the algorithm for solving the algebraic system.

(a) Time-schemes leading to the Stokes problem

We consider the general three-level scheme presented in Section 4.4.1 and already discussed (Section 6.3.1.b) in the case of the vorticity-streamfunction equations. Among the schemes (4.46) we select the subclass defined by $\theta_1 = \theta = 1 - \theta_2$, so that the family of schemes depends only on the two parameters ε and θ when applied to the time-dependent Stokes equations. The semidiscrete equations are then

$$\frac{(1+\varepsilon) \hat{u}_k^{n+1} - 2\varepsilon \hat{u}_k^n - (1-\varepsilon) \hat{u}_k^{n-1}}{2\Delta t} - \nu \left(\partial_{yy} - k^2\right) \left[\theta \hat{u}_k^{n+1} + (1 - \theta) \hat{u}_k^n\right]$$

$$+ \underline{i} k \hat{p}_k^{n+\tau} = \theta \hat{f}_{u,k}^{n+1} + (1 - \theta) \hat{f}_{u,k}^n \tag{7.29}$$

$$\frac{(1 + \varepsilon)\,\hat{v}_k^{n+1} - 2\varepsilon\hat{v}_k^n - (1 - \varepsilon)\,\hat{v}_k^{n-1}}{2\Delta t} - \nu\left(\partial_{yy} - k^2\right)\left[\theta\hat{v}_k^{n+1} + (1 - \theta)\,\hat{v}_k^n\right]$$

$$+\;\partial_y\hat{p}_k^{n+\tau} = \theta\hat{f}_{v,k}^{n+1} + (1 - \theta)\,\hat{f}_{v,k}^n \tag{7.30}$$

$$ik\hat{u}_k^{n+1} + \partial_y\hat{v}_k^{n+1} = 0 \tag{7.31}$$

$$\hat{u}_k^{n+1}(\pm1) = \left(\hat{u}_{\Gamma,k}^\pm\right)^{n+1} \tag{7.32}$$

$$\hat{v}_k^{n+1}(\pm1) = \left(\hat{v}_{\Gamma,k}^\pm\right)^{n+1}, \tag{7.33}$$

where $\hat{u}_k^n \cong \hat{u}_k\,(y, n\Delta t)$, $\hat{v}_k^n \cong \hat{v}_k\,(y, n\Delta t)$, and $\hat{p}_k^{n+\tau} \cong \hat{p}_k\,[y, (n+\tau)\,\Delta t]$. The number τ which characterizes the time $(n+\tau)\,\Delta t$ with $\tau > 0$ is arbitrary and must be chosen in order to ensure the better truncation error. This is due to the fact that no time derivative of the pressure appears in the equations so that the pressure is defined within an arbitrary function of time. As a matter of fact, it is the pressure gradient ∇p itself which should be fixed at the time-level $n+\tau$. In the above equations, the truncation error analysis shows that τ can be taken equal to θ, except in the case $\theta = 0$ (first-order scheme) where τ is arbitrary ($\tau > 0$) and, generally, is taken equal to 1. It must be noted that for $k = 0$, the pressure term disappears from Eq.(7.29) and the mode $\hat{p}_0$ appears only through its y-derivative in Eq.(7.30).

Therefore, the equations for determining the unknowns $(\hat{u}_k^{n+1}, \hat{v}_k^{n+1}, \hat{p}_k^{n+1})$ constitute a Stokes-type problem, also called a "generalized Stokes-problem". The classical Stokes problem corresponds to $\sigma = 0$. We use the simpler nomenclature "Stokes problem" even if $\sigma \neq 0$.

Now the discretization of Eqs.(7.29)-(7.33) with respect to y makes use of the Chebyshev collocation approximation. Therefore, the Fourier coefficients of u, v, and p at time $n\Delta t$ are approximated with polynomials of degree at most N (namely, $\hat{u}_{kN}^n$, $\hat{v}_{kN}^n$, and $\hat{p}_{kN}^n$), and Eqs. (7.29)-(7.33) are discretized using the collocation method (see Section 3.4.2) based on the Gauss-Lobatto points y_j [Eq.(6.26)]. The derivatives at each collocation point y_j are expressed in terms of the grid values by the usual expressions developed in Section 3.3.4.

By prescribing the equations (7.29)-(7.31) to be satisfied at all inner points y_j, $j = 1, \ldots, N$, and adding the boundary conditions, we get a system of $3N + 1$ complex equations for the $3\,(N + 1)$ complex unknowns (including the boundary values of the velocity). Hence, two equations must be added. One possibility is to add the normal momentum equation written for $j = 0$ and $j = N$. But, as shown by Orszag $et\ al.$ (1986), such a technique may be subject to instabilities which might be avoided by using a special form of the diffusive term (see Section 7.4.2.c). Another way to close the system is to add the incompressibility equation (7.31) prescribed at the boundaries $j = 0$ and $j = N$. We consider here this second option which is in common use. Therefore, we get an algebraic system of $3\,(N+1)$ equations for the $3\,(N+1)$ unknowns $\hat{u}_{kN}^{n+1}\,(y_j)$, $\hat{v}_{kN}^{n+1}\,(y_j)$, and $\hat{p}_{kN}^{n+1}\,(y_j)$, $j = 0, \ldots, N$.

For $k \neq 0$, it can be shown theoretically (Bernardi and Maday, 1992) and numerically (Bwemba, 1994) that the algebraic system admits a unique solution. For $k = 0$, the matrix of the system has two eigenvalues equal to zero. One corresponds to the fact that the pressure is defined up to a constant, but the other zero eigenvalue corresponds to a spurious pressure mode. This issue will be addressed at the end of the section, and up to that point we assume that $k \neq 0$.

The stability has been numerically studied by Bwemba (1994) in the case of the two-level scheme (θ-scheme) defined by $\varepsilon = 1$. By considering the homogeneous problem ($\hat{u}_{\Gamma,k}^{\pm} = \hat{v}_{\Gamma,k}^{\pm} = \hat{f}_{u,k} = \hat{f}_{v,k} = 0$), the algebraic system has the form

$$\mathcal{A}_1 \Phi^{n+1} + \mathcal{A}_0 \Phi^n = 0 \qquad (7.34)$$

where Φ^{n+1} is the $(3N-1)$-vector defined by

$$\Phi^{n+1} = \left(\hat{u}_{kN}^{n+1}\,(y_1), \ldots, \hat{u}_{kN}^{n+1}\,(y_{N-1}), \hat{v}_{kN}^{n+1}\,(y_1), \right.$$

$$\left. \ldots, \hat{v}_{kN}^{n+1}\,(y_{N-1}), \hat{p}_{kN}^{n+\tau}\,(y_0), \ldots, \hat{p}_{kN}^{n+\tau}\,(y_N) \right)^T.$$

The expression for the $(3N-1) \times (3N-1)$ matrices $\mathcal{A}_0$ and $\mathcal{A}_1$ is easily obtained from the Chebyshev differentiation matrix $\mathcal{D}$. A necessary condition for stability is that the spectral radius of $\mathcal{E} = \mathcal{A}_1^{-1} \mathcal{A}_0$ is not larger than 1. The eigenvalues μ of the matrix $\mathcal{E}$ are the roots of the equation

$$\det\left(\mathcal{A}_0 + \mu \mathcal{A}_1 \right) = 0 \qquad (7.35)$$

numerically calculated by means of a computer routine for generalized eigenvalue problems. Because the pressure appears only in $\mathcal{A}_1 \Phi^{n+1}$ (i.e., not in $\mathcal{A}_0 \Phi^n$) and the divergence equation only at level $n+1$, the equation (7.35) admits only $N-3$ nonzero roots. It is found that $\rho\,(\mathcal{E}) \leq 1$ for any Δt if $\theta \geq 1/2$. For $\theta < 1/2$, there exists a critical time-step Δt_* such that $\rho\,(\mathcal{E}) \leq 1$ if $\Delta t \leq \Delta t_*$. This result is in agreement with the usual result concerning the stability of the θ-scheme applied to the diffusion equation (Section 4.3.2), contrary to the case of the vorticity-streamfunction equations (Section 6.3.2.e).

For several reasons already discussed (Sections 4.4.2.a, 4.5.1.b), the three-level scheme defined by $\varepsilon = 2$ and $\theta = \tau = 1$ is preferred to the two-level Crank-Nicolson scheme defined by $\varepsilon = 1$ and $\theta = \tau = 1/2$. The apparent advantage of the Crank-Nicolson scheme concerning the number of time-levels involved disappears when considering the Navier-Stokes equations, since both schemes use a three-level Adams-Bashforth extrapolation for the nonlinear convective term. The solution of the algebraic system is addressed in the next section.

(b) Solution of the Stokes problem by the Uzawa method

As indicated, the implicit discretization of the time-dependent Stokes equations leads to the solution of a Stokes problem at each time-cycle. In this section, we discuss the solution of the discrete Stokes problem by means of the Uzawa method.

Let us consider Eqs. (7.29)-(7.33) in the case $\varepsilon = 2$ and $\theta = \tau = 1$. Then, for the sake of simplification, we set

$$\underline{i}\hat{u}_k^{n+1} = u, \quad \hat{v}_k^{n+1} = v, \quad \hat{p}_k^{n+1} = p, \quad \frac{3}{2\nu\Delta t} + k^2 = \sigma,$$

$$-\frac{i}{\nu}\hat{f}_{u,k}^{n+1} - \underline{i}\frac{4\hat{u}_k^n - 3\hat{u}_k^{n-1}}{2\Delta t} = f_u, \quad -\frac{1}{\nu}\hat{f}_{v,k}^{n+1} - \underline{i}\frac{4\hat{v}_k^n - 3\hat{v}_k^{n-1}}{2\Delta t} = f_v,$$

$$\underline{i}\left(\hat{u}_{\Gamma,k}^{\pm}\right)^{n+1} = \gamma_{\pm}, \quad \left(\hat{v}_{\Gamma,k}^{\pm}\right)^{n+1} = g_{\pm}.$$

Hence, the Stokes problem to be solved for each Fourier mode and at each time-cycle is

$$u'' - \sigma u + kp = f_u, \quad -1 < y < 1 \tag{7.36}$$

$$v'' - \sigma v - p' = f_v, \quad -1 < y < 1 \tag{7.37}$$

$$ku + v' = 0, \quad -1 < y < 1 \tag{7.38}$$

$$u(-1) = \gamma_-, \quad u(1) = \gamma_+ \tag{7.39}$$

$$v(-1) = g_-, \quad v(1) = g_+. \tag{7.40}$$

In the Chebyshev collocation method considered here, the momentum equations (7.36) and (7.37) are enforced at the inner collocation points y_j, $j = 1, \ldots, N-1$ [see Eq.(6.26)], while the incompressibility equation (7.38) is enforced at all collocation points, including the boundaries $y_0 = 1$ and $y_N = -1$. The velocity (u, v) and the pressure p are approximated with polynomials of degree at most N, namely, $u \cong u_N \in \mathbb{P}_N$, $v \cong v_N \in \mathbb{P}_N$, and $p \cong p_N \in \mathbb{P}_N$. Such an approximation, where velocity and pressure are both approximated in $\mathbb{P}_N$, is usually denoted by $\mathbb{P}_N - \mathbb{P}_N$, in contrast with the method where the pressure is approximated with a polynomial of degree $N - 2$ ($\mathbb{P}_N - \mathbb{P}_{N-2}$ approximation). This point will be discussed below and especially in Section 7.4.2.b.

We set $u_j = u_N(y_j)$, $v_j = v_N(y_j)$, $p_j = p_N(y_j)$, and we introduce the notations

$$D\phi_j = \sum_{l=0}^{N} d_{j,l}^{(1)}\phi_l, \quad D^2\phi_j = \sum_{l=0}^{N} d_{j,l}^{(2)}\phi_l. \tag{7.41}$$

The system (7.36)-(7.40) is discretized according to

$$\left(D^2 - \sigma\right)u_j + kp_j = f_{u,j}, \quad j = 1, \ldots, N-1, \tag{7.42}$$

$$\left(D^2 - \sigma\right)v_j - Dp_j = f_{v,j}, \quad j = 1, \ldots, N-1, \tag{7.43}$$

$$k\, u_j + D\, v_j = 0, \quad j = 0, \ldots, N, \tag{7.44}$$

$$u_0 = \gamma_+, \quad u_N = \gamma_-, \tag{7.45}$$

$$v_0 = g_+, \quad v_N = g_-. \tag{7.46}$$

It is convenient to write these equations in vector form. We introduce the vectors of the unknowns

$$U_\star = \begin{pmatrix} u_1 \\ \vdots \\ u_{N-1} \end{pmatrix}, \quad V_\star = \begin{pmatrix} v_1 \\ \vdots \\ v_{N-1} \end{pmatrix}, \quad P = \begin{pmatrix} p_0 \\ \vdots \\ p_N \end{pmatrix}$$

such that Eqs (7.42)-(7.44) become

$$\mathcal{H}\, U_\star + k\, \mathcal{I}_1\, P = S_u^\star \tag{7.47}$$

$$\mathcal{H}\, V_\star - \mathcal{D}_5\, P = S_v^\star \tag{7.48}$$

$$k\, \mathcal{I}_3\, U_\star + \mathcal{D}_6\, V_\star = S_Q, \tag{7.49}$$

with $\mathcal{H}$ defined by

$$\mathcal{H} = \mathcal{D}_2 - \sigma \mathcal{I},$$

where $\mathcal{D}_2$ is the $(N-1) \times (N-1)$ matrix defined in Section 4.4.2, namely,

$$\mathcal{D}_2 = \left[d_{i,j}^{(2)} \right], \quad i = 1, \ldots, N-1, \quad j = 1, \ldots, N-1. \tag{7.50}$$

The $(N-1) \times (N+1)$ matrix $\mathcal{D}_5$ is defined by

$$\mathcal{D}_5 = \left[d_{i,j}^{(1)} \right], \quad i = 1, \ldots, N-1, \quad j = 0, \ldots, N, \tag{7.51}$$

and the $(N+1) \times (N-1)$ matrix $\mathcal{D}_6$ by

$$\mathcal{D}_6 = \left[d_{i,j}^{(1)} \right], \quad i = 0, \ldots, N, \quad j = 1, \ldots, N-1. \tag{7.52}$$

Then $\mathcal{I}$ is the $(N-1) \times (N-1)$ identity matrix, $\mathcal{I}_1$ is the $(N-1) \times (N+1)$ matrix defined in Section 6.3.2.b [Eq.(6.33)], and $\mathcal{I}_3$ is the $(N+1) \times (N-1)$ matrix defined by

$$\mathcal{I}_3 = [\delta_{i-1,j}], \quad i = 0, \ldots, N, \quad j = 1, \ldots, N-1, \tag{7.53}$$

where $\delta_{i,j}$ is the Kronecker delta.

Lastly, the $(N-1)$-vectors, $S_u^\star = [s_{u,j}]$ and $S_v^\star = [s_{v,j}], j = 1, \ldots, N-1$, are, respectively, defined by

$$s_{u,j} = f_{u,j} - d_{j,0}^{(2)}\, \gamma_+ - d_{j,N}^{(2)}\, \gamma_-,$$

$$s_{v,j} = f_{v,j} - d_{j,0}^{(2)}\, g_+ - d_{j,N}^{(2)}\, g_-,$$

while the $(N+1)$-vector $S_Q = [s_{Q,j}]$, $j = 0, \ldots, N$, is defined by

$$s_{Q,0} = -k\,\gamma_+ - d^{(1)}_{0,0}\,g_+ - d^{(1)}_{0,N}\,g_-\,,$$

$$s_{Q,j} = -d^{(1)}_{j,0}\,g_+ - d^{(1)}_{j,N}\,g_-\,,\quad j = 1,\ldots,N-1\,,$$

$$s_{Q,N} = -k\,\gamma_- - d^{(1)}_{N,0}\,g_+ - d^{(1)}_{N,N}\,g_-\,.$$

The solution of (7.47)-(7.49) is obtained by the Uzawa method. The method consists of formally solving (7.47) and (7.48) with respect to $U_\star$ and $V_\star$:

$$U_\star = \mathcal{H}^{-1}\,(S^\star_u - k\mathcal{I}_1\,P)$$

$$V_\star = \mathcal{H}^{-1}\,(S^\star_v + \mathcal{D}_5\,P)\,,$$

which are then brought into Eq.(7.49) to give the equation

$$U\,P = G\,, \tag{7.54}$$

where U is the Uzawa operator defined by

$$U = \mathcal{D}_6\,\mathcal{H}^{-1}\,\mathcal{D}_5 - k^2\,\mathcal{I}_3\,\mathcal{H}^{-1}\,\mathcal{I}_1 \tag{7.55}$$

and

$$G = S_Q - k\mathcal{I}_3\,\mathcal{H}^{-1}\,S^\star_u - \mathcal{D}_6\,\mathcal{H}^{-1}\,S^\star_v\,.$$

Therefore, the direct Uzawa algorithm is as follows :

1. Calculate P by solving Eq.(7.54).
2. Calculate $U_\star$ and $V_\star$ by, respectively, solving Eqs.(7.47) and (7.48).

The implementation of the algorithm involves the solution of two algebraic systems with matrices U and $\mathcal{H}$, respectively. The eigenvalues of $\mathcal{H}$ are real and negative since this matrix is the discrete approximation to the Helmholtz equations with Dirichlet conditions (see Section 3.7.1). The matrix U has real positive eigenvalues except for $k = 0$, in which case two eigenvalues are zero (the case $k = 0$ is discussed below). For fixed N and k ($\neq 0$), the condition number $\kappa(\sigma)$ of U is a slowly varying function of σ. For example, for $N = 16$, $k = 1$, we find $\kappa(10) = 1.77 \times 10^{-3}$ and $\kappa(10^4) = 2.54 \times 10^{-4}$.

In the present one-dimensional case, the matrices U and $\mathcal{H}$ can be inverted and their inverse stored. Thus, at each time-step the above algorithm reduces to three matrix-vector products. On the other hand, in the multidimensional case, the inversions cannot be considered and the solution must be iterative. The crude iterative Uzawa procedure for solving the system (7.47)-(7.49) is the following :

$$P^{m+1} - P^m = -\alpha\,(k\mathcal{I}_3\,U^m_\star + \mathcal{D}_6\,V^m_\star - S_Q) \tag{7.56}$$

$$\mathcal{H} U_*^{m+1} = S_u^* - k \mathcal{I}_1 P^{m+1} \tag{7.57}$$

$$\mathcal{H} V_*^{m+1} = S_v^* + \mathcal{D}_5 P^{m+1}, \tag{7.58}$$

where m refers to the iteration and α is a relaxation parameter. The convergence of such an iterative procedure is very poor and must be improved by applying a preconditioning operator $\mathcal{A}_0$ to $P^{m+1} - P^m$ and calculating the relaxation parameter α in a dynamic way as described in Section 3.8. The choice of the preconditioner $\mathcal{A}_0$ is crucial and difficult. To our knowledge, the ideal preconditioner does not exist yet, mainly because of the poor properties of the Chebyshev collocation Uzawa operator.

Le Marec *et al.* (1996) have applied the preconditioner, originally proposed by Cahouet and Chabard (1988) for a finite-element method, to the Chebyshev collocation method. The idea is to examine what the Uzawa operator would be if the problem were also approximated in the y-direction by the Fourier rather than the Chebyshev method. For this it is convenient to come back to the continuous equations (7.36)-(7.3.21) and to consider their expressions in the Fourier space (with frequency λ) corresponding to y. We recall that $\sigma = k^2 + \sigma_0$ where k refers to the Fourier approximation in the x-direction and $\sigma_0 = 2/(3\nu\Delta t)$. Thus, the expression of the Uzawa operator in the (k, λ) Fourier space is

$$\frac{k^2 + \lambda^2}{\sigma_0 + k^2 + \lambda^2}$$

which may be considered as the Fourier approximation to the operator A such that

$$A_0 = \left[I - \sigma_0 \left(\partial_{xx} + \partial_{yy} \right)^{-1} \right]^{-1}.$$

This operator, supplemented with homogeneous Neumann boundary conditions, is proposed as a preconditioner by Cahouet and Chabard (1988). In the present case, the discrete form of A is the matrix

$$\mathcal{A}_0 = \left[\mathcal{I}_0 - \sigma_0 \left(\mathcal{D}_7 - k^2 \mathcal{I}_0 \right)^{-1} \right]^{-1},$$

where $\mathcal{I}_0$ is the $(N+1) \times (N+1)$ identity matrix and $\mathcal{D}_7$ is the $(N+1) \times (N+1)$ matrix approximating the second-order y-derivative operator with Neumann boundary conditions, that is,

$$\mathcal{D}_7 = \begin{pmatrix} d_{0,0}^{(1)} \cdots d_{0,N}^{(1)} \\ [\ \mathcal{D}_4\ ] \\ d_{N,0}^{(1)} \cdots d_{N,N}^{(1)} \end{pmatrix}, \tag{7.59}$$

where $\mathcal{D}_4$ is the $(N-1) \times (N+1)$ matrix defined in Eq.(6.36). Therefore, Eq.(7.56) is replaced by

$$\mathcal{A}_0 \left(P^{m+1} - P^m \right) = -\alpha_m \left(k \mathcal{I}_3 U_*^m + \mathcal{D}_6 V_*^m - S_Q \right), \tag{7.60}$$

where α_m is the relaxation parameter whose expression is specified below. The system consisting of Eqs.(7.57), (7.58) and (7.60) is solved iteratively according to the following algorithm :

1. *Initialization*
P^0 is given,

$$\mathcal{H} U_*^0 = S_u^* - k \mathcal{I}_1 P^0$$

$$\mathcal{H} V_*^0 = S_v^* + \mathcal{D}_5 P^0$$

$$R^0 = k \mathcal{I}_3 U_*^0 + \mathcal{D}_6 V_*^0 - S_Q,$$

and

$$-\left(\mathcal{D}_7 - k^2 \mathcal{I}_0\right) \Phi^0 = R^0$$

$$W^0 = R^0 + \sigma_0 \Phi^0.$$

2. *Current iteration* $(m \geq 0)$
P^m, U_*^m, V_*^m, R^m, and W^m are known,

$$\mathcal{H} \overline{U}_*^m = -k \mathcal{I}_1 W^m$$

$$\mathcal{H} \overline{V}_*^m = \mathcal{D}_5 W^m$$

$$\overline{R}^m = k \mathcal{I}_3 \overline{U}_*^m + \mathcal{D}_6 \overline{V}_*^m$$

then

$$P^{m+1} = P^m - \alpha_m W^m$$

$$U_*^{m+1} = U_*^m - \alpha_m \overline{U}_*^m$$

$$V_*^{m+1} = V_*^m - \alpha_m \overline{V}_*^m$$

$$R^{m+1} = R^m - \alpha_m \overline{R}^m,$$

and

$$-\left(\mathcal{D}_7 - k^2 \mathcal{I}_0\right) \overline{\Phi}^m = \overline{R}^m$$

$$W^{m+1} = W^m - \alpha_m \left(\overline{R}^m + \sigma_0 \overline{\Phi}^m\right).$$

The algorithm is stopped when the divergence measured by R^m is sufficiently small, as well as $\overline{U}_*^m$ and $\overline{V}_*^m$.

Le Marec *et al.* (1996) have tested two types of relaxation parameters. The parameter associated with the steepest descent method

$$\alpha_m = \frac{(R^m, W^m)}{\left(\overline{R}^m, W^m\right)}$$

and the one corresponding to the minimal residual method

$$\alpha_m = \frac{\left(R^m, \overline{R}^m\right)}{\left(\overline{R}^m, \overline{R}^m\right)},$$

where $(\cdot, \cdot)$ designates the Euclidian scalar product. Both parameters give roughly the same rate of convergence τ. This rate of convergence is defined as

$$\tau = \left(\frac{\|\nabla \cdot \vec{V}^m\|}{\|\nabla \cdot \vec{V}^0\|}\right)^{1/m},$$

where $\| \cdot \|$ is the discrete L^∞-norm. Numerical experiments performed by Le Marec *et al.* (1996) show that $\tau \simeq 0.52 - 0.68$ for two- and three-dimensional Cartesian computations. On the other hand, the rate deteriorates for three-dimensional axisymmetric computations since it is about $0.69 - 0.75$ according to the values of σ_0. These results show that, at least in the Cartesian case, the number of iterations needed to reach a reasonable convergence may be relatively small if the initial field is not too far from solenoidal. This happens in a time-integration process when the iterative Uzawa procedure is initialized with the solution at the previous time-cycle.

The above Uzawa algorithm needs only the solution of Helmholtz or Poisson problems with Dirichlet or Neumann conditions. In the multidimensional case, the algebraic systems can be solved by the matrix-diagonalization method described in Section 3.7.

Now we consider the special case $k = 0$. For $k = 0$, the matrix $\mathcal{U}$ is found to have two eigenvalues equal to zero. This behaviour was already mentioned. One of the zero eigenvalues expresses the fact that the pressure is defined up to a constant. The second null eigenvalue corresponds to the so-called spurious mode of pressure associated with the collocation discretization. This comes from the fact that, when $k = 0$, the pressure appears in Eqs.(7.36)-(7.40) only through its derivative $p'(y)$ and that it is the same in the approximate problem. More precisely, let us consider the Chebyshev approximation $p_N(y)$ to the pressure $p(y)$, namely,

$$p_N(y) = \hat{p}_0 T_0 + \hat{p}_N T_N(y) + \sum_{k=1}^{N-1} \hat{p}_k T_k(y)$$

with $T_0 = 1$. Then, by differentiation, we get

$$p'_N(y) = \hat{p}_N T'_N(y) + \sum_{k=1}^{N-1} \hat{p}_k T'_k(y).$$

Therefore, the mode $\hat{p}_0$ does not appear in the discrete derivative. This mode is not spurious since it has a physical meaning. On the other hand, since $T'_N(y_j) = 0$ for $j = 1, \ldots, N - 1$, the mode $\hat{p}_N$ does not also appear in the discrete Stokes problem, which does not involve the pressure nor the boundary values of its derivative since the discrete system is closed by pre-scribing the incompressibility equation (7.39) at the boundaries. Therefore, the mode $\hat{p}_N$, which is not controlled by the problem is spurious, its value remains undetermined. If we assume the algebraic system to be solved by a special technique (Section 7.4.1.b), the resulting pressure remains con-taminated by the spurious mode and may exhibit oscillations. The velocity field, however, is not contaminated and is correctly computed. Hence, if the pressure field is not of physical interest, the presence of spurious modes is unimportant. If not, the possible remedies are :

(i)   filter the spurious mode,

(ii)   calculate the correct pressure from the Poisson equation,

(iii) use a staggered mesh,

(iv)   use the single-grid $I\!P_N - I\!P_{N-2}$ approximation.

This question of spurious modes will again be discussed in the two-dimensional case (Section 7.4) where the problem is more delicate than here. In fact, for $k = 0$, the Stokes problem (7.36)-(7.40) considerably sim-plifies and becomes

$$u'' - \sigma u = f_u, \quad -1 < y < 1 \tag{7.61}$$

$$u(-1) = \gamma_-, \quad u(1) = \gamma_+, \tag{7.62}$$

and

$$v'' - \sigma v - p' = f_v, \quad -1 < y < 1 \tag{7.63}$$

$$v' = 0, \quad -1 < y < 1 \tag{7.64}$$

$$v(-1) = g_-, \quad v(1) = g_+. \tag{7.65}$$

This problem is easily solved. The first two equations give u. The conser-vation of the total flow rate imposes $g_+ = g_- = g$, so that $v(y) = g$; then the v-equation gives, up to a constant,

$$p = -\sigma g y - \int^y f_v(\eta) \, d\eta.$$

Note that, for $k = 0$, the stability of the time-discretization scheme (7.29)-(7.33) reduces to the stability of the diffusion equation for $\hat{u}_k^{n+1}$ and $\hat{v}_k^{n+1}$.

To close this section, we mention that the consideration of the stress-free boundary conditions (6.14) does not change the algorithm described above. As done in Section 3.7.1, the value of the tangential velocity at the boundary $y = 1$, namely, u_0, is eliminated thanks to the boundary con-dition $D u_0 = 0$. In this way, we again obtain Eqs. (7.47)-(7.49) with a redefinition of $\mathcal{H}$, $\mathcal{I}_3$, S_u^* and S_Q.

(c) Solution of the Stokes problem by the influence matrix method

The Uzawa method, just described, losses some efficiency in the multidimensional case because of the necessity to employ an iterative procedure whose convergence may be slow. In the context of direct method, an alternative is constituted by the influence matrix method. The method is commonly used in two-dimensional problems, but its application to the three-dimensional case cannot be considered due to the huge size of the influence matrix, whose order is determined by the number of collocation points belonging to the domain boundary.

The influence matrix method for solving the Stokes problem in primitive variables has been introduced by Kleiser and Schumann (1980). They were interested in the calculation of three-dimensional channel flows with two periodic directions. In the third direction, associated with no-slip boundary conditions, Kleiser and Schumann derive a Poisson equation for the pressure and determine the boundary values of the pressure ensuring that the velocity divergence is zero. This problem is solved by means of the Chebyshev tau method. This method extended to the two-dimensional configuration has been developed by Le Quéré and Alziary de Roquefort (1985), then it was improved by Tuckerman (1989). This improvement ensures that the polynomial approximating the divergence of the velocity is identically zero, as obtained, in the one-dimensional case, by Kleiser and Schumann (1980). The extension of the method to the Chebyshev collocation approximation was proposed by Canuto *et al.* (1988) in the one-dimensional case and by Madabhushi *et al.* (1993) in the two-dimensional case. The mathematical analysis of the influence matrix technique associated with the tau method has been done by Canuto and Sacchi-Landriani (1986) for the Legendre polynomial approximation and by Sacchi-Landriani (1986) in the Chebyshev polynomial case.

To begin with, we present the influence matrix method in its simplified version, which does not ensure the velocity divergence to be exactly zero. The reason for this deficiency will be discussed, and the remedy resulting from this discussion will be presented.

c1. The "simplified" influence matrix method

As commonly done when dealing with the Stokes or Navier-Stokes equations, we begin to construct a Poisson equation for the pressure. As a matter of fact, in Fourier space, as considered in Eqs.(7.36)-(7.40), the pressure is a solution of a Helmholtz equation. This equation is obtained by considering the linear combination of Eq.(7.36) multiplied by k, and Eq.(7.37) differentiated with respect to y, in order to make apparent the divergence of the velocity $Q = k\,u + v'$. We obtain

$$(Q'' - \sigma\,Q) - (p'' - k^2\,p) = k\,f_u + f_v'\,.$$

Then, from Eq.(7.38), we have simply

$$p'' - k^2 p = f_p,$$ (7.66)

where

$$f_p = -k f_u - f_v'.$$ (7.67)

Therefore, if p satisfies Eq.(7.66) and if u and v are, respectively, determined by Eq.(7.36) and (7.37), the divergence Q satisfies the Helmholtz equation

$$Q'' - \sigma Q = 0, \quad -1 < y < 1.$$ (7.68)

Consequently $Q = 0$ everywhere in $-1 \le y \le 1$ if Q satisfies the conditions

$$Q(-1) = 0, \quad Q(1) = 0.$$ (7.69)

Therefore, the $\mathcal{P}$-*Problem* to be solved is decomposed into two problems. The first one determines p and v :

$$p'' - k^2 p = f_p, \quad -1 < y < 1$$ (7.70)

$$v'' - \sigma v - p' = f_v, \quad -1 < y < 1$$ (7.71)

$$v(\pm 1) = g_\pm,$$ (7.72)

$$v'(\pm 1) = -k g_\pm = h_\pm,$$ (7.73)

and the second one determines u once p is known

$$u'' - \sigma u = f_u - k p, \quad -1 < y < 1,$$ (7.74)

$$u(\pm 1) = \gamma_\pm.$$ (7.75)

This simple Helmholtz problem is easily solved.

One may observe that the problem for p and v is similar to the one previously encountered for ω and ψ in Section 6.3, so that the solution algorithm for calculating (p, v) will follow the same outlines. This leads to inevitable repetitions, but is necessary for an understanding of the developments which follow.

The Chebyshev collocation method, applied to the solution of the two problems above, leads to algebraic systems determining the unknowns

$$u(y_j) \cong u_N(y_j) = u_j, \quad v(y_j) \cong v_N(y_j) = v_j, \quad p(y_j) \cong p_N(y_j) = p_j,$$

where u_N, v_N, and p_N are polynomials of degree at most N, and y_j, $j = 0, \ldots, N$, are the Gauss-Lobatto points (6.26).

The system determining (p_j, v_j), $j = 0, \ldots, N$, is

$$L_{p,j} \equiv (D^2 - k^2) p_j - f_{p,j} = 0, \quad j = 1, \ldots, N-1$$ (7.76)

$$L_{v,j} \equiv \left(D^2 - \sigma\right) v_j - D\,p_j - f_{v,j} = 0, \quad j = 1, \ldots, N-1 \qquad (7.77)$$

$$v_0 = g_+ \qquad (7.78)$$

$$v_N = g_- \qquad (7.79)$$

$$D\,v_0 = h_+ \qquad (7.80)$$

$$D\,v_N = h_-, \qquad (7.81)$$

and the system determining u_j, $j = 0, \ldots, N$, is

$$L_{u,j} \equiv \left(D^2 - \sigma\right) u_j + k\,p_j - f_{u,j} = 0, \quad j = 1, \ldots, N-1 \qquad (7.82)$$

$$u_0 = \gamma_+ \qquad (7.83)$$

$$u_N = \gamma_-. \qquad (7.84)$$

This last system is easily solved once as p_j is known from the first system. It is convenient to write this first system in vector form. Let us consider the vectors P and V containing the unknowns, and the $2(N+1)$-vector Φ, respectively, defined by

$$P = \begin{pmatrix} p_0 \\ \vdots \\ p_N \end{pmatrix}, \quad V = \begin{pmatrix} v_0 \\ \vdots \\ v_N \end{pmatrix}, \quad \Phi = \begin{pmatrix} P \\ V \end{pmatrix},$$

then the vector form of the system (7.76)-(7.81) is

$$\mathcal{A}\,\Phi = S. \qquad (7.85)$$

The $2(N+1) \times 2(N+1)$ matrix $\mathcal{A}$ is

$$\mathcal{A} = \begin{pmatrix} \mathcal{B}_2 & \mathcal{O}_1 \\ -\mathcal{D}_5 & \mathcal{B}_1 \\ \mathcal{O}_2 & \mathcal{I}_2 \\ \mathcal{O}_2 & \mathcal{D}_3 \end{pmatrix},$$

where $\mathcal{B}_1$ and $\mathcal{B}_2$ are $(N-1) \times (N+1)$ matrices such that

$$\mathcal{B}_1 = \mathcal{D}_4 - \sigma \mathcal{I}_1, \quad \mathcal{B}_2 = \mathcal{D}_4 - k^2 \mathcal{I}_1$$

with $\mathcal{D}_4$ and $\mathcal{I}_1$ defined in Section 6.3.2.b by Eqs.(6.36) and (6.33), respectively. The matrices $\mathcal{O}_1$, $\mathcal{O}_2$, and $\mathcal{I}_2$ were also defined in the same section. The matrix $\mathcal{D}_5$ is defined by Eq.(7.51). Finally, the $2(N+1)$-vector S is

$$S = (f_{p,1}, \ldots, f_{p,N-1}, f_{v,1}, \ldots, f_{v,N-1}, g_+, g_-, h_+, h_-)^T.$$

As for the ω-ψ problem, the large $2(N+1) \times 2(N+1)$ system (7.85) is solved by an algorithm leading to the solution of six smaller $(N+1) \times (N+1)$ systems. This algorithm is the discrete form of the influence matrix technique usually presented on the continuous problem, and which will be described later.

The solution of (7.85) is sought in the form

$$\Phi = \tilde{\Phi} + \xi_1 \overline{\Phi} + \xi_2 \overline{\Phi}_2 , \tag{7.86}$$

where ξ_1 and ξ_2 are scalars and $\tilde{\Phi}$, $\overline{\Phi}_1$ and $\overline{\Phi}_2$ are respectively the vector solution of

$$\hat{\mathcal{A}} \tilde{\Phi} = \tilde{S} \tag{7.87}$$

$$\hat{\mathcal{A}} \overline{\Phi}_1 = \overline{S}_1 \tag{7.88}$$

$$\hat{\mathcal{A}} \overline{\Phi}_2 = \overline{S}_2 , \tag{7.89}$$

where

$$\hat{\mathcal{A}} = \begin{pmatrix} \mathcal{B}_2 & \mathcal{O}_1 \\ -\mathcal{D}_5 & \mathcal{B}_1 \\ \mathcal{O}_2 & \mathcal{I}_2 \\ \mathcal{I}_2 & \mathcal{O}_2 \end{pmatrix} .$$

Therefore, $\hat{\mathcal{A}}$ is the matrix which would result from the discrete problem (7.76)-(7.81) if the Neumann conditions on v [Eqs.(7.80) and (7.81)] were replaced by Dirichlet conditions on p. These Dirichlet conditions are homogeneous so that the vector $\tilde{S}$ has the form

$$\tilde{S} = (f_{p,1} , \dots , f_{p,N-1} , f_{v,1} , \dots , f_{v,N-1} , g_+ , g_- , 0 , 0)^T .$$

Then, the vectors $\overline{S}_1$ and $\overline{S}_2$ are

$$\overline{S}_1 = (0 , \dots , 0 , 0 , 1)^T , \quad \overline{S}_2 = (0 , \dots , 0 , 1 , 1)^T .$$

From this it results that Φ, as defined by Eq.(7.86), is the solution to

$$\hat{\mathcal{A}} \Phi = S_\star \tag{7.90}$$

with

$$S_\star = (f_{p,1} , \dots , f_{p,N-1} , f_{v,1} , \dots , f_{v,N-1} , g_+ , g_- , \xi_2 , \xi_1)^T .$$

Now it is easy to check that the first $2N$ equations of system (7.90) coincide with the $2N$ equations (7.76)-(7.79). In other words, the vector Φ defined by Eq.(7.86) satisfies these $2N$ equations whatever the constants ξ_1 and ξ_2. These constants are then determined such that the last two equations (7.80) and (7.81), namely, the Neumann conditions on v, are satisfied. Thus, we impose

$$\mathcal{C} \left(\tilde{\Phi} + \xi_1 \overline{\Phi}_1 + \xi_2 \overline{\Phi}_2 \right) = E \tag{7.91}$$

where the $2 \times 2(N+1)$ matrix $\mathcal{C}$ and the vector E are, respectively, defined by

$$\mathcal{C} = [\mathcal{O}_2 \quad \mathcal{D}_3] , \quad E = (h_+ , h_-)^T .$$

Now we consider the vectors $\tilde{V}$, $\overline{V}_1$, and $\overline{V}_2$ such that

$$\tilde{\Phi} = \begin{pmatrix} \tilde{P} \\ \tilde{V} \end{pmatrix}, \quad \overline{\Phi}_l = \begin{pmatrix} \overline{P}_l \\ \overline{V}_l \end{pmatrix}, \quad l = 1, 2$$

with

$$\tilde{P} = (\tilde{p}_0, \ldots, \tilde{p}_N)^T, \quad \overline{P}_l = (\overline{p}_{l,0}, \ldots, \overline{p}_{l,N})^T, \quad l = 1, 2$$

$$\tilde{V} = (\tilde{v}_0, \ldots, \tilde{v}_N)^T, \quad \overline{V}_l = (\overline{v}_{l,0}, \ldots, \overline{v}_{l,N})^T, \quad l = 1, 2.$$

Because of the special structure of $\mathcal{C}$, the 2×2 system (7.91) reduces to

$$\xi_1 \mathcal{D}_3 \overline{V}_1 + \xi_2 \mathcal{D}_3 \overline{V}_2 = \tilde{E}$$

where

$$\tilde{E} = E - \mathcal{D}_3 \tilde{V}. \tag{7.92}$$

This system is written as

$$\mathcal{M} \Xi = \tilde{E} \tag{7.93}$$

where $\Xi = (\xi_1, \xi_2)^T$ and the influence (or "capacitance") matrix $\mathcal{M}$ is the 2×2 matrix

$$\mathcal{M} = [\mathcal{D}_3 \overline{V}_1 \quad \mathcal{D}_3 \overline{V}_2] = [m_{i,j}], \quad i = 0, N, \quad j = 1, 2$$

such that

$$m_{i,j} = \sum_{l=0}^{N} d_{i,l}^{(1)} \, \overline{v}_{j,l}.$$

The solution of the system (7.93) gives the constants ξ_1 and ξ_2, so that the solution Φ is completely determined. The matrix $\mathcal{M}$ is invertible (as discussed later) except for $k = 0$. For this value of k, the matrix $\mathcal{M}$ has a null eigenvalue. This corresponds to the fact that the pressure is determined up to a constant. In the present one-dimensional case, the mode $k = 0$ is easily calculated as explained in the preceding section.

Now the important point is that the decomposition described above allows us to replace the solution of the large initial $2(N + 1) \times 2(N + 1)$ system by that of the smaller $(N + 1) \times (N + 1)$ systems. This is obtained thanks to the special structure of the matrix $\hat{A}$. Thus, the system (7.87) is split into the following two systems. The first one determines $\tilde{P}$:

$$\hat{A}_1 \tilde{P} = \tilde{S}_1 \tag{7.94}$$

where

$$\hat{A}_1 = \begin{pmatrix} \mathcal{B}_2 \\ \mathcal{I}_2 \end{pmatrix}$$

and

$$\tilde{S}_1 = (f_{p,1}, \ldots, f_{p,N-1}, 0, 0)^T.$$

Then $\tilde{P}$ having been calculated, the second system determines $\tilde{V}$, namely

$$\hat{A}_2 \tilde{V} = \tilde{S}_2 + \tilde{A}_3 \tilde{P} \tag{7.95}$$

with

$$\hat{A}_2 = \begin{pmatrix} B_1 \\ I_2 \end{pmatrix}, \quad \hat{A}_3 = \begin{pmatrix} D_5 \\ O_2 \end{pmatrix}$$

and

$$\tilde{S}_2 = (f_{v,1}, \ldots, f_{v,N-1}, g_+, g_-)^T.$$

In the same way, each of the systems (7.88) and (7.89) is split into two smaller systems. The first system determines $\overline{P}_l$, $l = 1, 2$:

$$\hat{A}_1 \overline{P}_l = \overline{S}_{l,1}, \quad l = 1, 2 \tag{7.96}$$

with

$$\overline{S}_{1,1} = (0, \ldots, 0, 0, 1)^T, \quad \overline{S}_{2,1} = (0, \ldots, 0, 1, 0)^T$$

and the second system determines $\overline{V}_l$, $l = 1, 2$, namely

$$\hat{A}_2 \overline{V}_l = \hat{A}_3 \overline{P}_l, \quad l = 1, 2. \tag{7.97}$$

The dimension of each of these smaller systems is $(N+1) \times (N+1)$. The matrices $\hat{A}_l$, $l = 1, 2$, are inverted once and for all before the start of the time-integration and their inverses are stored. In the same way, the elementary solutions $\overline{P}_l$ and $\overline{V}_l$, $l = 1, 2$, are calculated in the preprocessing stage, as well as the matrix $\mathcal{M}$ and its inverse $\mathcal{M}^{-1}$. The elementary solutions and the matrix $\mathcal{M}^{-1}$ are stored, so that the solution at each time-cycle is obtained through *Algorithm A* :

1. Calculate $\tilde{P}$ and $\tilde{V}$ by

$$\tilde{P} = \hat{A}_1^{-1} \tilde{S}_1, \quad \tilde{V} = \tilde{A}_2^{-1} \left(\tilde{S}_2 + \tilde{A}_3 \tilde{P} \right).$$

2. Calculate $\tilde{E}$ by Eq.(7.92).
3. Calculate Ξ by $\Xi = \mathcal{M}^{-1} \tilde{E}$.
4. Calculate P and V by Eq.(7.86).
5. Calculate $U = (u_0, \ldots, u_N)^T$ by solving the system (7.82)-(7.84).

We point out that the resolvability of the initial system (7.85) (for $k \neq 0$) is a consequence of the resolvability of the various systems (7.94)-(7.97) corresponding to Dirichlet problems for Helmholtz equations and the resolvability of the system (7.93).

As for the vorticity-streamfunction equations, the algorithm just described for solving the discrete system (7.76)-(7.81) can also be obtained from the Chebyshev collocation approximation of a set of differential problems derived from Eqs.(7.70)-(7.73). As a matter of fact, this is the usual way to introduce the influence matrix method, as done by Kleiser and Schumann (1980). This method consists of searching the boundary values of the pressure ensuring that the divergence of the velocity at the boundary is zero. From a more general point of view, the advantage of the influence matrix method has already been discussed in Sections 3.7.2 (Remark 1) and 6.3.2.c.

Returning to problem (7.70)-(7.73), we look for its solution according to the decomposition

$$p = \tilde{p} + \overline{p}, \quad v = \tilde{v} + \overline{v}. \tag{7.98}$$

The pair $(\tilde{p}, \tilde{v})$ is the solution of the $\tilde{\mathcal{P}}$-*Problem* defined by

$$\tilde{p}'' - k^2 \tilde{p} = f_p, \quad -1 < y < 1$$

$$\tilde{p}(-1) = 0, \quad \tilde{p}(1) = 0,$$

and

$$\tilde{v}'' - \sigma \tilde{v} = f_v + \tilde{p}', \quad -1 < y < 1$$

$$\tilde{v}(-1) = g_-, \quad \tilde{v}(1) = g_+.$$

Thus, the determination of $(\tilde{p}, \tilde{v})$ reduces to the successive solution of two Helmholtz equations with Dirichlet conditions.

Then the pair $(\overline{p}, \overline{v})$ is the solution of the $\overline{\mathcal{P}}$-*Problem* :

$$\overline{p}'' - k^2 \overline{p} = 0, \quad -1 < y < 1 \tag{7.99}$$

$$\overline{v}'' - \sigma \overline{v} - \overline{p}' = 0, \quad -1 < y < 1 \tag{7.100}$$

$$\overline{v}(-1) = 0, \quad \overline{v}(1) = 0 \tag{7.101}$$

$$\overline{v}'(-1) = h_- - \tilde{v}'(-1), \quad \overline{v}'(1) = h_+ - \tilde{v}'(1). \tag{7.102}$$

Now this problem coupling $\overline{p}$ and $\overline{v}$ is transformed into a problem in which the boundary values $\overline{p}(\pm 1)$ must be determined such that the Neumann condition on $\overline{v}$ is satisfied. This defines the $\overline{\mathcal{P}}_0$-*Problem* :

$$\overline{p}'' - k^2 \overline{p} = 0, \quad -1 < y < 1$$

$$\overline{v}'' - \sigma \overline{v} - \overline{p}' = 0, \quad -1 < y < 1$$

$$\overline{p}(-1) = \xi_1, \quad \overline{p}(1) = \xi_2$$

$$\overline{v}(-1) = 0, \quad \overline{v}(1) = 0,$$

where the unknown quantities ξ_1 and ξ_2 are determined such that the boundary conditions on $\bar{v}'(\pm 1)$ are satisfied. Now we take advantage of the linearity of the $\overline{\mathcal{P}}_0$-problem to express its solution as the linear combination

$$\bar{p} = \xi_1\bar{p}_1 + \xi_2\bar{p}_2, \quad \bar{v} = \xi_1\bar{v}_1 + \xi_2\bar{v}_2, \qquad (7.103)$$

where the elementary solutions $(\bar{p}_l, \bar{v}_l)$, $l = 1, 2$, satisfy the $\overline{\mathcal{P}}_l$-*Problems*, namely,

$\overline{\mathcal{P}}_1$-*Problem* :

$$\bar{p}_1'' - k^2\bar{p}_1 = 0, \quad -1 < y < 1$$

$$\bar{p}_1(-1) = 1, \quad \bar{p}_1(1) = 0,$$

and

$$\bar{v}_1'' - \sigma\bar{v}_1 = \bar{p}_1', \quad -1 < y < 1$$

$$\bar{v}_1(-1) = 0, \quad \bar{v}_1(1) = 0.$$

$\overline{\mathcal{P}}_2$-*Problem* :

$$\bar{p}_2'' - k^2\bar{p}_2 = 0, \quad -1 < y < 1$$

$$\bar{p}_2(-1) = 0, \quad \bar{p}_2(1) = 1,$$

and

$$\bar{v}_2'' - \sigma\bar{v}_2 = \bar{p}_2', \quad -1 < y < 1$$

$$\bar{v}_2(-1) = 0, \quad \bar{v}_2(1) = 0.$$

Therefore, the linear combination (7.103) satisfies the equations (7.99), (7.100), and the boundary conditions (7.101) whatever ξ_1 and ξ_2. These quantities are then determined by imposing the conditions (7.102), namely,

$$\bar{v}_1'(-1)\,\xi_1 + \bar{v}_2'(-1)\,\xi_2 = h_- - \tilde{v}'(-1)$$

$$\bar{v}_1'(1)\,\xi_1 + \bar{v}_2'(1)\,\xi_2 = h_+ - \tilde{v}'(1).$$

This system is written

$$\mathcal{M}\,\Xi = \tilde{E} \qquad (7.104)$$

where $\Xi = (\xi_1, \xi_2)^T$, $\tilde{E} = (h_- - \tilde{v}'(-1), h_+ - \tilde{v}'(1))^T$, and

$$\mathcal{M} = \begin{pmatrix} \bar{v}_1'(-1) & \bar{v}_2'(-1) \\ \bar{v}_1'(1) & \bar{v}_2'(1) \end{pmatrix} \qquad (7.105)$$

is the influence (or "capacitance") matrix. Thus, the knowledge of ξ_1 and ξ_2 determines the solution p, v and, finally, the determination of u by Eqs.(7.74)-(7.75) achieves the solution of the Stokes problem.

The various Dirichlet ($\tilde{\mathcal{P}}$, $\overline{\mathcal{P}}_1$, and $\overline{\mathcal{P}}_2$) problems are solved using the Chebyshev (collocation or tau) method. Considering the collocation approximation, it is easy to see that the present method is identical to the algorithm above for solving the algebraic system. More precisely, the discretized $\tilde{\mathcal{P}}$-problem identifies with systems (7.94) and (7.95) ; then the discretized $\overline{\mathcal{P}}_l$-problems, $l = 1, 2$, identify with systems (7.96) and (7.97). Finally, the system (7.104) identifies with (7.93). The continuous formulation of the influence matrix method, as just described, is much easier to conceive than the preceding algebraic algorithm and, in fact, it constitutes the guide to construct this algorithm.

The $\overline{\mathcal{P}}_l$-problems, $l = 1, 2$, are time-independent and can be solved once and for all at a stage performed before the start of the time-integration. Moreover, the elementary solutions possess the symmetry properties

$$\overline{p}_2(y) = \overline{p}_1(-y) , \quad \overline{v}_2(y) = -\overline{v}_1(-y) ,$$

so that only the $\overline{\mathcal{P}}_1$-problem has to be solved, and only the elementary solution $(\overline{p}_1, \overline{v}_1)$ is stored. Also the influence matrix $\mathcal{M}$ is calculated, inverted and stored. Hence, at each time-cycle, the solution of the Stokes problem (7.70)-(7.75) is obtained through *Algorithm B* :

1. Solve the $\tilde{\mathcal{P}}$-problem.
2. Calculate $\tilde{E}$.
3. Calculate Ξ by $\Xi = \mathcal{M}^{-1} \tilde{E}$.
4. Calculate (p, v) by Eq.(7.98).
5. Calculate u by solving Eqs.(7.74)-(7.75).

Since the $\overline{\mathcal{P}}_1$-problem is time-independent and very simple, one may be tempted to solve it analytically. This is not recommended and leads to a serious loss of accuracy. The reason is that, in such a situation, the equivalence between the Chebyshev collocation approximation to the sets of differential $\tilde{\mathcal{P}}$- and $\overline{\mathcal{P}}$-problems and the algebraic system (7.85) is lost. As a matter of fact, the exact solution of the $\overline{\mathcal{P}}_l$-problems exhibits a boundary layer behaviour when σ and k are large, so that the polynomial approximation to $(\overline{p}_l, \overline{v}_l)$ is very far from the exact solution, unless N is sufficiently large. However, the accuracy of the polynomial representation of $(\overline{p}_l, \overline{v}_l)$ is unnecessary, since it constitutes only an intermediate step in the procedure, and the real accuracy [that of the algebraic system (7.85)] is recovered when the final combination (7.98) is reconstituted. Note that the error obtained by the use of the analytical solution of $\overline{\mathcal{P}}_l$ is diminished if N is sufficiently large to represent correctly the boundary layer behaviour of $(\overline{p}_l, \overline{v}_l)$. However, this could necessitate a much larger value of N than effectively needed to represent the final solution.

Thanks to the property $\overline{v}_2(y) = -\overline{v}_1(-y)$, the matrix $\mathcal{M}$ can be expressed in terms of $\overline{v}_1'(\pm 1)$ only, and its eigenvalues are $\lambda_\pm = \overline{v}_1'(-1) \pm \overline{v}_1'(1)$. From the exact calculation of $\overline{v}_1$, it is found that these eigenvalues are nonzero except for the case $k = 0$ where $\lambda_+ = 0$. This is a consequence of the property of the pressure which is determined to within a constant. This result, concerning the invertibility of $\mathcal{M}$, however, does not apply to the approximate problem for any value of N. As explained above, whatever the Chebyshev method, collocation or tau, as discussed above, the accuracy of the polynomial approximation to the elementary solutions $(\bar{p}_l, \bar{v}_l)$, $l = 1, 2$, is very poor, except if N is very large. Therefore, the reasoning based on the continuous differential problem gives information only for large N. An analysis of the eigenvalues of $\mathcal{M}$, in the Chebyshev tau approximation, was done by Sacchi-Landriani (1986).

Finally, taking into account that ξ_1 and ξ_2 are the boundary values ensuring that the Neumann condition is satisfied, we may avoid the storage of the K solutions $(\bar{p}_1, \bar{v}_1)$, where K is the cut-off frequency of the Fourier series. This is obtained by replacing *step 4* of Algorithm B by the successive solution of the following problems, deduced from the $\overline{\mathcal{P}}_0$-*Problem* :

$$p'' - k^2 p = f_p, \quad -1 < y < 1$$

$$p(-1) = \xi_1, \quad p(1) = \xi_2,$$

and

$$v'' - \sigma v = f_v + p', \quad -1 < y < 1$$

$$v(-1) = g_-, \quad v(1) = g_+,$$

which determines the final solution (p, v). This variant is not really necessary in the one-dimensional case, but is currently used in two-dimensional problems because of memory requirement.

Of course, this variant has its counterpart in the algorithm developed earlier for the solution of the full algebraic system. Hence, *step* 4 of Algorithm A is replaced by

$$\hat{A}_1 P = G,$$

where

$$G = (f_{p,1}, \cdots, f_{p,N-1}, \xi_2, \xi_1)^T$$

and

$$\hat{A}_1 V = H$$

with

$$H = (f_{v,1}, \cdots, f_{v,N-1}, g_+, g_-)^T.$$

Now we discuss the important point concerning the constraint on the divergence of the velocity. From the continuous point of view, the divergence $Q = k u + v'$ is identically zero since it satisfies the homogeneous Helmholtz equation (7.68) with the homogeneous Dirichlet conditions (7.69). But, in the discrete case, the fact that the momentum equation (7.71) is not satisfied at the boundaries $y = \pm 1$ prevents the grid values of the approximate divergence

$$Q_j = k\,u_j + D\,v_j \qquad (7.106)$$

to be exactly zero at the inner collocation points (it is prescribed zero at the boundaries). This can be seen in the following two equivalent ways.

First, consider the equations (7.70)-(7.75) approximated with the Chebyshev collocation method. It has been shown (Section 3.4.3) that the polynomial approximation to the solution of a differential problem satisfies, in fact, a modified equation called the "error equation" with the same boundary conditions. More precisely, the polynomials u_N, v_N, and p_N are solutions to

$$u_N'' - \sigma\,u_N + k\,p_N = f_{uN} + e_{uN}, \quad -1 < y < 1$$

$$v_N'' - \sigma\,v_N - k\,p_N' = f_{vN} + e_{vN}, \quad -1 < y < 1$$

$$p_N'' - k^2\,p_N = f_{pN} + e_{pN}, \quad -1 < y < 1$$

$$u(\pm 1) = \gamma_\pm, \quad v(\pm 1) = g_\pm, \quad v'(\pm 1) = -k\,g_\pm = h_\pm,$$

where f_{uN}, f_{vN}, and f_{pN} are the polynomials interpolating, respectively, f_u, f_v, and f_p at the collocation points. The polynomials (of degree N at most) e_{uN}, e_{vN}, and e_{pN} are of the general form

$$e_N = (\lambda\,y + \mu)\,T_N'(y),$$

where λ and μ are directly connected to the residuals of the corresponding equation at the boundaries [see Eqs.(3.93) and (3.94)]. Now, if we construct, from the above equations, the Helmholtz equation satisfied by the polynomial Q_N approximating the divergence Q we obtain the problem

$$Q_N'' - \sigma\,Q_N = k\,e_{uN} + e_{vN}' - e_{pN} \equiv E_N \qquad (7.107)$$

$$Q_N(-1) = 0, \quad Q_N(1) = 0. \qquad (7.108)$$

Taking the expressions of e_{uN}, e_{vN}, and e_{pN} into account, it is easy to see that the polynomial E_N at the inner collocation points

$$E_N(y_j) = e_{vN}'(y_j) = (\lambda_v\,y_j + \mu_v)\,T_N''(y_j)$$

is not zero since $T_N''(y_j) = (-1)^{j+1}\,N^2/(1 - y_j^2) \neq 0$ for $j = 1, \ldots, N - 1$. Thus, the Helmholtz equation satisfied by Q_N is not homogeneous and, consequently, Q_N cannot be zero, even at the collocation points. From Eqs.(3.93) and (3.94), the magnitude of λ_v and μ_v and, consequently of $Q_N(y_j)$ depend on the way in which the v-equation is satisfied at the boundaries $y = \pm 1$. Thus, $Q_N(y_j) \to 0$ when $N \to \infty$ but is not identically zero for arbitrary N. Similar behaviour exists in the case of the Chebyshev tau method (Kleiser and Schumann, 1980), the error coming from the last two Galerkin equations that are not satisfied (Section 3.4.3.a).

The second way consists of directly analyzing the discrete problem (7.76)-(7.84). We construct the algebraic system satisfied by the grid values $Q_j = Q_N(y_j)$ of the velocity divergence defined by Eq.(7.106). For that we combine Eqs.(7.82) and (7.77) according to

$$k\,L_{u,j} + \sum_{l=1}^{N-1} d_{j,l}^{(1)}\,L_{v,l} = 0\,, \quad j = 1,\ldots,N-1. \tag{7.109}$$

To make apparent the velocity divergence Q_j in this equation, we must commute the summations in the second term. This necessitates having the same summation limits, namely, $l = 0$ to $l = N$, in both sums. Consequently, we add and subtract the terms corresponding to $l = 0$ and $l = N$ in the second term of Eq.(7.109). We get

$$\left(D^2 - \sigma\right)Q_j - d_{j,0}^{(1)}L_{v,0} - d_{j,N}^{(1)}L_{v,N} - \left[\left(D^2 - k^2\right)p_j + kf_{u,j} + Df_{v,j}\right] = 0$$

$$j = 1,\ldots,N-1. \tag{7.110}$$

Now, taking into account Eq.(7.67) expressing that

$$k\,f_{u,j} + D\,f_{v,j} = -f_{p,j}\,,$$

we observe that the term in square brackets is nothing other than $L_{p,j}$ which is zero as prescribed by Eq.(7.76). Therefore, the algebraic system satisfied by Q_j is

$$\left(D^2 - \sigma\right)Q_j = d_{j,0}^{(1)}\,L_{v,0} + d_{j,N}^{(1)}\,L_{v,N}\,, \quad j = 1,\ldots,N-1 \tag{7.111}$$

$$Q_0 = 0\,, \quad Q_N = 0\,. \tag{7.112}$$

Since $L_{v,0}$ and $L_{v,N}$ are not zero, the above system is not homogeneous and its solution is not identically zero. It is easy to check that the right-hand-side of Eq.(7.111) coincides with $E_N(y_j)$ determined above.

From the analysis above, it can be deduced that the divergence Q_j could be made equal to zero for $j = 0,\ldots,N$ if the pressure equation

$$L_{p,j} = 0\,, \quad j = 1,\ldots,N-1$$

was replaced by the corrected equation

$$L_{p,j} = -\left(d_{j,0}^{(1)}\,L_{v,0} + d_{j,N}^{(1)}\,L_{v,N}\right)\,, \quad j = 1,\ldots,N-1 \tag{7.113}$$

so that Eq.(7.111) becomes an homogeneous equation ensuring, with the boundary conditions (7.112), that $Q_j = 0$ for $j = 0,\ldots,N$.

Since the correction term in Eq.(7.113) depends on the solution itself, it remains unknown and must be determined as part of the global solution. This may be obtained through a correction method which is now described.

c2. The "rigorous" influence matrix method

This method is presented by Canuto et al. (1988) for the one-dimensional collocation approximation. Its application to the two-dimensional case has been developed by Madabhushi et al. (1993) and by Sabbah and Pasquetti (1998). The method is called the "extended influence matrix method" in the last two works. We prefer the designation "rigorous influence matrix method," or simply "influence matrix method," because it constitutes the correct way to understand and apply the method, as was done originally by Kleiser and Schumann (1980) in the case of the tau method.

Since the source of the difficulty is of a discrete nature, it is advisable to discuss the correction method in the discrete framework.

By introducing the notations

$$A = D^2 - k^2, \quad B = D^2 - \sigma^2,$$

we write the discrete $\mathcal{P}$-Problem to be solved as

$$A p_j = f_{p,j} - D b_j, \quad j = 1, \ldots, N-1 \tag{7.114}$$

$$B v_j - D p_j = f_{v,j} + b_j, \quad j = 0, \ldots, N \tag{7.115}$$

$$v_0 = g_+, \quad v_N = g_- \tag{7.116}$$

$$D v_0 = h_+, \quad D v_N = h_-, \tag{7.117}$$

where the correction term b_j (which is determined as a part of the solution) is zero except at the boundaries $y = \pm 1$, namely ,

$$b_j = 0, \quad j = 1, \ldots, N-1.$$

It is clear that the correction term $-D b_j$ in the pressure equation (7.114) is nothing other than the right-hand side of Eq.(7.113).

As previously, the solution is decomposed according to

$$p_j = \tilde{p}_j + \overline{p}_j, \quad v_j = \tilde{v}_j + \overline{v}_j. \tag{7.118}$$

The couple $(\tilde{p}_j, \tilde{v}_j)$, $j = 0, \ldots, N$, is determined as previously by the $\tilde{\mathcal{P}}$-Problem defined by

$$A \tilde{p}_j = f_{p,j}, \quad j = 1, \ldots, N-1$$

$$\tilde{p}_0 = 0, \quad \tilde{p}_N = 0,$$

and

$$B \tilde{v}_j = f_{v,j} + D \tilde{p}_j, \quad j = 1, \ldots, N-1$$

$$\tilde{v}_0 = g_+, \quad \tilde{v}_N = g_-.$$

l	$b_{l,0}$	$b_{l,N}$	$q_{l,0}$	$q_{l,N}$
1	0	0	0	1
2	0	0	1	0
3	0	1	0	0
4	1	0	0	0

Table 7.1. Boundary values $b_{l,j}$ and $q_{l,j}$.

When $\tilde{p}_j$ and $\tilde{v}_j$ have been calculated, the residual of the $\tilde{v}$-equation at the boundaries may be determined, that is,

$$\tilde{b}_j = B\,\tilde{v}_j - D\,\tilde{p}_j - f_{v,j}\,, \quad j = 0, N\,.$$

Then the couple $(\bar{p}_j, \bar{v}_j)$, $j = 0, \ldots, N$, together with b_0 and b_N, are determined by the $\overline{\mathcal{P}}$-*Problem* :

$$A\bar{p}_j = -D\,b_j\,, \quad j = 1, \ldots, N-1 \tag{7.119}$$

$$B\bar{v}_j - D\bar{p}_j = 0\,, \quad j = 1, \ldots, N-1 \tag{7.120}$$

$$\bar{v}_0 = 0\,, \quad \bar{v}_N = 0 \tag{7.121}$$

$$D\bar{v}_0 = h_+ - D\tilde{v}_0\,, \quad D\bar{v}_N = h_- - D\tilde{v}_N \tag{7.122}$$

$$b_0 = \tilde{b}_0 + \bar{b}_0\,, \quad b_N = \tilde{b}_N + \bar{b}_N\,, \tag{7.123}$$

where

$$\bar{b}_j = B\bar{v}_j - D\bar{p}_j\,, \quad j = 0, N\,.$$

Now the $\overline{\mathcal{P}}$-problem is solved by means of the superposition of elementary solutions of the form

$$\phi_j = \sum_{l=1}^{4} \xi_l\,\phi_{l,j} \tag{7.124}$$

with $\phi_j = \bar{p}_j, \bar{v}_j, \bar{b}_j, b_j$. Therefore, we begin by solving the $\overline{\mathcal{P}}_l$-*Problems*, $l = 1, \ldots, 4$, defined by

$$A\bar{p}_{l,j} = -D\,b_{l,j}\,, \quad j = 1, \ldots, N-1$$

$$\bar{p}_{l,0} = q_{l,0}\,, \quad \bar{p}_{l,N} = q_{l,N}$$

and

$$B\bar{v}_{l,j} = D\bar{p}_{l,j}\,, \quad j = 1, \ldots, N-1$$

$$\bar{v}_{l,0} = 0\,, \quad \bar{v}_{l,N} = 0\,,$$

where again $b_{j,l} = 0$ for $j = 1, \ldots, N-1$, and every l. The boundary values $b_{l,j}$ and $q_{l,j}$, for $j = 0, N$, required for the solution of the $\overline{\mathcal{P}}_l$-problems, are given in Table 7.1.

After $(\bar{p}_{l,j}, \bar{v}_{l,j})$, $l = 1, \ldots, 4$, $j = 0, \ldots, N$, have been calculated, the residuals $\bar{b}_{l,0}$ and $\bar{b}_{l,N}$ are determined by

$$\bar{b}_{l,j} = A\bar{v}_{l,j} - D\bar{p}_{l,j}, \quad l = 1, \ldots, 4, \quad j = 0, N.$$

It is easy to see that the combination (7.124) satisfies the first three equations (7.119)-(7.121) of the $\overline{\mathcal{P}}$-problem, whatever the constants ξ_l, $l = 1, \ldots, 4$. Therefore, these constants are determined by requiring the last equations (7.122) and (7.123) to be satisfied by the combination (7.124), that is

$$\sum_{l=1}^{4} \xi_l D\bar{v}_{l,0} = \gamma_+ - D\tilde{v}_0$$

$$\sum_{l=1}^{4} \xi_l D\bar{v}_{l,N} = \gamma_- - D\tilde{v}_N$$

$$\sum_{l=1}^{4} \xi_l \left(b_{l,0} - \bar{b}_{l,0} \right) = \tilde{b}_0$$

$$\sum_{l=1}^{4} \xi_l \left(b_{l,N} - \bar{b}_{l,N} \right) = \tilde{b}_N.$$

The 4×4 matrix $\mathcal{M}$ of this system is the "extended influence matrix". It is invertible except for $k = 0$ (Madabhushi et al., 1993). For $k = 0$, the matrix $\mathcal{M}$ has two eigenvalues equal to zero. This behaviour is identical to that encountered previously for the Uzawa operator $\mathcal{U}$. This result is easily understood since the addition of the correction term b, to ensure the solenoidal character of the approximate velocity field, effectively recovers the original algebraic system (7.42)-(7.46). Note that, for $k = 0$, the solution of the Stokes problem is easily calculated, as shown at the end of Section 7.3.2.b.

The $\overline{\mathcal{P}}_l$-problems are time-independent. Hence, they can be solved once and for all in the preprocessing stage of the time-integration process, and their solutions are stored. Also, the matrix $\mathcal{M}$ is calculated and inverted at this stage and its inverse is stored. Therefore, the solution of the $\overline{\mathcal{P}}$-problem [Eqs.(7.114)-(7.117)] at each time-cycle is obtained through an algorithm similar to the above Algorithm B.

c3. Shear-stress-free boundary

We close this section by considering the case where the no-slip condition at $y = 1$ is replaced by the stress-free boundary condition (6.14). In this case, the unknowns p and v are coupled to u through the incompressibility equation at $y = 1$, since Eq.(7.75) is replaced by

$$u(-1) = 0, \quad u'(1) = 0$$

and Eq.(7.73) by

$$v'(-1) = 0, \quad k\,u(1) + v'(1) = 0.$$

The influence matrix method applies with a small change due to this coupling. The decomposition (7.98) is also done for u, namely,

$$u = \tilde{u} + \bar{u}.$$

In the $\tilde{\mathcal{P}}$-problem the velocity $\tilde{u}$ is calculated from the knowledge of $\tilde{p}$ as it is for $\tilde{v}$. On the other hand, the $\bar{\mathcal{P}}$-*Problem* in the simplified version of the method is now

$$\bar{u}'' - \sigma\bar{u} + k\bar{p} = 0, \qquad -1 < y < 1$$

$$\bar{v}'' - \sigma\bar{v} - \bar{p}' = 0, \qquad -1 < y < 1$$

$$\bar{p}'' - k^2\bar{p} = 0, \qquad -1 < y < 1$$

$$\bar{u}(-1) = 0, \quad \bar{u}'(1) = 0,$$

$$\bar{v}(-1) = 0, \quad \bar{v}(1) = 0,$$

$$\bar{v}'(-1) = -\tilde{v}'(-1), \quad k\bar{u}(1) + \bar{v}'(1) = -k\tilde{u}(1) - \tilde{v}'(1).$$

Then the algorithm follows the same lines, except that $\bar{u}$ is also decomposed according to Eq.(7.103) and $\bar{u}_1$ and $\bar{u}_2$ have to be determined in order to construct the influence matrix on the same principles as that above. Note that such changes in the algorithm also apply for the rigorous influence matrix method.

(d) Time-schemes leading to the Darcy problem

The time-discretization considered here belongs to the general class of splitting (or fractional step) methods. At the first step, a provisional value $\tilde{\mathbf{V}}^{n+1}$ of the velocity $\mathbf{V}$, that does not satisfy the incompressibility equation, is calculated. Then, at the second step, this velocity $\hat{\mathbf{V}}^{n+1}$ is corrected in such a way that the final velocity $\mathbf{V}^{n+1}$ and the pressure p^{n+1} satisfy the complete Navier-Stokes equations. This is the basic principle of the projection method introduced by Chorin (1968) and Temam (1969). Since then, the method has largely been used in association with various spatial approximations. At the same time, some improvements have been introduced in order to increase the accuracy of the method.

To introduce time-splitting in a general way it is convenient to consider the time-dependent Stokes equations written in vector form

$$\partial_t \mathbf{V} - \nu\,\nabla^2\mathbf{V} + \nabla p = \mathbf{f} \quad \text{in } \Omega \tag{7.125}$$

$$\nabla \cdot \mathbf{V} = 0 \quad \text{in } \Omega, \qquad (7.126)$$

associated with the initial condition $\mathbf{V} = \mathbf{V}_0$ and the boundary condition

$$\mathbf{V} = \mathbf{V}_\Gamma \quad \text{on } \Gamma = \partial\Omega. \qquad (7.127)$$

In the *first step* ("diffusion step"), the provisional velocity $\tilde{\mathbf{V}}^{n+1}$ is calculated according to the three-level scheme :

$$\frac{(1+\varepsilon)\,\tilde{\mathbf{V}}^{n+1} - 2\,\varepsilon\mathbf{V}^n - (1-\varepsilon)\,\mathbf{V}^{n-1}}{2\,\Delta t} - \nu\nabla^2\left[\theta\,\tilde{\mathbf{V}}^{n+1} + (1-\theta)\,\mathbf{V}^n\right]$$

$$+ \nabla\left(\lambda_1\,p^n\right) = \theta\,\mathbf{f}^{n+1} + (1-\theta)\,\mathbf{f}^n \quad \text{in } \Omega. \qquad (7.128)$$

The associated boundary condition is generally

$$\tilde{\mathbf{V}}^{n+1} = \mathbf{V}_\Gamma^{n+1} \quad \text{on } \Gamma \qquad (7.129)$$

since $\tilde{\mathbf{V}}^{n+1}$, as given by Eq.(7.128) with $\lambda_1 = 1$, may be seen as an approximation to $\mathbf{V}^{n+1}$. However, if $\lambda_1 = 0$, as it was in the original projection method, the equation (7.128) is no longer consistent with (7.125) and one may question the best boundary value of $\tilde{\mathbf{V}}^{n+1}$ to improve the accuracy (see, e.g., Fortin *et al.*, 1971 ; Kim and Moin, 1985).

The *second step* ("projection step") determines the solution $(\mathbf{V}^{n+1}, p^{n+1})$ by

$$(1+\varepsilon)\,\frac{\mathbf{V}^{n+1} - \tilde{\mathbf{V}}^{n+1}}{2\,\Delta t} + \nabla\left(\lambda_2\,p^{n+1} + \lambda_3\,p^n\right) = \mathbf{0} \quad \text{in } \Omega \qquad (7.130)$$

$$\nabla \cdot \mathbf{V}^{n+1} = 0 \quad \text{in } \Omega \qquad (7.131)$$

$$\mathbf{V}^{n+1} \cdot \mathbf{n} = \mathbf{V}_\Gamma^{n+1} \cdot \mathbf{n} \quad \text{on } \Gamma. \qquad (7.132)$$

The parameters ε, θ, λ_1, λ_2, and λ_3 characterize the scheme and determine its accuracy. Note that $\lambda_2 \neq 0$ is required to ensure the presence of the pressure at level $n+1$. The elimination of $\tilde{\mathbf{V}}^{n+1}$ from (7.128) and (7.130) gives the semidiscrete equation actually approximating the momentum equation (7.125), that is,

$$\frac{(1+\varepsilon)\,\mathbf{V}^{n+1} - 2\,\varepsilon\,\mathbf{V}^n - (1-\varepsilon)\,\mathbf{V}^{n-1}}{2\,\Delta t} - \nu\nabla^2\left[\theta\,\mathbf{V}^{n+1} + (1-\theta)\,\mathbf{V}^n\right]$$

$$+ \nabla\left[\lambda_2\,p^{n+1} + (\lambda_1 + \lambda_3)\,p^n\right] - \frac{2\theta\,\Delta t}{1+\varepsilon}\nu\nabla^2\left[\nabla\left(\lambda_2\,p^{n+1} + \lambda_3\,p^n\right)\right]$$

$$= \theta\,\mathbf{f}^{n+1} + (1-\theta)\,\mathbf{f}^n. \qquad (7.133)$$

As usual, the consistency and accuracy are analyzed by Taylor's expansion with respect to time. The scheme (7.133) is consistent with the momentum equation (7.125) and is accurate to $O(\Delta t)$ if the condition

$$\lambda_1 + \lambda_2 + \lambda_3 = 1$$

is satisfied. It is accurate to second order $O(\Delta t^2)$ if, moreover, the following conditions are satisfied

$$\frac{\varepsilon}{2} = \theta = 1 - (\lambda_1 + \lambda_3),$$

$$\frac{\theta}{1 + \varepsilon}(\lambda_2 + \lambda_3) = 0.$$

In the original projection method (Chorin, 1968 ; Temam, 1969) the pressure p^n does not appear in the first step, namely, $\lambda_1 = 0$. From the above conditions, we easily see that such a splitting scheme cannot be $O(\Delta t^2)$. The first-order explicit scheme ($\varepsilon = 1$, $\theta = 0$, $\lambda_1 = 0$, $\lambda_2 = 1$, $\lambda_3 = 0$) associated with the spatial approximation, referred to as Option (2) below, has been studied and applied to flow computations by Phillips and Roberts (1993), who examined the constraint on the time-step required for stability.

In the incremental projection method (Goda, 1979 ; Van Kan, 1986) the pressure p^n is involved in the first step, namely, $\lambda_1 = 1$. The resulting second-order schemes are then defined by $\lambda_3 = -\lambda_2$ and $\varepsilon/2 = \theta = \lambda_2$. The Crank-Nicolson (CN) scheme corresponds to $\varepsilon = 1$, $\theta = 1/2$, and $\lambda_2 = -\lambda_3 = 1/2$. The Backward-Differentiation BDI2 scheme (see Section 4.4) is defined by $\varepsilon = 2$, $\theta = 1$, and $\lambda_2 = -\lambda_3 = 1$. As discussed several times, the BDI2 scheme is preferred to the CN scheme. Therefore, the BDI2 scheme is considered in the sequel.

The determination of $\tilde{\mathbf{V}}^{n+1}$ at the first step requires the solution of two non coupled Helmholtz equations with Dirichlet conditions for the components $\tilde{u}$ and $\tilde{v}$ of $\tilde{\mathbf{V}}^{n+1}$. This is done by using the Chebyshev collocation method based on Gauss-Lobatto points (Sections 3.4. and 3.7).

The second step (7.130)-(7.132) corresponds to the Darcy problem, also called the "div-grad problem". Note that this problem is well-posed (see, e.g., Azaiez et al., 1994a) if the normal velocity only is prescribed at the boundary Γ. The tangential velocity is then determined from the global solution. The effective approximate solution of problem (7.130)-(7.132) strongly depends on the closure of the discrete problem. This will be made more precise later.

Up to now, we have discussed the accuracy of the splitting method by considering only the discretization with respect to time of the equations of motion. As a matter of fact, the spatial approximation technique and especially the treatment of boundary conditions play a fundamental role in the overall accuracy of the discrete method. Therefore, we are discussing now the full discrete problem by returning to the one-dimensional equations

(7.23)-(7.25) satisfied by the Fourier components $\hat{u}_k$, $\hat{v}_k$, $\hat{p}_k$ of the solution. By setting

$$\underline{i}\,\hat{u}_k = u\,, \quad \hat{v}_k = v\,, \quad \hat{p}_k = p\,, \quad \underline{i}\,\hat{f}_{u,k} = f_u\,, \quad \hat{f}_{v,k} = f_v\,,$$

equations (7.23)-(7.25) are

$$\partial_t u - \nu\left(\partial_{yy} - k^2\right)u - k\,p = f_u\,, \tag{7.134}$$

$$\partial_t v - \nu\left(\partial_{yy} - k^2\right)v + \partial_y p = f_v\,, \tag{7.135}$$

$$k\,u + \partial_y v = 0\,, \tag{7.136}$$

to be solved in $-1 < y < 1$ with the initial condition

$$u\left(y,0\right) = u_0\left(y\right)\,, \quad v\left(y,0\right) = v_0\left(y\right)\,, \tag{7.137}$$

and the boundary conditions

$$u\left(\pm 1, t\right) = \gamma_\pm\,, \quad v\left(\pm 1, t\right) = g_\pm\,. \tag{7.138}$$

Then we introduce the time-discretization and polynomial approximation so that

$$u\left(y_j, n\,\Delta t\right) \cong u_N^n\left(y_j\right) = u_j^n\,,$$

$$v\left(y_j, n\,\Delta t\right) \cong v_N^n\left(y_j\right) = v_j^n\,,$$

$$p\left(y_j, n\,\Delta t\right) \cong p_N^n\left(y_j\right) = p_j^n$$

where y_j are the collocation points defined by Eq.(6.26). The y-derivatives are defined by Eq.(7.41). With these notations, the splitting scheme approximating the above equations is

Step 1

$$\frac{1}{2\,\Delta t}\left(3\,\tilde{u}_j^{n+1} - 4\,u_j^n + u_j^{n-1}\right) - \nu\left(D^2 - k^2\right)\tilde{u}_j^{n+1} - k\,p_j^n = f_{u,j}^{n+1}\,,$$

$$j = 1,\ldots,N-1$$

$$\tag{7.139}$$

$$\tilde{u}_0^{n+1} = \gamma_+^{n+1}\,, \quad \tilde{u}_N^{n+1} = \gamma_-^{n+1}\,, \tag{7.140}$$

$$\frac{1}{2\,\Delta t}\left(3\,\tilde{v}_j^{n+1} - 4\,v_j^n + v_j^{n-1}\right) - \nu\left(D^2 - k^2\right)\tilde{v}_j^{n+1} + D\,p_j^n = f_{v,j}^{n+1}\,,$$

$$j = 1,\ldots,N-1$$

$$\tag{7.141}$$

$$\tilde{v}_0^{n+1} = g_+^{n+1}\,, \quad \tilde{v}_N^{n+1} = g_-^{n+1}\,. \tag{7.142}$$

Each of the above systems is easily solved using classical algorithms. The start of the time-integration ($n = 0$) requires the knowledge of u_j^{-1}, v_j^{-1}

and p_j^0. Starting schemes for multistep methods have been discussed in Section 4.5.1.d and will again be considered for the Navier-Stokes equations in Section 7.4.3, in which the determination of the initial pressure p_j^0 will also be addressed.

Step 2

$$\frac{3}{2\,\Delta t}\left(u_j^{n+1} - \tilde{u}_j^{n+1}\right) - k\left(p_j^{n+1} - p_j^n\right) = 0, \quad j = 1,\ldots,N-1 \qquad (7.143)$$

$$\frac{3}{2\,\Delta t}\left(v_j^{n+1} - \tilde{v}_j^{n+1}\right) + D\left(p_j^{n+1} - p_j^n\right) = 0, \quad j = 1,\ldots,N-1 \qquad (7.144)$$

$$k\,u_j^{n+1} + D\,v_j^{n+1} = 0, \quad j = 1,\ldots,N-1 \qquad (7.145)$$

$$v_0^{n+1} = g_+^{n+1}, \quad v_N^{n+1} = g_-^{n+1}. \qquad (7.146)$$

Note that the above equations have been collocated only at the inner points, $j = 1,\ldots,N-1$. This furnishes [with Eq.(7.3.124)] $3N - 1$ equations for the $3N + 3$ unknowns u_j^{n+1}, v_j^{n+1}, and p_j^{n+1}, $j = 0,\ldots,N$. Consequently, four equations have to be added. We obtain two more equations by prescribing either Eq.(7.145) or Eq.(7.144) at the boundaries $j = 0$ and $j = N$. As will be seen below, the first choice leads to an Uzawa operator to calculate the pressure, while the second choice leads to a Helmholtz operator. This being done, we need two more equations. One possibility is to prescribe Eq.(7.143) at $j = 0, N$. This choice is in agreement with the mathematical properties of the Darcy problem but introduces an error in the tangential velocity at the boundaries. However, if we consider that the above Darcy problem is not isolated, but results from the splitting of unsteady equations with given boundary conditions, we may contemplate the possibility of adding the exact boundary conditions

$$u_0^{n+1} = \gamma_+^{n+1}, \quad u_N^{n+1} = \gamma_-^{n+1}, \qquad (7.147)$$

instead of Eq.(7.143) at $j = 0, N$.

To summarize, we have the choice between four options to close the system (7.143)-(7.146) :

Option (1) : Use Eq.(7.145) and Eq.(7.143) at $j = 0, N$.
Option (2) : Use Eq.(7.145) at $j = 0, N$ and Eq.(7.147).
Option (3) : Use Eq.(7.144) and Eq.(7.143) at $j = 0, N$.
Option (4) : Use Eq.(7.144) at $j = 0, N$ and Eq.(7.147).

Now, we analyze the truncation error associated with each of these options. For this purpose, we eliminate the intermediate quantities $\tilde{u}_j^{n+1}$ and $\tilde{v}_j^{n+1}$, in order to derive the discrete equations effectively solved by $\left(u^{n+1}, v^{n+1}, p^{n+1}\right)$. We successively consider the four options.

Option (1). We get the system

$$\frac{1}{2\,\Delta t}\left(3\,u_j^{n+1} - 4\,u_j^n + u_j^{n-1}\right)$$

$$-\nu\left(D^2 - k^2\right)u_j^{n+1} - k\,p_j^{n+1} + \Delta t\,E_{u,j}^{n+1} = f_{u,j}^{n+1}\,, \quad j = 1,\ldots,N-1 \tag{7.148}$$

$$\frac{1}{2\,\Delta t}\left(3\,v_j^{n+1} - 4\,v_j^n + v_j^{n-1}\right)$$

$$-\nu\left(D^2 - k^2\right)v_j^{n+1} + D\,p_j^{n+1} + \Delta t\,E_{v,j}^{n+1} = f_{v,j}^{n+1}\,, \quad j = 1,\ldots,N-1 \tag{7.149}$$

$$k\,u_j^{n+1} + D\,v_j^{n+1} = 0\,, \quad j = 0,\ldots,N \tag{7.150}$$

$$u_0^{n+1} - \frac{2\,\Delta t}{3}\,k\left(p_0^{n+1} - p_0^n\right) = \gamma_+^{n+1}\,, \quad u_N^{n+1} - \frac{2\,\Delta t}{3}\,k\left(p_N^{n+1} - p_N^n\right) = \gamma_-^{n+1} \tag{7.151}$$

$$v_0^{n+1} = g_+^{n+1}\,, \quad v_N^{n+1} = g_-^{n+1}\,, \tag{7.152}$$

where

$$E_{u,j}^{n+1} = \frac{2}{3}\nu\,k\left[\sum_{l=0}^{N} d_{j,l}^{(2)}\left(p_l^{n+1} - p_l^n\right) - k^2\left(p_j^{n+1} - p_j^n\right)\right]\,, \tag{7.153}$$

$$E_{v,j}^{n+1} = -\frac{2}{3}\nu\left[\sum_{l=1}^{N-1} d_{j,l}^{(2)}\,D\left(p_l^{n+1} - p_l^n\right) - k^2\,D\left(p_j^{n+1} - p_j^n\right)\right]\,. \tag{7.154}$$

The terms $\Delta t\,E_{u,j}^{n+1}$ and $\Delta t\,E_{v,j}^{n+1}$ in Eqs.(7.148) and (7.149), respectively, are error terms due to splitting. With Taylor's expansion in time around $(n+1)\Delta t$, we evaluate these terms to be $O\left(\Delta t^2\right)$, that is, the same order of the whole discrete equations. Note that these splitting error terms vanish at steady state. The incompressibility equation (7.150) is obviously free of any temporal truncation error. On the other hand, the boundary equations (7.151) exhibit an error $O\left(\Delta t^2\right)$ for the tangential velocity at the boundary since, by Taylor's expansion, we get

$$u_{0,N}^{n+1} = \gamma_\pm^{n+1} + \frac{2\,\Delta t^2}{3}\,k\,(\partial_t p)_{0,N}^{n+1} + \ldots\,. \tag{7.155}$$

The error term in Δt^2 is the "slip velocity" associated with the splitting method.

The pressure solution of Eqs.(7.148)-(7.152) is exempt from spurious modes when $k \neq 0$, since the pressure appears in nondifferentiated form. On the other hand, for $k = 0$, the resulting system admits the spurious pressure mode $T_N(y)$ (we recall that the constant mode T_0 is physical and not spurious) and this for the same reason that was explained in Section 7.3.2.b. As a matter of fact, it is easy to see that the above system is similar (except the error terms $\Delta t\, E_{u,j}^{n+1}$ and $\Delta t\, E_{v,j}^{n+1}$) to the system (7.42)-(7.46).

Option (2). We get the same equations (7.148)-(7.150) with

$$E_{u,j}^{n+1} = \frac{2\nu}{3} k \left[\sum_{l=1}^{N-1} d_{j,l}^{(2)} \left(p_l^{n+1} - p_l^n \right) - k^2 \left(p_j^{n+1} - p_j^n \right) \right] \qquad (7.156)$$

and $E_{v,j}^{n+1}$ is again given by Eq.(7.154), that is, the inner truncation error of the momentum equations is again $O\left(\Delta t^2\right)$. On the other hand, the presence of the slip velocity in Eq.(7.155) is now avoided since Eq.(7.151) is replaced by Eq.(7.147). Finally, for $k = 0$, the system suffers from the same spurious mode as Option (1).

As already mentioned, a possible way to avoid the presence of spurious pressure modes is constituted by the $\mathbb{P}_N - \mathbb{P}_{N-2}$ approximation method which, moreover, simplifies the question associated with the closure of the algebraic system. This method, of special interest to multidimensional problems, is described in Section 7.4.2.

Option (3). We get Eqs. (7.148) and (7.149) with $E_{u,j}^{n+1}$ given by Eq.(7.153) and $E_{v,j}^{n+1}$ by

$$E_{v,j}^{n+1} = -\frac{2\nu}{3} \left[\sum_{l=0}^{N} d_{j,l}^{(2)} D \left(p_l^{n+1} - p_l \right) - k^2 D \left(p_j^{n+1} - p_j^n \right) \right] . \qquad (7.157)$$

Therefore, the inner truncation error is $O\left(\Delta t^2\right)$. Of course, Eq.(7.150) is restricted to inner collocation points $j = 1, \ldots, N-1$. Therefore the divergence of the approximate velocity is not the null polynomial as in Options (1) and (2). It is zero at the inner collocation points only. Equation (7.151) is considered in Option (3) as in Option (1), so that the tangential velocity at the boundaries is again subject to an erroneous slip velocity as shown by Eq.(7.155). The system is closed by the boundary equations (7.152) and the equations (7.144) prescribed at $j = 0, N$, that is,

$$v_0^{n+1} + \frac{2\Delta t}{3} D \left(p_0^{n+1} - p_0^n \right) = \tilde{v}_0^{n+1}, \quad v_N^{n+1} + \frac{2\Delta t}{3} D \left(p_N^{n+1} - p_N^n \right) = \tilde{v}_N^{n+1}. \qquad (7.158)$$

Since $\tilde{v}_0^{n+1} = v_0^{n+1} = g_+^{n+1}$ and $\tilde{v}_N^{n+1} = v_N^{n+1} = g_-^{n+1}$, these equations reduce to

$$D \left(p_j^{n+1} - p_j^n \right) = 0, \quad j = 0, N. \qquad (7.159)$$

The Chebyshev collocation solution obtained in this Option (3) is free from spurious modes whatever k (the constant mode T_0 is not considered as spurious). The reason lies in the fact that Eq. (7.159) involves the derivative $T_N'(\pm 1)$ which is not zero, while $T_N'(y_j) = 0$ for $j = 1, \ldots, N-1$ (see the discussion on spurious modes in Section 7.3.2.b).

Equation (7.159) means that

$$D p_j^{n+1} = D p_j^n = \ldots = D p_j^0, \quad j = 0, N.$$

Thus, the normal pressure gradient at the boundary retains its initial value. Obviously, except for very special cases, this is not correct. Numerical experiments (Hugues and Randriamampianina, 1998 ; Raspo et al., 2002) on the Chebyshev collocation approximation to the Navier-Stokes equations show that the errors on the pressure, as well as the velocity, are quite large (see Section 7.4.4). These errors prevent the use of such a strong formulation of this projection method as pointed out by Guermond and Quartapelle (1998) who recommend implementing the method in a weak framework.

Note that, in the original projection method ($\lambda_1 = 0$), equation (7.159) is replaced by

$$D p_j^{n+1} = 0, \quad j = 0, N \qquad (7.160)$$

which has no physical meaning either (see, e.g., Fortin et al., 1971 ; Peyret and Taylor, 1983 ; Temam, 1991) and is subject to the same conclusion as above concerning its use in a strong framework. For theoretical results concerning the convergence of the projection method and its implementation we refer to Guermond (1996, 1999).

Hugues and Randriamampianina (1998) have proposed a Chebyshev collocation projection method that removes the drawback concerning the "incorrect" Neumann condition on the pressure. This method considers, as a first step, instead of p^n, a predicted pressure $\bar{p}^{n+1}$ suitably defined. The method, which has been found very accurate, is described in Section 7.4.

Option (4). Equations (7.148) and (7.149) are again valid with $E_{u,j}^{n+1}$ given by Eq.(7.156) and $E_{v,j}^{n+1}$ given by Eq.(7.157). Therefore, the inner truncation error associated with the momentum equations is again $O\left(\Delta t^2\right)$. Equation (7.150) is restricted to the inner collocation points $j = 1, \ldots, N$ as in Option (3), therefore with the same result on the velocity divergence. Equation (7.151) is replaced by Eq.(7.147), so that the slip velocity is avoided. Lastly, the boundary equations (7.158) have to be added to Eq.(7.152) to close the system. Therefore, as for Option (3), the normal pressure gradient satisfies Eq.(7.159). Here also, the solution is free of spurious modes.

To close this section, we have to address the question of stability. There does not exist a complete stability analysis of the discrete problems represented by the various options considered above. For Option (1), Heinrichs (1998) checks numerically the stability of the method. However, he uses relatively small time-steps so that he does not verify if the stability is unconditional or not. Hugues and Randriamampianinia (1998), and Raspo et al. (2002) have tested the incremental projection method [Options (3) and (4)] applied to the Navier-Stokes equations. They found that the critical time-step is of the magnitude usually associated with the AB/BDI2 scheme. They did not consider, however, the possible unconditional stability of the discrete time-dependent Stokes equations.

(e) Solution of the Darcy problem

In this section, we describe the algorithms for solving the algebraic system derived in the previous section. We remind ourselves that the calculation of the provisional velocity is the same for all four options considered above. The determination of $(\tilde{u}^{n+1}, \tilde{v}^{n+1})$ reduces to the solution of the algebraic system resulting from the Chebyshev collocation approximation to the Dirichlet problem for the Helmholtz equation. This has been largely discussed in Chapter 3.

Therefore, the task here is to solve the algebraic system (7.143)-(7.146) supplemented with four boundary equations which differ according to the considered option. In a general way, the technique is to eliminate the velocity in order to obtain a system determining the pressure. With Options (1) and (2) this system is of Uzawa type while it is of Helmholtz type in the case of Options (3) and (4).

To simplify the notations we set

$$u_j^{n+1} = u_j, \quad v_j^{n+1} = v_j, \quad p_j^{n+1} = p_j, \quad \sigma = 3/(2\,\Delta t),$$

such that Eqs. (7.143)-(7.145) simply become

$$\sigma\,u_j - k\,p_j = \varphi_{u,j} \tag{7.161}$$

$$\sigma\,v_j + D\,p_j = \varphi_{v,j} \tag{7.162}$$

$$k\,u_j + D\,v_j = 0 \tag{7.163}$$

$$v_0 = g_+, \quad v_N = g_-, \tag{7.164}$$

where

$$\varphi_{u,j} = \sigma\,\tilde{u}_j^{n+1} - k\,p_j^n, \quad \varphi_{v,j} = \sigma\,\tilde{v}_j^{n+1} + D\,p_j^n.$$

We define the following vectors containing the unknowns

$$U = \begin{pmatrix} u_0 \\ \vdots \\ u_N \end{pmatrix}, \quad V = \begin{pmatrix} v_0 \\ \vdots \\ v_N \end{pmatrix}, \quad P = \begin{pmatrix} p_0 \\ \vdots \\ p_N \end{pmatrix},$$

and

$$U_\star = \begin{pmatrix} u_1 \\ \vdots \\ u_{N-1} \end{pmatrix}, \quad V_\star = \begin{pmatrix} v_1 \\ \vdots \\ v_{N-1} \end{pmatrix}.$$

Note that $U_\star = \mathcal{I}_1 U$ and $V_\star = \mathcal{I}_1 V$ where $\mathcal{I}_1$ is the $(N-1) \times (N+1)$ matrix defined in Section 6.3.2.b [Eq.(6.33)].

Now, we consider successively the four options defined in the previous section.

Option (1). The system is constituted by Eq.(7.161) for $j = 0, \ldots, N$, Eq.(7.162) for $j = 1, \ldots, N-1$, Eq.(7.163) for $j = 0, \ldots, N$, and Eq.(7.164). Therefore, the vectors of unknowns are U, $V_\star$, and P, they are determined by

$$\sigma U - k P = S_u \qquad (7.165)$$

$$\sigma V_\star + \mathcal{D}_5 P = S_v^\star \qquad (7.166)$$

$$k U + \mathcal{D}_6 V_\star = S_Q . \qquad (7.167)$$

The $(N-1) \times (N+1)$ matrix $\mathcal{D}_5$ and the $(N+1) \times (N-1)$ matrix $\mathcal{D}_6$ are defined, respectively, by Eqs.(7.51) and (7.52). The vectors S_u, $S_v^\star$, and S_Q are, respectively, defined by

$$S_u = (\varphi_{u,0}, \ldots, \varphi_{u,N})^T , \quad S_v^\star = (\varphi_{v,1}, \ldots, \varphi_{v,N-1})^T ,$$

$$S_Q = (\varphi_{Q,0}, \ldots, \varphi_{Q,N})^T , \quad \varphi_{Q,j} = - \left(d_{j,0}^{(1)} g_+ + d_{j,N}^{(1)} g_- \right) , \quad j = 0, \ldots, N .$$

By eliminating U and $V_\star$ from Eqs.(7.165)-(7.167) we obtain the system determining P :

$$\left(\mathcal{D}_6 \mathcal{D}_5 - k^2 \mathcal{I}_0 \right) P = -\sigma S_Q + k S_u + \mathcal{D}_6 S_v , \qquad (7.168)$$

where $\mathcal{I}_0$ is the $(N+1) \times (N+1)$ identity matrix. The operator $\mathcal{U}_1 = \mathcal{D}_6 \mathcal{D}_5 - k^2 \mathcal{I}_0$ is an Uzawa operator where $\mathcal{D}_6 \mathcal{D}_5$ could be qualified as "pseudo-discrete-Laplacian" since the jth element of the vector $\mathcal{D}_6 \mathcal{D}_5 P$ is

$$\sum_{l=0}^{N} \left[d_{j,l}^{(2)} - \left(d_{j,0}^{(1)} d_{0,l}^{(1)} + d_{j,N}^{(1)} d_{N,l}^{(1)} \right) \right] p_l ,$$

making apparent the second-order derivative $\sum_{l=0}^{N} d_{j,l}^{(2)} p_l$.

After P has been calculated as the solution of (7.168), U and $V_\star$ are determined by (7.165) and (7.166), respectively. For $k \neq 0$, the matrix $\mathcal{U}_1$ is invertible. On the other hand, for $k = 0$, the matrix $\mathcal{U}_1$ has two eigenvalues equal to zero. One of these corresponds to the fact that pressure is defined up to a constant. The second null eigenvalue corresponds to the spurious mode $T_N(y)$ as discussed several times. In the present one-dimensional case, the solution of (7.161)-(7.164) for $k = 0$ is easily calculated (see Section 7.3.2.b). In the two-dimensional problem approximated by Chebyshev collocation in both directions, the corresponding Uzawa matrix has four eigenvalues equal to zero. In this situation, it is possible to avoid these spurious modes by considering the so-called $\mathbb{P}_N - \mathbb{P}_{N-2}$ approximation (Maday *et al.*, 1987 ; Heinrichs, 1993 ; Azaiez *et al.*, 1994b ; Botella, 1997) in which the pressure is approximated with a polynomial of degree $N - 2$

while the velocity is approximated with a polynomial of degree N. Such an approximation will be discussed in detail in Section 7.4.2.b.

Option (2). The system is made with Eqs.(7.161) and (7.162) for $j = 1, \ldots, N - 1$, Eq.(7.163) for $j = 0, \ldots, N$, Eq.(7.164), and the boundary conditions $u_0 = \gamma_+$, $u_- = \gamma_-$. The system determining the unknown vectors $U_\star$, $V_\star$, and P is

$$\sigma U_\star - k \mathcal{I}_1 P = S_u^\star \tag{7.169}$$

$$\sigma V_\star + \mathcal{D}_5 P = S_v^\star \tag{7.170}$$

$$k \mathcal{I}_3 U_\star + \mathcal{D}_6 V_\star = S'_Q , \tag{7.171}$$

where $\mathcal{I}_3$ is the $(N+1) \times (N-1)$ matrix defined by Eq.(7.53) and $S_u^\star = \mathcal{I}_1 S_u$. Finally, the vector S'_Q is defined by

$$S'_Q = (-k\gamma_+ + \varphi_{Q,0}, \varphi_{Q,1}, \ldots, \varphi_{Q,N-1}, -k\gamma_- + \varphi_{Q,N})^T .$$

The elimination of $U_\star$ and $V_\star$ gives the system determining P, that is,

$$\left(\mathcal{D}_6 \mathcal{D}_5 - k^2 \mathcal{I}_3 \mathcal{I}_1\right) P = -\sigma S'_Q + k \mathcal{I}_3 S_u^\star + \mathcal{D}_6 S_v^\star . \tag{7.172}$$

The Uzawa operator $\mathcal{U}_2 = \mathcal{D}_6 \mathcal{D}_5 - k^2 \mathcal{I}_3 \mathcal{I}_1$ has the same properties as $\mathcal{U}_1$. When P is calculated, Eq.(7.169) and (7.170) give, respectively, $U_\star$ and $V_\star$.

Option (3). The system is made with Eqs.(7.161) and (7.162) for $j = 0, \ldots, N$, Eq.(7.163) for $j = 1, \ldots, N - 1$, and Eq.(7.164). Therefore, the unknowns vectors are U, $V_\star$, and P, that are determined by the equations

$$\sigma U - k P = S_u \tag{7.173}$$

$$\sigma \mathcal{I}_3 V_\star + \mathcal{D} P = S'_v \tag{7.174}$$

$$k \mathcal{I}_1 U + \mathcal{D}_1 V_\star = S_Q^\star , \tag{7.175}$$

where $\mathcal{D}$ is the $(N+1) \times (N+1)$ matrix defined by Eq.(3.51) and $\mathcal{D}_1$ is the $(N-1) \times (N-1)$ matrix defined by Eq.(4.66). The vector S'_v is defined by

$$S'_v = (-\sigma g_+ + \varphi_{v,0}, \varphi_{v,1}, \ldots, \varphi_{v,N-1}, -\sigma g_- + \varphi_{v,N})^T$$

and $S_Q^\star = \mathcal{I}_1 S_Q$.

To derive the system determining P we have to make a combination of Eqs.(7.173) and (7.174) in order to exhibit the left-hand side of Eq.(7.175). This technique is similar to the one developed previously for the first two options, but it has to take into account the fact that the incompressibility equation (7.175) contains only $N - 1$ equations. Thus, Eq.(7.173) is multiplied by $k \mathcal{I}_1$ and Eq.(7.174) is multiplied by $\mathcal{D}_5$, in order to extract

$(N-1)$ equations from the $(N+1)$-equation systems (7.173) and (7.174), respectively,

$$\sigma\, k\, \mathcal{I}_1\, U - k\, \mathcal{I}_1\, P = k\, \mathcal{I}_1\, S_u = k\, S_u^*$$

$$\sigma\, \mathcal{D}_1\, V_\star + \mathcal{D}_4\, P = \mathcal{D}_5\, S_v'\,,$$

where we have taken into account that $\mathcal{D}_5\, \mathcal{I}_3 = \mathcal{D}_1$ and $\mathcal{D}_5\, \mathcal{D} = \mathcal{D}_4$ with $\mathcal{D}_4$ defined by Eq.(6.36). Now, by adding these two equations and using Eq.(7.175), we get the system

$$\left(\mathcal{D}_4 - k^2\, \mathcal{I}_1\right) P = -\sigma\, S_Q^* + k\, S_u^* + \mathcal{D}_5\, S_v' \equiv S_P\,. \tag{7.176}$$

This system contains $(N-1)$ equations. It is completed with the first and last equations of the system (7.174), that is, Eq.(7.162) for $j = 0$ and $j = N$, so that

$$\left(\mathcal{D}_7 - k^2\, \mathcal{I}_4\right) P = S_P'\,, \tag{7.177}$$

where the $(N+1) \times (N+1)$ matrix $\mathcal{D}_7$ has been defined in Eq.(7.59). The $(N+1)$-element vector S_P' is such that

$$S_P' = \left(\varphi_{v,0},\, S_P^T,\, \varphi_{v,N}\right)^T$$

and $\mathcal{I}_4$ is the $(N+1) \times (N+1)$ matrix deduced from the identity matrix $\mathcal{I}_0$ by setting to zero the first and last elements of the main diagonal. When P is known, U and $V_\star$ are calculated from Eqs.(7.173) and (7.174), respectively.

It is easy to see that Eq.(7.177) represents the discrete approximation to a Helmholtz equation with Neumann boundary conditions. Obviously, this is nothing but the discrete counterpart of the projection method commonly presented in continuous form. To be more precise, let us consider the differential problem (7.134)-(7.138). The splitting scheme based on the BDI2 scheme for the predictor step leads to the following equations :

Step 1

$$\tilde{u}'' - \left(\sigma_1 + k^2\right)\tilde{u} = \tilde{f}_u\,, \quad -1 < y < 1$$

$$\tilde{v}'' - \left(\sigma_1 + k^2\right)\tilde{v} = \tilde{f}_v\,, \quad -1 < y < 1$$

$$\tilde{u}\,(\pm 1) = \gamma_\pm\,, \quad \tilde{v}\,(\pm 1) = g_\pm\,.$$

Step 2

$$\sigma\, u - k\, q = \sigma\, \tilde{u}\,, \quad -1 < y < 1$$

$$\sigma\, v + q' = \sigma\, \tilde{v}\,, \quad -1 < y < 1$$

$$k\, u + v' = 0\,, \quad -1 < y < 1$$

$$v\,(\pm 1) = g_\pm\,,$$

where $\tilde{u} = \tilde{u}^{n+1}$, $\tilde{v} = \tilde{v}^{n+1}$, $u = u^{n+1}$, $v = v^{n+1}$, $q = p^{n+1} - p^n$, $\sigma = 2/(3\,\Delta t)$ and $\sigma_1 = \sigma/\nu$. Now, considering Step 2, we obtain the equation for q by adding the first equation multiplied by k and the second one after differentiation, namely,

$$\sigma \left(k\,u + v'\right) + \left(q'' - k^2\,q\right) = \sigma \left(k\,\tilde{u} + \tilde{v}'\right).$$

Then, taking into account the incompressibility equation, we finally obtain

$$q'' - k^2\,q = \sigma\left(k\,\tilde{u} + \tilde{v}'\right), \quad -1 < y < 1, \tag{7.178}$$

to be solved with the boundary conditions deduced from the second equation evaluated at $y = \pm1$, namely,

$$q'\left(\pm1\right) = \sigma\left[\tilde{v}\left(\pm1\right) - v\left(\pm1\right)\right] = 0. \tag{7.179}$$

When q is calculated, the two first equations of the above system give, respectively, u and v.

The Chebyshev collocation solution to the above Neumann problem for q coincides with the system (7.177) with only the (formal) difference of unknown : q instead of p^{n+1}. For $k = 0$, Eq.(7.178) becomes a Poisson equation and it is easy to verify that the compatibility condition is automatically satisfied since the Poisson equation and the Neumann condition come both from the same set of differential equations. It was obviously the same in the discrete presentation of the projection method described previously.

Option (4). The system is constituted by Eq.(7.161) for $j = 1, \ldots, N-1$, Eq.(7.162) for $j = 0, \ldots, N$, Eq.(7.163) for $j = 1, \ldots, N-1$, Eq.(7.164), and the boundary conditions $u_0 = \gamma_+$ and $u_N = \gamma_-$. The unknown vectors are $U_\star$, $V_\star$, and P. They are determined by the system

$$\sigma\,U_\star - k\,\mathcal{I}_1\,P = S_u^\star \tag{7.180}$$

$$\sigma\,\mathcal{I}_3\,V_\star + \mathcal{D}\,P = S_v' \tag{7.181}$$

$$k\,U_\star + \mathcal{D}_1\,V_\star = S_Q^\star. \tag{7.182}$$

Proceeding as in Option (3), we again get the system (7.177) for determining P. Then Eqs.(7.180) and (7.181) give $U_\star$ and $V_\star$, respectively. The only difference with Option (3) lies in the determination of u_j^{n+1} at the boundaries.

7.3.3 Navier-Stokes equations

In this section, we consider the solution of the Navier-Stokes equations (7.1)-(7.2) in the domain $\{0 \le x \le 2\pi,\ -1 \le y \le 1\}$, assuming the solution to be 2π-periodic in x and verifying the boundary conditions (7.18) and the initial condition (7.17). The difference with the time-dependent

Stokes equations considered in the previous section lies in the presence of the nonlinear convective term $\mathbf{A}(\mathbf{V}, \mathbf{V})$. Generally, this term is treated explicitly in order to avoid costly iterative procedures. However, it may happen, in some circumstances, that such an implicit treatment induces too strong a constraint on the size of the time-step. In this case, it may be preferable to consider an implicit treatment of the term $\mathbf{A}(\mathbf{V}, \mathbf{V})$. These two types of time-discretization will now be addressed, in a way similar to that developed in Chapter 6 for the vorticity-streamfunction equations.

(a) Semi-implicit scheme

Whatever the time-discretization scheme considered in Section 7.3.2, the term $\mathbf{A}(\mathbf{V}, \mathbf{V})$ is evaluated in an explicit way through the Adams-Bashforth extrapolation. Thus, for the three-level BDI2 scheme defined by $\varepsilon = 2$ and $\theta = 1$, the approximation to $\mathbf{A}(\mathbf{V}, \mathbf{V})$ is

$$\mathbf{A}^{n+1} \cong 2\,\mathbf{A}^n - \mathbf{A}^{n-1}, \tag{7.183}$$

where $\mathbf{A}^n = \mathbf{A}(\mathbf{V}^n, \mathbf{V}^n)$. This term is included either in Eqs.(7.29)-(7.30) or in Eq.(7.128) according to the considered time-scheme. The evaluation of $\mathbf{A}^n$ follows the development in Section 6.3.3 for the analogous term B^n, making use of the pseudospectral technique (Sections 2.8, 3.5.2, and 6.5.3) for the calculation of the Fourier coefficients of the various products at the collocation points y_j. The influence of the form of the nonlinear term $\mathbf{A}(\mathbf{V}, \mathbf{V})$ on the stability will be analyzed in Section 7.4.5.

For some types of flows, the convective term $\mathbf{A}$ can be treated partly implicitly, while avoiding the need for an iterative solution procedure, in a way similar to that mentioned for the Burgers equation in Section 3.5.2. More precisely, let us assume the existence of a constant velocity $\mathbf{U}$ such that $\mathbf{V} = \mathbf{U} + \mathbf{W}$ where $\mathbf{W}$ is sufficiently "small" with respect to $\mathbf{U}$. The unknown vector in Eqs.(7.29)-(7.32) is then $\mathbf{W}^{n+1}$ and the approximation (7.183) is replaced by

$$\mathbf{A}^{n+1} = \left(\mathbf{V}^{n+1} \cdot \nabla\right)\mathbf{V}^{n+1}$$

$$\cong \left(\mathbf{U} \cdot \nabla\right)\mathbf{W}^{n+1} + 2\left(\mathbf{W}^n \cdot \nabla\right)\mathbf{W}^n - \left(\mathbf{W}^{n-1} \cdot \nabla\right)\mathbf{W}^{n-1}.$$

Of course, such a decomposition also applies to the splitting scheme (7.122). Therefore, for both time-discretization schemes, we are led to the solution, at each time-cycle, of advection-diffusion-type equations instead of Helmholtz equations. In the multi-dimensional case, the algebraic system resulting from the Chebyshev collocation approximation can be solved by the Schur decomposition technique described in Section 3.7. This semi-implicit time-scheme has been used by Forestier *et al.* (2000a,b) for the calculation of two- and three-dimensional wake flows.

(b) Fully implicit scheme

The fully implicit BDI2 scheme applied to the Navier-Stokes equations (7.1)-(7.2) leads to the following nonlinear problem, to be solved at each time-cycle

$$L\left(\mathbf{V}^{n+1}\right) + \mathbf{A}\left(\mathbf{V}^{n+1}, \mathbf{V}^{n+1}\right) + \nabla p^{n+1} = \mathbf{G}^{n+1} \quad \text{in } \Omega \qquad (7.184)$$

$$\nabla \cdot \mathbf{V}^{n+1} = 0 \quad \text{in } \Omega \qquad (7.185)$$

$$\mathbf{V}^{n+1}(x, \pm 1) = \left(\mathbf{V}_{\Gamma}^{\pm}\right)^{n+1}, \qquad (7.186)$$

where L is the linear operator defined by

$$L\left(\mathbf{V}^{n+1}\right) = \left(\sigma I - \nu \nabla^2\right) \mathbf{V}^{n+1} \qquad (7.187)$$

with $\sigma = 3/(2\Delta t)$. The forcing term $\mathbf{G}^{n+1}$ is defined by

$$\mathbf{G}^{n+1} = \mathbf{f}^{n+1} - \left(4\mathbf{V}^n - \mathbf{V}^{n-1}\right)/(2\Delta t) .$$

The nonlinear problem (7.184)-(7.186) is solved iteratively by using, for example, the preconditioned relaxation procedure (characterized by the superscript m) considered by Fröhlich et al. (1991) :

$$L\left(\overline{\mathbf{V}}^{m+1}\right) + \nabla p^{n+1,m+1} = \mathbf{G}^{n+1} - \mathbf{A}\left(\mathbf{V}^{n+1,m}, \mathbf{V}^{n+1,m}\right) - L\left(\mathbf{V}^{n+1,m}\right) \quad \text{in } \Omega \qquad (7.188)$$

$$\nabla \cdot \overline{\mathbf{V}}^{m+1} = 0 \quad \text{in } \Omega \qquad (7.189)$$

$$\overline{\mathbf{V}}^{m+1}(x, \pm 1) = \left(\mathbf{V}_{\Gamma}^{\pm}\right)^{n+1} - \mathbf{V}^{n+1,m}(x, \pm 1) , \qquad (7.190)$$

and

$$\mathbf{V}^{n+1,m+1} = \mathbf{V}^{n+1,m} + \alpha \overline{\mathbf{V}}^{m+1} . \qquad (7.191)$$

Therefore, at each iteration, we are faced with the solution of a Stokes problem. This is done by using one of the Chebyshev collocation (Uzawa or influence matrix) methods described in Sections 7.3.3.b and 7.3.3.c. The iterative procedure is started with

$$\mathbf{V}^{n+1,0} = \mathbf{V}^n .$$

The relaxation parameter α was taken to be constant by Fröhlich et al. (1991), but improvement of the convergence could be obtained by using a dynamic procedure (at the price of the calculation of norms) as indicated in Section 3.8. The efficiency of the method has been estimated by Fröhlich et al. (1991) for the steady Rayleigh-Bénard problem considered in Section 6.5.1. The spatial approximation makes use of the Chebyshev collocation method based on a staggered mesh. The unknowns are the values of the velocity at the Gauss-Lobatto points. The momentum equation and incompressibility equation are enforced at, respectively, the Gauss-Lobatto points and at the Gauss points. This type of approximation leads to an Uzawa equation determining the pressure in a way similar to the one described in Section 7.3.2.b. The Uzawa operator is inverted once and for all in a preprocessing stage. The conclusions concerning the properties of the iterative method, in comparison with the semi-implicit AB/BDI2 method, are quite similar to the ones drawn for the vorticity-streamfunction equations in Section 6.5.1.a.

In the case of the splitting method discussed in Section 7.3.3.d, the nonlinear term $\mathbf{A}\left(\tilde{\mathbf{V}}^{n+1}, \tilde{\mathbf{V}}^{n+1}\right)$ is included in the first step of the scheme [Eq.(7.128) with $\varepsilon = 2$, $\theta = 1$]. The second step remains unchanged. Thus, the iterative procedure has to be applied to the calculation of the provisional velocity $\tilde{\mathbf{V}}^{n+1}$ only, that is, to the solution of two-dimensional Burgers equations. This can be done by using a preconditioned relaxation procedure similar to the one described above.

7.4 Chebyshev-Chebyshev method

In this section we address the solution of the two-dimensional Navier-Stokes equations (7.1)-(7.2) in a square domain using a Chebyshev collocation approximation in both directions. The time-discretization schemes are similar to those discussed in the preceding section, except that higher-order accuracy is considered, thanks to the multistep AB/BDIk scheme.

First we consider time-schemes leading to the Stokes problem, and we discuss its solution using the influence matrix method. Then we consider the time-splitting schemes of projection-type leading to the solution of the Darcy problem. For this solution, we describe two methods, each of which corresponds to a special type of spatial approximation, namely, $I\!\!P_N - I\!\!P_N$ and $I\!\!P_N - I\!\!P_{N-2}$ approximations.

Every method discussed in this section has been applied to the computation of complex, unsteady, two- or three-dimensional flows. An application of the $I\!\!P_N - I\!\!P_N$ projection method is described in Section 7.5. Examples of the application of the $I\!\!P_N - I\!\!P_{N-2}$ projection method and the influence matrix method are given in Sections 8.4.2.b and 9.6.3, respectively.

The methods are described in the two-dimensional case only for the sake of simplicity. There is no fundamental obstacle to their three-dimensional extension. The use of the influence matrix method is, however, restricted to two Chebyshev directions (and, possibly, periodic in the third one) because of the huge size of the influence matrix whose order is equal to the number of collocation points belonging to the boundary.

Therefore, the Navier-Stokes equations (7.1)-(7.3) are solved in the domain $\Omega = (-1, 1)^2$, with the boundary condition

$$\mathbf{V} = \mathbf{V}_\Gamma \quad \text{on } \Gamma = \partial\Omega \tag{7.192}$$

and the initial condition

$$V = V_0 = (u_0, v_0) \quad \text{at } t = 0, \tag{7.193}$$

where V_Γ and V_0 are assumed to satisfy the compatibility and regularity conditions mentioned in Section 5.1. In particular, V_Γ is assumed to be regular enough to ensure the solution to be sufficiently differentiable.

7.4.1 The influence matrix method

(a) Time-discretization

The time-discretization scheme makes use of the finite-difference multistep AB/BDIk scheme described in Section 4.5.1, namely

$$\frac{1}{\Delta t} \sum_{j=0}^{k} a_j V^{n+1-j} + \sum_{j=0}^{k-1} b_j A^{n-j} + \nabla p^{n+1} - \nu \nabla^2 V^{n+1} = f^{n+1} \tag{7.194}$$

$$\nabla \cdot V^{n+1} = 0, \tag{7.195}$$

where $A^{n-j} = A\left(V^{n-j}, V^{n-j}\right)$. The coefficients a_j and b_j are given in Table 4.4 for schemes of order $2 \leq k \leq 4$. Therefore, at each time-cycle, the following Stokes problem has to be solved

$$\nabla^2 V - \sigma V - \nabla p = F \quad \text{in } \Omega \tag{7.196}$$

$$\nabla \cdot V = 0 \quad \text{in } \Omega \tag{7.197}$$

$$V = V_\Gamma \quad \text{on } \Gamma, \tag{7.198}$$

where

$$V = V^{n+1}, \quad p = p^{n+1}/\nu, \quad \sigma = a_0/(\nu \Delta t),$$

$$F = -f^{n+1} + \frac{1}{\Delta t} \sum_{j=1}^{k} a_j V^{n+1-j} + \sum_{j=0}^{k-1} b_j A^{n-j}.$$

The question concerning the calculation of the first time-cycles with a multistep scheme like (7.194) is discussed in Section 4.5.1.c.

(b) Solution of the Stokes problem

The influence matrix method for solving the one-dimensional Stokes problem has largely been discussed in Section 7.3.2.c2. Therefore, we restrict ourselves here to presenting the method and to discussing the peculiar points characteristic of the two-dimensional configuration.

To begin with, the Poisson equation satisfied by the pressure is derived by applying the divergence operator to Eq.(7.196) and taking Eq.(7.197) into account, we obtain

$$\nabla^2 p = -\nabla \cdot F.$$

This implies that $Q = \nabla \cdot \mathbf{V}$ satisfies the Helmholtz equation

$$\nabla^2 Q - \sigma Q = 0, \tag{7.199}$$

so that $Q = 0$ everywhere if $Q = 0$ on the boundary Γ, which gives the missing boundary conditions. Therefore, the problem (7.196)-(7.198) is replaced by

$$\nabla^2 p = -\nabla \cdot \mathbf{F} \quad \text{in } \Omega \tag{7.200}$$

$$H\mathbf{V} - \nabla p = \mathbf{F} \quad \text{in } \Omega \tag{7.201}$$

$$\mathbf{V} = \mathbf{V}_\Gamma \quad \text{on } \Gamma \tag{7.202}$$

$$\nabla \cdot \mathbf{V} = 0 \quad \text{on } \Gamma, \tag{7.203}$$

where H is the Helmholtz operator

$$H = \nabla^2 - \sigma I. \tag{7.204}$$

The problem (7.200)-(7.203) is approximated by the Chebyshev collocation method based on the Gauss-Lobatto points

$$x_i = \cos \frac{\pi i}{N_x}, \quad i = 0, \ldots, N_x, \quad y_j = \cos \frac{\pi j}{N_y}, \quad j = 0, \ldots, N_y.$$

We denote by Ω_N the set of collocation points in the interior of Ω,

$$\Omega_N = \{x_i, y_j; \; i = 1, \ldots, N_x - 1, \; j = 1, \ldots, N_y - 1\},$$

by $\overline{\Omega}_N$, the set of all collocation points,

$$\overline{\Omega}_N = \{x_i, y_j; \; i = 0, \ldots, N_x, \; j = 0, \ldots, N_y\},$$

and by $\overline{\Omega}_N^I$ the set $\overline{\Omega}_N$ minus the four corners. We denote by Γ_N the set of collocation points belonging to the boundary Γ, and by Γ_N^I this same set minus the four corners. Finally, we recall that $I\!\!P_N$ represents the space of polynomials of degree at most equal to N_x in x and to N_y in y. Therefore, the solution $[\mathbf{V} = (u, v), p]$ of problem (7.200)-(7.203) is approximated by $[\mathbf{V}_N = (u_N, v_N), p_N]$, where u_N, v_N, and p_N belong to $I\!\!P_N$. Such an approximation, where the velocity components and the pressure are in the same polynomial space, is denoted by "$I\!\!P_N - I\!\!P_N$".

The Chebyshev collocation approximation to the problem (7.200)-(7.203) is

$$\nabla^2 p_N = -\nabla \cdot \mathbf{F}_N \quad \text{in } \Omega_N \tag{7.205}$$

$$H\mathbf{V}_N - \nabla p_N = \mathbf{F}_N \quad \text{in } \Omega_N \tag{7.206}$$

$$\mathbf{V}_N = \mathbf{V}_\Gamma \quad \text{on } \Gamma_N^I \tag{7.207}$$

$$\nabla \cdot \mathbf{V}_N = 0 \quad \text{on } \Gamma_N^I. \tag{7.208}$$

The various derivatives occurring in these equations are expressed by formulas of type (3.45). Thus, Eqs. (7.205)-(7.206) constitute a system of $[3\,(N_x - 1)\,(N_y - 1) + 4\,(N_x + N_y - 2)]$ scalar equations for the same number of scalar unknowns, namely, $(N_x - 1)\,(N_y - 1)$ for each velocity component $u_N\,(x_i, y_j)$ and $v_N\,(x_i, y_j)$, $(x_i, y_j) \in \Omega_N$, and $[N_x\,N_y + (N_x + N_y - 3)]$ values of the pressure $p_N\,(x_i, y_j)$, $(x_i, y_j) \in \Omega_N^I$. Note that the forcing term $\mathbf{F}_N$ is evaluated either by matrix products or by the pseudospectral technique (Section 3.5.2).

As shown in Section 7.3.2.c, the velocity field $\mathbf{V}_N$ defined by (7.205)-(7.208) is not divergence-free because the momentum equation normal to the boundary is not satisfied. Therefore, it is necessary to introduce a correction term $\nabla \cdot \mathbf{B}$ into the pressure equation in order to recover an exactly divergence-free velocity field. The correction vector $\mathbf{B}$ (whose components are polynomials belonging to $I\!\!P_N$) is zero in Ω_N and its normal component B_n at the boundary Γ_N^I is the residual of the momentum equation normal to the boundary.

Therefore, we have to solve the $\mathcal{P}$-*Problem* defined by the algebraic system

$$\nabla^2 p_N = -\nabla \cdot \mathbf{F}_N - \nabla \cdot \mathbf{B} \quad \text{in } \Omega_N \tag{7.209}$$

$$H\,\mathbf{V}_N - \nabla p_N = \mathbf{F}_N \quad \text{in } \Omega_N \tag{7.210}$$

$$\mathbf{V}_N = \mathbf{V}_\Gamma \quad \text{on } \Gamma_N^I \tag{7.211}$$

$$B_n = (H\,\mathbf{V}_N - \nabla p_N - \mathbf{F}_N)\,.\mathbf{n} \quad \text{on } \Gamma_N^I, \tag{7.212}$$

where $\mathbf{n}$ is the unit normal vector to Γ and $B_n = \mathbf{B} \cdot \mathbf{n}$. As noted above, the vector $\mathbf{B}$ vanishes in Ω_N, namely,

$$\mathbf{B} = \mathbf{0} \quad \text{in } \Omega_N. \tag{7.213}$$

Note that $\mathbf{B}_t$, the component of $\mathbf{B}$ tangential to Γ, is not involved in the algorithm since $\mathbf{B}$ appears in Eq.(7.209) only through its divergence.

Now the task is to solve the $\mathcal{P}$-problem in an efficient way. This is done through the influence matrix technique. As usual, we pose

$$p_N = \tilde{p}_N + \overline{p}_N, \quad \mathbf{V}_N = \tilde{\mathbf{V}}_N + \overline{\mathbf{V}}_N. \tag{7.214}$$

The grid values of the part $\left(\tilde{p}_N, \tilde{\mathbf{V}}_N\right)$ are determined by the discrete $\tilde{\mathcal{P}}$-*Problem* constituted by

$$\nabla^2 \tilde{p}_N = -\nabla \cdot \mathbf{F}_N \quad \text{in } \Omega_N$$

$$\tilde{p}_N = 0 \quad \text{on } \Gamma_N^I,$$

and by

$$H\,\tilde{\mathbf{V}}_N = \mathbf{F}_N + \nabla \tilde{p}_N \quad \text{in } \Omega_N$$

$$\tilde{\mathbf{V}}_N = \mathbf{V}_\Gamma \quad \text{on } \Gamma_N^I.$$

The solution of these systems gives, successively, $\tilde{p}_N$ and $\tilde{\mathbf{V}}_N$, from which we deduce

$$\tilde{B}_n = \left(H\,\tilde{\mathbf{V}}_N - \nabla\tilde{p}_N - \mathbf{F}_N \right).\mathbf{n} \quad \text{on } \Gamma_N^I ,$$

where $\tilde{B}_n$ is the normal component of vector $\tilde{\mathbf{B}}$ such that

$$\tilde{\mathbf{B}} = \mathbf{0} \quad \text{in } \Omega_N$$

and $\tilde{\mathbf{B}}_t = 0$ on Γ_N^I. Then the grid values of the part $\left(\overline{p}_N, \overline{\mathbf{V}}_N \right)$ are a solution to the discrete $\overline{\mathcal{P}}$-Problem :

$$\nabla^2 \overline{p}_N = -\nabla \cdot \mathbf{B} \quad \text{in } \Omega_N \tag{7.215}$$

$$H\,\overline{\mathbf{V}}_N - \nabla\overline{p}_N = \mathbf{0} \quad \text{in } \Omega_N \tag{7.216}$$

$$\overline{\mathbf{V}}_N = \mathbf{0} \quad \text{on } \Gamma_N^I \tag{7.217}$$

$$\nabla \cdot \overline{\mathbf{V}}_N = -\nabla \cdot \tilde{\mathbf{V}}_N \quad \text{on } \Gamma_N^I , \tag{7.218}$$

and

$$B_n = \tilde{B}_n + \overline{B}_n \quad \text{on } \Gamma_N^I , \tag{7.219}$$

where

$$\overline{B}_n = \left(H\,\overline{\mathbf{V}}_N - \nabla\overline{p}_N \right) \cdot \mathbf{n} \quad \text{on } \Gamma_N^I .$$

As in the one-dimensional case, the $\overline{\mathcal{P}}$-problem is solved through the superposition of elementary solutions according to

$$\phi = \sum_{l=1}^{2\,L} \xi_l\,\phi_l , \tag{7.220}$$

where $\phi = \overline{p}_N, \overline{\mathbf{V}}_N, \overline{B}_n, \mathbf{B}$. The integer L is the number of collocation points on Γ_N^I, namely $L = 2\,(N_x + N_y - 2)$ and ξ_l, $l = 1, \ldots, 2\,L$, are constants. The elementary solutions ϕ_l are obtained from a sequence of $\overline{\mathcal{P}}_l$-Problems, $l = 1, \ldots, 2\,L$, defined as follows:

$\overline{\mathcal{P}}_l$-Problem, $l = 1, \ldots, L$:

$$\nabla^2 \overline{p}_l = 0 \qquad \text{in } \Omega_N$$

$$\overline{p}_{l\,|\eta_m} = \delta_{m,l} , \qquad \eta_m \in \Gamma_N^I ,$$

and

$$H\,\overline{\mathbf{V}}_l = \nabla\overline{p}_l \qquad \text{in } \Omega_N$$

$$\overline{\mathbf{V}}_l = \mathbf{0} \qquad \text{on } \Gamma_N^I ,$$

and

$$\mathbf{B}_l = \mathbf{0} \qquad \text{in } \overline{\Omega}_N ,$$

where η_m, $m = 1, \ldots, L$, refers to the collocation points on Γ_N^I. Again, when $\bar{p}_l$ and $\overline{\mathbf{V}}_l$ are determined, we calculate the residual $\overline{B}_{l\,n}$ of the normal momentum equation at the boundary

$$\overline{B}_{l\,n} = \left(H\,\overline{\mathbf{V}}_l - \nabla\bar{p}_l\right) \cdot \mathbf{n} \quad \text{on } \Gamma_N^I. \tag{7.221}$$

$\overline{\mathcal{P}}_l$-Problem, $l = L+1, \ldots, 2\,L$:

$$\nabla^2 \bar{p}_l = -\nabla \cdot \mathbf{B}_l \qquad \text{in } \Omega_N$$

$$\bar{p}_l = 0 \qquad \text{on } \Gamma_N^I$$

$$B_{l\,n\,|\eta_m} = \delta_{L+m,l}, \qquad \eta_m \in \Gamma_N^I.$$

Since

$$\mathbf{B}_l = 0 \qquad \text{in } \Omega_N$$

the above system completely determines $\bar{p}_l$ in $\overline{\Omega}_N^I$. Then,

$$H\,\overline{\mathbf{V}}_l = \nabla\bar{p}_l \qquad \text{in } \Omega_N,$$

$$\overline{\mathbf{V}}_l = 0 \qquad \text{on } \Gamma_N^I,$$

and the residual $\overline{B}_{l\,n}$ on Γ_N^I is again determined by Eq.(7.221).

It is not complicated to verify that the combination (7.220) satisfies Eqs.(7.215)-(7.217) of the $\overline{\mathcal{P}}$-problem, whatever the constants ξ_l, $l = 1, \ldots,$ $2\,L$. These constants are then determined by prescribing the combination (7.220) to also satisfy Eqs.(7.218) and (7.219), namely,

$$\sum_{l=1}^{2\,L} \xi_l \left(\nabla \cdot \overline{\mathbf{V}}_l\right)_{\eta_m} = -\left(\nabla \cdot \tilde{\mathbf{V}}_l\right)_{\eta_m} \qquad \eta_m \in \Gamma_N^I$$

$$\sum_{l=1}^{2\,L} \xi_l \left(B_{l\,n} - \overline{B}_{l\,n}\right)_{\eta_m} = \tilde{B}_{n\,|\eta_m} \qquad \eta_m \in \Gamma_N^I,$$

or, in vector form,

$$\mathcal{M}\,\Xi = \tilde{E} \tag{7.222}$$

with $\Xi = (\xi_1, \ldots, \xi_{2\,L})^T$. The matrix $\mathcal{M}$ is the influence matrix. This matrix has four eigenvalues equal to zero. One of them has a physical meaning: it is associated to the property of the pressure field to be defined up to a constant. The other null eigenvalues are associated to the spurious pressure modes, which will be characterized later. Therefore, the solution Ξ of Eq.(7.222) is not determined in a unique way. It suffices, however, to have one solution $(\xi_1, \ldots, \xi_{2\,L})$ which defines a set of boundary values for p_N and B_n ensuring that Eqs.(7.215)-(7.219) are satisfied. The procedure proposed by Tuckerman (1989) consists of replacing the matrix $\mathcal{M}$ in

Eq.(7.222) by a matrix $\mathcal{M}_0$ defined as follows : first, calculate the eigenvalues λ_j, $j = 1, \ldots, 2L$, of the matrix $\mathcal{M}$ and the associated eigenvectors, and let $\mathcal{P}$ be the matrix made with these eigenvalues ; then construct the matrix

$$\mathcal{M}_0 = \mathcal{P} \Lambda_0 \mathcal{P}^{-1},$$

where Λ_0 is the diagonal matrix made with the eigenvalues λ_j except that the null eigenvalues are replaced by any nonzero value γ, for example, $\gamma = 1$. Then $\mathcal{M}_0$ can be inverted, so that

$$\Xi = \mathcal{M}_0^{-1} \tilde{E}. \tag{7.223}$$

Note that the matrix $\mathcal{M}_0$ acts like $\mathcal{M}$ except on its null space, on which it acts like the identity. It may be observed that this regularization procedure is similar to the one proposed in Section 3.7.1 for the solution of the Neumann problem for the Poisson equation using the matrix-diagonalization method. In this case there exists one null eigenvalue. The technique described, which consists of setting to zero the component of $\tilde{V}$ corresponding to the null eigenvalue, would correspond to the special choice $\gamma = \infty$. Now, Ξ having been determined, it is easy to reconstruct the solution $(p_N, \mathbf{V}_N)$ thanks to Eqs.(7.214) and (7.220). However, in order to avoid the storage of the elementary solutions as well as the summations (7.220), it is generally more efficient to calculate the boundary values of p_N and B_n and, then, to solve again an algebraic system. More precisely, from the expressions

$$p_N = \tilde{p}_N + \sum_{l=1}^{2L} \xi_l \, \bar{p}_l, \quad \mathbf{B} = \sum_{l=1}^{2L} \xi_l \, \mathbf{B}_l$$

and the definition of the boundary values of $\bar{p}_l$ and $B_{l\,n}$ used in Problems $\overline{\mathcal{P}}_l$, we easily get

$$p_{N\,|\,\eta_m} = \xi_m, \quad B_{n\,|\,\eta_m} = \xi_{L+m}, \quad \eta_m \in \Gamma_N^I,$$

and we may calculate $\nabla \cdot \mathbf{B}$ taking Eq.(7.213) into account. Then the solution $(p_N, \mathbf{V}_N)$ is calculated by solving, successively, the two systems :

$$\nabla^2 p_N = -\nabla \cdot \mathbf{F}_N - \nabla \cdot \mathbf{B} \quad \text{in } \Omega_N \tag{7.224}$$

$$p_{N\,|\eta_m} = \xi_m, \quad \eta_m \in \Gamma_N^I, \tag{7.225}$$

and

$$H \mathbf{V}_N = \mathbf{F}_N - \nabla p_N \quad \text{in } \Omega_N \tag{7.226}$$

$$\mathbf{V}_N = \mathbf{V}_\Gamma \quad \text{on } \Gamma_N^I. \tag{7.227}$$

The various algebraic systems are solved by means of the matrix-diagonalization technique described in Section 3.7.2. The $\overline{\mathcal{P}}_l$-problems are time-independent and can be solved in the preprocessing stage performed before starting the time-integration. So, the matrix $\mathcal{M}$ is calculated (using its symmetry properties) during this stage, and the matrix $\mathcal{M}_0$ is calculated, inverted, and its inverse is stored. Generally, the elementary solutions are not stored. Therefore the algorithm is the following :

A. Preprocessing stage
A.1. Solve the $\overline{\mathcal{P}}_l$-problems, $l = 1, \ldots, 2L$.
A.2. Construct the matrix $\mathcal{M}_0$. Invert $\mathcal{M}_0$ and store $\mathcal{M}_0^{-1}$.

B. At each time-cycle
B.1. Calculate the right-hand side $\mathbf{F}_N$ in Eq.(7.200) by matrix products or by the pseudo-spectral technique (FFT).
B.2. Solve the $\tilde{\mathcal{P}}$-problem, calculate $\tilde{E}$.
B.3. Calculate Ξ by Eq.(7.223).
B.4. Calculate the final solution $(p_N, \mathbf{V}_N)$ by solving Eqs.(7.224)-(7.227).

Remarks
To conclude the discussion, we make two remarks. The first one concerns the characterization and removal of the spurious pressure modes and the second remark is relative to the compatibility of the boundary conditions.

1. *Spurious pressure modes*
As mentioned in Section 7.3, the existence of spurious pressure modes is connected to the fact that some Chebyshev modes do not appear in the collocation at Gauss-Lobatto points. This is discussed now in more detail.

As for the one-dimensional case, the application of the influence matrix method with the correction term $\mathbf{B}$ amounts to solving the discrete problem

$$\nabla^2 \mathbf{V}_N - \sigma \mathbf{V}_N + \nabla p_N = \mathbf{F}_N \quad \text{in } \Omega_N \tag{7.228}$$

$$\nabla \cdot \mathbf{V}_N = 0 \quad \text{in } \overline{\Omega}_N^I \tag{7.229}$$

$$\mathbf{V}_N = \mathbf{V}_\Gamma \quad \text{on } \Gamma_N^I . \tag{7.230}$$

Therefore the pressure appears only through its gradient evaluated at the inner collocation points. The polynomial $p_N(x, y)$ can be expressed as

$$p_N(x, y) = \sum_{k=0}^{N_x} \sum_{l=0}^{N_y} \hat{p}_{k,l} T_k(x) T_l(y) . \tag{7.231}$$

Now it is not difficult to detect the modes which are not involved in the gradient ∇p_N evaluated at points belonging to Ω_N, namely, those for which $\nabla p_N(x_i, y_j) = 0$ for $(x_i, y_j) \in \Omega_N$. Thus, we find :

(i)   the constant mode $T_0(x) T_0(y) = 1$,
(ii)   the line mode $T_{N_x}(y) T_0(y) = T_{N_x}(x)$,
(iii) the column mode $T_0(x) T_{N_y}(y) = T_{N_y}(y)$, and
(iv)   the checkboard mode $T_{N_x}(x) T_{N_y}(y)$.

Therefore, the Chebyshev coefficients $\hat{p}_{0,0}$, $\hat{p}_{N_x,0}$, $\hat{p}_{0,N_y}$ and $\hat{p}_{N_x,N_y}$ do not occur in the collocation approximation to the problem (7.228)-(7.230). This explains why some eigenvalues of $\mathcal{M}$ are zero. The constant mode has a physical meaning, the others are spurious.

Moreover, since the algebraic system (7.228)-(7.230) does not involve the pressure values at the corners of the computational domain, there exists another set of special modes, the so-called "corner modes" defined by

$$(1 \pm x)(1 \pm y) T'_{N_x}(x) T'_{N_y}(y) . \tag{7.232}$$

Each of these modes vanishes in $\overline{\Omega}_N$ except at one corner. Moreover, the gradient of these modes is zero at any collocation point of $\overline{\Omega}_N$ except on the two sides of the boundary adjacent to the associated corner. The values of the pressure at the corners cannot be determined by imposing the incompressibility equation at these points since the divergence of the velocity at a corner is completely defined (as zero, like here) by the knowledge of $\mathbf{V}_\Gamma$. The corner modes are generally qualified as "spurious," although their nature and their effect are slightly different from the other ones. More precisely, they are not a possible source of null eigenvalues of the influence matrix or, more generally, of the algebraic system (7.228)-(7.230). The value of the pressure at the corners has no effect on the pressure values of the other collocation points, while the presence of the other spurious modes contaminates the pressure solution in the whole domain. Thus, the knowledge of the corner values of the pressure is unnecessary as long as the formal expression of the polynomial $p_N(x, y)$ is not of interest. The existence of the spurious pressure modes was detected by Morchoisne (1983) and their theoretical analysis (inf-sup conditions) has been done by Bernardi et al. (1988).

An equivalent way to detect the spurious modes associated with the influence matrix method (Balachandar and Madabhushi, 1994) is to examine directly the system (7.209)-(7.213). Let us set $\phi = \phi^I + \phi^{II}$, where $\phi = p_N, \mathbf{V}_N, \mathbf{B}$ and ϕ^I represents the solution of (7.209)-(7.213) free of spurious modes. The spurious part ϕ^{II} is such that $\mathbf{V}_N^{II} = \mathbf{0}$ since these modes have no effect on the velocity. Therefore, the spurious part $(p_N^{II}, \mathbf{B}^{II})$ is any nonzero solution to the homogeneous system

$$\nabla^2 p_N^{II} + \nabla \cdot \mathbf{B}^{II} = 0 \quad \text{in } \Omega_N$$

$$\nabla p_N^{II} = 0 \quad \text{in } \Omega_N$$

$$\mathbf{B}^{II} = \mathbf{0} \quad \text{in } \Omega_N$$

$$\left(\nabla p_N^{II} + \mathbf{B}^{II}\right) \cdot \mathbf{n} = 0 \quad \text{on } \Gamma_N^I .$$

A careful examination of the system shows that we have the following four nonzero solutions

$$p_N^{II} = \text{constant}, \qquad \mathbf{B}^{II} = (0,0),$$

$$p_N^{II} = T_{N_x}(x), \qquad \mathbf{B}^{II} = \left(-T'_{N_x}(x), 0\right),$$

$$p_N^{II} = T_{N_y}(y), \qquad \mathbf{B}^{II} = \left(0, -T'_{N_y}(y)\right),$$

$$p_N^{II} = T_{N_x}(x) T_{N_y}(y), \quad \mathbf{B}^{II} = \left(-T'_{N_x}(x) T_{N_y}(y), -T_{N_x}(x) T'_{N_y}(y)\right),$$

exhibiting the spurious modes.

In the case where the correction term $\mathbf{B}$ is not considered (the "simplified" influence matrix method), the above system reduces to

$$\nabla^2 p_N^{II} = 0, \quad \nabla p_N^{II} = 0 \quad \text{in } \Omega_N.$$

These equations do not admit a nonzero solution (except the constant) : there are no spurious modes. Of course, in both cases, the corner modes are always present.

It is important to remember that spurious modes do not affect the velocity field. Therefore, their presence is harmless if the pressure field is not of interest. However, in a number of physical problems the knowledge of the pressure is indispensable. In these conditions, it is necessary to recover an accurate pressure field. In the following, we propose two different ways to fulfill this requirement :
(i) by filtering the spurious modes, or
(ii) by determining the pressure as the solution of a Poisson equation with a Neumann boundary condition.

First, let us consider the filtering technique. This consists of the following operations :

1. From $p_N(x_i, y_j)$, $i = 0, \ldots, N_x$, $j = 0, \ldots, N_y$, calculate the Chebyshev coefficients $\hat{p}_{k,l}$, $k = 0, \ldots, N_x$, $l = 0, \ldots, N_y$.

2. Set to zero the coefficients $\hat{p}_{N_x,0}$, $\hat{p}_{0,N_y}$, and $\hat{p}_{N_x,N_y}$.

3. From the filtered spectrum, calculate the new (filtered) pressure field $\overline{p}_N(x_i, y_j)$, $i = 0, \ldots, N_x$, $j = 0, \ldots, N_y$, defined to within a constant.

The implementation of this algorithm necessitates the knowledge of the values of the pressure p_N at the corners. These values p_l^c, $l = 1, \ldots, 4$, are not given by the collocation solution (corner modes) and, hence, their determination necessitates a special treatment. We describe, here, the technique proposed by Sabbah and Pasquetti (1998) which is based on a least-squares procedure. The pressure $p_N(x, y)$ is expressed as

$$p_N = p_N^0 + \sum_{l=1}^{4} p_l^c C_l, \tag{7.233}$$

where p_N^0 is the calculated pressure field in $\overline{\Omega}_N^I$ supplemented with the value zero at the corners and C_l, $l = 1,\ldots,4$, are the polynomials constructed from (7.232) and normalized such that $C_l = 1$ at the associated corner. Now the field (7.233) is filtered as explained above, to become

$$\overline{p}_N = \overline{p}_N^0 + \sum_{l=1}^{4} p_l^c \, \overline{C}_l \,. \tag{7.234}$$

Then, with the objective of getting a smooth field $\overline{p}_N$, Sabbah and Pasquetti (1998) propose to determine the values p_l^c, $l = 1,\ldots,4$, from the minimization of the functional

$$J\left(p_l^c\right) = \int_{\Omega} |\nabla \overline{p}_N|^2 \, w \, dx \, dy \,, \tag{7.235}$$

where $w = \left(1 - x^2\right)^{-1/2} \left(1 - y^2\right)^{-1/2}$ is the Chebyshev weight. Thus, by taking Eq.(7.234) into account in Eq.(7.235) and by differentiating J with respect to p_m^c, $m = 1,\ldots,4$, we get the least-squares equations

$$\int_{\Omega} \left[\sum_{l=1}^{4} p_l^c \, \nabla \overline{C}_l \cdot \nabla \overline{C}_m + \nabla \overline{p}_N^0 \cdot \nabla \overline{C}_m\right] w \, dx \, dy = 0 \,, \quad m = 1,\ldots,4 \,.$$

This integral is evaluated by means of the Gauss-Lobatto quadrature [see Eq.(3.14)], so that the system determining p_l^c, $l = 1,\ldots,4$, is

$$\mathcal{A} P^c = S \,, \tag{7.236}$$

where $P^c = (p_1^c, \ldots, p_4^c)$, $\mathcal{A}$ is the matrix, and S the vector whose elements are, respectively,

$$a_{m,l} = \sum_{x_i, y_j} \frac{1}{\overline{c}_i \, \overline{c}_j} \left(\nabla \overline{C}_l \cdot \nabla \overline{C}_m\right)_{i,j} \,, \quad (x_i, y_j) \in \Gamma_N^I \,,$$

$$s_m = -\sum_{x_i, y_j} \frac{1}{\overline{c}_i \, \overline{c}_j} \left(\nabla p_0' \cdot \nabla \overline{C}_m\right)_{i,j} \,, \quad (x_i, y_j) \in \Gamma_N^I \,,$$

with $\overline{c}_k$ defined by Eq.(3.16). It must be noticed that the summations are taken on the boundary only, because the gradient $\nabla \overline{C}_l$ vanishes in Ω_N. This property can easily be proven. Let $\overline{C}_l = C_l - C_l^*$ where C_l^* is a linear combination of the spurious modes. The result is that $\nabla \overline{C}_l = \nabla C_l - \nabla C_l^*$ vanishes in Ω_N because of the properties of the spurious and corner modes.

The second way to recover the correct pressure field is to define it as the solution of the Neumann problem (5.30)-(5.31) with $\mathbf{V}$ replaced by the calculated velocity field $\mathbf{V}_N$ which is exactly solenoidal owing to the application of the influence matrix method. This way is much more accurate than the filtering technique described above (see Section 7.4.4.a), whose weakness is connected to the necessary determination of the pressure values at the corners. Moreover, the extra-cost in computing time induced by the solution of the Neumann problem remains unimportant since this latter is calculated (as post-processing) only when the pressure is really needed. Finally, it must be remarked that, if needed, the values of the pressure at the corners can be calculated as for any boundary value problem of Robin type (Section 3.7.2). More precisely, let us consider the side $-1 \leq x \leq 1$, $y = 1$. The derivatives $\partial_x p_N$ at the corners $(-1, 1)$ and $(1, 1)$ may be taken equal to the derivatives $\partial_y p_N$ normal to the sides $x = -1$ and $x = 1$, respectively, extended up to the corners. Then, from the first-order differentiation formula (3.45)-(3.46), associated to the known values of $\partial_x p_N (\pm 1, 1)$ and $p_N (x_i, 1)$, $i = 1, \ldots, N - 1$, we easily deduce the corner values $p_N (\pm 1, 1)$.

To close this remark, we mention the comparative study made by Schumack *et al.* (1991) on ways to avoid the spurious modes of pressure in the solution of the Stokes problem. Some of them have already been mentioned in Section 7.3. In the following sections we describe two projection methods exempt of spurious modes, one is based on the $I\!P_N - I\!P_{N-2}$ approximation (Section 7.4.2) and the other makes use of a Neumann boundary condition for the pressure (Section 7.4.3).

2. Total flow rate condition

The influence matrix method provides an approximate velocity field whose divergence is the null polynomial. Therefore, consistency requires that the polynomial approximation $\mathbf{V}_{\Gamma N} = (u_{\Gamma N}, v_{\Gamma N})$ to the boundary velocity $\mathbf{V}_\Gamma$ satisfies exactly the flow rate condition

$$\int_\Gamma (\mathbf{V}_{\Gamma N} \cdot \mathbf{n}) \, d\Gamma = 0. \tag{7.237}$$

If the spatial resolution is sufficiently high, this condition is satisfied to machine accuracy. If not, the deviation from the zero total flow rate, even as small as 10^{-6}, may induce long-time instabilities. The remedy is to slightly modify $\mathbf{V}_{\Gamma N}$ on some part of the boundary in order to recover condition (7.237). This can be done in the following way.

Let Γ_1 be the part of Γ where $\mathbf{V}_{\Gamma N}$ is replaced by $\mathbf{V}_{\Gamma N}^\star$ such that

$$\mathbf{V}_{\Gamma N}^\star \cdot \mathbf{n} = \begin{cases} \mathbf{V}_{\Gamma N} \cdot \mathbf{n} & \text{at the extremities of } \Gamma_1. \\ \mathbf{V}_{\Gamma N} \cdot \mathbf{n} + \varepsilon & \text{at the inner points of } \Gamma_1, \end{cases}$$

where ε is determined by requiring

$$\int_{\Gamma_1} \mathbf{V}_{\Gamma N}^\star \cdot \mathbf{n} \, d\Gamma + \int_{\Gamma \backslash \Gamma_1} \mathbf{V}_{\Gamma N} \cdot \mathbf{n} \, d\Gamma = 0.$$

Another approach is the one used by Forestier *et al.* (2000b) in which $\mathbf{V}_{\Gamma N}^\star$ is represented on a polynomial basis satisfying the flux rate condition and is determined by the least-squares minimization of $|\mathbf{V}_{\Gamma N}^\star \cdot \mathbf{n} - \mathbf{V}_{\Gamma N} \cdot \mathbf{n}|^2$ on Γ_1.

7.4.2 The projection method

The projection method has been presented in Section 7.3.2, in the Fourier-Chebyshev case, associated with first- and second-order time-discretization schemes. Here, considering higher-order time schemes, we describe two commonly used projection methods based on a Chebyshev collocation approximation in both spatial directions (or in three, in the general case). The first one considers polynomial approximation of different degree for velocity and pressure but using the same grid of collocation points ($I\!P_N - I\!P_{N-2}$ method). The second method makes use of the same polynomial space for velocity and pressure ($I\!P_N - I\!P_N$ method) and is based, as is classical for projection methods, on a Neumann boundary condition for the pressure.

(a) The general semidiscrete projection method

As is known, the projection method consists of two steps. In the first step, a provisional value $\tilde{\mathbf{V}}^{n+1}$ of the velocity is calculated without taking into account the incompressibility equation. Then, in the second step, this provisional value is corrected by determining the pressure ensuring that the final velocity field is solenoidal. The scheme is as follows:

Step 1 :

$$\frac{1}{\Delta t}\left(a_0\tilde{\mathbf{V}}^{n+1} + \sum_{j=1}^{k} a_j \mathbf{V}^{n+1-j}\right) + \sum_{j=0}^{k-1} b_j\,\mathbf{A}^{n-j} \tag{7.238}$$
$$+ \nabla\overline{p}^{n+1} - \nu\nabla^2\tilde{\mathbf{V}}^{n+1} = \mathbf{f}^{n+1}\quad\text{in }\Omega$$

$$\tilde{\mathbf{V}}^{n+1} = \mathbf{V}_\Gamma^{n+1}\quad\text{on }\Gamma, \tag{7.239}$$

with $\mathbf{A}^{n-j} = \mathbf{A}\left(\mathbf{V}^{n-j}, \mathbf{V}^{n-j}\right)$.

Step 2 :

$$\frac{a_0}{\Delta t}\left(\mathbf{V}^{n+1} - \tilde{\mathbf{V}}^{n+1}\right) + \nabla\left(p^{n+1} - \overline{p}^{n+1}\right) = \mathbf{0}\quad\text{in }\Omega \tag{7.240}$$

$$\nabla\cdot\mathbf{V}^{n+1} = 0\quad\text{in }\Omega \tag{7.241}$$

$$\mathbf{V}^{n+1}\cdot\mathbf{n} = \mathbf{V}_\Gamma^{n+1}\cdot\mathbf{n}\quad\text{on }\Gamma. \tag{7.242}$$

The coefficients a_j and b_j for $2 \le k \le 4$ are given in Table 4.4 of Section 4.5.1.a. They ensure, respectively, the material derivative of $\mathbf{V}$ and the nonlinear term $\mathbf{A}$ to be approximated at time $(n+1)\,\Delta t$ with an error $O\left(\Delta t^k\right)$. Moreover, the global accuracy of the scheme depends on the definition of the pressure $\overline{p}^{n+1}$ as discussed now.

By eliminating the provisional velocity $\tilde{\mathbf{V}}^{n+1}$ from Eqs.(7.238) and (7.240), we obtain the semidiscrete momentum equation actually solved, namely,

$$\frac{1}{\Delta t}\sum_{j=0}^{k} a_j \, \mathbf{V}^{n+1-j} + \sum_{j=0}^{k-1} b_j \, \mathbf{A}^{n-j} + \nabla p^{n+1} - \nu \, \nabla^2 \mathbf{V}^{n+1} + \Delta t \, \mathbf{E}^{n+1} = \mathbf{f}^{n+1}$$

$$(7.243)$$

with

$$\mathbf{E}^{n+1} = -\frac{\nu}{a_0}\nabla^2 \left[\nabla \left(p^{n+1} - \overline{p}^{n+1}\right)\right] . \qquad (7.244)$$

Obviously, when the problem (7.238)-(7.242) is approximated by a Chebyshev collocation, the discrete form of $\mathbf{E}^{n+1}$ depends on the type of the polynomial approximation ($I\!\!P_N - I\!\!P_N$ or $I\!\!P_N - I\!\!P_{N-2}$) and on the way in which the algebraic system is closed. In any case, however, the discrete operator in $\mathbf{E}^{n+1}$ acts on the polynomial approximation of the quantity $q^{n+1} = p^{n+1} - \overline{p}^{n+1}$, so that the order of the error in time of Eq.(7.243) is unchanged.

The examination of Eqs.(7.243) and (7.244) shows that the truncation error of this equation is $O\left(\Delta t^k\right)$ if $p^{n+1} - \overline{p}^{n+1} = O\left(\Delta t^{k-1}\right)$. Therefore, the choice of $\overline{p}^{n+1}$ determines the accuracy of the method :

1. The original method (Chorin, 1968 ; Temam, 1969) simply uses

$$\overline{p}^{n+1} = 0$$

and the truncation error is $O\left(\Delta t\right)$.

2. The incremental method (Goda, 1979) makes use of

$$\overline{p}^{n+1} = p^n$$

so that the truncation error is $O\left(\Delta t^2\right)$.

3. A general high-order approximation is obtained (Karniadakis *et al.*, 1991 ; Botella, 1997) from the extrapolation formula

$$\overline{p}^{n+1} = \sum_{j=0}^{k-2} c_j \, p^{n-j}, \quad k \geq 2, \qquad (7.245)$$

with $c_0 = 1$ for $k = 2$, $c_0 = 2$, $c_1 = -1$ for $k = 3$ and, more generally, $c_j(k) = b_j(k-1)$. Formula (7.245) leads to a truncation error of $O\left(\Delta t^k\right)$ for Eq.(7.243).

4. Hugues and Randriamampianina (1998) determine $\overline{p}^{n+1}$ as the solution of a Neumann problem for a Poisson equation, such that $p^{n+1} - \overline{p}^{n+1} = O\left(\Delta t^2\right)$. This technique is described in Section 7.4.2.b.

In *Step 1*, Eqs.(7.238) and (7.239) define a Dirichlet problem for the Helmholtz equation satisfied by each component of $\tilde{\mathbf{V}}^{n+1}$. These Helmholtz problems are easily solved by means of the Chebyshev collocation method associated with the matrix-diagonalization technique (Section 3.7.2) for the solution of the associated algebraic system. Note that the nonlinear convective term in Eq.(7.238) is evaluated through matrix products or using the pseudospectral technique (Section 3.5.2).

Step 2 corresponds to a Darcy problem (or div-grad problem) whose solution is the subject of the next two sections.

The Darcy problem is written as

$$\sigma \mathbf{V}^{n+1} + \nabla q^{n+1} = \sigma \tilde{\mathbf{V}}^{n+1} \quad \text{in } \Omega \tag{7.246}$$

$$\nabla \cdot \mathbf{V}^{n+1} = 0 \quad \text{in } \Omega \tag{7.247}$$

$$\mathbf{V}^{n+1} \cdot \mathbf{n} = \mathbf{V}_\Gamma^{n+1} \cdot \mathbf{n} \quad \text{on } \Gamma, \tag{7.248}$$

and

$$p^{n+1} = q^{n+1} + \overline{p}^{n+1}, \tag{7.249}$$

where $\sigma = a_0/\Delta t$.

Now let us assume that the velocity components u^{n+1} and v^{n+1}, as well as the pressure p^{n+1} (and consequently q^{n+1}) are approximated with polynomials of the same degree ($I\!P_N - I\!P_N$ method). Then suppose Eq.(7.246) to be enforced on Ω_N and Eq.(7.247) on $\overline{\Omega}_N^I$ (see Section 7.4.1.b for the definition of Ω_N and $\overline{\Omega}_N^I$). The resulting pressure solution is contaminated by spurious modes, since q^{n+1} appears only through its gradient at inner collocation points. Spatial approximations avoiding spurious modes may be of two types, which are now described.

1. The first type of method consists of approximating the pressure with a polynomial whose degree is lower than the one used for the velocity components, so that the pressure gradient does not vanish at points where the momentum equation is enforced. This can be done in either of the following approximation techniques :

(i) Use a staggered grid of MAC type where the velocity components and the pressure are computed on different grids (see, e.g., Peyret and Taylor, 1983). This is relatively simple to implement in the one-dimensional case (Malik *et al.*, 1985 ; Zang and Hussaini, 1985 ; Canuto *et al.*, 1988 ; Fröhlich and Peyret, 1990). On the other hand, in the two-dimensional case (Bernardi and Maday, 1988 ; Le Quéré, 1989) the implementation is appreciably more complicated since it involves three different grids and appropriate interpolation. For these reasons, the staggered grid method is not of common use.

(ii) Use a unique collocation grid (Gauss-Lobatto points) but approximate the velocity components and polynomials of different degree. This $I\!P_N - I\!P_{N-2}$ method, introduced by Bernardi *et al.* (1990) and developed in several works (e.g., Heinrichs, 1993 ; Azaiez *et al.*, 1994b; Botella, 1997), is described in detail in Section 7.4.2.b.

2. The second type of method consists of using the normal momentum equation at the boundary to close the discrete system rather than the incompressibility equation. In this case, the velocity components as well as the pressure are approximated with polynomials of the same degree ($I\!P_N - I\!P_N$ method). There are no spurious pressure modes since the normal gradient of the pressure at the boundary involves the derivative of the high-degree Chebyshev polynomial which does not vanish at ± 1 while it does at the inner collocation points. The $I\!P_N - I\!P_N$ method is described in detail in Section 7.4.2.c.

(b) The $I\!P_N - I\!P_{N-2}$ projection method

The pressure $\bar{p}^{n+1}$ in Eq.(7.238) is evaluated through the extrapolation formula (7.245). Let us define $I\!P_N$ as the space of polynomials of degree at most N_x in x and N_y in y. The velocity components are approximated with polynomials belonging to the $I\!P_N$-space. Thus, the provisional velocity $\tilde{\mathbf{V}}^{n+1}$ is approximated by $\tilde{\mathbf{V}}_N^{n+1} = (\tilde{u}_N^{n+1}, \tilde{v}_N^{n+1})$ and the velocity $\mathbf{V}^{n+1}$ by $\mathbf{V}_N^{n+1} = (u_N^{n+1}, v_N^{n+1})$. The pressure is approximated with a polynomial of degree less than two in both directions, namely, $p^{n+1} \cong p_{N-2}^{n+1} \in I\!P_{N-2}$ and, consequently, $q^{n+1} \cong q_{N-2}^{n+1} \in I\!P_{N-2}$. Equations (7.246) and (7.247) are enforced at the Gauss-Lobatto points $(x_i, y_j) \in \Omega_N$. This necessitates the evaluation of the pressure gradient at the same points. Hence, one must define the Lagrange interpolation polynomial of degree $N_x - 2$ (or $N_y - 2$) based on the points $x_i = \cos(\pi i / N_x), i = 1, \ldots, N_x-1$, [or $y_j = \cos(\pi j / N_y)$, $j = 1, \ldots, N_y - 1$].

Let us consider the function $\phi(x)$ defined in $[-1, 1]$ and the inner Gauss-Lobatto points $x_i = \cos \pi i / N, i = 1, \ldots, N-1$. The Lagrange interpolation polynomial based on this set of points is defined by

$$\phi_{N-2}(x) = \sum_{j=1}^{N-1} \phi(x_j)\, \hat{h}_j(x)\,, \tag{7.250}$$

where $\hat{h}_j(x)$ is the polynomial of degree $N - 2$ such that

$$\hat{h}_j(x_i) = \delta_{i,j}\,, \quad i, j = 1, \ldots, N - 1.$$

The polynomial $\hat{h}_j(x)$ has the expression

$$\hat{h}_j(x) = \frac{(-1)^{j+1}\left(1 - x_j^2\right) T_N'(x)}{N^2(x - x_j)}\,, \quad j = 1, \ldots, N - 1. \tag{7.251}$$

This expression is obtained by using the fact that the Gauss-Lobatto points $x_i, i = 1, \ldots, N - 1$, are the zeros of $T_N'(x)$ and by taking into account that

$$\frac{T_N'(x)}{x - x_j} \to (-1)^{j+1}\frac{N^2}{1 - x_j^2} \quad \text{for } x \to x_j, \quad j = 1, \ldots, N - 1.$$

We observe that

$$\hat{h}_j(x) = \frac{1 - x_j^2}{1 - x^2} h_j(x),$$

where $h_j(x)$ is the Lagrange interpolation coefficient associated with the $I\!P_N$ approximation and given by Eq.(3.44). The differentiation formula associated with the approximation (7.250) is written

$$\phi'_{N-2}(x_i) = \sum_{j=1}^{N-1} \hat{d}_{i,j}^{(1)} \phi(x_j), \quad i = 1, \ldots, N-1 \qquad (7.252)$$

with

$$\hat{d}_{i,j}^{(1)} = \hat{h}'_j(x_i) = \frac{1 - x_j^2}{1 - x_i^2} h'_j(x_i) + \frac{2 x_i}{1 - x_i^2} \delta_{i,j},$$

or

$$\hat{d}_{i,j}^{(1)} = \frac{(-1)^{j+1} (1 - x_j^2)}{(1 - x_i^2)(x_i - x_j)}, \quad i, j = 1, \ldots, N-1, \quad i \neq j$$

$$\qquad (7.253)$$

$$\hat{d}_{i,i}^{(1)} = \frac{3 x_i}{2 (1 - x_i^2)}, \quad i = 1, \ldots, N-1.$$

Now the collocation method applied to the solution of problem (7.238)-(7.239) determines the value of the approximate provisional velocity $\tilde{\mathbf{V}}_N^{n+1}$ at the Gauss-Lobatto points, namely,

$$\sigma \tilde{\mathbf{V}}_N^{n+1} - \nu \nabla^2 \tilde{\mathbf{V}}_N^{n+1} = \mathbf{F}_N^{n+1} \quad \text{in } \Omega_N \qquad (7.254)$$

$$\tilde{\mathbf{V}}_N^{n+1} = \mathbf{V}_\Gamma^{n+1} \quad \text{on } \Gamma_N^I, \qquad (7.255)$$

where

$$\mathbf{F}_N^{n+1} = \mathbf{f}_N^{n+1} - \frac{1}{\Delta t} \sum_{j=1}^{k} a_j \mathbf{V}_N^{n+1-j} - \sum_{j=0}^{k-1} b_j \mathbf{A}_N^{n-j} - \sum_{j=0}^{k-2} c_j \nabla p_{N-2}^{n-j}$$

and $\sigma = a_0/\Delta t$. The gradient ∇p_{N-2}^{n-j} is evaluated at the Gauss-Lobatto points by means of formulas of type (7.252)-(7.253). The nonlinear term in $\mathbf{F}_N^{n+1}$ is evaluated by matrix products or by the pseudospectral technique (Section 3.5.2). Equations (7.254) and (7.255) define an algebraic system for each component $\tilde{u}_N^{n+1}$ and $\tilde{v}_N^{n+1}$, that is easily solved by using the matrix-diagonalization technique (Section 3.7.2).

Then the collocation approximation of the Darcy problem (7.246)-(7.249) yields the algebraic system

$$\sigma \mathbf{V}_N^{n+1} + \nabla q_{N-2}^{n+1} = \sigma \tilde{\mathbf{V}}_N^{n+1} \quad \text{in } \Omega_N \qquad (7.256)$$

$$\nabla \cdot \mathbf{V}_N^{n+1} = 0 \quad \text{in } \Omega_N \qquad (7.257)$$

$$\mathbf{V}_N^{n+1} \cdot \mathbf{n} = \mathbf{V}_\Gamma^{n+1} \cdot \mathbf{n} \quad \text{on } \Gamma_N^I, \tag{7.258}$$

and

$$p_{N-2}^{n+1} = q_{N-2}^{n+1} + \overline{p}_{N-2}^{n+1} \quad \text{in } \Omega_N, \tag{7.259}$$

where the divergence operator acting on $\mathbf{V}_N^{n+1}$ is defined by the usual differentiation formulas (3.45)-(3.46), while the gradient ∇q_{N-2}^{n+1} is expressed by formulas like (7.252)-(7.253).

It is observed that the system (7.256)-(7.259) contains sufficient algebraic equations to completely determine the velocity and pressure fields in Ω_N. Therefore, one of the main advantages of this method is that it does not require the prescription of boundary conditions for the pressure. On the boundary Γ_N^I, the normal component of the velocity is defined by (7.258). On the other hand, the tangential component is not involved in the above system. Therefore, this component is not deduced from the solution of the Darcy problem (as it should be in the continuous case) but is simply defined by the boundary condition (7.192) of the Navier-Stokes problem.

We can estimate the error in time of the method by analyzing the truncation error of the discrete equations obtained after the elimination of the provisional velocity $\tilde{\mathbf{V}}_N^{n+1}(x_i, y_j)$. Thus, the discrete form of Eq.(7.243) involves the term

$$\mathbf{E}_{N-2}^{n+1}(x_i, y_j) = -\frac{\nu}{a_0} \left[\sum_{l=1}^{N_x-1} d_{i,l}^{(2)} \mathbf{G}_{N-2}^{n+1}(x_l, y_j) + \sum_{l=1}^{N_y-1} d_{j,l}^{(2)} \mathbf{G}_{N-2}^{n+1}(x_i, y_l) \right]$$

where $\mathbf{G}_{N-2}^{n+1}$ is the discrete gradient of q_{N-2}^{n+1} defined by differentiation formulas of type (7.252). The discrete equations resulting from Eq.(7.243) are applied to the inner points $(x_i, y_j) \in \Omega_N$, supplemented by the incompressibility equation $\nabla \cdot \mathbf{V}_N = 0$ enforced on Ω_N and the boundary conditions $\mathbf{V}_N^{n+1} = \mathbf{V}_\Gamma^{n+1}$ on Γ_N^I. Therefore it is clear that the truncation error of the overall discrete problem is $O(\Delta t^k)$ if $\overline{p}^{n+1}$ is defined by Eq.(7.245). Moreover, the splitting-error term $\Delta t \, \mathbf{E}_{N-2}^{n+1}$ vanishes at steady state.

Equations (7.256)-(7.259) are written in matrix form by introducing the $(N_x - 1) \times (N_y - 1)$ matrices containing the grid values of the unknowns

$$\mathcal{U} = \left[u_N^{n+1}(x_i, y_j) \right], \quad i = 1, \ldots, N_x - 1, \quad j = 1, \ldots, N_y - 1$$

$$\mathcal{V} = \left[v_N^{n+1}(x_i, y_j) \right], \quad i = 1, \ldots, N_x - 1, \quad j = 1, \ldots, N_y - 1$$

$$\mathcal{Q} = \left[q_{N-2}^{n+1}(x_i, y_j) \right], \quad i = 1, \ldots, N_x - 1, \quad j = 1, \ldots, N_y - 1.$$

The system is written

$$\sigma \mathcal{U} + \hat{\mathcal{D}}_x \mathcal{Q} = \sigma \tilde{\mathcal{U}} \tag{7.260}$$

$$\sigma \mathcal{V} + \mathcal{Q} \hat{\mathcal{D}}_y^T = \sigma \tilde{\mathcal{V}} \tag{7.261}$$

$$\mathcal{D}_x \mathcal{U} + \mathcal{V} \mathcal{D}_y^T = \mathcal{S}, \tag{7.262}$$

where $\mathcal{D}_x$ and $\mathcal{D}_y$ are, respectively, the $(N_x - 1) \times (N_x - 1)$ and $(N_y - 1) \times (N_y - 1)$ matrices analogous to the matrix $\mathcal{D}_1$ defined in Eq.(4.66) ; $\hat{\mathcal{D}}_x$ and $\hat{\mathcal{D}}_y$ are similar matrices made with the coefficients $\hat{d}_{i,j}^{(1)}$. The matrices $\tilde{\mathcal{U}}$ and $\tilde{\mathcal{V}}$ are defined by

$$\tilde{\mathcal{U}} = \left[\tilde{u}_N^{n+1}(x_i, y_j)\right], \qquad i = 1, \ldots, N_x - 1, \qquad j = 1, \ldots, N_y - 1$$

$$\tilde{\mathcal{V}} = \left[\tilde{v}_N^{n+1}(x_i, y_j)\right], \qquad i = 1, \ldots, N_x - 1, \qquad j = 1, \ldots, N_y - 1$$

and $\mathcal{S}$ is the $(N_x - 1) \times (N_y - 1)$ matrix containing the boundary values of the velocity, namely,

$$\mathcal{S} = -\left(\overline{\mathcal{D}}_x \mathcal{U}_\Gamma + \mathcal{V}_\Gamma \overline{\mathcal{D}}_y^T\right),$$

where

$$\mathcal{U}_\Gamma = \left[u_\Gamma^{n+1}(x_i, y_j)\right], \qquad i = 0, N_x, \quad j = 1, \ldots, N_y - 1$$

$$\mathcal{V}_\Gamma = \left[v_\Gamma^{n+1}(x_i, y_j)\right], \qquad i = 1, \ldots, N_x - 1, \quad j = 0, N_y$$

and

$$\overline{\mathcal{D}}_x = \left[d_{i,j}^{(1)}\right], \qquad i = 1, \ldots, N_x - 1, \quad j = 0, N_x$$

$$\overline{\mathcal{D}}_y = \left[d_{i,j}^{(1)}\right], \qquad i = 1, \ldots, N_y - 1, \quad j = 0, N_y.$$

The solution of the large system (7.260)-(7.262) is obtained through the elimination of $\mathcal{U}$ and $\mathcal{V}$ to get the system determining $\mathcal{Q}$, let

$$\mathcal{D}_x \hat{\mathcal{D}}_x \mathcal{Q} + \mathcal{Q}\left(\mathcal{D}_y \hat{\mathcal{D}}_y\right)^T = -\sigma\left(\mathcal{S} - \mathcal{D}_x \tilde{\mathcal{U}} - \tilde{\mathcal{V}}\mathcal{D}_y^T\right). \qquad (7.263)$$

The operator acting on $\mathcal{Q}$ is of Uzawa type. The eigenvalues of the matrix $\mathcal{D}_x \hat{\mathcal{D}}_x$ (or $\mathcal{D}_y \hat{\mathcal{D}}_y$) are real, distinct, and negative except one and only one which is zero, corresponding to the constant mode (Heinrichs, 1993). Moreover, when $N_x \to \infty$, the smaller eigenvalue $|\lambda|_{min} \to \pi^2/4$ (discarding the null eigenvalue) and the larger one $|\lambda|_{max} \cong 0.047 N_x^4$. This is typically the behaviour of the eigenvalues of the second-order derivative operator with Dirichlet conditions (see Table 3.1). Therefore, the Uzawa operator acting on $\mathcal{Q}$ in (7.263) may be characterized as a "pseudo-discrete-Laplacian".

System (7.263) is solved by means of the matrix-diagonalization procedure (Section 3.7.2) using, for the null eigenvalue, the technique proposed in Section 3.7.1 for solving the Neumann problem for the Poisson equation. Then, when $\mathcal{Q}$ is known, Eqs.(7.260) and (7.261) give $\mathcal{U}$ and $\mathcal{V}$, respectively. Note that in the three-dimensional case, the system determining $\mathcal{Q}$ is put in the tensor product form and the solution procedure follows the one described in Section 3.7.3.

Note that the value of the pressure at the boundary is not known from the discrete system (7.260)-(7.262). If this value is needed, it is rather simple to deduce an approximation of it from the Lagrange interpolation polynomial of type (7.250) evaluated at the extremities ± 1.

Moreover, we recall that the incompressibility equation is enforced at the inner collocation points but not on the boundary. The result is that the approximate divergence velocity $Q_N = \nabla \cdot \mathbf{V}_N$ is not the null polynomial as it is in the case of the influence matrix method. This drawback may be seen as an advantage is some situations where the solution exhibits a singularity at the boundary. Moreover, the method is less sensitive to the way in which the total flow rate condition is satisfied (see Remark 2 of Section 7.4.1.b).

Returning to the time-discretization of the Navier-Stokes equations, we are faced with the usual problem associated with multistep schemes, namely the calculation of the solution of the first time-cycles. More precisely, the calculation of $\left(\mathbf{V}^1, p^1\right)$ for the second-order scheme $(k = 2)$, $\left(\mathbf{V}^1, p^1\right)$ and $\left(\mathbf{V}^2, p^2\right)$ for the third-order scheme $(k = 3)$, and so on.

As mentioned in Section 4.5.1.d, the starting scheme may be one order less accurate than the current scheme without effect on the accuracy of the approximate solution. Therefore, for the second-order scheme one can set $\mathbf{V}^{-1} = \mathbf{V}^0 = \mathbf{V}_0$ and $p^0 = 0$, that leads to a first-order scheme provided the time-step is taken equal to $3\Delta t/2$.

For the third-order scheme, Botella and Peyret (2001) suggest the calculation of $\left(\mathbf{V}^1, p^1\right)$ and $\left(\mathbf{V}^2, p^2\right)$ by means of a second-order scheme derived from the RK2/CN scheme (4.85):

Step 1a : Calculation of $\mathbf{V}_1$:

$$\frac{\mathbf{V}_1 - \mathbf{V}^n}{\Delta t} + \mathbf{A}\left(\mathbf{V}^n, \mathbf{V}^n\right) + \nabla p^n - \nu \nabla^2 \mathbf{V}^n = \mathbf{f}^n \qquad \text{in } \Omega$$

$$\mathbf{V}_1 = \mathbf{V}_\Gamma^{n+1} \qquad \text{on } \Gamma.$$

Step 1b : Calculation of $\tilde{\mathbf{V}}^{n+1}$:

$$\frac{2\,\tilde{\mathbf{V}}^{n+1} - \mathbf{V}_1 - \mathbf{V}^n}{\Delta t} + \mathbf{A}\left(\mathbf{V}_1, \mathbf{V}_1\right) + \nabla p^n - \nu \nabla^2 \tilde{\mathbf{V}}^{n+1} = \mathbf{f}^{n+1} \qquad \text{in } \Omega$$

$$\tilde{\mathbf{V}}^{n+1} = \mathbf{V}_\Gamma^{n+1} \qquad \text{on } \Gamma.$$

Step 2 : Projection :

$$\frac{\mathbf{V}^{n+1} - \tilde{\mathbf{V}}^{n+1}}{2\,\Delta t} + \frac{1}{2}\nabla\left(p^{n+1} - p^n\right) = 0 \qquad \text{in } \Omega$$

$$\nabla \cdot \mathbf{V}^{n+1} = 0 \qquad \text{in } \Omega$$

$$\mathbf{V}^{n+1} \cdot \mathbf{n} = \mathbf{V}_\Gamma^{n+1} \cdot \mathbf{n} \qquad \text{on } \Gamma.$$

It is possible to show (Botella, 1998) that this scheme is equivalent at $O\left(\Delta t^2\right)$ to the classical Crank-Nicolson scheme. The scheme is applied to $n = 0$ and $n = 1$. Therefore, the knowledge of an initial pressure field p^0 is required. As shown by Heywood and Rannacher (1982) and by Temam (1982), the initial pressure field (which is not data of the exact continuous problem) can be determined from the Neumann problem (5.30)-(5.31) with $\mathbf{V} = \mathbf{V}_0$, taking into account that the initial velocity field $\mathbf{V}_0$ satisfies the compatibility condition (5.11). We refer to Marion and Temam (1998) for a general discussion on the initial conditions. Examples of the application of the $I\!\!P_N - I\!\!P_{N-2}$ method are presented in Sections 7.6.4 and 8.4.2.b.

(c) The predicted $I\!\!P_N - I\!\!P_N$ projection method

The $I\!\!P_N - I\!\!P_N$ projection method was defined as Option (3) in Section 7.3.2.d, and the weakness of the method concerning the Neumann condition for the pressure, in the framework of the strong collocation approximation, was discussed. This drawback, however, can be removed by using, for the pressure $\overline{p}^{n+1}$ occurring in Eq.(7.235), the solution of a Poisson equation with Neumann boundary conditions. This method, proposed by Hugues and Randriamampianina (1998), is now presented associated with the second-order time-scheme [$k = 2$ in Eqs.(7.238)-(7.242)], but higher-order schemes could be used as long as stability is ensured.

The pressure $\overline{p}^{n+1}$ is calculated, in a predicted step, from Eqs.(5.30)-(5.31) using an explicit formulation, namely,

$$\nabla^2 \overline{p}^{n+1} = \nabla \cdot \left[\mathbf{f}^{n+1} - \left(2\mathbf{A}^n - \mathbf{A}^{n-1}\right)\right] \quad \text{in } \Omega \qquad (7.264)$$

$$\partial_n \overline{p}^{n+1} = \left[\mathbf{f}^{n+1} - \left(2\mathbf{A}^n - \mathbf{A}^{n-1}\right) + \nu\left(2\mathbf{D}^n - \mathbf{D}^{n-1}\right)\right.$$
$$\left. - \frac{1}{2\Delta t}\left(3\mathbf{V}_\Gamma^{n+1} - 4\mathbf{V}_\Gamma^n + \mathbf{V}_\Gamma^{n-1}\right)\right] \cdot \mathbf{n} \quad \text{on } \Gamma. \qquad (7.265)$$

The use of the classical form of the diffusive term $\mathbf{D} = \nabla^2\mathbf{V}$ leads to instability. On the other hand, the form

$$\mathbf{D} = \nabla\left(\nabla \cdot \mathbf{V}\right) - \nabla \times \left(\nabla \times \mathbf{V}\right) = -\nabla \times \left(\nabla \times \mathbf{V}\right) \qquad (7.266)$$

considered by Orszag et al. (1986) and by Karniadakis et al. (1991) is found to be stable. As a matter of fact, because the constraint $\nabla \cdot \mathbf{V} = 0$ is not enforced at the boundary, the compatibility condition of the Neumann problem (7.264)-(7.265) would not be satisfied if the term $\nabla\left(\nabla \cdot \mathbf{V}\right)$ was retained in $\mathbf{D}$, as it is when using $\mathbf{D} = \nabla^2\mathbf{V}$. Formally (and numerically checked) the value $\overline{p}^{n+1}$ is an approximation to p^{n+1} with an error $O\left(\Delta t^2\right)$. Therefore, the term $\Delta t\, E^{n+1}$ in Eq.(7.243) is $O\left(\Delta t^3\right)$.

It should be recalled that, in the $I\!\!P_N - I\!\!P_N$ approximation method, the velocity components and the pressure are approximated with polynomials of the same degree, namely, u_N^{n+1}, v_N^{n+1}, p_N^{n+1}, $\overline{p}_N^{n+1}$ (and $q_N^{n+1} = p_N^{n+1} - \overline{p}_N^{n+1}$) are polynomials of degree at most N_x in x and N_y in y.

The Chebyshev-collocation solution of the Neumann problem (7.264)-(7.265) gives $\bar{p}_N^{n+1}$. Then, when $\bar{p}_N^{n+1}$ has been calculated, the polynomial approximation $\tilde{\mathbf{V}}_N^{n+1} = \left(\tilde{u}_N^{n+1}, \tilde{v}_N^{n+1}\right)$ to the polynomial velocity $\tilde{\mathbf{V}}^{n+1} = \left(\tilde{u}^{n+1}, \tilde{v}^{n+1}\right)$ is determined from the solution of Eqs.(7.238)-(7.239) with $k = 2$, namely,

$$\sigma \tilde{\mathbf{V}}^{n+1} - \nu \nabla^2 \tilde{\mathbf{V}}^{n+1} = \mathbf{F}^{n+1} \quad \text{in } \Omega \tag{7.267}$$

$$\nabla \cdot \tilde{\mathbf{V}}^{n+1} = 0 \quad \text{in } \Omega \tag{7.268}$$

$$\tilde{\mathbf{V}}^{n+1} = \mathbf{V}_\Gamma^{n+1} \quad \text{on } \Gamma, \tag{7.269}$$

with $\sigma = 3/(2\,\Delta t)$ and

$$\mathbf{F}^{n+1} = \mathbf{f}^{n+1} - \frac{4\,\mathbf{V}^n - \mathbf{V}^{n-1}}{2\,\Delta t} - \left(2\,\mathbf{A}^n - \mathbf{A}^{n-1}\right) - \nabla \bar{p}^{n+1}. \tag{7.270}$$

The collocation approximation to Eqs.(7.267)-(7.269) gives an algebraic system for the grid values of each component of $\tilde{\mathbf{V}}^{n+1}$.

Now it remains to solve the Darcy problem (7.246)-(7.249). As explained in Section 7.3.2.e this is done by deriving a Poisson equation for q^{n+1} supplemented with a Neumann boundary condition. This can be done either directly from the discrete system resulting from the Chebyshev collocation approximation to problem (7.246)-(7.249) or from the continuous equations themselves. This latter derivation is simpler in the multidimensional case. Therefore, the Poisson equation is obtained by applying the divergence operator to Eq.(7.246) and taking into account the incompressibility equation (7.247). The associated Neumann condition is derived by projecting Eq.(7.246) normal to the boundary. We obtain the Neumann problem

$$\nabla^2 q^{n+1} = \sigma \nabla \cdot \tilde{\mathbf{V}}^{n+1} \quad \text{in } \Omega \tag{7.271}$$

$$\partial_n q^{n+1} = \sigma \left(\tilde{\mathbf{V}}^{n+1} - \mathbf{V}^{n+1}\right) \cdot \mathbf{n} = 0 \quad \text{on } \Gamma, \tag{7.272}$$

which automatically satisfies the compatibility condition. Equation (7.272) shows that p^{n+1} satisfies the Neumann condition

$$\partial_n p^{n+1} = \partial_n \bar{p}^{n+1} \quad \text{on } \Gamma,$$

that is consistent with the Navier-Stokes problem and accurate to second-order. Problem (7.271)-(7.272) is solved by means of the Chebyshev collocation method. Once q_N^{n+1} is known, Eq.(7.246) applied to $\overline{\Omega}_N^I$ determines the approximate velocity field $\mathbf{V}_N^{n+1}$ whose divergence is zero at the inner collocation points.

In summary, the algorithm is the following :

1. Solve Eqs.(7.264)-(7.265) for determining $\bar{p}_N^{n+1}$ in $\overline{\Omega}_N^I$.
2. Solve Eqs.(7.267)-(7.269) for determining $\tilde{\mathbf{V}}_N^{n+1}$ in $\overline{\Omega}_N^I$.

3. Solve Eqs.(7.271)-(7.272) for determining q_N^{n+1} in $\overline{\Omega}_N^I$.

4. Calculate p_N^{n+1} in $\overline{\Omega}_N^I$ from Eq.(7.249), namely, $p_N^{n+1} = q_N^{n+1} + \overline{p}_N^{n+1}$.

5. Calculate $\mathbf{V}_N^{n+1}$ in $\overline{\Omega}_N^I$ from Eq.(7.246), namely,

$$\mathbf{V}_N^{n+1} = \tilde{\mathbf{V}}_N^{n+1} - \frac{1}{\sigma} \nabla q_N^{n+1} .$$

The calculation of $\left(\mathbf{V}^1, p^1\right)$ at the first time-cycle ($n = 0$) with the three-level scheme ($k = 2$) is performed by setting $\mathbf{V}^{-1} = \mathbf{V}^0 = \mathbf{V}_0$ and the time-step equal to $3 \Delta t/2$, $\overline{p}^1$ being determined by Eqs.(7.264)-(7.265).

Note that the boundary condition on the tangential velocity component is not exactly satisfied since this velocity is obtained from Eq.(7.246), namely,

$$\mathbf{V}_N^{n+1} \cdot \mathbf{t} = \mathbf{V}_\Gamma^{n+1} \cdot \mathbf{t} - \frac{\Delta t}{a_0} \left[\nabla \left(p_N^{n+1} - \overline{p}_N^{n+1}\right)\right] \cdot \mathbf{t} \quad \text{on } \Gamma_N^I . \quad (7.273)$$

The evaluation of the tangential gradient of $q_N^{n+1} = p_N^{n+1} - \overline{p}_N^{n+1}$ on Γ_N^I necessitates the knowledge of the values of q_N^{n+1} at the corners. These values are calculated by using the Neumann condition $\partial_n q_N^{n+1} = 0$ according to the technique described in Section 3.7.2 in the general case of Robin conditions and applied to the calculation of the pressure at the corners in Remark 1 of Section 7.4.1.

The error term ("slip velocity") in the right-hand side of Eq.(7.273) is $O\left(\Delta t^3\right)$ since $p_N^{n+1} - \overline{p}_N^{n+1} = O\left(\Delta t^2\right)$ as numerically checked (Hugues and Randriamampianina, 1998; Raspo et al., 2002). Therefore, the effect of this slip velocity on the global accuracy $O\left(\Delta t^2\right)$ is negligible. However, the slip velocity does not vanish at steady state because the difference $p_N^{n+1} - \overline{p}_N^{n+1}$ is not zero at steady state. In order to avoid the slip velocity, one might be tempted to define the tangential velocity from the exact boundary condition (7.192) rather than from Eq.(7.273). This is the Option (4) defined in Section 7.3.2.d. Unfortunately, such a procedure appeared unstable in some applications, while it is stable in the case where $\overline{p}^{n+1} = p^n$.

Lastly, one could wonder whether it is quite necessary to calculate the predicted pressure $\overline{p}^{n+1}$. One would then restrict oneself to set $\overline{p}^{n+1} = p^n$, namely, $q^{n+1} = p^{n+1} - p^n$, and to solve the Poisson equation (7.271) with a Neumann boundary condition similar to Eq.(7.265). However, the compatibility condition associated with this Neumann problem is satisfied at $O\left(\Delta t\right)$ only, whatever the form of $\mathbf{D}$, and the method appears to be unstable. On the other hand, a stable projection method involving such a Neumann condition can be constructed (Karniadakis et al., 1991) by inverting the two steps of the procedure. In the first step the pressure field and a solenoidal velocity field are calculated by discarding the diffusive term. The pressure field p^{n+1} is a solution of the Neumann problem (7.264)-(7.265) which satisfies the compatibility condition. In the second step, the final velocity is calculated by taking the diffusion into account. Although providing accurate results, this method presents the drawback of not controlling the divergence Q_N^{n+1} of the approximation of the final velocity field $\mathbf{V}^{n+1}$. The divergence Q_N^{n+1} satisfies a Helmholtz equation, as in the influence matrix method, but without the boundary condition ensuring that Q_N^{n+1} is effectively zero. However, it must be pointed out that the polynomial Q_N^{n+1} tends toward zero when the spatial resolution is increased. A similar splitting scheme has been proposed by Batoul *et al.* (1994), associated with the $\mathbb{P}_N - \mathbb{P}_{N-2}$ approximation, avoiding the need for a boundary condition for the pressure. This method presents the same behaviour concerning the divergence of the velocity.

7.4.3 Shear-stress-free boundary

When a part of the boundary Γ is stress-free, the no-slip condition is replaced by

$$V_n = 0, \quad \partial_n (V_t) = 0, \tag{7.274}$$

where V_n and V_t are, respectively, the normal and tangential components of the velocity vector $\mathbf{V}$ with respect to the boundary Γ. Of course, it is assumed here that the boundary velocity $\mathbf{V}_\Gamma$ on the complementary part of the stress-free boundary is compatible with condition (7.274). In the following, and this to fix the ideas, the side $\Gamma_1 = \{y = 1, -1 < x < 1\}$ will be supposed to be stress-free, that is, Eq.(7.274) becomes

$$v = 0, \quad \partial_y u = 0 \quad \text{on } \Gamma_1. \tag{7.275}$$

(a) Influence matrix method
The implementation of the boundary conditions (7.275) in the influence matrix method associated with the Fourier-Chebyshev approximation has been discussed at the end of Section 7.3.2.c. The extension to the Chebyshev-Chebyshev case is straightforward.

(b) $\mathbb{P}_N - \mathbb{P}_{N-2}$ projection method
In *step 1* of the projection method, the calculation of the provisional velocity $\tilde{\mathbf{V}}^{n+1}$ is done by simply replacing the no-slip condition on Γ_1 by

$$\tilde{v}^{n+1} = 0, \quad \partial_y \tilde{u}^{n+1} = 0 \quad \text{on } \Gamma_1. \tag{7.276}$$

The solution of the Darcy problem in *step 2* requires a little more work. For the normal component on Γ_1, there is no difficulty in prescribing

$$v^{n+1} = 0 \quad \text{on } \Gamma_1. \tag{7.277}$$

The difficulty appears in determining the tangential velocity u^{n+1} on Γ_1 such that it satisfies the second condition (7.275). This can be done by differentiating the tangential part of Eq.(7.246) and, taking the second equation (7.275) into account, we get the discrete equation

$$\partial_y u_N^{n+1} = -\frac{1}{\sigma} \partial_{yx} q_N^{n+1} \quad \text{on } \Gamma_{1\,N}^I \tag{7.278}$$

where $\Gamma^I_{1\,N}$ refers to the inner collocation points belonging to Γ_1. The term $\partial_{yx}q_N^{n+1}(x_i, 1)$, $i = 1, \ldots, N_x - 1$, is evaluated from the values $q_N^{n+1}(x_i, y_j)$, $i = 1, \ldots, N_x - 1$, $j = 1, \ldots, N_y - 1$, that is,

$$\partial_x q_N^{n+1}(x_i, y_j) = \sum_{k=1}^{N_x-1} \hat{d}^{(1)}_{i,k} q_N^{n+1}(x_k, y_j)$$

then

$$\partial_x q_N^{n+1}(x_i, y_j) = \sum_{l=1}^{N_y-1} \left[\partial_x q_N^{n+1}(x_i, y_l)\right] \hat{h}_l(y)$$

and

$$\partial_{yx} q_N^{n+1}(x_i, 1) = \sum_{l=1}^{N_y-1} \left[\partial_x q_N^{n+1}(x_i, y_l)\right] \hat{h}'_l(1),$$

where $\hat{h}'_l(1)$ is calculated from Eq.(7.251).

Now the derivative $\partial_y u_N^{n+1}(x_i, 1)$ is evaluated through the usual differentiation formula using the known values $u_N^{n+1}(x_i, y_j)$, $j = 1, \ldots, N_y$. Thus, Eq.(7.278) determines the boundary value $u_N^{n+1}(x_i, 1)$.

Note that Eq.(7.278) exhibits the error term $(\Delta t/a_0)\partial_{yx}q_N^{n+1}$ which is formally $O(\Delta t^2)$. This is the analog of the slip velocity occurring in the $I\!P_N - I\!P_N$ projection method [Eq.(7.273)].

(c) $I\!P_N - I\!P_N$ projection method
The implementation of the stress-free conditions is similar to that described for the $I\!P_N - I\!P_{N-2}$ method. It is, however, simpler since the quantity q_N^{n+1} is known on $\Gamma^I_{1\,N}$ from the solution of Eqs.(7.271)-(7.272) and is calculated at the corners as explained above. It may be useful to observe that the pressure satisfies the homogeneous Neumann condition $\partial_n p = 0$ on a stress-free boundary provided that $\mathbf{f} \cdot \mathbf{n} = 0$.

7.4.4 Assessment of methods

The purpose of this section is to assess and compare the various methods by considering simple examples : analytical solutions and regularized cavity flow. The considered methods are based on the AB/BDIk scheme with $k = 2$ or $k = 3$ and the nonlinear term $\mathbf{A}(\mathbf{V}, \mathbf{V})$ is considered in the convective form (see Section 7.4.5 for a discussion on the various forms of this term). These methods are :

1. The AB/BDI2, $I\!P_N - I\!P_N$, influence matrix method [Section 7.4.1] : IM.

2. The AB/BDI2, $I\!P_N - I\!P_N$, original projection method [Section 7.4.2, $\bar{p}^{n+1} = 0$] : OP.

3. The AB/BDI2, $I\!P_N - I\!P_N$, incremental projection method [Section

7.4.2, $\bar{p}^{n+1} = p^n]$: IP.

4. The AB/BDI2, $I\!P_N - I\!P_N$, predicted projection method [Section 7.4.2, $\bar{p}^{n+1}$ solution of Eqs.(7.264)-(7.265)] : PP.

5. The AB/BDI2, $I\!P_N - I\!P_{N-2}$, projection method [Section 7.4.2, $\bar{p}^{n+1} = p^n]$: P2.

6. The AB/BDI3, $I\!P_N - I\!P_{N-2}$, projection method [Section 7.4.2, $\bar{p}^{n+1} = 2p^n - p^{n-1}]$: P3.

The following points will be discussed :

(a) Comparison of the accuracy of the pressure calculated by the IM and P2 methods. In the case of the IM method, the pressure is recovered either by filtering or by the solution of a Poisson equation.

(b) Evaluation of the effect of $\bar{p}^{n+1}$ on the accuracy of the $I\!P_N - I\!P_N$ projection methods (OP, IP and PP).

(c) Comparison of the accuracy in time of the IM, PP, P2 and P3 methods.

(d) Comparison of the stability of the IM, PP, P2 and P3 methods.

(e) Comparison of the spatial accuracy of the IM, PP and P2 methods.

(f) Comparison of the cost of the IM, PP and P2 methods.

These questions are now successively addressed.

(a) Accuracy of the pressure in IM and P2 methods

The application of the influence matrix method with the correction term **B** produces spurious pressure modes which have to be removed to get an accurate pressure field. As discussed in Section 7.4.1.b, this can be done either by a filtering process or by the solution of a Poisson equation with a Neumann condition. These two techniques are evaluated and their results are compared to those given by the P2 method. This is done by considering the calculation of the analytical steady solution in $\Omega = \{-1 < x, y < 1\}$:

$$u = \sin(\pi x/2) \cos(\pi y/2) , \quad v = -\cos(\pi x/2) \sin(\pi y/2) ,$$
$$p = (\cos \pi x + \cos \pi y)/4 + 10(x + y) .$$

(7.279)

This defines the forcing term **f** in Eq.(7.1) and the boundary value V_Γ. The initial condition is the exact solution (7.279), and the time-integration (with $\nu = 10^{-1}$) is pursued until the residual of the velocity is stabilized (at less than 2×10^{-12}). The arbitrary constant in the pressure field is determined by matching the calculated and exact pressures at the center point of the domain Ω_N. The calculations (Botella, 1998; Sabbah, 2000) are done with $N_x = N_y = N$. The accuracy of a quantity ϕ is evaluated through the discrete L^2-norm based on Ω_N^I :

$$E_\phi = \left\{ \frac{1}{(N-1)^2} \sum_{i,j=1}^{N-1} \left[\phi_N^{n+1}(x_i, y_j) - \phi(x_i, y_j, t_{n+1}) \right]^2 \right\}^{1/2}$$

(7.280)

where ϕ_N is the computed solution and ϕ is the exact one. Table 7.2 gives the errors on the u-component of the velocity (error $\overline{E}_u$) and on the pressure (error $\overline{E}_p$). The error $\overline{E}_p^I$ refers to the pressure field obtained by the filtering process and the error $\overline{E}_p^{II}$ to the one computed from the Poisson equation. The second technique is much more accurate and its accuracy is similar to the one associated with the P2 method (note that different computers have been used for the IM and P2 methods). The last column in Table 7.2 gives the maximal value Q_{max} of $\nabla \cdot \mathbf{V}_N$ on the boundary given by the P2 method.

N	IM method $\overline{E}_u$	$\overline{E}_p^I$	$\overline{E}_p^{II}$
12	5.26×10^{-11}	1.20×10^{-5}	2.80×10^{-8}
16	1.08×10^{-14}	2.00×10^{-6}	2.27×10^{-12}

N	P2 method $\overline{E}_u$	$\overline{E}_p$	Q_{max}
12	1.99×10^{-11}	5.16×10^{-7}	2.39×10^{-9}
16	3.75×10^{-13}	7.46×10^{-11}	1.29×10^{-11}

Table 7.2. Errors given by the IM and P2 methods for the steady solution (7.279).

(b) Influence of $\overline{p}^{n+1}$ in OP, IP and PP methods

This example is intended to evaluate the effect of the choice of $\overline{p}^{n+1}$ on the accuracy in time of the $I\!P_N - I\!P_N$ projection methods. This is done by considering a three-dimensional unsteady solution (Raspo et al., 2002) in a cylindrical coordinate system (r, θ, z) with $\mathbf{V} = (u, v, w)$. The annular domain Ω is defined by $1 \leq r \leq 2, 0 \leq \theta \leq 2\pi, -1 \leq z \leq 1$. The coordinate r is changed in $\rho = r - 2$ so that $-1 \leq \rho \leq 1$. The numerical tests are done with the analytical time-periodic solution

$$u = (1/2\pi)\left(1 + \cos^2 4\pi t\right) \sin^2 \pi \rho \cdot \sin 2\pi z \cdot \cos \theta$$

$$v = -(1/2\pi)\left(1 + \cos^2 4\pi t\right) \sin^2 \pi\rho \cdot \sin 2\pi z \cdot \sin \theta$$

$$w = -(1/2\pi)\left(1 + \cos^2 4\pi t\right) \sin 2\pi\rho \cdot \sin^2 \pi z \cdot \cos \theta \tag{7.281}$$

$$p = (\cos \pi\rho + \cos \pi z) \cos \theta + 10\left(x + y\right) \cos \theta \cdot \cos 4\pi t,$$

which defines the forcing term $\mathbf{f}$ in Eq.(7.1), the boundary value $\mathbf{V}_\Gamma$, and the initial condition $\mathbf{V}_0$.

The spatial approximation makes use of the Fourier Galerkin method in the θ-direction and Chebyshev collocation in the ρ- and z-directions (see Section 7.5). We denote by K the cut-off frequency of the truncated Fourier series, and by N_ρ and N_z the degree of the polynomial approximation in ρ- and z-directions, respectively. The calculations are done with $\nu = 2 \times 10^{-3}$ and the resolution $2K = N_\rho = N_z = 40$. This resolution is expected to be sufficient to ensure that the measured error is due only to the error in time for the considered values of the time-step Δt. However, as will be observed, the presence of a rather large spatial error associated with the normal pressure gradient at the boundary may require higher resolutions.

The interest in using a time-periodic solution like (7.281) is that the error itself is periodic after a transient stage. Thus, it is recommended integrating over a large number of periods in order to check the stability of the method, characterized by a constant maximal error defined by

$$E_\phi = \max_t \overline{E}_\phi(t) \qquad (7.282)$$

where $\overline{E}_\phi$ is defined by an equation similar to (7.280) based on the collocation points (ρ_i, θ_j, z_k). The error E_ϕ characterizes the accuracy in time of the method. Figures 7.1 and 7.2 show, respectively, the errors E_u and E_p calculated on the inner collocation points. Figure 7.3 shows the error E_p^Γ calculated on the points belonging to the boundary of the computational domain. It may be observed that the spatial error of the IP method [curve (2)] becomes preponderant as soon as $\Delta t < 10^{-3}$. This threshold is lowered by increasing the resolution to $2K = N_\rho = N_z = 54$ [curve (3)]. On the other hand, the temporal error of the OP method [curve (1)] is too large to make apparent the spatial error for the values of Δt considered. Lastly, the error associated with the PP method [curve (4)] shows a perfect decrease in Δt^2. Results concerning the rate of decrease of the errors are reported in Table 7.3. This table also gives the behaviour of the slip velocity V_s exhibited by Eq.(7.273).

Method	$\overline{E}_u$	V_s	$\overline{E}_p$	$\overline{E}_p^\Gamma$
OP	Δt	Δt	$\Delta t^{0.88}$	$\Delta t^{0.74}$
IP	Δt^2	Δt^2	$\Delta t^{1.65}$	Δt
PP	Δt^2	Δt^3	Δt^2	Δt^2

Table 7.3. Rate of decrease of the errors given by the OP, IP, and PP methods for the time-periodic solution (7.281).

These various numerical results clearly show the weaknesses of the OP and IP methods compared to the PP method. This latter constitutes an efficient way to apply the concept of projection associated with the $\mathbb{P}_N - \mathbb{P}_N$ approximation in the strong collocation framework.

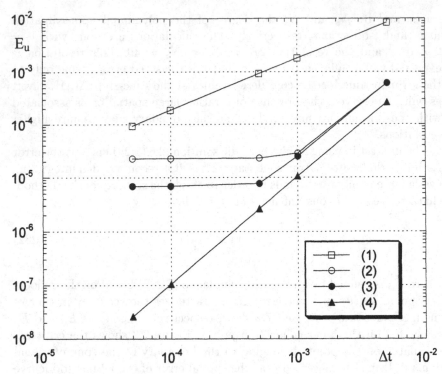

FIGURE 7.1. Error E_u for the solution (7.281) given by the projection methods :
(1) OP method ($40 \times 40 \times 40$ resolution), (2) IP method ($40 \times 40 \times 40$ resolution),
(3) IP method ($54 \times 54 \times 54$ resolution), (4) PP method ($40 \times 40 \times 40$ resolution).

(c) Accuracy in time of the IM, PP, P2, and P3 methods

The purpose of this test case is to compare the accuracy in time of the IM, PP, P2, and P3 methods. The comparison is based on the calculation of the two-dimensional time-periodic solution similar to the steady solution (7.279), namely,

$$u = \cos{(5t)} \sin{(\pi x/2)} \cos{(\pi y/2)}, \quad v = -\cos{(5t)} \cos{(\pi x/2)} \sin{(\pi y/2)} ,$$

$$p = \cos^2{(5t)} (\cos{\pi x} + \cos{\pi y}) /4 + 10 (x + y) \cos{(5t)} ,$$

$$\tag{7.283}$$

which defines the forcing term $\mathbf{f}$ in Eq.(7.1), the boundary value $\mathbf{V}_\Gamma$, and the initial condition $\mathbf{V}_0$.

Numerical tests (Hugues and Randriamampianina, 1998 ; Botella, 1997; Sabbah, 2000) are performed for $\nu = 10^{-2}$, $N_x = N_y = 32$, and Δt varying between 10^{-4} and 10^{-2}. The three second-order IM, PP and P2 methods give essentially the same errors E_u and E_p [defined by Eq.(7.282)]: they behave like Δt^2 and their magnitude is nearly identical, namely, $E_u \simeq 2.1 \times 10^{-4}$ and $E_p \simeq 3.6 \times 10^{-3}$ for $\Delta t = 10^{-2}$. For the P3 method, we

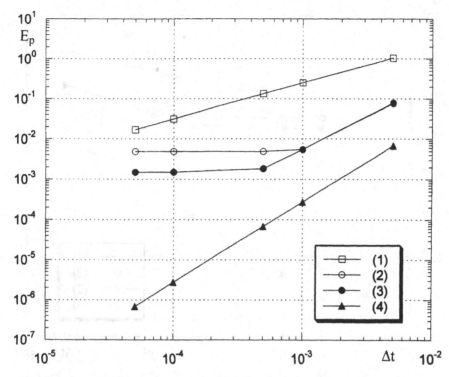

FIGURE 7.2. Error E_p, at inner points, for the solution (7.281) given by the projection methods : (1) OP method ($40 \times 40 \times 40$ resolution), (2) IP method ($40 \times 40 \times 40$ resolution), (3) IP method ($54 \times 54 \times 54$ resolution), (4) PP method ($40 \times 40 \times 40$ resolution).

obtain the Δt^3-behaviour with the typical values $E_u \simeq 5.2 \times 10^{-5}$ and $E_p \simeq 3.1 \times 10^{-4}$ for $\Delta t = 10^{-2}$.

(d) Stability of the methods

The stability of the various methods is compared by evaluating the maximum allowable time-steps in the calculation of the regularized driven cavity flow. This flow, which takes place in the square $0 \le X, Y \le 1$, is defined by the boundary conditions

$$u(X, 1) = -16X^2 (1 - X)^2 , \quad v(X, 1) = 0 ,$$

on the side $Y = 1$ and $u = 0$, $v = 0$ on the other three sides. The forcing term $\mathbf{f}$ in Eq.(7.1) is zero. The initial condition is $\mathbf{V}_0 = \mathbf{0}$. The Reynolds number $Re = \nu^{-1}$, based on the side of unit length, is $Re = 400$. The application of the Chebyshev method necessitates the change of variable in order to transform the square $[0, 1] \times [0, 1]$ into the square $[-1, 1] \times [-1, 1]$.

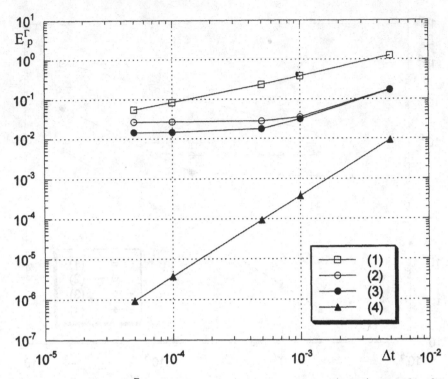

FIGURE 7.3. Error E_p^Γ, at the boundary, for the solution (7.281) given by the projection methods : (1) OP method ($40 \times 40 \times 40$ resolution), (2) IP method ($40 \times 40 \times 40$ resolution), (3) IP method ($54 \times 54 \times 54$ resolution), (4) PP method ($40 \times 40 \times 40$ resolution).

N	IM	PP	P2	P3	ω-ψ
16	0.082	0.055	0.081	0.049	0.067
24	0.049	0.033	0.047	0.032	0.047
32	0.032	0.027	0.030	0.023	0.033

Table 7.4. Critical time-step Δt_* for the regularized cavity flow at $Re = 400$.

Table 7.4. gives the critical time-step Δt_* (estimated within an error of $\pm 10^{-3}$) in the case of the IM, PP, P2, P3 methods and the AB/BDI2 influence matrix method applied to the vorticity-streamfunction equations (Section 6.4). The calculations are done by using the same polynomial degree in both directions $N_x = N_y = N$. The results show that the PP method requires time-steps slightly smaller than the other methods, the difference diminishing when the resolution is increased. Also, we may observe the decrease of the critical time-step when the order of accuracy increases (P2 and P3 methods).

N	IM	PP	P2	ω-ψ
16	24.6799 (0.60)	25.0387 (0.60)	24.7799 (0.60)	25.2329 (0.60)
24	24.9144 (0.63)	24.9180 (0.63)	24.9157 (0.63)	24.9344 (0.63)
32	24.7845 (0.65)	24.7844 (0.65)	24.7845 (0.65)	24.7845 (0.65)

Table 7.5. Maximal value M_1 of the vorticity on $Y = 1$ for the regularized cavity flow at $Re = 400$. The abscissa X_M of the point where the maximum is reached is given in parentheses.

N	IM	PP	P2	ω-ψ
16	25.0476 (0.63)	25.3854 (0.62)	25.1604 (0.625)	25.4675 (0.62)
24	24.9135 (0.63)	24.9170 (0.63)	24.9148 (0.63)	24.9333 (0.63)
32	24.9109 (0.63)	24.9107 (0.63)	24.9110 (0.63)	24.9110 (0.63)
40	24.9109 (0.63)	-	24.9109 (0.63)	-

Table 7.6. Maximal value M_2 of the vorticity on $Y = 1$ for the regularized cavity flow at $Re = 400$. The abscissa X_M of the point where the maximum is reached is given in parentheses.

(e) Spatial accuracy of the methods

The convergence of the solution with respect to the polynomial degree N is assessed by considering the steady regularized cavity flow at $Re = 400$, already considered in the previous section. The convergence is measured by examining a sensitive quantity such as the maximal value of the vorticity $\omega = \partial_X v - \partial_Y u$ on the side $Y = 1$. Thus, we may define M_1 as the maximum of ω_N on the collocation points belonging to $Y = 1$. Such a quantity, however, is not really significant due to the unequal distribution of the Gauss-Lobatto points and the large variations of the boundary vorticity. So, it is much more significant to consider the maximal value of the polynomial $\omega_N(x, 1)$ on $x \in [-1, 1]$. From the knowledge of the grid values $\omega_N(x_i, 1)$, $i = 0, \ldots, N$, the polynomial $\omega_N(x, 1)$ is reconstructed either through the Lagrange interpolation polynomial (3.43) or through the Chebyshev expansion (3.17) after having calculated the Chebyshev coefficients with Eq.(3.36). Therefore, we denote by M_2 the maximal value of $\omega_N(x, 1)$ taken on a large number of equispaced points, say 201. Tables 7.5 and 7.6, respectively, display the values of M_1 and M_2 obtained with the IM, PP, P2 methods and with the influence matrix method applied to the ω-ψ equation (Section 6.4).

A comparison between the behaviour of M_1 and M_2 is very instructive. The evolution of M_1, when N increases, does not give any indication on the convergence of the solution, while the results for M_2 exhibit a perfectly monotonic convergence. In particular, the relative difference between M_2 obtained for $N = 32$ and the converged value ($M_2 = 24.9109$) is less than 1.6×10^{-4} for any method. This illustrates not only the high accuracy of the

Chebyshev method but also its superiority over local methods like finite-differences concerning the information which can be extracted from the numerical results. Indeed, the finite-difference method gives the solution only at some points of the computational domain. The spectral method gives much more, since it gives the approximate solution in the form of a polynomial valid for any location in the computational domain.

(f) Cost of the IM, PP and P2 methods

For every method the CPU time needed by the calculations in the prepro-cessing stage is negligible compared to the time spent during the integration in time, even for the IM method which necessitates important precalcula-tions. For each of the IM, PP, and P2 methods, the computations performed at each time-cycle are essentially : (1) the evaluation of the right-hand sides, and (2) the solution of the Helmholtz problems. We assume that the right-hand sides are evaluated by matrix-matrix products and that the algebraic systems associated with the Helmholtz problems are solved by the matrix-diagonalization procedure. Therefore, the computational effort amounts to matrix-matrix products.

The cost of a method, in terms of CPU time, depends on :

(1) the number of time-cycles needed to reach some prescribed time, with a prescribed accuracy in time,

(2) the number of matrix-matrix products to be done at each time-cycle, and

(3) the spatial resolution needed to reach some prescribed accuracy, since the size of the matrices has an effect on the CPU time needed to perform these products.

From the results discussed in previous Sections, it appears that the tem-poral and spatial accuracy is comparable for the three IM, PP, and P2 methods, whereas the size of the time-step required for stability is slightly smaller for the PP method (1.2-1.5 times). Moreover, the number of matrix-matrix products done at each time-cycle is 34 for the IM method, 28 for the PP method, and 20 for the P2 method. In conclusion, the P2 method is less costly whereas the PP method is generally more costly.

7.4.5   The form of the nonlinear term

In Section 7.2.2, the influence of the form of the nonlinear term of the Navier-Stokes equations on the stability of large Reynolds number time-dependent solutions has been discussed for the Fourier method. It was pointed out that, in the absence of the forcing term $\mathbf{f}$, the semiconservation properties of the Navier-Stokes equations (i.e., conservation with $\nu = 0$) were preserved for the Fourier Galerkin approximation. Therefore, Fourier Galerkin or, equivalently, dealiased Fourier collocation, approximation of the nonlinear term does not introduce instability. On the other hand, the same properties of semi-conservation do not hold for the Chebyshev approximation (Canuto *et al.*, 1988). Since a theoretical analysis for the Chebyshev method is not available, we must have recourse to numerical experiments in order to understand the influence of the form of the nonlinear term.

Wilhelm and Kleiser (2000) have studied the case of a spectral-element method based on Legendre polynomials and staggered mesh. They consider the four forms of the nonlinear terms, namely, the convective, conservative, skew-symmetric, and rotational form (Section 5.1) and also the alternative rotational form where $\mathbf{A}(\mathbf{V}, \mathbf{V}) = \omega \times \mathbf{V} + \nabla(|\mathbf{V}|^2/2)$ while retaining ∇p. Their numerical results, for channel flow at a Reynolds number equal to 7500, show that only the convective form and the above alternative rotational forms are stable, whereas the conservative, the skew-symmetric, and the usual rotational forms are unstable, whatever the time-discretization scheme, time-step, and spatial resolution. The argument put forward by Wilhelm and Kleiser (2000) to explain the different behaviour between the various forms is the lack of verification of the constraint $\nabla \cdot \mathbf{V} = 0$ at the Gauss-Lobatto points where the nonlinear term is evaluated. However, it is likely that aliasing also has an influence.

From the numerical experiments (Botella *et al.*, 2001) reported below it seems that, in Chebyshev collocation methods, two effects are responsible for instability : aliasing and defective verification of the incompressibility condition. These experiments have been performed on the problem of the regularized cavity flow (Sections 7.4.4.d and 7.4.4.e) by means of the influence matrix method (IM-method, Section 7.4.1) and the $\mathbb{P}_N - \mathbb{P}_{N-2}$ projection method (P2-method, Section 7.4.2). In both methods, the same second-order AB/BDI2 time-scheme is used and the nonlinear term is evaluated without aliasing removal. This term is considered in the four classical forms : convective, conservative, skew-symmetric, and rotational defined in Section 5.1.

In the IM method, the divergence of the approximate velocity $\mathbf{V}_N$ is exactly zero everywhere, namely, $Q_N = \nabla \cdot \mathbf{V}_N$ is the null polynomial, while in the P2 method Q_N is zero only at the inner collocation points. At the collocation points belonging to the boundary Γ_N^I, Q_N is not zero but tends toward zero when the resolution is increased. This situation is different from the one considered by Wilhelm and Kleiser since a single collocation mesh is used here to enforce both momentum and incompressibility equations. For a given form of the nonlinear term, the aliasing error at inner collocation points is the same in both the IM and P2 methods. The only difference between both methods, concerning the nonlinear term, lies in the value of Q_N at the boundary.

For small values of Re (say $Re = 400$) all the forms are stable, whatever the method, because viscous effects are preponderant. Different behaviours appear for larger values of Re.

For example, Table 7.7 displays the results obtained for $Re = 2000$. For the considered time-steps and spatial resolutions ($N_x = N_y = N$) it is observed that :

Form	Method	$N = 32$		$N = 50$		$N = 80$	
		$10^2 \Delta t = 1$	0.5	1	0.05	0.5	0.25
CONV	IM	S	S	S	S	S	S
	P2	S	S	S	S	S	S
CONS	IM	S	S	S	S	S	S
	P2	U	U	U	U	U	U
S-S	IM	S	S	S	S	S	S
	P2	U	U	S	S	S	S
ROT	IM	S	S	S	S	S	S
	P2	S	S	S	S	S	S

Table 7.7. Stability properties, in the case $Re = 2000$, associated with the various forms of the nonlinear term (CONV = convective, CONS = conservative, S-S = skew-symmetric, ROT = rotational, for the influence matrix method (IM) and the $\mathbb{P}_N - \mathbb{P}_{N-2}$ projection method (P2), S = stable, U = unstable).

Form	Method	
	IM	P2
CONV	S	S
CONS	U	U
S-S	S	U
ROT	S	S

Table 7.8. Stability properties, in the case $Re = 7500$, $N = 50$, and $\Delta t = 5 \times 10^{-3}$, associated with the various form of the nonlinear term (CONV = convective, CONS = conservative, S-S = skew-symmetric, ROT = rotational, for the influence matrix method (IM), and the $\mathbb{P}_N - \mathbb{P}_{N-2}$ projection method (P2), S = stable, U = unstable).

1. The IM method is stable whatever the form of the nonlinear term.
2. For the P2 method, the conservative form is unstable.
3. For the P2 method, the skew-symmetric form is unstable for low resolution ($N = 32$).

Further computations performed with the P2 method and with the conservative form show that stability is obtained with $N = 160$ and $\Delta t = 2.5 \times 10^{-3}$. The fact that the calculations are stable for these values, while they are unstable for the same time-step but with $N = 80$, makes evident that instability is not a result of the choice of too large time-steps. The nature of the instability is connected to the spatial approximation itself as was observed in the Chebyshev approximation to the advection-diffusion (Section 4.2.3) but the reasons are obviously different. From these results, it seems that, for $Re = 2000$, the observed instability is most likely due more to the defective verification of the incompressibility condition at the boundary than the inner aliasing effects. By the way, it is interesting to point out that the same cavity flow at $Re = 2000$ has been calculated using the vorticity-streamfunction equations and the collocation influence matrix method given in Section 6.4. The computations done for $N = 32$ and $\Delta t = 2 \times 10^{-2}$ are stable whatever the form of the nonlinear term [Eqs.(5.16) and (5.17)]. This is in agreement with the above argument since the computed velocity field is automatically solenoidal.

Other computations done with a higher Reynolds number, namely, $Re = 7500$ with $N = 50$ and $\Delta t = 5 \times 10^{-3}$, reported in Table 7.8, show that the conservative form is unstable even with the IM method. A possible explanation is that viscous effects are no longer sufficient to prevent the influence of the aliasing error. This is strengthened by the fact that stability is recovered by increasing the resolution to $N = 80$ having, for effect, to reduce the aliasing terms.

In conclusion, it is clear that, for the considered Chebyshev collocation methods, the convective and rotational forms are the more stable. However, it must be pointed out that, at the present state of knowledge, the explanations put forward above must be taken with caution. Definitive conclusions could only be drawn after justification of the numerical results by theoretical analyses.

7.5 Example of application: three-dimensional flow in a rotating annulus

In this section we present an application of the $\mathbb{P}_N - \mathbb{P}_N$ predicted projection method (PP method) presented in Section 7.4.2. The application (Serre *et al.*, 2001) concerns three-dimensional flow in a rotating annulus with forced flux. This configuration has already been considered in Section 6.5.2 in the axisymmetric case characterized by a relatively low value of the dimensionless flow rate : $C_w \leq 120$. Here we consider the three-dimensional flow which appears for larger values of C_w. The problem is made dimensionless by means of the following characteristic quantities : h for length, Ω^{-1} for time, $\Omega^{-1} R_1$ for velocity, and $(R_m + 1) L^2 \Omega^2 h^2$ for pressure. We refer to Section 6.5.2 for the definition of these various quantities. The dimensionless equations, considered in the rotating frame of reference (ρ, θ, z) with $\mathbf{V} = (u, v, w)$, are

$$\partial_t u + A_u + \partial_\rho p$$

$$-E\left[\nabla^2 u - \frac{u}{L^2 (R_m + \rho)^2} - \frac{2}{L^2 (R_m + \rho)^2}\partial_\theta v\right] - 2v = 0 \qquad (7.284)$$

$$\partial_t v + A_v + \frac{1}{R_m + \rho}\partial_\theta p$$

$$-E\left[\nabla^2 v - \frac{v}{L^2\left(R_m + \rho\right)^2} + \frac{2}{L^2\left(R_m + \rho\right)^2}\partial_\theta u\right] + 2\,u = 0 \tag{7.285}$$

$$\partial_t w + A_w + L\,\partial_z p - E\nabla^2 w = 0 \tag{7.286}$$

$$\partial_\rho u + \frac{u}{R_m + \rho} + \frac{1}{R_m + \rho}\partial_\theta v + L\,\partial_z w = 0, \tag{7.287}$$

where (A_u, A_v, A_w) is the convective term, ∇^2 is the Laplacian operator,

$$\nabla^2 = \frac{1}{L^2}\partial_{\rho\rho} + \frac{1}{L^2\left(R_m + \rho\right)}\partial_\rho + \frac{1}{L^2\left(R_m + \rho\right)^2}\partial_{\theta\theta} + \partial_{zz}, \tag{7.288}$$

and E is the Ekman number defined by Eq.(6.149).

The above equations are solved in the domain $(-1 \le \rho \le 1, 0 \le \theta \le 2\pi, -1 \le z \le 1)$. The boundary conditions prescribed at $\rho = \pm 1$ are deduced from the Ekman boundary layer solution, namely,

$$u = V_g\left[e^{Z_+}\sin Z_+ - e^{Z_-}\sin Z_-\right],$$

$$v = -V_g\left[e^{Z_+}\cos Z_+ - e^{Z_-}\cos Z_-\right],$$

$$w = 0,$$

where $V_g = C_w E^{1/2}/\left[2\pi L\left(R_m + \rho\right)\right]$ and $Z_\pm = -E^{-1/2}\left(1 \pm z\right)$. The no-slip conditions $u = 0$, $v = 0$, $w = 0$ are prescribed at $z = \pm 1$. The initial condition will be specified later. The problem being periodic in the azimuthal direction, the spatial approximation makes use of truncated Fourier series in θ. Thus, each dependent variable $\phi = u, v, w, p$ is approximated according to

$$\phi_K\left(\rho, \theta, z, t\right) = \sum_{k=-K+1}^{K}\hat{\phi}_k\left(\rho, z, t\right)e^{i\,k\,\theta}. \tag{7.289}$$

Then the Galerkin equations (see Section 6.2.1) determining the Fourier coefficients $\hat{u}_k$, $\hat{v}_k$, $\hat{w}_k$, and $\hat{p}_k$ are

$$\partial_t\hat{u}_k + \hat{A}_{u,k} + \partial_\rho\hat{p}_k$$

$$-E\left[\overline{\nabla}^2\hat{u}_k - \frac{k^2 + 1}{L^2\left(R_m + \rho\right)^2}\hat{u}_k - i\frac{2\,k}{L^2\left(R_m + \rho\right)^2}\hat{v}_k\right] - 2\,\hat{v}_k = 0 \tag{7.290}$$

$$\partial_t\hat{v}_k + \hat{A}_{v,k} + \frac{i\,k}{R_m + \rho}\hat{p}_k$$

$$-E\left[\overline{\nabla}^2\hat{v}_k - \frac{k^2 + 1}{L^2\left(R_m + \rho\right)^2}\hat{v}_k + i\frac{2\,k}{L^2\left(R_m + \rho\right)^2}\hat{u}_k\right] + 2\,\hat{u}_k = 0 \tag{7.291}$$

$$\partial_t \hat{w}_k + \hat{A}_{w,k} + L\partial_z \hat{p}_k - E\left[\overline{\nabla}^2 \hat{w}_k - \frac{k^2}{L^2 (R_m + \rho)^2}\hat{w}_k\right] = 0 \qquad (7.292)$$

$$\partial_\rho \hat{u}_k + \frac{\hat{u}_k}{R_m + \rho} + i\frac{k\,\hat{v}_k}{R_m + \rho} + L\partial_z \hat{w}_k = 0, \qquad (7.293)$$

with associated boundary conditions at $\rho = \pm 1$, $z = \pm 1$. The operator $\overline{\nabla}^2$ is the Laplacian operator (7.288) without the θ-term. The problem is solved by using the second-order version of the splitting scheme (7.238)-(7.242) and the $I\!P_N - I\!P_N$ projection method described in Section 7.4.3.e. Thus, the solution method is as follows :

1. The predicted pressure $\overline{p}^{n+1}$ is calculated from equations similar to Eqs.(7.264)-(7.265).

2. The provisional velocity $\tilde{\mathbf{V}}_k^{n+1} = (\tilde{\hat{u}}_k^{n+1}, \tilde{\hat{v}}_k^{n+1}, \tilde{\hat{w}}_k^{n+1})$ is calculated from equations similar to Eqs.(7.238)-(7.239). The nonlinear term $(\hat{A}_{u,k}, \hat{A}_{v,k}, \hat{A}_{w,k})$ is calculated through the pseudospectral technique (Section 6.3.3) based on even collocation (Section 2.4) defined by the collocation points $\theta_j = 2\pi j/N_\theta$, $j = 0, \ldots, N_\theta$, with $N_\theta = 2K$ and $\rho_i = \cos(\pi i/N_\rho)$, $i = 0, \ldots, N_\rho$, $z_k = \cos(\pi k/N_z)$, $k = 0, \ldots, N_z$. To avoid the coupling of Eqs.(7.290) and (7.291) through the terms $i\,k\,\hat{v}_k$ and $i\,k\,\hat{u}_k$, Serre *et al.* (2001) evaluated these terms in an explicit way using the Adams-Bashforth extrapolation. Note that, if the equations were considered in a fixed frame of reference [the Coriolis term $(-2\,\hat{v}_k, \hat{u}_k)$ being absent], the coupling could be avoided, while considering an implicit discretization, by introducing (Orszag and Patera, 1983) the new dependent variables

$$\tilde{\lambda}_k^{n+1} = \tilde{\hat{u}}_k^{n+1} + i\tilde{\hat{v}}_k^{n+1}, \quad \tilde{\mu}_k^{n+1} = \tilde{\hat{u}}_k^{n+1} - i\tilde{\hat{v}}_k^{n+1}.$$

Then uncoupled equations for $\tilde{\lambda}_k^{n+1}$ and $\tilde{\mu}_k^{n+1}$ are obtained by addition and subtraction of the two momentum equations.

3. The solution $(\hat{u}_k^{n+1}, \hat{v}_k^{n+1}, \hat{p}_k^{n+1})$ is obtained from the projection step (7.240)-(7.242) solved through a Neumann problem analogous to (7.271)-(7.272).

Serre *et al.* (2001) did a detailed study of the problem by varying the flow rate C_w and the curvature parameter R_m. Here we want to illustrate the application of the PP projection method by considering a typical case, namely, $C_w = 530$, $E = 2.24 \times 10^{-3}$, $L = 3.37$, and $R_m = 5$. The calculations are done with $N_\rho = N_z = 48$, $K = 32$, and $\Delta t = 4 \times 10^{-3}$. This flow is computed by using, as an initial condition, the axisymmetric flow at $C_w = 460$. The resulting flow is time-periodic and again axisymmetric but it becomes three-dimensional after an instantaneous small disturbance was applied to the azimuthal velocity near the entrance. This three-dimensional

FIGURE 7.4. Instantaneous iso-surface of the axial velocity component ($w = -3.5 \times 10^{-4}$) of the time-periodic three-dimensional flow for $C_w = 530$, $E = 2.24 \times 10^{-3}$, $L = 3.37$ and $R_m = 5$.

flow is illustrated in Figure 7.4 which displays the instantaneous iso-surface of the axial velocity component ($w = -3.5 \times 10^{-4}$). The iso-surface exhibits a spiral pattern whose number of arms (twelve) depends on the wavelength of the imposed disturbance. Figure 7.5 displays the representation of the same iso-surface pattern in a Cartesian frame.

Remark

The axis singularity

In the example considered above, the computational domain does not contain the axis $r = 0$, that is, the coordinate singularity associated with the axis is avoided. In the general cylindrical case (r, θ, z) with $0 \leq r \leq R$, the behaviour of the solution as $r \rightarrow 0$ must be taken into account. The requirement for analyticity of the physical (scalar or vector) quantities $\phi(r, \theta, z, t)$ implies some conditions in the form of the approximation in the r-direction (Orszag and Patera, 1983 ; Tuckerman, 1989 ; Pasquetti and Bwemba, 1994; Priymak, 1995). In practice, only some of these conditions are actually prescribed, for example the Laplacian of ϕ is required to be nonsingular. The weaker form of the regularity condition is the requirement that ϕ is single-valued at $r = 0$. This implies the vanishing of the azimuthal derivative

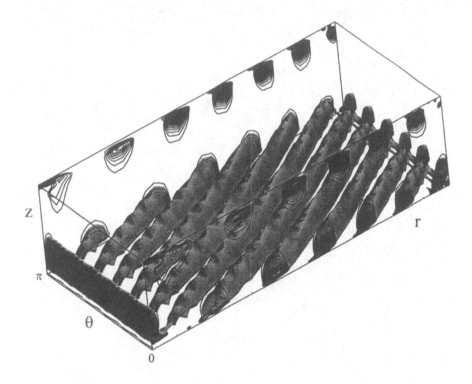

FIGURE 7.5. Representation in a cartesian frame of the instantaneous iso-surface of the axial velocity component ($w = -3.5 \times 10^{-4}$).

$\partial_\theta \phi$ at $r = 0$. The result is that the Fourier coefficients $\hat{\phi}_k\,(r, z, t)$ are such that $\hat{\phi}_k\,(0, z, t) = 0$ for $k \neq 0$, but the value $\hat{\phi}_0\,(0, z, t)$ remains undetermined.

Various ways are possible to avoid having to prescribe boundary conditions on the axis :

(1) Consider the interval $-R \leq r \leq R$ and the Gauss-Lobatto points $\rho_i = r_i/R, i = 0, \ldots, N_r$, where N_r is even, so that $r = 0$ is not a collocation point (Pulicani and Ouazzani, 1991 ; Pasquetti and Bwemba, 1994).

(2) Consider the interval $0 \leq r \leq R$ and the Gauss-Radau points (see Appendix A), so that the origin $r = 0$ does not belong to the set of collocation points (Le Marec et al., 1996 ; Manna and Vacca, 1999).

(3) Consider the interval $0 \leq r \leq R$ and the Gauss-Lobatto points, but introduce the change of dependent variable $\tilde{\phi} = r\,\phi$ so that $\tilde{\phi}$ satisfies the boundary condition $\tilde{\phi}\,(0, \theta, z, t) = 0$ (Pulicani and Ouazzani, 1991 ; Serre and Pulicani, 2001). The value of ϕ on the axis $r = 0$ is computed from the value of $\partial_r \phi$ at $r = R$. Because this later value depends on θ, the continuity of ϕ at $r = 0$ is obtained only approximately and needs the resolution to be large enough.

Part III

Special topics

8

Stiff and singular problems

This chapter discusses some techniques for handling stiff and singular problems, using Chebyshev methods. The solution of a stiff problem is regular but exhibits large variations in a region of small extent. For convergence reasons, the collocation points cannot be clustered arbitrarily in the rapid variation region, so that an appropriate distribution of the collocation points must be obtained through a coordinate transformation, preferably self-adaptive. Another way is to use a domain decomposition method which can be combined with the preceding one. The presentation of these various approaches is the subject of the first part of this chapter. The second part is devoted to the calculation of singular solutions, which have only a very small number of bounded derivatives. For such functions, the rate of convergence of the Chebyshev approximation is only algebraic and, in the case of a strong singularity, the polynomial representation is subject to oscillations. Two ways to recover high accuracy will be discussed. The first way consists of using a domain decomposition method so that the singularity is shifted to a corner of a subdomain. The second way, which is more efficient, consists of subtracting the most singular part of the solution, so that the computed solution is less singular and can be accurately represented by the Chebyshev approximation.

8.1   Introduction

As discussed in Sections 2.1.2 and 3.6, the approximation error associated with the spectral (Fourier or Chebyshev) method depends on the regularity of the solution. This supposes, however, a sufficiently high resolution characterized by the cut-off frequency K for the Fourier method or by the degree N of the approximation polynomial for the Chebyshev method. Even for regular functions, the resolution (i.e., K or N) required for an accurate representation may become very large in the case of some stiff problems. This has been numerically observed by Basdevant *et al.* (1986) who compare the results of various spectral methods (and finite-differences) for the solution of the Burgers equation exhibiting a large gradient.

A stiff problem is a problem whose solution is characterized by two or more space-scales and/or time-scales of different orders of magnitude. In the spatial case considered here, common stiff problems in fluid mechanics are boundary layer, shear layer, viscous shock, interface, flame front, etc. In all these problems there exists a region of small extent (with respect to the global characteristic length) in which the solution exhibits a large variation.

In the Chebyshev case considered in the following, a stiff solution is more easily represented (i.e., needs a lower polynomial degree N) when the large gradient region is located near a boundary rather than inside the domain of computation. For example, Fig.8.1 shows the Chebyshev spectrum $\{\hat{u}_k\}$ of two functions $u(x)$ having the same slope at $x = 1$ and $x = 0$, respectively. Curve (1) represents the spectrum of

$$u(x) = -1 - 2\tanh[5(x-1)] \qquad (8.1)$$

and curve (2) represents the spectrum of

$$u(x) = -\tanh(10\,x) . \qquad (8.2)$$

In both cases, the coefficients $\hat{u}_k$ are calculated by the FFT using the Gauss-Lobatto points [see Eq.(3.37)]. The difference between the rate of decrease of the Chebyshev coefficients clearly illustrates the property mentioned above. Note that, as pointed out by Solomonikoff and Turkel (1989), this property is not necessarily connected to the clustering of the Gauss-Lobatto points near the boundaries, and is a consequence of the global character of the approximation. It also holds for the Lagrange interpolation polynomial based on uniformly spaced points.

It is advisable to take advantage of the above property in the case of moderately thin boundary layers (Section 6.5.2). On the other hand, for the correct representation of very thin boundary layers, and mainly of inner layers, it is necessary to resort to appropriate techniques in order to keep the polynomial degree N within reasonable bounds. The usual approach to stiff problems makes use of adapted coordinate transformations. The domain

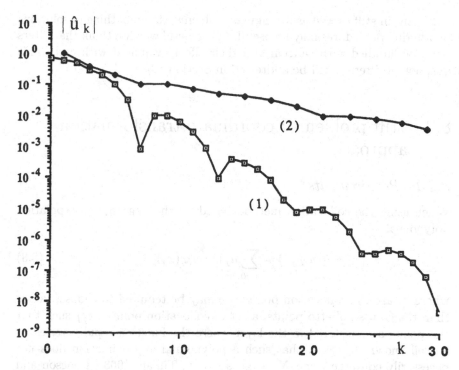

FIGURE 8.1. Chebyshev spectrum $\{\hat{u}_k\}$ of functions : $u(x) = -1 - 2\tanh[5(x-1)]$ (curve (1)) and $u(x) = -\tanh(10x)$ (curve (2)).

decomposition method is an alternate way of dealing with such problems. Moreover, when possible, the combination of these two techniques, taking advantage of each of them constitutes an efficient tool for the solution of stiff problems, especially when the function to be represented exhibits several regions of rapid variation. Lastly, the efficiency of these various approaches is greatly enhanced when they are associated to a self-adaptation procedure. These different points will be addressed in the next two sections.

The computation of singular solutions with a spectral method is rather difficult. Contrary to the stiff solutions discussed above, a singular solution has only a limited number of bounded derivatives, and sometimes none ! In such cases, the spectral accuracy is lost and the rate of convergence of the spectral approximation is only algebraic (see Sections 2.1.2 and 3.6). Moreover, when the singularity is strong, the Gibbs oscillations make the approximation completely unusable.

In flow problems, even for incompressible viscous flows, the presence of singularities is not exceptional, but their effect on the overall solution is often sufficiently weak to be neglected. On the other hand, when this is not the case, the computation of singular solutions with a spectral method necessitates a special treatment. This point will be the subject of Section 8.4.

Finally, in stiff as well as in singular problems, the smoothing of solutions by filtering procedures may be useful. In a general way, we think that filters must be handled with caution and their effect evaluated with care. This question of filtering will be addressed in Section 8.5.

8.2 Stiff problems : coordinate transformation approach

8.2.1 Requirements

When using the collocation method based on the Lagrange interpolation polynomial

$$u_N(x) = \sum_{j=0}^{N} h_j(x) \, u_N(x_j) \, , \tag{8.3}$$

where x_j are the collocation points, we may be tempted to choose, rather than the Gauss-Lobatto points, a set of collocation points $\{x_j\}$ such that the points are clustered in the layer where the function rapidly varies. It is well known, however, that such a polynomial approximation does not necessarily converge when $N \to \infty$ (see, e.g., Timan, 1963 ; Isaacson and Keller, 1966 ; Funaro, 1992). The fact that the distribution of collocation points cannot be arbitrary is illustrated by the famous Runge example. This example states that the polynomial approximation $u_N(x)$ to $u(x) = 1/(1 + x^2)$ in $[-5, 5]$, based on the equally spaced points, is such that the error $|u(x) - u_N(x)|$ becomes arbitrarily large if N is sufficiently large.

Therefore, it is preferable to introduce a coordinate transformation

$$x = f(\xi) \tag{8.4}$$

in the equations to be solved and then to use the Chebyshev method (tau or collocation) in the transformed domain. The mapping $x = f(\xi)$ must meet some requirements. First, it must be one-to-one and easy to invert. Then the choice of $f(\xi)$ is made on the following equivalent arguments.

Considering the function $u(x)$ to be approximated in $[-1, 1]$, the mapping $x = f(\xi)$ has to be chosen such that $u(f(\xi))$ is sufficiently smooth to be represented by a low degree Chebyshev approximation, namely,

$$u(f(\xi)) = v(\xi) \cong v_N(\xi) = \sum_{k=0}^{N} \hat{v}_k \, T_k(\xi) \, . \tag{8.5}$$

In an equivalent way, this means that the basis $T_k(f^{-1}(x))$ is better adapted to represent the function $u(x)$ than the original Chebyshev basis

$T_k(x)$. As an example, let us consider the one-to-one cubic transformation (Basdevant *et al.*, 1986) :

$$x = f(\xi, a) = (1 - a)\,\xi^3 + a\,\xi, \tag{8.6}$$

where a is a parameter such that $0 \le a < 1$. Figure 8.2 compares the three Chebyshev polynomials, $T_1 = x$, $T_2 = 2x^2 - 1$, and $T_3 = 4x^3 - 3x$, and the corresponding transformed basis functions $T_1\left(f^{-1}(x)\right)$, $T_2\left(f^{-1}(x)\right)$, and $T_3\left(f^{-1}(x)\right)$ for $a = 0.15$. It is clear that the transformed basis is more appropriate to represent a function having a steep slope at the origin than the original basis.We mention the works by Boyd (1987a,b) in which the mapping procedure (for infinite and semi-infinite domains), considered as defining a new set of basis functions, is analyzed.

Finally, another equivalent requirement of the coordinate transformation $x = f(\xi)$ is to cluster the points $x_i = f(\xi_i)$ with $\xi_i = \cos \pi i/N$, $i = 0, \ldots, N$, in the region of large gradient, namely,

$$x_{i+1} - x_i = f(\xi_{i+1}) - f(\xi_i) = (\xi_{i+1} - \xi_i)\,f'(\xi_i) + O[(\xi_{i+1} - \xi_i)^2].$$

Therefore, it is required that

$$|f'(\xi_i)| \ll 1 \tag{8.7}$$

for ξ_i belonging to the large gradient region. Then the derivative $v'(\xi_i) = u'(x_i)\,f'(\xi_i)$ keeps a reasonable bounded value ($|v'(\xi_i)| \approx 1$) if $|f'(\xi_i)|$ is sufficiently small.

8.2.2 Some typical mappings

In a general way, the coordinate transformation is considered in the form

$$x = f(\xi, a_1, a_2) \equiv f(\xi, A), \tag{8.8}$$

where $A = (a_1, a_2)$. The parameter a_1 characterizes the location $x = a_1$ where the function to be approximated exhibits a rapid variation. The parameter a_2 controls the magnitude of the coordinate contracting near $x = a_1$. It is desirable to recover, more or less exactly, the identity for some value of A. The two-parameter family (8.8) supposes the existence of a single region of rapid variation. Coordinate transformations involving more than two parameters are needed for the representation of solutions with multiple layers (Bayliss *et al.*, 1995a).

The interval $x \in [-1, 1]$ is mapped onto $\xi = [-1, 1]$. This requires

$$f(-1, A) = -1, \quad f(1, A) = 1, \tag{8.9}$$

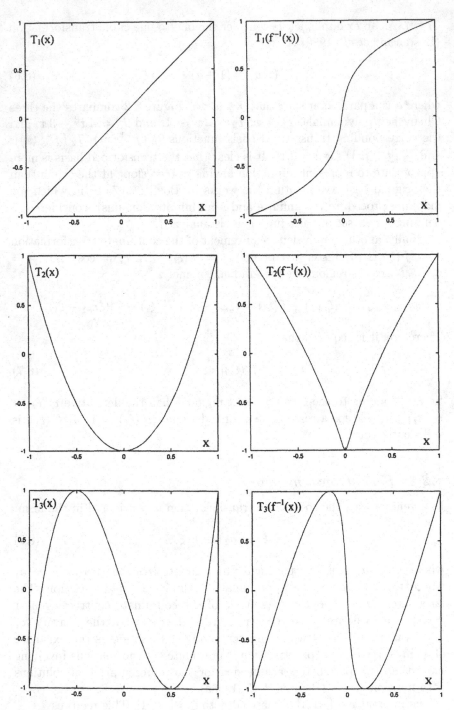

FIGURE 8.2. Chebyshev polynomials $T_1(x)$, $T_2(x)$, $T_3(x)$ and the transformed functions $T_1\left(f^{-1}(x)\right)$, $T_2\left(f^{-1}(x)\right)$, $T_3\left(f^{-1}(x)\right)$.

and the transformation is one-to-one, such that

$$\xi = f^{-1}\left(x, A\right) \equiv g\left(x, A\right) \tag{8.10}$$

is univalent.

Various mappings have been used. For example, the cubic transformation (8.6) has been employed (Basdevant et al., 1986 ; Guillard and Peyret, 1988; Guillard et al., 1992) to represent functions with rapid variation at the origin $x = 0$ for which the Chebyshev approximation is not well-adapted. To illustrate, we show some results obtained by Basdevant et al. (1986) for the stiff solution of the Burgers equation defined by

$$\partial_t u + u\,\partial_x u - \nu\,\partial_{xx} u = 0, \quad -1 < x < 1, \quad t > 0$$
$$u\left(-1, t\right) = 0, \quad u\left(1, t\right) = 0 \tag{8.11}$$
$$u\left(x, 0\right) = -\sin \pi x.$$

For the small chosen value of the viscosity ($\nu = 1/100\pi$) the solution develops into a sawtooth wave at the origin from about the time $t = 1/\pi$. Figure 8.3 compares the results obtained with $N = 64$ without mapping (Fig.8.3.a) and with the mapping (8.6) where $a = 1/25$ (Fig.8.3.b). The improvement is clearly seen and is easily explained by the shape of the curves in the transformed coordinate system ξ (Fig.8.4).

When the region of rapid variation is not located at the origin but at (or near) $x = a_1$, the cubic transformation is

$$x = f\left(\xi, A\right) = \left(1 - a_2\right) \xi^3 - a_1\,\xi^2 + a_2\,\xi + a_1. \tag{8.12}$$

The point $x = a_1$ is transformed onto $\xi = 0$ and $f'\left(0, A\right) = a_2$. Therefore the condition (8.7) implies $0 < a_2 \ll 1$. This transformation is one-to-one if $3\,a_2\left(1 - a_2\right) - a_1^2 \geq 0$, namely its application is restricted to the cases where the large gradient is not too far from the origin.

A similar polynomial approximation, but of higher degree (Malé, 1992), is constructed by combining the first transformation

$$x = F\left(X, a_1\right) = -a_1\,X^2 + X + a_1 \tag{8.13}$$

and the cubic mapping

$$X = f\left(\xi, a_2\right) = \left(1 - a_2\right) \xi^3 + a_2\,\xi. \tag{8.14}$$

Therefore, the mapping (8.13) brings $x = a_1$ to $X = 0$ and (8.14) clusters the points near $X = 0$ if $0 < a_2 < 1$. Note that the transformation (8.13) is one-to-one if $-1/2 < a_1 < 1/2$ while (8.14) is one-to-one whatever a_2. The compound transformation $X = f \circ F$ is easily inverted since each component is easily inverted.

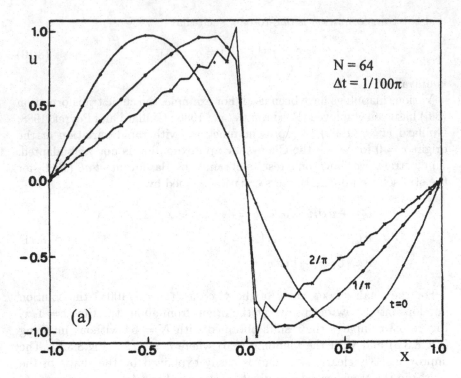

FIGURE 8.3.a. Solution of the Burgers equation : (a) without mapping.

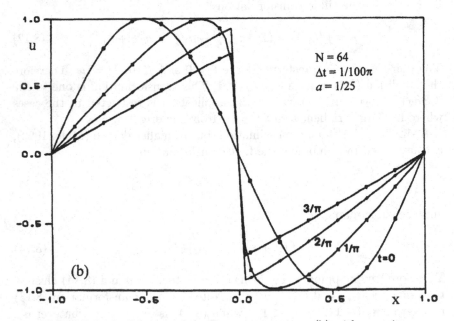

FIGURE 8.3.b. Solution of the Burgers equation : (b) with mapping.

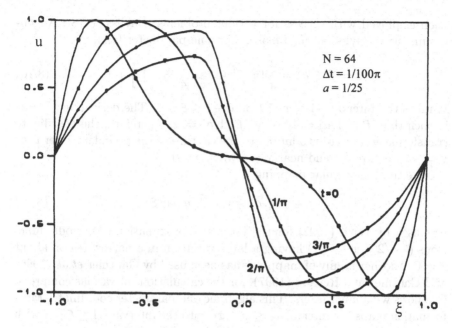

FIGURE 8.4. Solution of the Burgers equation in the transformed coordinate ξ.

Bayliss and Turkel (1992) have compared several coordinate transformations. Among these, the mapping based on the tangent function seems to be the best. It reads

$$x = f(\xi, A) = a_1 + a_2 \tan\left[\lambda(\xi - \xi_0)\right] \qquad (8.15)$$

with

$$\xi_0 = \frac{\kappa - 1}{\kappa + 1}, \quad \kappa = \frac{\tan^{-1}\left[(1 + a_1)/a_2\right]}{\tan^{-1}\left[(1 - a_1)/a_2\right]},$$

and

$$\lambda = \frac{\tan^{-1}\left[(1 - a_1)/a_2\right]}{1 - \xi_0}.$$

In the region of rapid variation, characterized by $x = a_1$ and $\xi = \xi_0$, the derivative $f'(\xi_0, a_1, a_2) \sim \pi a_2/2$ if $a_2 \to 0$. Note that, under these conditions, the inverse $\xi = g(x, A)$ gives an approximation of a quasi-step function with a near-discontinuity at $x = a_1$. On the other hand, for large values of a_2, the transformation (8.15) is close to the identity.

The Chebyshev polynomial approximation is naturally well-adapted to the representation of boundary layers. It may happen, however, that the thickness of the boundary layer is so small that a high-degree polynomial is necessary to capture it accurately. In this case, it is advisable to introduce an adapted mapping. The coordinate transformation (8.15) is singular for the boundary layers, namely for $a_1 = \pm 1$, and an alternative mapping has been employed with success for various problems (Bayliss, 1978 ; Augenbaum, 1989; Bayliss et al., 1995b). This one-parameter family is

$$x = f(\xi, a) = \frac{4}{\pi} \tan^{-1}\left[a \tan \frac{\pi}{4}(\xi - 1)\right] + 1. \tag{8.16}$$

It maps the interval $-1 \le x \le 1$ onto $-1 \le \xi \le 1$. The derivative $f'(\xi, a)$ is such that $f'(-1, a) = 1/a$ and $f'(1, a) = a$. Therefore, the coordinate transformation (8.16) is adapted to the capture of the boundary layer near $x = -1$ when $a > 1$ and near $x = 1$ when $a < 1$.

Note that the similar mapping

$$x = f(\xi, a) = \tan^{-1}(a \tan \xi) \tag{8.17}$$

was used by Boyd (1992) for the Fourier approximation of periodic functions $(-\pi/2 \le x \le \pi/2)$ having a large variation in a narrow region about $x = 0$. Another family of mappings has been used by Gauthier et al. (1996) and Gauthier and Renaud (1997), for the calculations of various compressible flows with inner layers. This algebraic one-parameter coordinate transformation maps the interval $x_1 \le x \le x_2$ onto the interval $-1 \le \xi \le 1$ with correspondance between $x_0 = (x_1 + x_2)/2$ and $\xi = 0$. It is of the form

$$x = f(\xi, x_1, x_2, a) = \frac{1}{2}(x_1 + x_2) + \frac{a\xi}{(1 + b^2 - \xi^2)^{1/2}} \tag{8.18}$$

with $b = 2a/(x_2 - x_1)$. At $\xi = 0$, the derivative of the transformation is

$$f'(0, x_1, x_2, a) = a(1 + b^2)^{-1/2}$$

so that the points are clustered near $\xi = 0$ (or, equivalently, $x = x_0$) if $a < 1$. The identity is recovered when $a \to \infty$ and a quasi-uniform distribution of points x_i, corresponding to the Gauss-Lobatto ξ_i points, is obtained if $a = 0.3277(x_2 - x_1)$. The case of the infinite domain is obtained by setting $x_1 \to -\infty$, $x_2 \to +\infty$, and $x_1 + x_2 = 0$. The resulting mapping

$$x = f(\xi, a) = \frac{a\xi}{(1 - \xi^2)^{1/2}} \tag{8.19}$$

has been discussed by Boyd (1982).

The mapping of the semi-infinite domain $0 \le x \le \infty$ onto the finite domain $-1 \le \xi \le 1$ may be done either with an algebraic transformation

$$x = f(\xi, a) = a\frac{1 + \xi}{1 - \xi} \tag{8.20}$$

or with an exponential one

$$x = f(\xi, a) = \frac{1}{a} \tanh^{-1}\left(\frac{1 + \xi}{2}\right). \tag{8.21}$$

Grosch and Orszag (1977) and Boyd (1982, 1987b) have compared algebraic and exponential mappings. They found that the algebraic mapping is preferable to exponential mappings, although such transformations have been used with some success in various problems (Hussaini and Zang, 1982; Guillard and Peyret, 1988 ; Raspo, 2001).

We refer to the various references quoted above, and to Canuto *et al.* (1988) and to Kosloff and Tal-Ezer (1993) for other types of mapping. In particular, these latter authors consider the one-parameter family

$$x = f\left(\xi, a\right) = \frac{\sin^{-1}\left(a\,\xi\right)}{\sin^{-1} a} \tag{8.22}$$

with $0 \leq a < 1$. The identity is recovered when $a \to 0$. The interest in this mapping is to transform the Gauss-Lobatto ξ_i points into equally spaced x_i points when $a \to 1$. This limit, however, is singular so that the optimal use of the mapping (8.22) corresponds to values of a close to 1, of the form $a = \cos\left(k\,\pi/N\right)$ where N is the polynomial degree and $k \geq 1$. As pointed out by these authors, and experienced by Renaut and Fröhlich (1995), the mapping (8.22) has a beneficial effect on the stability of time-integration schemes, but care must be taken concerning accuracy when choosing the value of k.

8.2.3   Self-adaptive coordinate transformation

(a) The basic principles

When solving a differential problem, whose solution exhibits rapid variation in a narrow region, the use of a coordinate transformation is of great interest in order to obtain accurate results while restricting the degree of the polynomial approximation within reasonable bounds. As introduced in the previous section, the considered coordinate transformation is of the form (8.8). In most of the problems of physical interest, the location of the steep gradient and its thickness are unknown and, moreover, they may evolve with time. Thus, the optimal value of the parameter vector A may change with time. In such a case, it is necessary to design an automatic procedure for determining the parameter A. This is the problem of self-adaptation commonly encountered in computational fluid dynamics.

The principle of the method is to write the equations in the transformed ξ-domain through the parametrized mapping (8.8). Then the transformed equations are solved by means of a standard spectral method. We shall address, here, only the case of Chebyshev approximation, but the Fourier method could also be considered (Augenbaum, 1989 ; Boyd, 1992). When necessary, the mapping has to be adapted to the evolving solution. This is done by determining the parameter A which minimizes some functional of the solution (Bayliss and Matkowsky, 1987 ; Guillard and Peyret, 1988). It is clear that such a functional has to be connected with the approximation error. In Section 3.6 it has been stated that, for a fixed polynomial degree N, the error (in weighted Sobolev norm) of the polynomial approximation is bounded by the norm $\|u\|_{H_w^m(-1,1)}$, where u is the function under consideration. Denoting by $v(\xi)$ the expression of $u(x)$ through the mapping (8.8), the parameter A is determined such that the norm $\|v\|_{H_w^m(-1,1)}$ calculated in the ξ-domain, namely,

$$\left[\int_{-1}^{1}\sum_{p=0}^{m}\left|\partial_\xi^p v\right|^2 w\,d\xi\right]^{1/2} \quad, \quad w = \left(1-\xi^2\right)^{-1/2}, \tag{8.23}$$

is minimal. Numerical experiments show that the best results are obtained with the H_w^2-norm. The use of H_w^1 gives bad results (Bayliss *et al.*, 1989; Guillard *et al.*, 1992) whereas the use of H_w^3 does not bring any improvement with respect to H_w^2. Moreover, the nondifferentiated term $|v|^2$ in (8.23) may be discarded without damage since its magnitude is independent of the mapping. Therefore, the resulting functional is

$$J_2(v, A) = \left[\int_{-1}^{1}\left(\left|\partial_\xi v\right|^2 + \left|\partial_{\xi\xi} v\right|^2\right) w\,d\xi\right]^{1/2}. \tag{8.24}$$

Bayliss *et al.* (1989) have proposed an alternative criterion based on the minimization of the maximum norm of the approximation error. They show that the error $v(\xi) - v_N(\xi)$, where $v_N(\xi)$ is the Chebyshev Galerkin (i.e., projection) approximation to $v(\xi)$, satisfies the inequality

$$\|v - v_n\|_{L^\infty(-1,1)} \le \frac{I_2(v, A)}{(N+1)^2} + O\left(\frac{1}{N^3}\right), \tag{8.25}$$

where $I_2(v, A)$ is derived from the general functional

$$I_{2j}(v, A) = \left\{\int_{-1}^{1}\left[\left(\sqrt{1-\xi^2}\partial_\xi\right)^{2j} v\right]^2 w\,d\xi\right\}^{1/2}. \tag{8.26}$$

The estimate (8.25) can be extended on the maximum norm of the Chebyshev collocation error with the loss of at most a factor of $\ln N$. Besides, Guillard *et al.* (1992) have shown that the functional I_{2j} gives an upper bound of the projection error in the L_w^2-norm, that is,

$$\|v - v_N\|_{L_w^2(-1,1)} \le \frac{1}{N^{2j}} I_{2j}(v, A).$$

It is easy to see that $I_2(v, A) \le J_2(v, A)$, therefore I_2 constitutes a tighter upper bound on the error. Bayliss *et al.* (1989) report that I_2 is more sensitive to variations in the behaviour of the solution than J_2, so that its use permits an accurate representation of the solution with

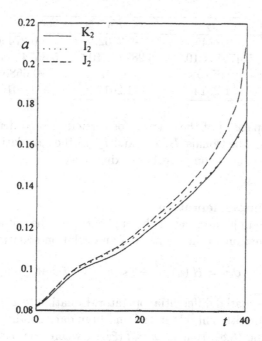

FIGURE 8.5. Evolution of the parameter a for various functionals.

fewer collocation points. It seems that additional numerical experiments, performed on various flow problems, are needed to confirm the superiority of I_2. To illustrate, we mention the results obtained by Guillard *et al.* (1992) concerning the calculation of the two-dimensional subsonic (Mach number = 0.8) viscous (Reynolds number = 400) mixing layer (see Section 8.3.4). By using the cubic transformation (8.6) with 53 collocation points, they compared the functionals J_2, I_2, and K_2 similar to J_2 but without the weight w. The argument of the functional is chosen as the streamwise component of the velocity. Figure 8.5 shows the evolution of the parameter a for the three functionals. It can be seen that, although the evolution of a is similar for the three different cases, the value of a given by I_2 is always the lowest, while the largest is given by the weighted Sobolev norm J_2. This can be explained by considering that I_2 and K_2 give less weight to the boundaries than J_2. Thus, this latter functional tends to give larger importance to rather small variations appearing far from the middle of the domain. Note that, at $t =$ 40, when the difference between the values of a becomes relatively large, the vorticity thickness is significantly large, and the distributions of the collocation points in the physical domain are almost the same. This is the reason why the numerical values of the flow quantities (vorticity ω and density ρ) at $t = 40$ given in Table 8.1 display very small differences.

	J_2	K_2	I_2
ω_{min}	-0.27455	-0.27622	-0.27456
ω_{max}	5.4751×10^{-3}	5.2829×10^{-3}	5.8574×10^{-3}
ρ_{min}	0.65798	0.65814	0.65802
ρ_{max}	1.2014	1.2015	1.2016

Table 8.1. Comparison of the extrema of vorticity ω and density ρ, given by the three functionals J_2, K_2, and I_2, in the calculation of the compressible mixing layer.

(b) Practical implementation

We now describe in greater detail the application of the self-adaptive coordinate transformation to the solution of an evolution equation of the type

$$\partial_t u = H(u), \quad -1 < x < 1, \ t > 0, \tag{8.27}$$

where H is some spatial differential operator. Equation (8.27) is assumed to be supplemented with suitable initial and boundary conditions. We introduce the mapping (8.8), that is, $x = f(\xi, A)$, where the parameter vector $A = (a_1, a_2)$ has to be determined such that the solution in the ξ-domain is accurately represented by a polynomial of reasonable degree N. Equation (8.27) is transformed by the above mapping according to

$$\partial_t v = \overline{H}(v), \quad -1 < \xi < 1, \ t > 0, \tag{8.28}$$

where $v(\xi, t) = u(x, t)$ and $\overline{H}$ is the transformed operator. The x-space is called the physical space while the ξ-space is called the computational space. Equation (8.28) associated with initial and boundary conditions, is solved by the Chebyshev collocation method (or, alternatively, by the tau method). Let $v_N^n(\xi)$ be the polynomial approximation to $v(\xi, t)$ at time $n \Delta t$. We assume that, at this time, the value of the parameter vector involved in the mapping is A and we denote by $F(v)$ the functional to be minimized, namely, J_2 or I_2. The approximate solution $v_N^n(\xi)$ can be expressed as

$$v_N^n(\xi) = \sum_{k=0}^{N} \hat{v}_k^n T_k(\xi) \tag{8.29}$$

and, in terms of transformed Chebyshev polynomials in the physical space, by

$$u^n(x) = v_N^n\left(f^{-1}(x, A)\right) = \sum_{k=0}^{N} \hat{v}_k^n T_k\left(f^{-1}(x, A)\right). \tag{8.30}$$

Note that $u^n(x)$ is a polynomial only if $f^{-1}(x, A)$ is itself a polynomial, this is not generally the case.

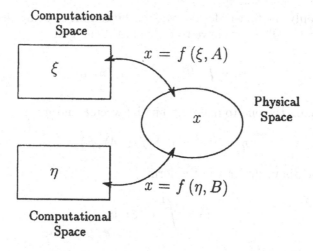

FIGURE 8.6. Coordinate transformations.

We then have to compute the approximate solution at time $(n+1)\Delta t$. The necessity of changing the parameter A is tested by the criterion

$$\frac{F(v_N^n) - F_{ref}}{F_{ref}} < \varepsilon, \qquad (8.31)$$

where $\varepsilon \ll 1$ and F_{ref} is the minimum of the functional found at the last adaptation, when A was chosen. The nonverification of the criterion (8.31) means that the coordinate transformation is no longer adapted and that a new value of the parameter has to be determined. Let us denote by B this new value, by $x = f(\eta, B)$ the new mapping (see Fig.8.6), and by $V^n(\eta)$ the expression of $u^n(x)$ in the new computational space, namely,

$$V^n(\eta) = u^n(f(\eta, B)) = v_N^n\left(f^{-1}(f(\eta, B), A)\right). \qquad (8.32)$$

For each η the quantity $V^n(\eta)$ can be computed according to the following expression

$$V^n(\eta) = \sum_{k=0}^{N} \hat{v}_k^n T_k\left(f^{-1}(f(\eta, B), A)\right). \qquad (8.33)$$

Note that $V^n(\eta)$ is not a polynomial in η except if $B = A$. Then the parameter B is determined by the minimization of the functional $F(V^n)$. A crucial point of the adaptive method is the actual computation of this functional. Even if $V^n(\eta)$ is known for every η from Eq.(8.33), the exact calculation of $F(V^n)$ is not possible and, in practice, we have to use a quadrature formula to evaluate the integral. An efficient way for this approximation is now described (Guillard et al. 1992). For clarity of the presentation, we describe below the calculation of the part of the functional containing only the first-order derivative, the technique being the same for the other terms. Thus, we have to evaluate the integral

$$F_1 = \int_{-1}^{1} [\partial_\eta V^n (\eta)]^2 \, w (\eta) \, d\eta \, . \tag{8.34}$$

By introducing the one-to-one mapping between ξ and η :

$$\eta = \varphi (\xi) = f^{-1} (f (\xi, A), B) \, ,$$

we write the above integral in the form

$$F_1 = \int_{-1}^{1} G (\xi) \, d\xi$$

with

$$G (\xi) = \frac{[\partial_\xi v_N^n (\xi)]^2}{\partial_\xi \varphi (\xi)} w (\varphi (\xi)) \, .$$

Since the values of the polynomial $v_N^n (\xi)$ at the Gauss-Lobatto points $\xi_i = \cos \pi i/N$, $i = 0, \ldots, N$ are known from the solution of Eq.(8.28), we can easily calculate $\partial_\xi v_N^n (\xi_i)$, thanks to the usual differentiation formula. Then, using the quadrature formula (3.14), we get

$$F_1 = \int_{-1}^{1} G (\xi) \, d\xi = \int_{-1}^{1} \frac{G (\xi)}{w (\xi)} w (\xi) \, d\xi \cong \frac{\pi}{N} \sum_{i=0}^{N} \frac{1}{\bar{c}_i} \frac{G (\xi_i)}{w (\xi_i)} \, . \tag{8.35}$$

An analogous calculation is carried out for the second-order derivative. Note that the integral F_1 could also be evaluated by means of the Clenshaw-Curtis formula (Davis and Rabinowitz, 1984) given in Appendix A.

The minimization of the functional $F (V^n)$ determines the new parameter B. Then the solution $v_N^n (\xi)$ is projected onto the new computational space using formula (8.33). Finally, from this approximate solution $V^n (\eta_i)$, the advancement in time of Eq.(8.28) determines the solution at time $(n + 1) \Delta t$.

Remarks

1. The minimization of the functional F is obtained thanks to classical algorithms (Press et al., 1992 ; Polak, 1971 ; Dennis and Schnabel, 1983). For the case where the mapping depends on a single scalar parameter, we recommend the use of the "golden section search" which is simple and efficient. For the vector case, specific algorithms (e.g., Powell's method or conjugate gradient methods) have to be considered. However, the recource to the two-parameter minimization is avoided when the location of the rapid variation (i.e., the parameter a_1) can be detected with sufficient accuracy. This is done generally by searching for the maximum of some

characteristic quantity, for example, the reaction rate in combustion problems or the vorticity in mixing layer problems. Therefore, the parameter a_1 being determined in this way, it remains to calculate a_2 by a one-parameter minimization technique.

2. In the computation of $F(V^n)$, it may be useful to apply a smoothing filter to the function $v_N^n(\xi)$, namely,

$$\bar{v}_N^n(\xi) = \sum_{k=1}^N \sigma_k \, \hat{v}_k^n \, T_k(\xi) \,,$$

where the filter σ_k may be the raised cosine filter $\sigma_c(k/N)$ defined by Eq.(8.67). This filtering only serves to define the new mapping and is not used to calculate the unknowns at subsequent times. The utility of smoothing appears by considering the behaviour of $v_N^n(\xi)$. It is clear from Eq.(8.33) that if the function $v_N^n(\xi)$ is not well-behaved, then the function $V^n(\eta)$ will be also ill-behaved for some values of the parameter B. Therefore, for these values, the computation of the functional $F(V^n)$ may be very inaccurate, leading, for example, to the existence of several local minima. As a consequence, the minimization algorithm may be unsuccessful or may converge to a bad value. But it is precisely when $v_N^n(\xi)$ becomes ill-behaved that a new change of variable is calculated. The smoothing avoids this kind of inconvenience.

3. In addition to the adaptive algorithm described above, the functional can also be used to monitor the number of collocation points by requiring the error to be lower than a prescribed level (Bayliss *et al.*, 1989). More precisely, after having determined the new parameter B, if the criterion

$$\frac{F(V^n)}{N^2} < \varepsilon' \ll 1 \tag{8.36}$$

is not satisfied, then the number N is increased.

4. In practice, Bayliss and Matkowsky (1987) and Bayliss *et al.* (1995a) add a penalty term to the functional (J_2 or I_2) in order to prevent the Jacobian of the transformation from becoming too large. The purpose is to avoid a very strong accumulation of points in the region of rapid variation to the detriment of the others. The same phenomenon was encountered by Guillard and Peyret (1988) who limit the range of variation of the parameter of the mapping (8.6). On the other hand, such a limitation was not found necessary by Guillard *et al.* (1992) who used a different way to compute the functional J_2.

5. The transformed equation (8.28) has nonconstant coefficients which change when the parameter A is changed. As a consequence, the use of an implicit scheme, associated with a direct solution method (matrix inversion or diagonalization), is not possible and an iterative procedure has to be considered. Another possibility is to use an explicit scheme (Guillard *et al.*, 1992).

8.3   Stiff problem: domain decomposition approach

A complementary way to deal with stiff problems is the domain decomposition approach. We refer to the next chapter for a detailed description of the domain decomposition method and its application to the Navier-Stokes equations. In the present section we only want to discuss the various ways, to apply the domain decomposition concept to the calculation of solutions, which vary rapidly.

The domain decomposition method consists of splitting the computational domain into subdomains. The equation under consideration is solved in each subdomain and the global solution is obtained by prescribing suitable transmission conditions through the interface separating two adjacent subdomains. A straightforward way to apply the domain decomposition approach is to adapt the size of the domains and the degree of the polynomial approximation to the local level of stiffness of the solution. However, taking into account the fact that the domain decomposition method is often implemented on a parallel computer (each subdomain being assigned to one processor), the use of a different resolution in each subdomain makes the algorithm rather inefficient. In this case, it is better to vary on the size of the subdomains (i.e., the location of the interfaces) and/or on adapted local coordinate transformations. According to the various possibilities we may distinguish four levels of sophistication :

level 1 : Fixed subdomains and nonadapted coordinate systems.
level 2 : Fixed subdomains and adapted coordinate transformations.
level 3 : Adapted subdomains and nonadapted coordinate systems.
level 4 : Adapted subdomains and adapted coordinate transformations.

These various options are now successively discussed.

8.3.1   *Fixed subdomains and nonadapted coordinate systems*

In this simple case, the distribution of subdomains is chosen once and for all and no adapted coordinate transformation is considered. When the location of the rapid variation region is known (and fixed with respect to time) two strategies are possible :
(1) locate the interface within the layer so as to benefit from the ability of the Chebyshev polynomial approximation to represent accurately a boundary layer solution, or
(2) locate the interface inside a subdomain of small extent and high polynomial degree (Pulicani, 1988 ; Peyret, 1990 ; Bayliss and Turkel, 1992).
On the other hand, if the location of the rapid variation is not known with precision (or if it varies with time), it is appropriate to enclose the inner layer into one (or several) subdomain(s) of small extent and high polynomial degree (Lacroix *et al.*, 1988).

As an example, we consider the solution of the one-dimensional Helmholtz problem (3.124)-(3.125). The domain $-1 \leq x \leq 1$ is divided into two or three subdomains Ω_j defined by $\alpha_{j-1} \leq x \leq \alpha_j$, $j = 1, \ldots, J$ (with $J = 2$ or $J = 3$) so that $\alpha_0 = -1$ and $\alpha_J = 1$. The solution u_j in each subdomain is calculated using continuity of the function and its derivative at an interface between two adjacent subdomains (see Section 9.2.1). The results given in Tables 8.2 and 8.3 have been obtained by using the Chebyshev tau method, N_j being the degree of the polynomial approximation in Ω_j.

The first example (Pulicani, 1988) considers the solution

$$u(x) = \frac{1}{2}[\tanh(20x) + 1] \qquad (8.37)$$

that determines f in Eq.(3.124) and $g_\pm$ in Eq.(3.125). This solution exhibits a relatively large derivative at the origin. The error $\overline{E}$ in the discrete L^2-norm [Eq.(3.126)] is given in Table 8.2 for various decompositions, but with the same total resolution characterized by N (see also Fig.9.2).

N	1-Domain	2-Domain* $\alpha_1 = 0$	3-Domain**	
			$\alpha_1 = -\alpha_2 = -0.2$	$\alpha_1 = -\alpha_2 = -0.1$
34	2.22×10^{-2}	1.36×10^{-3}	3.49×10^{-3}	2.02×10^{-4}
54	3.98×10^{-3}	1.87×10^{-5}	1.55×10^{-4}	9.29×10^{-7}
74	4.94×10^{-4}	1.12×10^{-7}	4.85×10^{-6}	1.86×10^{-8}

Table 8.2. Error $\overline{E}$ for the inner layer solution (8.37)
(* $N_1 = N_2 = N/2$, ** $N_1 = N_3 = [N/3]$, $N_2 = N - 2N_1$).

In the two-domain case $N_1 = N_2 = N/2$, and in the three-domain case $N_1 = N_3 = [N/3]$ and $N_2 = N - 2N_1$. The following conclusions can be drawn :

1. The superiority of the multidomain method over the one-domain method is obvious.

2. The accuracy associated with the three-domain method strongly depends on the extent of the inner domain which includes the region of rapid variation. Note that the chosen values of N_j are such that the polynomial degree is roughly the same in each subdomain. It is possible that another distribution of N_j would give better results.

The second example (Peyret, 1990) considers the calculation of the boundary layer solution (3.127) using the one-domain and two-domain methods. If the boundary layer is thick (say $\nu = 10^{-2}$) the one-domain solution is more accurate than the two-domain solution. This point will be discussed in Section 9.2.4. On the other hand, in the case of a very thin boundary layer (say $\nu = 10^{-4}$) the advantage of the two-domain method is obvious as shown by the results displayed in Table 8.3. However, because of the global character of the spectral approximation, the resolution in the domain Ω_1

in which the solution is smooth, must be kept at a certain level in order to preserve the overall accuracy.

N	1-Domain	2-Domain, $\alpha_1 = 0.8$	
		$N_1 = 10, N_2 = N - 10$	$N_1 = 8, N_2 = N - 8$
20	6.13×10^{-2}	1.28×10^{-3}	1.56×10^{-4}
25	1.53×10^{-2}	3.82×10^{-6}	2.71×10^{-7}
30	1.57×10^{-3}	5.49×10^{-10}	6.57×10^{-10}

Table 8.3. Error for the boundary layer solution (3.127) with $\nu = 10^{-4}$ (the boundary layer is located at $x = 1$).

8.3.2   Fixed subdomains and adapted coordinate transformations

The crude domain decomposition technique can be improved by associating it with a self-adaptive coordinate transformation in every subdomain (Garbey, 1994 ; Bayliss et al., 1995b). The two strategies considered by these authors require some knowledge of the location of the rapid variation layer.

In the first strategy the subdomains are chosen so that the layer lies at the center of a subdomain. Then the mapping (8.15) with $a_1 = 0$ is employed and the parameter a_2 is determined by minimizing the functionals I_2 or I_4 (Garbey, 1994).

The second strategy consists of choosing the subdomains so that the interface is located inside the layer of rapid variation. Thus, this latter becomes a boundary layer and the mapping (8.16) is used in the considered subdomain with the parameter a determined as above. Numerical experiments on typical examples show that the second strategy is more efficient.

8.3.3   Adapted subdomains and nonadapted coordinate systems

An alternative to the previous approach consists of adapting the interfaces between the subdomains while keeping a nonadapted coordinate system. Considering the case of a two-domain method, Bayliss et al. (1992) propose two ways for the determination of the location α_1 of the interface between the two subdomains Ω_1 and Ω_2. In the first method, the interface is determined such that the error within each subdomain is comparable. This is done by minimizing the difference $|I_2(u_1) - I_2(u_2)|$ where u_j refers to the solution in Ω_j. This minimization is obtained through a bisection technique. The second method is based on the minimization of the largest error within the different subdomains. Therefore, the interface α_1 is chosen such that the maximum of $\{I_2(u_1), I_2(u_2)\}$ is minimal. These methods can be extended to the case of several subdomains. The authors found that the first method, which tends to equalize the errors, is somewhat more flexible and robust, particularly in the case of several subdomains.

8.3.4 Adapted subdomains and adapted coordinate transformations

In this method, developed by Renaud and Gauthier (1997), the location of the interfaces between the subdomains, as well as the coordinate transformation within each subdomain, are simultaneously adapted in an automatic way. The mapping used by these authors is the one-parameter family (8.18) and the functional F is J_2 defined by Eq.(8.24).

Let us describe the adaptation procedure in the two-domain case. The subdomains Ω_j, $j = 1, 2$, are separated by the interface $x = \alpha_1$. Assume that, starting from an initial configuration (which may correspond to some time-cycle in a time-dependent problem) defined by α_1 and the parameters a_1 and a_2 corresponding, respectively, to Ω_1 and Ω_2, we want to find a new configuration $(\alpha_1^\star, a_j^\star, j = 1, 2)$ better adapted to the solution. The procedure is as follows :

1. The best location $\alpha_1^\star$ of the interface is searched for in a set of K discrete values defined by

$$\alpha_{1,k} = \left(\alpha_1 - \frac{\Delta x}{2}\right) + \frac{k-1}{K-1}\Delta x, \quad k = 1, \dots, K, \tag{8.38}$$

where Δx is the length of the interval in which the interface is allowed to move and K is the number of discrete values (e.g., $K = 10$). The interface $\alpha_{1,k}$ defines the subdomains $\Omega_{j,k}$, $j = 1, 2$.

2. For each value $\alpha_{1,k}$, $k = 1, \dots, K$, the following two steps are carried out :

 2.a. For each subdomain $\Omega_{j,k}$, $j = 1, 2$, the best value of the parameter $a_{j,k}$ is determined by minimizing the functional F, namely, $F_{j,k}$.

 2.b. The total functional $F_{T,k} = F_{1,k} + F_{2,k}$ is calculated.

3. Thus, a set of K values of $F_{T,k}$, $k = 1, \dots, K$, has been defined. The minimal value of $F_{T,k}$ obtained for $k = k^\star$, namely, $F_{T,k^\star}$, defines the best location $\alpha_1^\star = \alpha_{1,k^\star}$ of the interface and the best local mapping defined by $a_j^\star = a_{j,k^\star}$, $j = 1, 2$.

In the case of several subdomains Ω_j, $j = 1, \dots, J$, the interfaces α_j and the parameters a_j defining the mappings are determined by an iterative procedure, described in detail by Renaud and Gauthier (1997).

To illustrate the application of the method, we briefly present some results concerning the temporal development of the two-dimensional com-

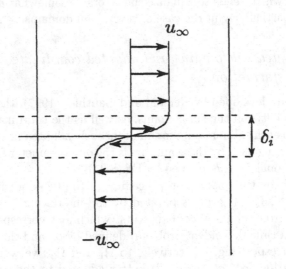

FIGURE 8.7. Initial configuration of the mixing layer.

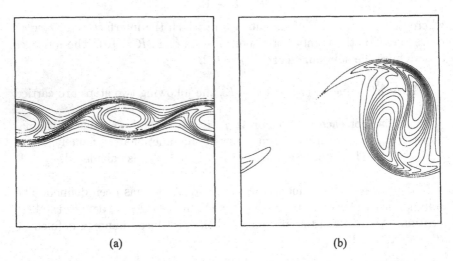

(a)   (b)

FIGURE 8.8. Instantaneous iso-vorticity lines : (a) before pairing, (b) after pairing.

pressible mixing layer already considered in Section 8.2.3. At initial time
the basic state has constant pressure and total enthalpy. Moreover, the y-
component v of the velocity is zero whereas the x-component u varies with
respect to y like a hyperbolic tangent function (Fig.8.7). The Reynolds
number (based on $2u_\infty$ and on the initial vorticity thickness δ_i) is equal
to 1000, the Mach number is 0.3, and the Prandtl number is 0.71. The
flow is assumed to be periodic in the x-direction. The length L of the
computational domain is sufficiently large to allow the presence of two vor-
tical structures, namely, L is equal to twice the most-amplified inviscid
wavelength predicted by the linear stability analysis (Blumen, 1970). This
problem is studied by solving the Navier-Stokes equations for compressible
flows using the Fourier Galerkin approximation in the x-direction and the
Chebyshev collocation approximation in the y-direction. The adaptive domain decomposition/coordinate transformation method is applied in the
y-direction, using the mean value of u in the x-direction as an argument of
the functional J_2.

Previous numerical experiments in a one-domain configuration (Malé,
1992 ; Guillard *et al.*, 1992), in which the basic state is disturbed in a random way, show initially the formation of a two-vortex structure (Fig.8.8.a)
and then the pairing of the two vortices, giving rise to only one (Fig.8.8.b).
In this case, the mode characteristic of the secondary instability and responsible for the pairing, is excited by the random perturbation. Another set of
experiments has been made (Gauthier *et al.*, 1996) using a deterministic
initial sine function disturbance such that only the mode corresponding to
the two-vortex structure is excited. Figure 8.9 shows the evolution of the
dimensionless vorticity thickness δ/δ_i obtained with the present adaptive
method and with the MUSCL and ENO finite-volume methods (see, e.g.,
Grasso and Meola, 1996 ; Shu, 1998). More precisely, the curves (1)-(4) in
Fig.8.9 correspond to :

(1) The second-order MUSCL finite-volume method (uniform mesh $251 \times$
    201).

(2) The third-order ENO finite-volume method (uniform mesh 201×201).

(3) The adaptive multidomain spectral method (100 Fourier modes in the
    x-direction, three subdomains in the y-direction with 51 collocation
    points in each one).

(4) The adaptive multidomain spectral method (128 Fourier modes in
    the x-direction, five subdomains in the y-direction with 51 collocation
    points in the three central domains and 31 in the farthest ones).

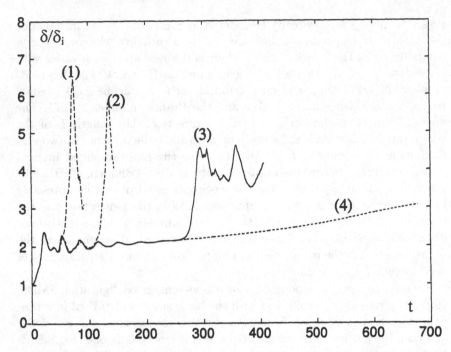

FIGURE 8.9. Evolution of the vorticity thickness given by various methods : (1) second-order MUSCL finite-volume method, (2) third-order ENO finite-volume method, (3) adaptive three-domain spectral method, (4) adaptive five-domain spectral method.

The peak of each curve characterizes the vortex pairing. In the present experiments the corresponding mode is not excited so that only numerical (approximation and round-off) errors are responsible for its formation. This explains why pairing appears later when the computation is more accurate. In particular, the solution given by the five-domain method shows no pairing, at least during the time of integration. It seems that the effect of round-off errors is not perceptible during this time and, because the flow is largely damped by dissipation, it is unlikely that the mode under consideration would significantly increase. In this computation we have obtained an approximate solution which reproduces faithfully the characteristics of the exact solution of the mathematical problem. Note that computations done with the five-domain method but with an initial random perturbation shows the vortex pairing.

This example is interesting since it clearly shows the possible existence of physically realistic behaviours due only to a numerical artefact. A low-order method can give results which seem to be physically realistic while they are wrong from the mathematical point of view! In a general way, we do not think that it is reasonable to rely upon the numerical errors to act as physical mechanisms.

8.4   Singular problems

In a large number of flow problems of practical interest, the solution is not regular and possesses only a small number of bounded derivatives. Even for incompressible viscous flows in a simple rectangular cavity, the solution may not be infinitely differentiable. A classical example is given by the existence of an infinite sequence of eddies in a corner with no-slip boundary conditions (Moffatt, 1964). In this situation, the second-order derivatives of the vorticity become infinite at the corner (Section 3.6), but this singularity is sufficiently weak so as not to necessitate special treatment.

In a general way, singularities of the solution of elliptic problems may be due to various reasons : (i) geometrical singularity of the boundary of the computational domain, (ii) discontinuity of the boundary conditions, (iii) abrupt change of the type of boundary conditions (e.g., Dirichlet/Neumann conditions), (iv) singularity of the forcing term, (v) incompatibility of the forcing term with the boundary conditions (e.g., the solution of $\nabla^2 u = -1$ in a corner with $u = 0$ as a boundary condition).

It is striking that singularities are often introduced by mathematical modelling of the flow and are not physically realistic. Let us consider, for example, the flow inside a cylinder chamber where one of its sides is compressed under the motion of a piston (Botella and Peyret, 2001). Its mathematical modelling introduces singularities at the boundary: more precisely, the velocity of the fluid is discontinuous at the contact points between the moving piston and the stationary wall chamber. This modelling leads to a physically unrealistic flow since, by following the analysis of Taylor (1962), an infinite force is required to move the piston and, in particular, the stress is infinite at the contact points. We may certainly think that, in reality, the piston does not make contact with the wall, and that there exists a small gap that enables an exchange of fluid between the chamber and the external cylinder. Nevertheless, the fluid leakage must be small enough so that the real-life problem is close to the modelled one. Analogous unrealistic boundary conditions are considered in the lid-driven cavity flow (Section 8.4.2) that, in spite of the presence of singularities, has become one of the reference problems for evaluating numerical methods in fluid dynamics. For this last problem, where the motion of fluid in a rectangular cavity is generated by the uniform sliding of one of its walls, the validity of the modelling has been justified by Hansen and Kelmanson (1994).

The presence of singularities makes the numerical solution much less accurate than expected with a spectral method. The "infinite" accuracy is lost and the rate of convergence, which depends on the regularity of the solution, is only algebraic (see Sections 2.1.2 and 3.6). For a solution with low regularity, the use of spectral methods shows little advantage over local approximation methods. Therefore, if one wants to use spectral methods, a suitable treatment of the singularities is mandatory for preserving, as far as possible, the high accuracy of the methods.

When the boundary conditions or the forcing term are discontinuous, a process of regularization is sometimes used. The technique consists of multiplying the boundary condition by a regularizing function in order to smooth the discontinuity. The regularizing function, however, must be sufficiently stiff so as not to modify excessively the solution to be computed. Therefore, the singular problem is replaced by a stiff one, with the difficulties discussed in the previous section. An evaluation of the regularization procedure in the case of the lid-driven cavity flow has been made by Batoul *et al.* (1994). Note that smoothing of the discontinuity can also be obtained by a filtering technique (Section 8.5). Although the technique of regularization may be useful in some problems, and may sometimes be considered as physically realistic, we suggest using it with caution.

The aim of the present section is to discuss two ways handling singular solutions with Chebyshev polynomial approximations. We are concerned with the usual situation where the location of the singularity is known. In a general way, the singular points belong to the boundary except in the special case (iv) mentioned above when the forcing term exhibits an inner singularity.

The first approach consists of using a domain decomposition method in order to locate the singular point at a corner of a subdomain. As a matter of fact, it is known that the rate of decay of the interpolation error for corner singularities is essentially twice as fast than for edge singularity. This doubling of the rate has been proven theoretically by Bernardi and Maday (1991) for the spectral-Legendre method. It has been numerically experienced by Botella and Peyret (2001) in the case of the Chebyshev method, and by Pathria and Karniadakis (1995) for the spectral-element Legendre method, following the theoretical results stated by Babuška and Suri (1987, 1994) for the $h - p$ method.

The second approach is based on the subtraction of the leading part of the asymptoptic expansion of the solution near the singular point. Therfore, the solution to be numerically computed is less singular than the original solution, even if it is not completely regular. The subtraction method requires the *a priori* knowledge of the asymptotic expansion. For nonhomogeneous boundary conditions, the leading term of the asymptotic expansion is completely determined and the application of the subtraction method is quite simple. On the other hand, when the boundary conditions are homogeneous the leading term itself is not completely determined by the local analysis and depends on some multiplicative constant that can only be determined by considering the global problem. In this case, the application of the subtraction method is much more complicated since it requires some connection between the local asymptotic solution and the global numerical one. This question has been the subject of a large number of studies in finite-difference and finite-element methods. Here, we expose a simple iterative procedure adapted to the Chebyshev

method. It must be pointed out that, even if the boundary conditions
are nonhomogeneous so that the leading term is completely determined,
the higher-order terms may depend on some undetermined constants. In
this case, the constants have to be determined as part of the global solution.

These two approaches are presented in the following sections : the Laplace
equation in Section 8.4.1 and the Navier-Stokes equations (in vorticity-
streamfunction and velocity-pressure formulations) in Section 8.4.2.

8.4.1 Singular solution of the Laplace equation

As a typical example of a singular solution to the Laplace equation we consider the case where the type of boundary conditions abruptly changes on
one side of the computational domain. The problem, known as the "problem of Motz," is defined by

$$\nabla^2 u = 0, \quad -1 < x < 1, \quad 0 < y < 1 \tag{8.39}$$

$$u(-1, y) = 500, \quad \partial_x u(1, y) = 0 \tag{8.40}$$

$$\partial_y u(x, 0) = 0 \quad \text{for } -1 < x < 0, \quad u(x, 0) = 0 \quad \text{for } 0 \le x < 1 \tag{8.41}$$

$$\partial_y u(x, 1) = 0. \tag{8.42}$$

The solution near the origin $(0, 0)$ behaves like $r^{1/2} \sin \theta/2$ where (r, θ) are
local polar coordinates such that $x = r \cos \theta$, $y = r \sin \theta$. Therefore, the
first-order derivatives of u become infinite at the origin O like $r^{-1/2}$. We
refer to Grisvard (1985) for the determination of asymptotic solutions near
singular points. Note that the solution of (8.39)-(8.42) is regular at the
corners.

(a) Classical solution method
Problem (8.39)-(8.42) is approximated by means of the Chebyshev collocation method associated with the influence matrix technique (Remark 1
of Section 3.7.2) allowing the use of the matrix-diagonalization procedure.
Note that an affine transformation is required in the y-direction to map the
interval $[0, 1]$ onto $[-1, 1]$. Computations are done with $N_x = N_y = 100$,
that is, with high resolution. Figure 8.10.a shows the graphs of $u(x, 0.005)$
and $\partial_x u(x, 0.005)$ calculated on 201 equispaced points using the polynomial representation of the solution (see Section 7.4.4.d). This representation
makes clear the presence of the Gibbs oscillations that destroy the accuracy
of the solution.

(b) Domain decomposition method
Now the problem (8.39)-(8.40) is solved by means of the domain decomposition method described in Chapter 9. The domain $\Omega = \{-1 < x < 1,$
$0 < y < 1\}$ is split into two square subdomains Ω_1 and Ω_2 whose interface is located at the singular point $x = 0$. In each subdomain Ω_j the

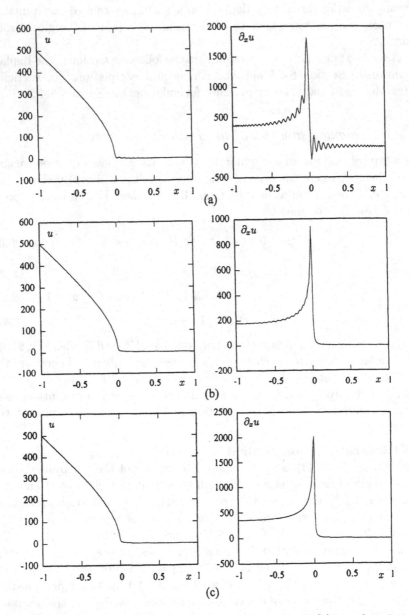

FIGURE 8.10. Problem of Motz : (a) one-domain method, (b) two-domain method, (c) one-domain subtraction method.

solution is approximated with polynomials of degree N_{xj} and N_y in the x-
and y-directions, respectively. The singularity is now located at a corner of
the subdomains. It is clear that the representation of the solution in the
x-direction by two polynomials, one in Ω_1 and the other in Ω_2, is better
adapted than the one using a single polynomial whose derivative should
approach infinity at the inner point $x = 0$ on the side $y = 0$. Figure 8.10.b
displays the results obtained with $N_{x1} = N_{x2} = 50$, $N_y = 100$. The improvement with regard to the classical one-domain solution is undeniable,
but some oscillations remain.

(c) Subtraction method

The asymptotic expansion near the singular point $(0,0)$ of the solution to
the problem (8.39)-(8.42) is of the form

$$u^\star = \sum_{k \geq 0} A_k \, r^{(2k+1)/2} \sin\left(\frac{2k+1}{2}\theta\right), \qquad (8.43)$$

where k is an integer. The constants A_k are unknown and must be determined by the global problem. We now give the application of the method
based on the subtraction of the first term

$$u_0^\star = A_0 \, r^{1/2} \sin\frac{\theta}{2} = \frac{A_0}{\sqrt{x^2 + y^2}} \sin\left(\frac{1}{2}\tan^{-1}\frac{y}{x}\right) \equiv A_0 v\,(x,y)\,. \quad (8.44)$$

This function $u_0^\star$ satisfies the equation (8.39) and the boundary conditions
(8.41) but, obviously, not the other boundary conditions.

We set

$$u = u_0^\star + \overline{u} = A_0 v + \overline{u} \qquad (8.45)$$

so that $\overline{u}$ is less singular than u since it behaves like $r^{3/2}$ near the origin
O. The part $\overline{u}$ is the solution of the transformed problem

$$\nabla^2 \overline{u} = 0, \quad -1 < x < 1, \ \ 0 < y < 1 \qquad (8.46)$$

$$\overline{u}\,(-1,y) = 500 - A_0 v\,(-1,y)\,, \quad \partial_x \overline{u}\,(1,y) = -A_0 \partial_x v\,(1,y) \qquad (8.47)$$

$$\partial_y \overline{u}\,(x,0) = 0 \ \text{for} \ -1 < x < 0, \quad \overline{u}\,(x,0) = 0 \ \text{for} \ 0 \leq x < 1 \qquad (8.48)$$

$$\partial_y \overline{u}\,(x,1) = -A_0 \partial_y v\,(x,1)\,. \qquad (8.49)$$

The constant A_0 is unknown and is determined as part of the global
solution through the following iterative procedure.

If A_0^m is assumed to be known, the solution $\overline{u}^{m+1}$, with $m \geq 0$, is determined by

$$\nabla^2 \overline{u}^{m+1} = 0, \quad -1 < x < 1, \ \ 0 < y < 1 \qquad (8.50)$$

$$\overline{u}^{m+1}\,(-1,y) = 500 - A_0^m v\,(-1,y)\,, \quad \partial_x \overline{u}^{m+1}\,(1,y) = -A_0^m \, \partial_x v\,(1,y) \qquad (8.51)$$

$$\partial_y \overline{u}^{m+1}\,(x,0) = 0 \ \text{for} \ -1 < x < 0, \quad \overline{u}^{m+1}\,(x,0) = 0 \ \text{for} \ 0 \leq x < 1 \qquad (8.52)$$

$$\partial_y \overline{u}^{m+1}(x,1) = -A_0^m \, \partial_y v(x,1) . \tag{8.53}$$

Then A_0^{m+1} is determined by requiring the solution

$$u^{m+1} = A_0^m v + \overline{u}^{m+1} \tag{8.54}$$

to coincide with the asymptotic solution $A_0^{m+1} v$ in some neighbourhood of the origin O. Since a single constant has to be determined, the neighbourhood could be reduced to one point chosen sufficiently close to O. This way can be efficient in some simple cases (Botella and Peyret, 2001). However, it is generally too much dependent on the choice of the selected point. It is much better to consider a least-squares method based on several points (X_l, Y_l), $l = 1, \ldots, L$, distributed on the semicircle γ of center O and radius $\rho \ll 1$. Therefore, A_0^{m+1} is defined as the value that minimizes the expression

$$I\left(A_0^{m+1}\right) = \sum_{l=1}^{L} \left(u_l^{m+1} - A_0^{m+1} v_l\right)^2 ,$$

where $u_l^{m+1} = u^{m+1}(X_l, Y_l)$ and $v_l = v(X_l, Y_l)$. Then by prescribing

$$\frac{dI\left(A_0^{m+1}\right)}{dA_0^{m+1}} = 0 ,$$

we get the expression

$$A_0^{m+1} = \frac{\sum_l u_l^{m+1} v_l}{\sum_l v_l} .$$

Now the elimination of u_l^{m+1}, taking Eq.(8.54) into account, yields

$$A_0^{m+1} = A_0^m + \frac{\sum_l \overline{u}_l^{m+1} v_l}{\sum_l v_l} . \tag{8.55}$$

In fact, the convergence of the iterative procedure is very much improved by introducing a relaxation parameter α whose choice depends on the value of the denominator $\sum_l v_l$. Therefore, Eq.(8.55) is replaced by

$$A_0^{m+1} = A_0^m + \alpha \frac{\sum_l \overline{u}_l^{m+1} v_l}{\sum_l v_l} . \tag{8.56}$$

The iterative procedure is stopped when

$$\frac{\left|A_0^{m+1} - A_0^m\right|}{\left|A_0^{m+1}\right|} < \varepsilon , \tag{8.57}$$

where ε is a given small number. This procedure is initiated by choosing A_0^0 as the value obtained by making $A_0^0\, v$ equal to the numerical solution u given by the classical method (without subtraction) at the first collocation point near $x = 0$ on the vertical axis.

The application of formula (8.56) necessitates the evaluation of $\overline{u}^{m+1}$ at points (X_l, Y_l) that are not collocation points. This is done by using the polynomial approximation to $\overline{u}^{m+1}$ defined for any (x,y) belonging to the domain Ω and, in particular, for any (X_l, Y_l) (see Section 7.4.4.d). This constitutes a very valuable property of spectral methods which may be used in various circumstances. In the present case, it allows us to take the semicircle γ as close to the singular point O as needed. Indeed, the accuracy of the determination of A_0 depends on the value of ρ since it is assumed that the asymptotic solution $A_0 v$ and the exact solution coincide at distance ρ. That is to say, the terms of the expansion (8.43) which are not considered in the algorithm are negligible. We must notice, however, that the accuracy of the final solution $u = A_0 v + \overline{u}$ is generally better than the accuracy with which A_0 is determined. The reason is that the consideration of $A_0 v$ serves uniquely to make the computed solution $\overline{u}$ less singular. A small inaccuracy in the calculation of A_0 means only that the most singular part of u has not been completely subtracted. The effect on the computed solution $\overline{u}$ is rather weak as observed in the numerical experiments, some results of which are reported now.

To illustrate the subtraction method, we report the results obtained by Raspo (private communication) using the Chebyshev-collocation method (with $N_x = 80$, $N_y = 40$) for the solution of problem (8.50)-(8.53). Computations were done with various values of the radius ρ, the number L of points on γ, and the relaxation parameter α. A preliminary solution obtained without subtraction gives the initial value $A_0^0 = 400.0754$. For the typical values $\rho = 10^{-4}$, $L = 9$, and $\alpha = 20$, the convergence with $\varepsilon = 10^{-8}$ is reached after 11 iterations. The constant A_0 is found to be equal to 401.2326. This value has to be compared with the value 401.163 obtained by Wigley (1988). This author, using a low-resolution finite-difference method, has computed the first 14 constants A_k, $k = 0, \ldots, 13$, by employing an accurate method based on the Green theorem. For comparison, values of the solution u at some typical points (x_n, y_n) are given in Table 8.4. The profiles of u and $\partial_x u$ along the line $y = 0.005$, displayed in Fig.8.10.c, are free of oscillations.

It must be remarked that the method also applies when more than one term of the asymptotic expansion are subtracted. The least-squares technique provides a system of algebraic equations for determining the constants A_k. In the present case, numerical results showed no significant change by also subtracting the second term $u_1^\star = A_1\, r^{3/2} \sin(3\theta/2)$.

Note that the subtraction method can be combined with the domain decomposition approach to benefit from the advantages of both approaches.

(x_n, y_n)	$\left(\dfrac{2}{7}, \dfrac{2}{7}\right)$	$\left(0, \dfrac{2}{7}\right)$	$\left(\dfrac{1}{28}, \dfrac{1}{28}\right)$	$\left(0, \dfrac{1}{28}\right)$	$\left(-\dfrac{1}{28}, \dfrac{1}{28}\right)$
Spectral method	78.559	141.559	33.587	53.182	83.671
Wigley (1988)	78.559	141.560	33.592	53.186	83.671

Table 8.4. Numerical solution to the problem of Motz at various points (x_n, y_n).

(d) Remark on corner singularities

The example discussed above concerns the case where the computational domain is a rectangle. However, as already mentioned, singularities are often due to irregularities in the shape of the boundary of the domain, essentially reentrant corners (e.g., L-shaped domain). In these circumstances, the use of a domain decomposition method is indispensable. Then the singularity may be subtracted as explained above.

Pathria and Karniadakis (1993) have adapted the "auxiliary mapping method" (Babuška and Oh, 1990) to the spectral-element approximation. The basic idea is to introduce an appropriate mapping in the elements containing a singular corner whose effect is to weaken the singularity. Then, if the resulting singularity is sufficiently weak, the transformed solution can be computed just as it is. If not, the subtraction method can be applied.

8.4.2   Navier-Stokes equations

In this section, we discuss the two approaches (domain decomposition and subtraction methods) for handling singularities in the solution of the Navier-Stokes equations. The first example considers the singularity due to the abrupt change (no-slip/stress-free) of the boundary condition on one side of the computational domain (the so-called "stick-slip" problem). The second example concerns the classical lid-driven cavity flow in which singularities are due to discontinuities of the velocity at two corners of the domain.

(a) Domain decomposition method : stick-slip problem

This example comes from the Czochralsky crystal growth technique (Raspo et al., 1996). The mathematical modelling of the problem makes use of the axisymmetric Navier-Stokes equations (5.33), (5.36), and (5.39) supplemented with the temperature equation (5.41) in the rectangular domain $0 < x < R$, $0 < z < H$ where x and z are cylindrical coordinates. The part $0 < x < a_1$ of the side $z = H$ is constituted by a no-slip wall (crystal) whereas the complementary part $a_1 \leq x < R$ is a stress-free boundary (free surface of the melt). Therefore, the vorticity ω is singular at the contact point B between the crystal and the free surface and it behaves like $r^{-1/2}$

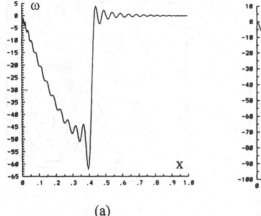

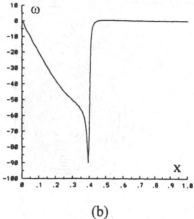

(a)   (b)

FIGURE 8.11. Vorticity profiles under the crystal-free surface for the stick-slip problem : (a) one-domain method, (b) two-domain method.

where r is the distance to B. Moreover, because the crystal is rotating, the azimuthal velocity v is singular at point B since its derivative behaves like $r^{-1/2}$. A continuous (linear) profile of temperature is prescribed on $z = H$. The rotation of the crystal is characterized by a Reynolds number Re and the magnitude of the heating by a Grashof number Gr.

The problem is solved by using the two-domain decomposition method described in Section 9.5.2. The interface is located at $x = a_1$ in order to transfer the singularity at the corner of the subdomains (see Fig.9.3.a). Figure 8.11 displays the vorticity profiles (interpolated on 201 equispaced points) along the second line of collocation points under the crystal-free surface boundary. The resolution is 65×65 for the one-domain solution (Fig.8.11.a) and 33×65 in each subdomain for the two-domain solution (Fig.8.11.b). The interest of using a domain decomposition method clearly appears when comparing the two profiles. Figure 8.12 shows the iso-stream-function lines (projection of the streamlines on the meridian plane) obtained for $Re = 2000$ and $Gr = 10^5$. It is observed that the flow patterns are completely different in the two solutions. In fact, for this kind of flow, there exists a threshold for which the convective regime changes type and the loss of accuracy of the one-domain solution has the effect of delaying the appearance of this change. Therefore, the loss of accuracy leads not only to oscillations but also to a shift in the threshold. This last point is of importance when physical conclusions are drawn from numerical results and illustrates, once more, the crucial effect of accuracy in some flow computations.

(b) Subtraction method : lid-driven cavity flow
This example concerns the well-known two-dimensional lid-driven cavity flow. The Navier-Stokes equations (5.1)-(5.2) with $\mathbf{f} = 0$ are solved in the

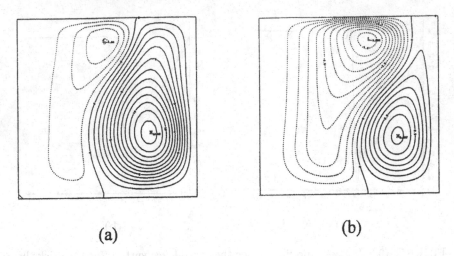

(a)   (b)

FIGURE 8.12. Iso-streamfunction lines for the stick-slip problem : (a) one-domain method, (b) two-domain method.

domain $0 \leq X, Y \leq 1$, with the boundary conditions : $\mathbf{V} = (u, v) = (-1, 0)$ on the side $Y = 1$, $0 < X < 1$, and $\mathbf{V} = \mathbf{0}$ on the other three sides. In this problem, the Reynolds number Re is based on the unit length of the square and on the sliding velocity, such that $\nu = Re^{-1}$ in Eq.(5.1). We consider here the case where the ultimate flow is steady, in particular, for $Re = 1000$, but the method applies as well to the calculation of unsteady flows obtained for larger values of Re (Botella and Peyret, 2001).

At corners $s = A = (0, 1)$ and $B = (1, 1)$, where the velocity is discontinuous, the solution is singular. The application of the subtraction method requires knowledge of the asymptotic expansion of the solution valid near the singular points. To determine this expansion it is convenient to introduce the streamfunction ψ and to consider a local coordinate system (r, θ) defined by $X = X_s + r \cos \theta$, $Y = Y_s + r \sin \theta$, where (X_s, Y_s) corresponds to the singular point s. The equation satisfied by the streamfunction (in the steady case) is

$$\nabla^4 \psi = Re\, \mathbf{V}.\nabla \left(\nabla^2 \psi \right) , \tag{8.58}$$

where $\mathbf{V} = (u, v)$ is connected to ψ by the relations $u = \partial_Y \psi$, $v = -\partial_X \psi$.

In the neighbourhood of the point s, the general solution to Eq.(8.58) can be written as the asymptotic expansion

$$\psi^s = \sum_{k \geq 1} \psi_k^s , \tag{8.59}$$

where each term ψ_k^s is less singular than the previous one ψ_{k-1}^s. For the present problem, we have

$$\psi_k^s = r^{\alpha_k} f_k (\theta) ,$$

where the functions f_k are smooth and the exponents α_k (which can be complex) are such that

$$1 \leq \mathcal{R}e\,(\alpha_1) < \mathcal{R}e\,(\alpha_2) < \dots.$$

Note that, for some problems, the asymptotic sequence r^{α_k} is not sufficient and logarithmic terms like $r^{\alpha_k} \ln^{\beta_k} r$ have to be added (Moffatt and Duffy, 1980 ; Hancock *et al.*, 1981 ; Botella and Peyret, 2001).

A systematic procedure (Botella and Peyret, 2001) allows us to determine the functions ψ_k, namely α_k and $f_k\,(\theta)$, so that the local boundary conditions near the corners s are satisfied. We find the expansion

$$\psi^s = r\,f_1\,(\theta) + r^2 f_2\,(\theta) + r^3 f_3\,(\theta) + r^{\lambda_1} f_4\,(\theta) + \dots. \tag{8.60}$$

The first term $\psi_1^s = r f_1$ is a solution to the Stokes problem ($Re = 0$), while the following two terms are associated with the Navier-Stokes equations and, consequently, depend on the Reynolds number Re. All these terms are completely determined by the local analysis. On the other hand, the fourth term $\psi_4^s = r^{\lambda_1} f_4$ (where $\lambda_1 \simeq 3.74 + 1.13i$), which is an eigenfunction of the Stokes problem, is determined up to a multiplicative constant. The subtraction of the first two terms suffices to give accurate results. The expression of $f_1\,(\theta)$ and $f_2\,(\theta)$ and the resulting expressions of the velocity and pressure are given by Gupta *et al.* (1981) and by Botella and Peyret (1998). The behaviour near s of the velocity $\mathbf{V}$ and the pressure p is easily deduced from (8.60) using the relations expressing $\mathbf{V}$ in terms of ψ, and the momentum equations for determining p.

At the corners $C = (1,0)$ and $D = (0,0)$, where the boundary conditions are continuous, since noslip conditions are prescribed on the corresponding sides, the behaviour of the streamfunction is given by the fourth term ψ_4^s discussed above [see Eq.(3.124)]. At these points, the second-order derivatives of the pressure and the third-order derivatives of the velocity are infinite. This singularity is sufficiently weak so that it need not to be subtracted, as for corners A and B.

Let $(\mathbf{V}^\star, p^\star)$ be the most singular part of the solution such that $\mathbf{V}^\star = \mathbf{V}^A + \mathbf{V}^B$, $p^\star = p^A + p^B$ where $(\mathbf{V}^s, p^s)$ refers to the first K terms of the asymptotic solution near the singular point $s = A$ or $s = B$ and is deduced from the expansion (8.59) or, equivalently, (8.60) :

$$\mathbf{V}^s = \sum_{k=1}^{K} \mathbf{V}_k^s, \quad p^s = \sum_{k=1}^{K} p_k^s.$$

Now the part $(\mathbf{V}^\star, p^\star)$ is subtracted, namely,

$$\mathbf{V} = \mathbf{V}^\star + \overline{\mathbf{V}}, \quad p = p^\star + \overline{p},$$

so that the resulting field $(\overline{\mathbf{V}}, \overline{p})$ is more regular than $(\mathbf{V}, p)$. The field $(\overline{\mathbf{V}}, \overline{p})$ satisfies the set of (dimensionless) equations

$$\partial_t \overline{\mathbf{V}} + \mathbf{N}\left(\overline{\mathbf{V}}, \mathbf{V}^\star\right) - \frac{1}{Re} \nabla^2 \overline{\mathbf{V}} + \nabla \overline{p} = \mathbf{F}^\star \qquad (8.61)$$

$$\nabla \cdot \overline{\mathbf{V}} = 0, \qquad (8.62)$$

that is, similar to the Navier-Stokes system. The term $\mathbf{N}$ comes from the nonlinear term $\mathbf{A}(\mathbf{V}, \mathbf{V})$. In the case where the convective form is used (as in the computations reported below), we have

$$\mathbf{N}\left(\overline{\mathbf{V}}, \mathbf{V}^\star\right) = \left(\overline{\mathbf{V}} \cdot \nabla\right) \overline{\mathbf{V}} + \left(\mathbf{V}^\star \cdot \nabla\right) \overline{\mathbf{V}} + \left(\overline{\mathbf{V}} \cdot \nabla\right) \mathbf{V}^\star.$$

The vector $\mathbf{F}^\star$ depends only on $\mathbf{V}^\star$. Equations (8.61) and (8.62) have to be solved in the domain $0 < X, Y < 1$, with the boundary condition

$$\overline{\mathbf{V}} = \mathbf{g}^\star \qquad (8.63)$$

deduced from the one imposed on $\mathbf{V}$. Since only the steady solution is of interest, the initial condition $\overline{\mathbf{V}} = \overline{\mathbf{V}}_0$ is any solenoidal vector field satisfying the compatibility condition of type (5.11). The results reported below have been obtained by Botella and Peyret (1998, 2001) by means of the Chebyshev-collocation $I\!P_N - I\!P_{N-2}$ projection method described in Section 7.4.2.b. We denote by NS0 the solution obtained without subtraction, by NS1 the solution obtained by subtracting the first (Stokes) term of the asymptotic expansion (8.60), and by NS2 that obtained by subtracting the first two terms. Near the singular point A, the computed solution $(\overline{\mathbf{V}}, \overline{p})$ behaves according to :
    NS1 solution :

$$\overline{\mathbf{V}} + \mathbf{V}^B \sim Re\, r\, \mathbf{F}_2(\theta), \quad \overline{p} + p^B \sim \ln r\, G_2(\theta),$$

    NS2 solution :

$$\overline{\mathbf{V}} + \mathbf{V}^B \sim Re^2\, r^2\, \mathbf{F}_3(\theta), \quad \overline{p} + p^B \sim Re\, r\, G_3(\theta),$$

the fields $\mathbf{V}^B$ and p^B being smooth in this region. Analogous behaviours are obviously observed near B.

    Figure 8.13 compares the iso-vorticity lines of the NS0 and NS2 solutions (on the scale of the figure, the NS1 solution is indistinguishable from the NS2 solution).

    When using a spectral method, a valuable indication on the accuracy of the solution is given by the decay of the expansion coefficients. Figure 8.14 displays the spectrum of the computed vorticity $\overline{\omega}$ (ω in the NS0 case) on the wall $X = 0$ (Fig.8.14.a) and on the vertical centerline $X = 0.5$

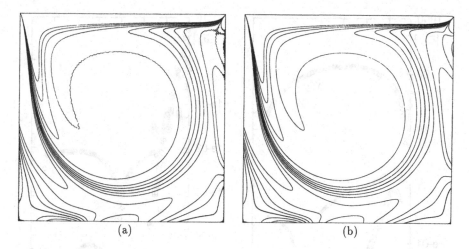

FIGURE 8.13. Iso-vorticity lines of the lid-driven cavity flow at $Re = 1000$:
(a) NS0 solution, (b) NS2 solution.

N	Pressure p		Vorticity ω	
	no filter	filter $\sigma_{v,12}$	no filter	filter $\sigma_{v,12}$
48	0.3561904	0.3570182	-84.24558	-83.86083
64	0.3556856	0.3553056	-83.47874	-83.82833
96	0.3551954	0.3552367	-83.70878	-83.82041
128	0.3552255	0.3552370	-83.88577	-83.82035
160	0.3552442	0.3552699	-83.82351	-83.82035

Table 8.5. Convergence of the pressure and vorticity at point $(0, 0.95)$
close to the singular point $A = (0, 1)$.

(Fig.8.14.b). The improvement brought by the subtraction method is man-
ifest. The tail of the spectra, however, exhibits a saturation indicating a pol-
lution effect due to the presence of the singularities. Nevertheless, Fig.8.14.b
shows clearly the exponential decay of the first part of the spectrum.

It is instructive to also examine the values of the solution near the singular points. Table 8.5 shows the values of the pressure and vorticity at point
$(0, 0.95)$ close to the singular point A. Note that convergence is improved
by using the filter $\sigma_{v,p}$ proposed by Vandeven (1991) and discussed in the
next section.

Now, taking the high-resolution NS2 solution $(\overline{\mathbf{V}}_N, \overline{p}_N)$, computed with
$N = 160$ as a reference solution $(\overline{\mathbf{V}}_{ref}, \overline{p}_{ref})$ so that

$$\mathbf{V}_{ref} = \mathbf{V}^\star + \overline{\mathbf{V}}_{ref}, p_{ref} = p^\star + \overline{p}_{ref},$$

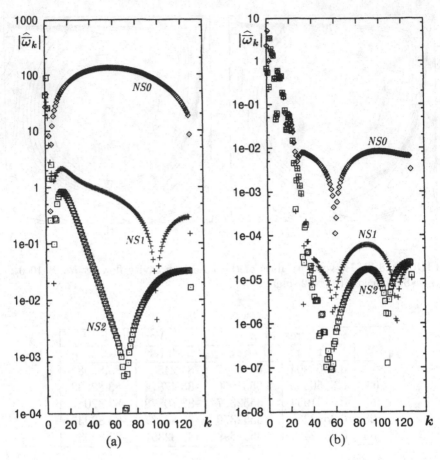

FIGURE 8.14. Spectrum of the vorticity for the lid-driven cavity flow at $Re = 1000$: (a) on $X = 0$, (b) on $X = 0.5$.

it is valuable to examine the rate of decay of the errors in the continuous $L_w^2(\Omega)$-norm

$$\overline{E}_{\mathbf{V}} = \|\mathbf{V}_N - \mathbf{V}_{ref}\| = \left(\|u_N - u_{ref}\|^2 + \|v_N - v_{ref}\|^2 \right)^{1/2}$$

and $\overline{E}_p = \|p_N - p_{ref}\|$, where $\mathbf{V}_N = \mathbf{V}^\star + \overline{\mathbf{V}}_N$ and $p_N = p^\star + \overline{p}_N$. The norm $L_w^2(\Omega)$ is calculated, as explained in Section 3.6, by constructing the polynomials u_N, v_N, and p_N and then by using the Gauss-Lobatto quadrature formula based on a large number of points. Figure 8.15 compares the rate of decay of the errors, making clear the improvement brought by the subtraction technique. This global rate remains algebraic in agreement with the theory. However, a much faster decay of the local error is observed for points located sufficiently far from the singular points. As an example, the error obtained on the maximum of u on the line $Y = 0.5$ is evaluated to be

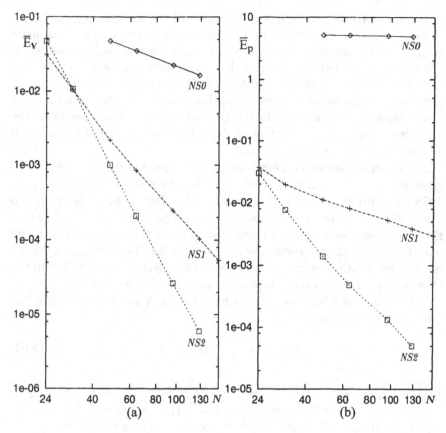

FIGURE 8.15. Errors obtained for the lid-driven cavity flow at $Re = 1000$: (a) error on velocity, (b) error on pressure.

$O\left(N^{-5}\right)$ for the NS0 solution and, respectively, $O\left(N^{-9}\right)$ and $O\left(N^{-17}\right)$ for the NS1 and NS2 solutions.

To close this section we mention the earlier Chebyshev application of the subtraction method, in which only the first (Stokes) term is subtracted, to the calculation of the lid-driven cavity flow. Thus, Schultz *et al.* (1989) have considered the Stokes and Navier-Stokes equations in the streamfunction formulation and Schumack *et al.* (1991) have considered the Stokes equations in the velocity-pressure variables.

8.5   Filtering technique

When using a spectral approximation to solve stiff or singular problems the approximate solution is often subject to oscillations. The presence of oscillations may have several causes, in particular, it may be due to an insufficient resolution or to the fundamental incapacity of the spectral approximation to represent correctly the function under consideration. A typical example is given by the representation of a discontinuous function by Fourier or Chebyshev series, exhibiting the well-known Gibbs phenomenon (see, e.g., Canuto et al., 1988 ; Gottlieb and Shu, 1997). The solutions of elliptic problems, with which we are concerned in this book, are exempt from interior discontinuities. But, as seen in the previous section, discontinuity in the boundary conditions at some point (or other reason) makes the solution singular at the point of discontinuity.

A number of works has been devoted to preserve the accuracy when representing a discontinuous function by Fourier series (e.g., Majda et al., 1978 ; Canuto et al., 1988 ; Vandeven, 1991 ; Boyd, 1996 ; Gottlieb and Shu, 1997). The general approach consists of a filtering technique, in which the expansion coefficients are multiplied by some factor, in order to damp out the high-frequency coefficients responsible for the oscillations. Filters devised for Fourier series apply as well to Chebyshev series owing to the close analogy between both approximations (see Section 3.3.1).

Let us consider a function $u(x)$ defined for $x \in \Lambda$ and represented by the truncated series expansion

$$u_M(x) = \sum_{k \in I_M} \hat{u}_k \, \varphi_k(x) . \tag{8.64}$$

In the Fourier case, $\Lambda = [0, 2\pi]$, $\varphi_k(x) = e^{ikx}$, $M = K$, and $I_M = \{-K, \ldots, K\}$. In the Chebyshev case, $\Lambda = [-1, 1]$, $\varphi_k(x) = T_k(x)$, $M = N$, and $I_M = \{0, \ldots, N\}$. The filtered expansion $\overline{u}_M(x)$ is defined by

$$\overline{u}_M(x) = \sum_{k \in I_M} \sigma\left(\frac{k}{M}\right) \hat{u}_k \, \varphi_k(x) \tag{8.65}$$

where the factor $\sigma(k/M)$ has, for effect, to damp out the higher modes while preserving, as far as possible, the accuracy of the spectral approximation. The filter $\sigma(\eta)$ must satisfy some general requirements. Following Vandeven (1991) and Gottlieb and Shu (1997), a real and even function $\sigma(\eta)$ is called a filter of order p if :

   (i) $\sigma(0) = 1$, $\sigma^{(m)} = 0$, $1 \le m \le p - 1$,
   (ii) $\sigma(\eta) = 0$, $|\eta| \ge 1$,
   (iii) $\sigma(\eta) \in C^{p-1}$, $\eta \in (-\infty, +\infty)$.
The last two conditions imply that :
   (iv) $\sigma^{(m)}(\pm 1) = 0$, $0 \le m \le p - 1$.

We list below some typical filters :
   1. The Lanczos filter (first-order filter) :

$$\sigma_l(\eta) = \frac{\sin(\pi \eta)}{\pi \eta} . \tag{8.66}$$

2. The raised cosine filter (second-order filter) :

$$\sigma_c (\eta) = \frac{1}{2} [1 + \cos (\pi \, \eta)] \, . \tag{8.67}$$

3. The sharpened raised cosine filter (eighth-order filter) :

$$\sigma_s (\eta) = \sigma_c^4 (\eta) \left[35 - 84 \, \sigma_c (\eta) + 70 \, \sigma_c^2 (\eta) - 20 \, \sigma_c^3 (\eta) \right] \, . \tag{8.68}$$

4. The exponential filter (pth-order filter, with even p) :

$$\sigma_{e,p} (\eta) = e^{-\alpha \, \eta^p} \quad (\alpha > 0) \, . \tag{8.69}$$

Note that condition (iv) is not exactly satisfied by this filter as $\sigma_{e,p} (\pm 1) = e^{-\alpha}$, but it can be satisfied numerically if the coefficient α is chosen sufficiently large.

5. The Vandeven filter (pth-order filter) :

$$\sigma_{v,p} = 1 - \frac{(2 p - 1)!}{(p - 1)!^2} \int_0^{|\eta|} t^{p-1} (1 - t)^{p-1} \, dt \, , \tag{8.70}$$

that is, after integration

$$\sigma_{v,p} = 1 + (-1)^p \frac{(2 p - 1)!}{(p - 1)!} \sum_{j=0}^{p-1} \frac{(-1)^j}{(p - 1 - j)! \, j!} \frac{|\eta|^{2 p - j - 1}}{2 p - j - 1} \, .$$

It is proven by Vandeven (1991) that the filtered expansion based on this filter recovers exponential accuracy away from the discontinuity provided p is chosen increasing with N. More precisely, assuming that $u (x)$ is discontinuous at $x = x_0$ and setting $d (x) = x - x_0$, the following estimate holds

$$|u (x) - \overline{u}_M (x)| \leq \frac{C}{d (x)^{p-1} \, N^{p-1}} \, , \tag{8.71}$$

where C is a constant. This inequality shows that error can decrease faster than any power of N provided that p is sufficiently large and $d (x)$ not too small.

Some of the above filters (and others) have been tested in several works (e.g., Gottlieb and Orszag , 1977 ; Kopriva, 1987 ; Vandeven, 1991 ; Boyd, 1996) and may be applied to a large variety of problems. For example, the raised cosine is used in Section 8.2.3.b to smooth the computation of a functional arising in adaptive meshing. The Vandeven filter is used in Section 8.4.2.b to improve the convergence of the Chebyshev approximation near the singular corners of the lid-driven cavity. In this last example, Botella and Peyret (1998) have evaluated numerically the spatial nonuniformity of the filter $\sigma_{v,p}$ as shown by Eq.(8.71). The presence of the term $d (x)^{p-1}$ involves the error increases when point x is drawn near to the singularity. It becomes even $O(1)$ when $d(x) = O(1/N)$. Furthermore, it is known that a high-order filter is expected to give worse results than a low-order one near the singularity. As a result, the optimal choice of the order p depends on the location of point x with respect to the singular point x_0 (Boyd, 1996). Thus, in the lid-driven cavity problem, it is found that at point $(0, 0.95)$, that is relatively close to the corner $(0, 1)$, the high-order filter $\sigma_{v,12}$ gives excellent results (seven-digit convergence for the vorticity), while low-order filters with $p \leq 6$ prove to be insufficient. On the other hand, for the point $(0, 0.999)$, much closer to the corner than the previous one, the low-order filters $\sigma_{v,2}$ or $\sigma_{v,4}$ are preferable.

This example shows that filters must be used with care and their effect evaluated with precision when tuning the parameters. Experience leads us to think that filtering should be used mainly as a post-processing tool for convergence acceleration or for a specific cosmetic purpose. The application of a filter in the course of a time integration (especially if it is often applied) may be dangerous because significant frequencies could be too much damped, modifying, in this way, the general behaviour of the solution.

Note that, for the Chebyshev basis, the filtering process changes the boundary values of the function under consideration. To avoid such a drawback, Boyd (1998) proposes a simple technique based on the consideration of an alternative basis constructed from the Chebyshev basis.

9

Domain Decomposition Method

The domain decomposition method for the solution of differential problems consists of dividing the computational domain into a set of subdomains in which the solution is calculated by taking into account some transmission conditions at the interfaces between the subdomains. This chapter addresses some usual Chebyshev domain decomposition methods for elliptic problems : influence matrix, iterative Dirichlet/Neumann, and spectral-element methods. The effect of the decomposition on the stability of time-dependent problems will be discussed. The application of the influence matrix method to the multidomain solution of the Stokes problem will be considered for both formulations : vorticity-streamfunction and velocity-pressure. Lastly, examples of application to the Navier-Stokes equations will be presented.

9.1 Introduction

In this chapter we discuss the domain decomposition approach to the Chebyshev approximation of elliptic equations and, especially, the Stokes equations resulting from the time-discretization of the Navier-Stokes equations. It is classical to distinguish these methods according to the nature of the decomposition : overlapping and nonoverlapping decompositions. The first type of decomposition, which leads to an iterative solution procedure (Schwarz algorithm), is not in common use in spectral methods. This case will not be addressed here and we consider the usual method in which the original computational domain is divided into non-overlapping subdomains. Then equations are solved in each subdomain with suitable transmission conditions at the interfaces.

The interest to consider a domain decomposition technique appears in several situations. The first one concerns the geometry of the computational domain. The properties of the Chebyshev polynomial approximation require this geometry to be rectangular and, consequently, restricts the versatility of the method. However, thanks to the domain decomposition concept, the applicability of the Chebyshev method may be extended to domains which can be decomposed into rectangular subdomains. When such a regular decomposition is impossible, the elementary nonrectangular quadrangular subdomains have to be transformed into square subdomains by suitable transformations. The latter approach has been chosen by Korczak and Patera (1986) who developed the iso-parametric spectral-element method. Triangular subdomains may also be considered (Dubiner, 1991 ; Sherwin and Karniadakis, 1995).

The second situation, where the domain decomposition approach may be of interest, is the case of stiff problems. As largely discussed in Chapter 8, the consideration of subdomains of different size and/or different resolution, according to the local degree of stiffness of the solution, allows an accurate representation of this solution while keeping the total resolution within reasonable bounds. Chapter 8 has also shown the ability of the method to handle singular solutions when the singularity can be transferred to the corner of a subdomain, thanks to a suitable decomposition.

A very interesting consequence of the domain decomposition approach is that the size of the matrices involved is relatively small and, in any case, much smaller than in a one-domain method. This constitutes an attractive property since the Chebyshev matrices are generally ill-conditioned and the conditioning deteriorates when their size increases.

Furthermore, the concept of domain decomposition constitutes the basis of the implementation of numerical methods on parallel computers. Thus, the design of algorithms for the solution of partial differential equations and, especially, the Navier-Stokes equations, must take the parallelization aspect into consideration.

Finally, the necessity for using a domain decomposition technique appears naturally when the representation of the physical phenomena requires the simultaneous consideration of different models in different regions of the computational domain. An example is given by the motion of a fluid due to thermal convection in an enclosure whose walls are subject to thermal diffusion : the Navier-Stokes equations with the transport-diffusion equations for temperature are solved in the fluid domain while the pure diffusion equation is solved in the wall (Larroudé et al., 1994).

Domain decomposition methods for spectral approximations appeared at the beginning of the 1980s. Since that time, much progress has been made but one may yet expect improvements, especially concerning the algorithms for the Navier-Stokes equations. Our aim, in this chapter, is to present the general principles of the domain decomposition approach and to discuss some methods which have been experimented on in various flow problems.

The first part of the chapter is devoted to one-dimensional problems by presenting successively the patching method and the spectral-element method. Then the application of the patching method to the solution of the two-dimensional Helmholtz equation is discussed by emphasizing the use of the influence matrix technique for handling the interface problem. Lastly, in the third part of the chapter, this technique is extended to the solution of the Stokes problem in vorticity-streamfunction as well velocity-pressure formulations, and typical applications concerning the time-dependent Navier-Stokes equations are presented.

9.2   Differential equation

Considering the boundary value problem for a second-order differential equation we present, in this section, two commonly used domain decomposition methods. The first one (patching method) is based on the enforcement of the continuity of the solution and its first-order derivative at the interface between two adjacent subdomains. The second method (spectral-element method) is variational so that the continuity of the solution is prescribed at the interface while the continuity of the derivative is obtained only when the degree N of the polynomial approximation tends toward infinity.

9.2.1   Patching method: two-domain solution

Let us consider the differential problem (3.56)-(3.58) written in the general form

$$L u \equiv -\nu u' + a u + b = f \quad \text{in } \Omega = (-1, 1) \tag{9.1}$$

$$B_- u(-1) = g_- \tag{9.2}$$

$$B_+ u(1) = g_+ , \tag{9.3}$$

where f is a smooth function. The domain Ω is divided in two subdomains $\Omega_1 = (-1, a_1)$ and $\Omega_2 = (a_1, 1)$ separated by the interface $x = a_1$. The patching method (Orszag, 1980) consists of solving Eq.(9.1) in each subdomain with the appropriate boundary conditions at $x = \pm 1$ and with continuity conditions at the interface. For the second-order differential equation (9.1) the continuity of the function u and its first-order derivative u' is prescribed at the interface (see, e.g., Canuto et al., 1988). It is easy to see that these conditions ensure, for the exact problem, continuity of the derivatives of any order (provided they exist). For example, the continuity of the second-order derivative comes from the verification of the differential equation at the interface. Then, for higher-order derivatives, the same result is obtained by differentiation of the equation (9.1). Of course, this is not the situation in the discrete case and discontinuities in second and higher-order derivatives are observed.

Let u_m, $m = 1, 2$, be the restriction of u to Ω_m, and the problem (9.1)-(9.3) is replaced by the $\mathcal{P}_{1-2}$-*Problem* :

$$L u_1 = f \qquad \text{in } \Omega_1 \qquad (9.4)$$

$$B_- u_1 (-1) = g_- , \qquad (9.5)$$

$$L u_2 = f \qquad \text{in } \Omega_2 \qquad (9.6)$$

$$B_+ u_2 (1) = g_+ , \qquad (9.7)$$

with the continuity conditions

$$u_1 (a_1) = u_2 (a_1) \qquad (9.8)$$

$$u_1' (a_1) = u_2' (a_1) . \qquad (9.9)$$

Thus, the solutions u_1 and u_2 are coupled by the continuity conditions, so that the discrete approximation to the $\mathcal{P}_{1-2}$-problem by the Chebyshev (tau or collocation) method produces a large algebraic system determining the unknowns (Chebyshev coefficients or grid values) in both subdomains. This system can be solved either by a direct method or by an iterative one. The basic idea, for both methods, is to uncouple, as far as possible, the solution in each subdomain. This is done by using either a superposition of elementary solutions (influence matrix method) or by using an iteration-by-subdomain technique (Dirichlet/Neumann method). These methods are now successively described by considering the continuous problem rather than its discrete version. This simplifies the presentation of the algorithms, but it must be clear, as already pointed out several times, that both approaches are equivalent provided the discretization is coherent (see Remark 2 of Section 6.3.2.d).

(a) Influence matrix method
The influence (or capacitance) matrix method is a powerful tool for the solution of linear problems. As shown in several places in this book, it can be applied to a large variety of problems. Its application to the spectral domain decomposition method has been considered by Pulicani (1988) and Lacroix *et al.* (1988). The basic idea is to determine, in a direct way, the value of $u_1(a_1) = u_2(a_1) = \xi$ at the interface which ensures the continuity of the derivative. Previously, the method had been proposed by Macaraeg and Streett (1986) to enforce a global flux balance rather than the pointwise condition (9.9).

We set

$$u_m = \tilde{u}_m + \overline{u}_m , \qquad m = 1, 2. \qquad (9.10)$$

The function $\tilde{u}_m$, $m = 1, 2$, is the solution to the problems

$\tilde{\mathcal{P}}_1$-*Problem* :

$L\tilde{u}_1 = f$ in Ω_1

$B_- \tilde{u}_1 (-1) = g_-$

$\tilde{u}_1 (a_1) = 0,$

$\tilde{\mathcal{P}}_2$-*Problem* :

$L\tilde{u}_2 = f$ in Ω_2

$\tilde{u}_2 (a_1) = 0$

$B_+ \tilde{u}_2 (1) = g_+ .$

Then the complementary part $\bar{u}_m$, $m = 1, 2$, is the solution to the $\overline{\mathcal{P}}_{1-2}$-*Problem* :

$L\bar{u}_1 = 0$ in Ω_1

$B_- \bar{u}_1 (-1) = 0$

$\bar{u}_1 (a_1) = \xi,$

$L\bar{u}_2 = 0$ in Ω_2

$\bar{u}_2 (a_1) = \xi$

$B_+ \bar{u}_2 (1) = 0,$

where ξ must be found such that condition (9.9) is satisfied, namely,

$$\bar{u}'_1 (a_1) - \bar{u}'_2 (a_1) = -\tilde{u}'_1 (a_1) + \tilde{u}'_2 (a_1) .$$

By setting

$$\bar{u}_m = \xi \bar{v}_m , \quad m = 1, 2, \tag{9.11}$$

one replaces the $\overline{\mathcal{P}}_{1-2}$-problem by the two problems

$\overline{\mathcal{P}}_1$-*Problem* :

$L\bar{v}_1 = 0$ in Ω_1

$B_- \bar{v}_1 (-1) = 0$

$\bar{v}_1 (a_1) = 1,$

$\overline{\mathcal{P}}_2$-*Problem* :

$L\bar{v}_2 = 0$ in Ω_2

$\bar{v}_2 (a_1) = 1$

$B_+ \bar{v}_2 (1) = 0,$

so that ξ is determined by the equation

$$[\bar{v}'_1 (a_1) - \bar{v}'_2 (a_1)] \xi = - [\tilde{u}'_1 (a_1) - \tilde{u}'_2 (a_1)] . \tag{9.12}$$

In summary, the $\mathcal{P}_{1-2}$-problem, coupling u_1 and u_2, has been replaced by a set of uncoupled $\left(\tilde{\mathcal{P}}_m, \overline{\mathcal{P}}_m, \quad m = 1, 2\right)$ problems in each subdomain. These problems are easily solved by using the Chebyshev (tau or collocation) method. After ξ has been determined by Eq.(9.12) the final solution u_m, $m = 1, 2$, is reconstituted thanks to Eqs.(9.11) and (9.10). In the simple one-dimensional two-domain case the "influence matrix" is simply the scalar $\bar{v}'_1 (a_1) - \bar{v}'_2 (a_1)$. This quantity is nonzero except the case where the problem (9.1)-(9.3) reduces to the Poisson-type equation $u'' = f$ with Neumann boundary conditions. In this case, we also have $\tilde{u}'_1 (a_1) - \tilde{u}'_2 (a_1) = 0$ (from the compatibility relation of the Neumann problem), so that Eq.(9.12) is satisfied for any value of the constant ξ, in agreement with the fact that the exact solution is defined within a constant. Note that, in the present case, the $\overline{\mathcal{P}}_{1-2}$-problem is too degenerate to be significant since the interface value may be arbitrarily fixed.

Concerning the implementation of the method we refer to Section 6.3.2.b and, especially, to Remarks 1 and 2 which also apply in the present situation.

(b) Iterative Dirichlet/Neumann method

The iterative Dirichlet/Neumann method has been developed for spectral approximations by Funaro *et al.* (1988). The idea is to devise the iterative procedure such that the boundary condition at the interface $x = a_1$ is of Dirichlet type for u_1 and of Neumann type for u_2. Let us denote by n the index of iteration. We assume that the value ξ^n of u_1^n at $x = a_1$ is known. The iterative procedure consists of solving, successively, the two problems:

$\mathcal{P}_1^n$-*Problem* :

$L u_1^n = f$ in Ω_1

$B_- u_1^n (-1) = g_-$

$u_1^n (a_1) = \xi^n,$

$\mathcal{P}_2^n$-*Problem* :

$L u_2^n = 0$ in Ω_2

$u_2^{n\prime} (a_1) = u_1^{n\prime} (a_1)$

$B_+ u_2^n (1) = g_+ .$

Then ξ^{n+1} is defined by the relaxation formula

$$\xi^{n+1} = \alpha u_2^n (a_1) + (1 - \alpha) \, \xi^n \,, \tag{9.13}$$

where α is the relaxation parameter. The process is initiated by choosing an arbitrary value of ξ^0. At convergence (assumed to be reached), $\xi^{n+1} = \xi^n = \xi$ and, hence, $u_2^n (a_1) = u_1^n (a_1) = \xi$. Since the continuity of the derivative is ensured at each step of the iteration, the converged pair (u_1, u_2) is a solution of the two-domain $\mathcal{P}_{1-2}$-problem defined by Eqs.(9.4)-(9.9).

Funaro *et al.* (1988) have examined the convergence of the iterative procedure for the continuous differential problem in the case of the Helmholtz equation ($a = 0$, $b/\nu = \sigma \geq 0$) with Dirichlet conditions. They found that, if no relaxation is applied ($\alpha = 0$), the process converges if and only if $a_1 > 0$. In the more general case, the process converges provided the relaxation parameter α satisfies

$$0 < \alpha < \alpha^\star (\sigma) \,, \tag{9.14}$$

where

$$\alpha^\star (\sigma) = \begin{cases} 2 \,(1 + \alpha_1) & \text{if } \sigma = 0 \,, \\[2mm] 2 \left[\dfrac{\tanh \left(\sqrt{\sigma} \,(1 - a_1) \right)}{\tanh \left(\sqrt{\sigma} \,(1 + a_1) \right)} \right]^{-1} & \text{if } \sigma > 0 \,. \end{cases} \tag{9.15}$$

Funaro *et al.* (1988) remark that, for the optimal value $\alpha_{opt} = \alpha^\star / 2$, the process converges after two iterations. They also propose the choice of a relaxation parameter α_n varying with the iteration, namely, α_n is defined by

$$\frac{1}{\alpha_n} = 1 - \frac{u_2^n (a_1) - u_2^{n-1} (a_1)}{u_1^n (a_1) - u_1^{n-1} (a_1)}, \quad n \geq 2, \tag{9.16}$$

and α_1 being prescribed. The choice (9.16) gives $\alpha_2 = \alpha_{opt}$, therefore the convergence is reached after three iterations whatever the value of α_1.

Similar results of convergence were obtained by Funaro *et al.* (1988) for the discrete problem resulting from the approximation by the Chebyshev collocation method, but the expression of α^* in Eq.(9.14) is much more complicated. Consequently, the results obtained above for the differential problem may also be used in the discrete case provided that the polynomial degree N is sufficiently large.

Returning to the general case where $a \neq 0$ and assuming $\nu > 0$, divergence of the iterative procedure may occur when $a > 0$ and $\nu/a \ll 1$, namely, when advection is dominating (see Carlenzoli and Quarteroni, 1995). This can be explained by observing that, in the limit $\nu \to 0$, the Dirichlet condition $u_1^n(a_1) = \xi^n$ produces (except if ξ^n has the exact value !) a boundary layer for u_1 near the interface $x = a_1$. The presence of this layer is the cause of the divergence of the iterative procedure. Therefore, the choice of a Dirichlet or Neumann condition at the interface should be dictated by the sign of a. More precisely, if $a > 0$ the Neumann condition must be associated to u_1 and the Dirichlet condition to u_2. The opposite situation must be prescribed when $a < 0$.

Finally, let us assume that Eq.(9.1) is the Poisson-type equation $u'' = f$ and that the boundary condition (9.3) is of Neumann type, that is $u'(1) = g_+$. In such a case, the compatibility condition associated with the Neumann $\mathcal{P}_2^n$-problem has no reason to be satisfied (except if ξ^n has the exact value !). The way to avoid this difficulty is to simply exchange the $\mathcal{P}_1^n$- and $\mathcal{P}_2^n$-problems by prescribing, at $x = a_1$, a Dirichlet condition for u_2 in Ω_2 and a Neumann condition for u_1 in Ω_1. On the other hand, this difficulty does not appear in the pure Neumann case $u'(\pm 1) = g_\pm$, thanks to the compatibility relation associated with the original problem. However, as previously, the iterative method for the one-dimensional two-domain problem is degenerate since ξ^n can be arbitrarily prescribed.

(c) Relationship between the influence matrix and Dirichlet/Neumann methods

Since the direct influence matrix method and the iterative Dirichlet/Neumann method (after discretization) are both intended to solve the same algebraic system, we may expect to have a close relationship between these two methods. Indeed, in the case of the two-dimensional Helmholtz equation with Dirichlet conditions, Quarteroni and Sacchi-Landriani (1988) have shown that the Dirichlet/Neumann method amounts to solving iteratively the system

$$\mathcal{M} \Xi = \tilde{E}, \tag{9.17}$$

where Ξ is the vector whose elements are the interface values of the unknown and $\mathcal{M}$ is the influence matrix. More precisely, they show that the iterative procedure is the PR method defined by

$$A_0 \left(\Xi^{n+1} - \Xi^n \right) = \alpha \left(\tilde{E} - \mathcal{M} \Xi^n \right) \tag{9.18}$$

where A_0 is the preconditioning operator such that $\mathcal{A} = -\mathcal{M}_2$ with $\mathcal{M}_2$ defined below in Eq.(9.60). In the present scalar case, it is not difficult to show that the Dirichlet/Neumann procedure defined by the $\mathcal{P}_1^n$- and $\mathcal{P}_2^n$-problems and Eq.(9.13) amounts to the solution of Eq.(9.12), namely,

$$M \xi = \tilde{E} \tag{9.19}$$

by the "preconditioned" iterative procedure

$$A_0 \left(\xi^{n+1} - \xi^n \right) = \alpha \left(\tilde{E} - M \xi^n \right) \tag{9.20}$$

with $A_0 = -\bar{v}_2'(a_1)$.

9.2.2 Patching method : multidomain solution

The algorithms described above in the two-domain case can be easily extended to an arbitrary number of subdomains. We briefly describe this extension.

The domain $\Omega = (-1, 1)$ is divided into M subdomains $\Omega_m = (a_{m-1}, a_m)$, $m = 1, \ldots, M$, with $a_0 = -1$ and $a_M = 1$. We denote by u_m the restriction of u to Ω_m. The multidomain problem associated to Eq.(9.1)-(9.3) is defined by :
In every subdomain Ω_m, solve the equation

$$L u_m = f, \quad m = 1, \ldots, M, \tag{9.21}$$

with the interface conditions

$$u_m(a_m) = u_{m+1}(a_m), \quad m = 1, \ldots, M-1, \tag{9.22}$$

$$u_m'(a_m) = u_{m+1}'(a_m), \quad m = 1, \ldots, M-1, \tag{9.23}$$

and, for Ω_1 and Ω_M, respectively, the boundary conditions

$$B_- u_1(-1) = g_-, \tag{9.24}$$

$$B_+ u_M(1) = g_+. \tag{9.25}$$

(a) Influence matrix method
The extension of the influence matrix method to the multidomain case may follow exactly the algorithm described in Section 9.2.1.a, by considering special problems in Ω_1 and Ω_M where boundary conditions at $x = \pm 1$ have to be prescribed. However, it is convenient (at least in the one-dimensional

case and for Robin boundary conditions) to make the method more systematic by including the enforcement of the boundary conditions into the definition of the influence matrix.

The solution u_m in each subdomain Ω_m is sought in the form

$$u_m = \tilde{u}_m + \overline{u}_m, \quad m = 1, \ldots, M.$$

The part $\tilde{u}_m$ satisfies the $\tilde{P}_m$-Problem :

$$L\tilde{u}_m = f \quad \text{in } \Omega_m, \quad m = 1, \ldots, M$$

$$\tilde{u}_m(a_{m-1}) = 0, \quad \tilde{u}_m(a_m) = 0,$$

and the part $\overline{u}_m$ is the solution to the $\overline{P}_m$-Problem :

$$L\overline{u}_m = 0 \quad \text{in } \Omega_m, \quad m = 1, \ldots, M$$

$$\overline{u}_m(a_{m-1}) = \xi_{m-1}, \quad \overline{u}_m(a_m) = \xi_m,$$

where the boundary values ξ_0 and ξ_M, and the interface values ξ_m, $m = 1, \ldots, M-1$, are determined so that the boundary conditions (9.24), (9.25), and the interface conditions (9.23), are satisfied. We set

$$\overline{u}_m = \xi_{m-1}\overline{u}_{m,1} + \xi_m \overline{u}_{m,2}, \quad m = 1, \ldots, M$$

where $\overline{u}_{m,l}$, $l = 1, 2$, is the solution to the $\overline{P}_{m,l}$-Problem :

$$L\overline{u}_{m,l} = 0 \quad \text{in } \Omega_m, \quad m = 1, \ldots, M, \quad l = 1, 2,$$

$$\overline{u}_{m,l}(a_{m-1}) = -l + 2, \quad \overline{u}_{m,l}(a_m) = l - 1.$$

The $M + 1$ parameters ξ_m, $m = 0, \ldots, M$, are determined by requiring the continuity and boundary conditions to be satisfied. The continuity conditions (9.23) give

$$\xi_{m-1}\overline{u}'_{m,1}(a_m) + \xi_m\left[\overline{u}'_{m,2}(a_m) - \overline{u}'_{m+1,1}(a_m)\right] - \xi_{m+1}\overline{u}'_{m+1,2}(a_m)$$
$$= -\tilde{u}'_m(a_m) + \tilde{u}'_{m+1}(a_m), \quad m = 1, \ldots, M - 1,$$

$$(9.26)$$

and the boundary conditions (9.24) and (9.25) give, respectively,

$$\xi_0 B_- \overline{u}_{1,1}(-1) + \xi_1 B_- \overline{u}_{1,2}(-1) = g_- - B_- \tilde{u}_1(-1), \quad (9.27)$$

$$\xi_{M-1} B_+ \overline{u}_{M,1}(1) + \xi_M B_+ \overline{u}_{M,2}(1) = g_+ - B_+ \tilde{u}_M(1). \quad (9.28)$$

In summary, the $M + 1$ equations (9.26)-(9.28) constitute the algebraic system

$$\mathcal{M} \,\Xi = \tilde{E} \tag{9.29}$$

determining the $M + 1$ elements of the vector Ξ such that

$$\Xi = (\xi_0, \ldots, \xi_M)^T \,.$$

The influence matrix $\mathcal{M}$ is tridiagonal and is invertible except when the differential problem under consideration is a Neumann problem for a Poisson-type equation. In such a case $\mathcal{M}$ is singular, expressing the fact that the exact solution is defined up to a constant. Therefore, in order to be invertible, the matrix $\mathcal{M}$ has to be previously transformed in the same way as that already described in Section 7.4.1.b for the influence matrix of the Stokes problem.

One of the advantages of this influence matrix method is that only Dirichlet problems have to be solved whatever the type of the original boundary conditions. The basic element of the numerical code is simply a Dirichlet problem solver.

(b) Iterative Dirichlet/Neumann method

The extension of the Dirichlet/Neumann method to an arbitrary number M of subdomains is straightforward when M is even. In the case where M is odd, a slight modification has to be considered since Dirichlet conditions must be attributed at the two boundaries of one subdomain. The iteration is initiated by fixing these interface values and solving the Dirichlet problem in this subdomain. This provides Neumann conditions for the adjacent subdomains and the iterative process is similar to the one described for the two-domain case.

9.2.3   Spectral-element method

The spectral-element method has been introduced by Patera (1984) and then largely developed and applied to various problems of fluid flow. The method, which presents some similarity to the p-version of the finite-element method (see Babuška and Suri, 1994), is based on a variational formulation of the problem using Lagrange interpolation polynomials on the Gauss-Lobatto points as trial and test functions.

The method was originally associated with Chebyshev polynomials but it is now generally associated with Legendre polynomials because of a better accuracy of quadratures. Nevertheless, in order to remain in the frame of Chebyshev approximation, we present below the basic principles of the method as it was proposed by Patera (1984) and described by Quarteroni (1987) or Canuto et al. (1988). We refer to more recent works (Maday and Patera, 1989 ; Karniadakis and Henderson, 1998 ; Karniadakis and

Sherwin, 1999) for a general presentation of the method in the Legendre case and applications to the Navier-Stokes equations.

For the sake of simplicity of presentation, we consider the case of the Helmholtz equation with Dirichlet conditions, that is,

$$u'' - \sigma u = f, \quad -1 < x < 1 \tag{9.30}$$

$$u(-1) = g_-, \quad u(1) = g_+, \tag{9.31}$$

where σ is a positive constant and f is assumed to be square integrable in $\Omega = (-1, 1)$. Now we introduce the weak formulation

$$\int_{-1}^{1} (u'' - \sigma u) v \, dx = \int_{-1}^{1} f v \, dx,$$

where $v \in H_0^1$, namely, v and its first-order derivative are square integrable in Ω and $v(x)$ satisfies $v(-1) = v(1) = 0$. As usual, an integration by parts yields the equation

$$-\int_{-1}^{1} (u' v' + \sigma u v) \, dx = \int_{-1}^{1} f v \, dx, \tag{9.32}$$

which has to be approximated.

As previously, we consider the partition of $\Omega = (-1, 1)$ in M subdomains (called "elements") $\Omega_m = (a_{m-1}, a_m)$, $m = 1, \ldots, M$, with $a_0 = -1$, $a_M = 1$, and we denote by u_m the restriction of u to Ω_m. Therefore, Eq.(9.32) may be written as

$$-\sum_{m=1}^{M} \int_{a_{m-1}}^{a_m} (u' v' + \sigma u v) \, dx = \sum_{m=1}^{M} \int_{a_{m-1}}^{a_m} f v \, dx. \tag{9.33}$$

For each element Ω_m, we introduce the local coordinate $\hat{x}^m$ defined by

$$\hat{x}^m = \Phi_m(x) = \frac{2}{l_m}(x - a_{m-1}) - 1 \tag{9.34}$$

with $l_m = a_m - a_{m-1}$, such that $-1 \leq \hat{x}^m \leq 1$. Then the restriction u_m of u to Ω_m is approximated by a polynomial $u_{mN}(x)$ of degree N_m. With the local Gauss-Lobatto points

$$\hat{x}_i^m = \cos \frac{\pi i}{N_m}, \quad i = 0, \ldots, N_m, \tag{9.35}$$

the unknowns of the problem are the values $u_{m,i}$ of the polynomial u_{mN} at points $x_i^m = \Phi_m^{-1}(\hat{x}_i^m)$.

Now let us denote by $\hat{u}_{mN}(\hat{x}^m)$ the expression of $u_{mN}(x)$ in the local coordinate system, it is represented by the Lagrange interpolation polynomial based on the grid values $u_{m,i} = u_{mN}(x_i^m)$, $i = 0, \ldots, N_m$, namely,

$$\hat{u}_{mN}(\hat{x}^m) = \sum_{i=0}^{N_m} u_{m,i} \, \hat{h}_{m,i}(\hat{x}^m). \tag{9.36}$$

The Lagrange coefficients $\hat{h}_{m,i}$ are polynomials of degree N_m such that

$$\hat{h}_{m,i}\left(\hat{x}_j^m\right) = \delta_{i,j} .$$

Their expression [see Eq.(3.16)] is

$$\hat{h}_{m,i}\left(\hat{x}^m\right) = \frac{2}{\bar{c}_i \, N_m} \sum_{k=0}^{N_m} \frac{1}{\bar{c}_k} T_k\left(\hat{x}_i^m\right) T_k\left(\hat{x}^m\right) , \quad i = 0, \ldots, N_m .$$

It remains to define the test functions v associated with each point of Ω_m. For that, we must distinguish the inner points $i = 1, \ldots, N_m - 1$ from the boundary point $i = 0$ or from $i = N_m$.

For any Ω_m, $m = 1, \ldots, M$, and for $i = 1, \ldots, N_m - 1$, the test function $v_{m,i}$ is defined by

$$v_{m,i}\left(x\right) = \begin{cases} h_{m,i}\left(x\right) & \text{if } a_{m-1} \le x \le a_m , \\ 0 & \text{otherwise} , \end{cases} \tag{9.37}$$

where $h_{m,i}\left(x\right) = \hat{h}_{m,i}\left(\Phi_m^{-1}\left(\hat{x}^m\right)\right)$. Then, by bringing the expressions (9.36) and (9.37) into Eq.(9.33) and denoting by $f_{m,i}$ the values of f at points x_i^m, we get the discrete equations for $i = 1, \ldots, N_m - 1$:

$$\sum_{k=0}^{N_m} C_{ik}^m \, u_{m,k} = - \sum_{k=0}^{N_m} B_{ik}^m \, f_{m,k} , \quad m = 1, \ldots, M , \tag{9.38}$$

where

$$C_{ik}^m = A_{ik}^m + \sigma \, B_{ik}^m$$

with

$$A_{ik}^m = \frac{8}{\bar{c}_i \, \bar{c}_k \, l_m \, N_m^2} \sum_{l,n=0}^{N_m} \frac{1}{\bar{c}_l \, \bar{c}_n} T_l\left(\hat{x}_i^m\right) T_n\left(\hat{x}_k^m\right) \alpha_{ln} ,$$

$$B_{ik}^m = \frac{2 \, l_m}{\bar{c}_i \, \bar{c}_k \, N_m^2} \sum_{l,n=0}^{N_m} \frac{1}{\bar{c}_l \, \bar{c}_n} T_l\left(\hat{x}_i^m\right) T_n\left(\hat{x}_k^m\right) \beta_{ln} ,$$

with α_{ln} and β_{ln} defined by

$$\begin{aligned}
\alpha_{ln} &= \int_{-1}^{1} T_l'\left(x\right) T_n'\left(x\right) dx \\
&= \begin{cases} 0 , & l + n \text{ odd} , \\ \dfrac{l_n}{2} \left[J\left(\dfrac{|l - n|}{2}\right) - J\left(\dfrac{|l + n|}{2}\right) \right] , & l + n \text{ even} , \end{cases}
\end{aligned}$$

$$J\left(k\right) = \begin{cases} 0 , & k = 0 , \\ -4 \displaystyle\sum_{p=1}^{k} \dfrac{1}{2 \, p - 1} , & k \ge 1 , \end{cases}$$

and

$$
\beta_{l\,n} = \int_{-1}^{1} T_l\left(x\right) T_n\left(x\right)\, dx = \begin{cases} 0, & l+n \text{ odd}, \\[2mm] \dfrac{1}{1-(l+n)^2} + \dfrac{1}{1-(l-n)^2}, & l+n \text{ even}. \end{cases}
$$

Now, for each Ω_m, $m = 2, \ldots, M$, we consider the boundary point $x = a_{m-1}$ corresponding to $i = N_m$. For such a point, the test function v extends to the two adjacent elements Ω_{m-1} and Ω_m, and has the expression

$$
v_{m,N_m}\left(x\right) = \begin{cases} h_{m-1,0}\left(x\right) & \text{if } a_{m-2} \le x \le a_{m-1}, \\[1mm] h_{m,N_m}\left(x\right) & \text{if } a_{m-1} \le x \le a_m, \\[1mm] 0 & \text{otherwise}. \end{cases} \tag{9.39}
$$

By bringing the expressions (9.36) and (9.39) into Eq.(9.33), we get the discrete equation coupling the solutions into Ω_{m-1} and Ω_m, that is,

$$
\begin{aligned}
\sum_{k=0}^{N_{m-1}} C_{0\,k}^{m-1}\, u_{m-1,k} &+ \sum_{k=0}^{N_m} C_{N_m\,k}^{m}\, u_{m,k} \\[2mm]
&= -\sum_{k=0}^{N_{m-1}} B_{0\,k}^{m-1}\, f_{m-1,k} - \sum_{k=0}^{N_m} B_{N_m\,k}^{m}\, f_{m,k}, \quad m = 2, \ldots, M.
\end{aligned} \tag{9.40}
$$

Then we must add the equations expressing the continuity of the solution at the interfaces, namely,

$$
u_{m-1,0} = u_{m,N_m}, \quad m = 2, \ldots, M, \tag{9.41}
$$

and, finally, the boundary conditions

$$
u_{1,N_1} = g_-, \quad u_{M,0} = g_+. \tag{9.42}
$$

In summary, taking Eqs.(9.41) and (9.42) into account, Eqs.(9.38) and (9.40) constitute an algebraic system of $\sum_{m=1}^{M} N_m - 1$ equations for the same number of unknowns, that is,

$$
\mathcal{A} U = F \tag{9.43}
$$

with

$$
U = \left(u_{1,1}, \ldots, u_{1,N_1-1}, u_{2,1}, \ldots, u_{2,N_2}, \ldots, u_{M,1}, \ldots, u_{M,N_M}\right)^T.
$$

The matrix $\mathcal{A}$ has a quasi-block-diagonal structure, that is, the blocks are coupled by a line coming from the matching equation (9.40). Moreover, the matrix $\mathcal{A}$ is positive-definite and symmetric.

Two strategies are generally considered for the solution of the algebraic system (9.43) and, more especially, for the solution of the analogous system

M	N	Error $\overline{E}$
1	12	1.77×10^{-8}
1	16	1.41×10^{-12}
1	18	3.52×10^{-14}
2	16	5.70×10^{-8}
2	20	1.93×10^{-10}
2	24	1.54×10^{-13}
4	16	1.08×10^{-3}
4	40	2.86×10^{-11}

Table 9.1. Error $\overline{E}$ for the smooth solution $u = \sin \pi x$, M is the number of subdomains and N is the total number of degrees of freedom.

obtained in the two-dimensional case. The first one consists of solving the global system by an iterative procedure like the preconditioned conjugate gradient method (Section 3.8). This is of common application, especially for the solution of the Stokes equations (Maday and Patera, 1989 ; Karniadakis and Sherwin, 1999). The second strategy (Patera, 1986) is based on the decoupling of the elements followed by the separate solution of the subsystems containing the grid values interior to the individual elements. The first stage of this algorithm amounts to solving the Schur complement problem determining the interface values (see Section 9.3). This problem can be solved either by a direct or by an iterative method with suitable preconditioning (Chan and Goovaerts, 1989/90; Couzy and Deville, 1995).

9.2.4 Numerical illustration

To evaluate the accuracy of the domain decomposition methods we consider the Helmholtz problem

$$u'' - \sigma u = f, \quad -1 < x < 1 \tag{9.44}$$

$$u(-1) = g_-, \quad u(1) = g_+ \tag{9.45}$$

in the following cases : smooth solution and stiff solution with large derivative at the origin.

The first example concerns the solution

$$u = \sin \pi x,$$

which defines f and $g_- = g_+ = 0$. The problem (9.44)-(9.45) with $\sigma = 100$ is solved using the influence matrix technique (Section 9.2.2.a) associated with the Chebyshev tau method. Results concerning the discrete L^2 error $\overline{E}$ calculated on the Gauss-Lobatto points [see Eq.(3.126)] are given in Table 9.1. In the one-domain solution ($M = 1$), the degree of the polynomial approximation is N, while it is N/M for the M-domain solution. It is

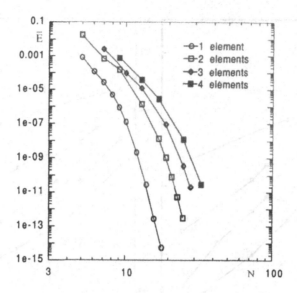

FIGURE 9.1. Error given by the spectral-element method for the solution $u = \cos \pi(x + 1/4)$; N is the total number of collocation points.

observed that, for the same total number of degrees of freedom, the best results are given by the one-domain solution. The same behaviour is observed in the second example constituted by the spectral-element solution of the problem (9.44)-(9.45) with $\sigma = 0$ in the case where

$$u = \cos \pi \left(x + 1/4 \right) ,$$

which defines f and $g_- = g_+ = -\sqrt{2}/2$. The evolution of the error with respect to the total number of degrees of freedom is shown in Fig.9.1. Note that, for the M-element solution, the local polynomial degree is again N/M.

As a matter of fact, this is a general observation : for the same total number of degrees of freedom, the multidomain representation of a smooth function is less accurate than its one-domain representation. A simple reasoning allows us to understand easily the cause of such behaviour. Let us assume $u(x)$ to be a polynomial of degree N, defined in $[-1, 1]$. In the domain $[-1, 1]$, the polynomial approximation of degree N obviously gives an exact representation of u. Now let us consider the two-domain decomposition $[-1, 0]$ and $[0, 1]$. It is clear that the polynomial $u(x)$ of degree N cannot be exactly represented in each subdomain if the degree of the local polynomial approximation is $N/2$.

The third example concerns a stiff solution having a large derivative at the origin, that is,

$$u = \frac{1}{2} \left[\tanh(20\,x) + 1 \right]$$

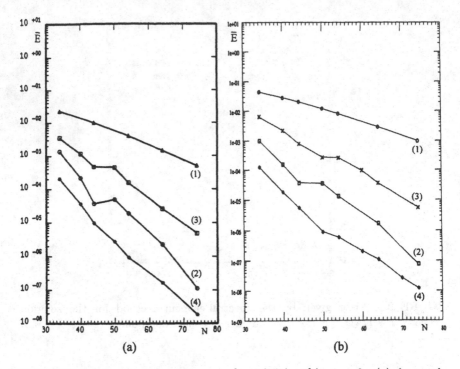

FIGURE 9.2. Error for the solution $u = [\tanh(20x)+1]/2$ given by (a) the patching tau method and (b) the spectral-element method : (1) $M = 1$, (2) $M = 2$, $a_1 = 0$, (3) $M = 3$, $a_2 = -a_1 = 0.2$, (4) $M = 3$, $a_2 = -a_1 = 0.1$. M is the number of subdomains or elements, a_1 and a_2 are the abscissae of the interfaces, N is the total number of degrees of freedom.

already considered in Section 8.3. Figure 9.2 compares the errors obtained with the tau influence matrix method (Fig.9.2.a) and the spectral-element method (Fig.9.2.b). The similarity between the behaviour of both sets of results is remarkable. In this example, the interest in using a multidomain method is clear.

9.3   Two-dimensional Helmholtz equation

The domain decomposition methods described in the previous section can be extended to multidimensional problems without fundamental difficulty. However, their implementation and, especially, the direct inversion of the influence matrix may be difficult, if not impossible, due to its size. Consequently, except in some simple configurations, iterative methods have to be considered, as already mentioned.

In the present section, we discuss the application of the patching method to the solution of the two-dimensional Helmholtz equation with Dirichlet

conditions in the case where the domain Ω is decomposed into two subdomains Ω_1 and Ω_2 (Fig.9.3.a). The case of more general boundary conditions does not pose any difficulty. For the advection-diffusion equation, the
application of the iterative Dirichlet/Neumann method in the case where
advection dominates must take into account the direction of the advective
velocity (Section 9.2.1.b).

Some examples of decomposition using more than two domains are shown
in Fig.9.3. The decomposition (b) is quite similar to the one-dimensional
configuration considered in Section 9.2. The decompositions (c) and (d),
although slightly different, can be handled in the same way except that, in
case (d), the corner C needs special treatment.

In the following, we begin by setting the patching problem which will
be solved by the Chebyshev collocation method. Then the influence matrix
method is briefly described and its relationship with the Schur-complement
problem is discussed.

9.3.1 Patching method

Let us consider the Helmholtz problem

$$Lu \equiv \nabla^2 u - \sigma u = f \quad \text{in } \Omega = (-1,1) \times (-1,1) \tag{9.46}$$

$$u = g \quad \text{on } \Gamma = \partial\Omega. \tag{9.47}$$

The domain Ω is decomposed into two subdomains Ω_m, $m = 1,2$, whose
interface γ is defined by $x = a_1$ (Fig.9.3.a). We denote by Γ_m the part of Γ
associated with Ω_m and by u_m the restriction of u to Ω_m. The two-domain
formulation of the problem (9.46)-(9.47) is defined by the $\mathcal{P}_{1-2}$-Problem:

$$L u_m = f \quad \text{in } \Omega_m, \quad m = 1,2 \tag{9.48}$$

$$u_m = g \quad \text{on } \Gamma_m, \tag{9.49}$$

with the transmission conditions

$$u_1 = u_2 \quad \text{on } \gamma \tag{9.50}$$

$$\partial_x u_1 = \partial_x u_2 \quad \text{on } \gamma, \tag{9.51}$$

ensuring the continuity of the function and its normal derivative at the
interface γ.

By the way, we note that these continuity conditions are also applied
in the multidomain decompositions (b)-(d). The determination of the solution at the corner C of decomposition (d) needs four equations. Three
equations are obtained by the continuity of u. The fourth one is obtained
by prescribing the continuity of either derivative $\partial_x u$ or $\partial_y u$. It is stated
by Canuto et al. (1988) that the jump in the other derivative is spectrally
small.

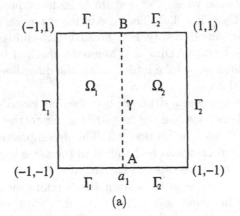

(a)

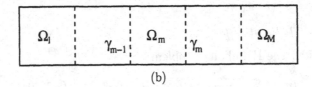

(b)

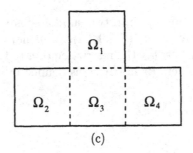

(c)

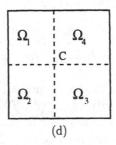

(d)

FIGURE 9.3. Examples of domain decomposition.

The approximation of the $\mathcal{P}_{1-2}$-problem by the Chebyshev collocation method requires the change of the x-coordinate in order to map each subdomain Ω_m on the square domain $(-1,1) \times (-1,1)$. This mapping is defined by $\hat{x}^m = \Phi_m(x)$ with Φ_m given by Eq.(9.34) and we denote by $\hat{L}$ the operator transformed from L. Let N_x^m be the degree of the polynomial approximation in the x-direction in the subdomain Ω_m. In the y-direction it is assumed that $N_y^1 = N_y^2 = N_y$. This assumption allows the collocation points to coincide at the interface γ. If not, interpolation should be done on γ to prescribe the continuity conditions. Note that, for the parallel implementation of a multidomain method, efficiency requires the number of collocation points to be the same in every subdomain.

The collocation points are defined by

$$\hat{x}_i^m = \cos \frac{\pi i}{N_x^m}, \qquad i = 0, \ldots, N_x^m, \quad m = 1, 2$$

$$y_j = \cos \frac{\pi j}{N_y}, \qquad j = 0, \ldots, N_y. \tag{9.52}$$

We denote by Ω_{mN} the set of inner collocation points in Ω_m, by Γ_{mN}^I the set of collocation points on Γ_m except the corners, and by γ_N^I the set of inner collocation points on γ. Finally, the polynomial approximation to u_m in Ω_m is denoted by $u_{mN}(\hat{x}^m, y)$.

9.3.2   Influence matrix method

This method has been described several times, so we restrict the discussion to a brief presentation of the discrete case. As usual, we introduce the decomposition

$$u_{mN} = \tilde{u}_{mN} + \bar{u}_{mN}, \quad m = 1, 2. \tag{9.53}$$

The grid values $\tilde{u}_{mN}(\hat{x}_i^m, y_j)$ of the polynomial $\tilde{u}_{mN}$ are the solution of the discrete $\tilde{\mathcal{P}}_m$-Problem :

$$\hat{L}\,\tilde{u}_{mN} = f \quad \text{in } \Omega_{mN} \tag{9.54}$$

$$\tilde{u}_{mN} = g \quad \text{on } \Gamma_{mN}^I \tag{9.55}$$

$$\tilde{u}_{mN} = 0 \quad \text{on } \gamma_N^I. \tag{9.56}$$

Then the part $\bar{u}_{mN}$ is sought in the form

$$\bar{u}_{mN} = \sum_{l=1}^{N_y-1} \xi_l \, \bar{u}_{m,l} \tag{9.57}$$

and the grid values $\bar{u}_{m,l}(\hat{x}_i^m, y_j)$ of the polynomial $\bar{u}_{m,l}$ are the solution of the discrete $\overline{\mathcal{P}}_{m,l}$-Problem :

$$\hat{L}\,\bar{u}_{m,l} = 0 \qquad \text{in } \Omega_{mN}$$

$$\bar{u}_{m,l} = 0 \qquad \text{on } \Gamma_{mN}^I$$

$$\bar{u}_{m,l|y_j} = \delta_{l,j} \qquad \text{for } y_j \in \gamma_N^I\,.$$

Note that $\xi_j, j = 1, \ldots, N_y-1$, is the common value $u_1(a_1, y_j) = u_2(a_1, y_j)$, that is to say, the continuity condition (9.50) is satisfied on γ_N^I whatever ξ_j. These quantities are determined so that condition (9.51) is satisfied on γ_N^I. We obtain the algebraic system

$$\sum_{l=1}^{N_y-1} \left[\partial_{\hat{x}^1}\bar{u}_{1,l}(1, y_j) - \partial_{\hat{x}^2}\bar{u}_{2,l}(-1, y_j)\right]\xi_l \tag{9.58}$$

$$= -\partial_{\hat{x}^1}\tilde{u}_1(1, y_j) + \partial_{\hat{x}^2}\tilde{u}_2(-1, y_j)\,, \quad j = 1, \ldots, N_y - 1\,,$$

which is written in vector form as

$$\mathcal{M}\,\Xi = \tilde{E}\,, \tag{9.59}$$

where $\Xi = (\xi_1, \ldots, \xi_{N_y-1})^T$. The influence (or capacitance) matrix $\mathcal{M}$ is such that $\mathcal{M} = \mathcal{M}_1 - \mathcal{M}_2$ where

$$\mathcal{M}_m = \left[M_{i,j}^m\right]\,, \quad m = 1, 2\,, \tag{9.60}$$

with

$$M_{i,j}^m = \partial_{\hat{x}^m}\bar{u}_{m,j}(3 - 2m, y_i)\,, \quad i, j = 1, \ldots, N_y - 1\,,$$

and $\tilde{E}$ is the vector whose components are defined by the right-hand side of Eq.(9.58).

In summary, the solution of the $\mathcal{P}_{1-2}$-problem is reduced to the solution of the two algebraic $\tilde{\mathcal{P}}_m$-systems and of the $2(N_y - 1)$ algebraic $\overline{\mathcal{P}}_{m,l}$-systems. We refer to Sections 6.3.2.d and 6.4.2 for general recommendations and remarks concerning the implementation of the influence matrix method. For unsteady problems, where the $\mathcal{P}_{1-2}$-problem has to be solved at each time-cycle, the efficiency of the method lies in the fact that the elementary $\overline{\mathcal{P}}_{m,l}$-problems can be solved once and for all before the start of the time-integration. Thus, the influence matrix is calculated at this stage, is inverted, and its inverse is stored. Moreover, in order to avoid the storage of the elementary solutions $\bar{u}_{m,l}$, it is preferable, after Ξ has been determined, to calculate the solution $u_{mN}, m = 1, 2$, from the discrete problem solved independently in each subdomain

$$\hat{L}\,u_{mN} = f \qquad \text{in } \Omega_{mN}$$

$$u_{mN} = g \qquad \text{on } \Gamma_{mN}^I$$

$$u_{mN|y_j} = \xi_j \qquad \text{for } y_j \in \gamma_N^I\,.$$

Now let us consider the multidomain decompositions (b)-(d) of Fig.9.3. The number of sets of elementary solutions involved in Eq.(9.57) is equal to the number of interfaces delimiting the subdomain Ω_m. For example, for decomposition (b), the linear combination (9.57) in the domain Ω_m, whose two sides are the interfaces γ_{m-1} and γ_m, has the general form

$$\bar{u}_{mN} = \sum_{l=1}^{N_y-1} \left(\xi_{m,l}^{(1)} \, \bar{u}_{m,l}^{(1)} + \xi_{m,l}^{(2)} \, \bar{u}_{m,l}^{(2)} \right) . \tag{9.61}$$

Then if $\bar{u}_{m,l}^{(1)}$ and $\bar{u}_{m,l}^{(2)}$ satisfy the boundary conditions

$$\bar{u}_{m,l|y_j}^{(1)} = \delta_{l,j} \, , \quad \bar{u}_{m,l|y_j}^{(2)} = 0 \qquad \text{for } y_j \in \gamma_{m-1,N}^I \, ,$$
$$\bar{u}_{m,l|y_j}^{(1)} = 0 \, , \quad \bar{u}_{m,l|y_j}^{(2)} = \delta_{l,j} \qquad \text{for } y_j \in \gamma_{mN}^I \, ,$$

the continuity of the solution at the interface γ_m implies $\xi_{m,l}^{(2)} = \xi_{m+1,l}^{(1)}$. The influence matrix $\mathcal{M}$ is block-tridiagonal, each block corresponding to an interface. If the number of subdomains is large the direct inversion of $\mathcal{M}$ is not possible, but the influence system analogous to (9.59) can be solved by the classical block-LU-decomposition (Isaacson and Keller, 1966) or by a block-cyclic-reduction algorithm adapted to parallel computers (Gallopoulos and Saad, 1989).

In the case of decomposition (c), the linear combination (9.54) in the subdomain Ω_3 is of the form

$$\bar{u}_{3N} = \sum_{l=1}^{N_y-1} \left(\xi_{3,l}^{(1)} \, \bar{u}_{3,l}^{(1)} + \xi_{3,l}^{(2)} \, \bar{u}_{3,l}^{(2)} \right) + \sum_{l=1}^{N_{x2}-1} \xi_{3,l}^{(3)} \, \bar{u}_{3,l}^{(3)} . \tag{9.62}$$

Such a decomposition will be considered in Section 9.5.3.b for the calculation of flows in a rotating system.

As already mentioned, iterative methods are indispensable for more complex configurations with a large number of subdomains. In this respect, it is recalled that the Dirichlet/Neumann procedure (Section 9.2.1.b) is equivalent to the iterative solution of the influence matrix system (9.59) with the preconditioner $\mathcal{A}_0 = -\mathcal{M}_2$ in the case of the two-domain decomposition. Other preconditioners are discussed by Quarteroni and Sacchi-Landriani (1988). Moreover, preconditioners devised for the Schur complement system (Bjørstadt and Widlund, 1986 ; Chan and Goovaerts, 1989/90 ; Chan and Mathew, 1994 ; Couzy and Deville, 1995 ; Quarteroni and Valli, 1999) may be used since the influence matrix coincides with the Schur complement as will be shown in the following.

To conclude this section, we mention a multidomain method proposed by Borchers et al. (1998) whose principle is close to the one on which the influence matrix technique is based. More precisely, this method (called

"Conjugate Gradient Boundary Iteration") consists of finding iteratively the value of the normal derivative of the solution at the interface ensuring the continuity of this solution, obtained by the minimization of some quadratic functional. Such a method constitutes, in a way, a dual version of the influence matrix technique. Note that dual Schur complement methods associated with finite-element approximations have been developed in structural mechanics (Fahrat and Roux, 1995) and applied to fluid mechanics (Vanderstraeten and Keunings, 1998).

9.3.3 Schur complement problem

The strategy that consists of constructing a separated problem for the interface values of the unknown is common in the field of domain decomposition. This leads to the Schur complement system whose construction is now described.

Returning to the P_{1-2}-problem (9.48)-(9.51), we approximate it by means of the Chebyshev collocation method. We denote by U_m, $m = 1, 2$, the $(N_{xm} - 1)(N_y - 1)$-component vector made with the inner grid values of the polynomial $u_{mN}(\hat{x}^m, y)$ in Ω_{mN}. The interface values $\xi_j = u_{1N}(1, y_j) = u_{2N}(-1, y_j)$, $j = 1, .., N_y$, constitute the $(N_y - 1)$-component vector Ξ. Then the discrete system approximating the P_{1-2}-problem may be written as

$$\begin{pmatrix} \mathcal{A}_{11} & \mathcal{O} & \mathcal{A}_{13} \\ \mathcal{O} & \mathcal{A}_{22} & \mathcal{A}_{23} \\ \mathcal{A}_{31} & \mathcal{A}_{32} & \mathcal{A}_{33} \end{pmatrix} \begin{pmatrix} U_1 \\ U_2 \\ \Xi \end{pmatrix} = \begin{pmatrix} F_1 \\ F_2 \\ 0 \end{pmatrix}, \qquad (9.63)$$

where $\mathcal{A}_{11}$ and $\mathcal{A}_{22}$ are the matrices associated with the discretization of the equation but involving only the inner grid values in Ω_1 and Ω_2, respectively; $\mathcal{A}_{13}$ and $\mathcal{A}_{23}$ are similar matrices but involving only the inner interface values ; $\mathcal{A}_{31}$, $\mathcal{A}_{32}$ and $\mathcal{A}_{33}$ are matrices associated with the discretization of the continuity condition of the normal derivative.

In finite-difference or finite-element methods, the system (9.63) would correspond to the global discretization of problem (9.46)-(9.47), in which the equation is also enforced at the interface γ [that is, transmission conditions like (9.50)-(9.51) are not considered]. In this case, we would have $\mathcal{A}_{31} = \mathcal{A}_{13}^T$ and $\mathcal{A}_{32} = \mathcal{A}_{23}^T$.

Equation (9.63) is written

$$\mathcal{A}_{11} U_1 + \mathcal{A}_{13} \Xi = F_1$$

$$\mathcal{A}_{22} U_2 + \mathcal{A}_{23} \Xi = F_2$$

$$\mathcal{A}_{31} U_1 + \mathcal{A}_{32} U_2 + \mathcal{A}_{33} \Xi = 0,$$

then, by eliminating U_1 and U_2, we get the equation determining Ξ, that is,

$$\left(-\mathcal{A}_{31}\mathcal{A}_{11}^{-1}\mathcal{A}_{13} - \mathcal{A}_{32}\mathcal{A}_{22}^{-1}\mathcal{A}_{23} + \mathcal{A}_{33}\right)\Xi = -\left(\mathcal{A}_{31}\mathcal{A}_{11}^{-1}F_1 + \mathcal{A}_{32}\mathcal{A}_{22}^{-1}F_2\right).$$
(9.64)

The matrix

$$\mathcal{C} = \mathcal{A}_{33} - \mathcal{A}_{31}\mathcal{A}_{11}^{-1}\mathcal{A}_{13} - \mathcal{A}_{32}\mathcal{A}_{22}^{-1}\mathcal{A}_{23} \qquad (9.65)$$

is the Schur complement of $\mathcal{A}_{33}$ in the matrix $\mathcal{A}$. In the next section, it is shown that $\mathcal{C}$ coincides with the influence matrix $\mathcal{M}$.

9.3.4   Identity between influence matrix and Schur complement

First of all, since the influence matrix method is nothing but an algorithm for solving the global algebraic system approximating the $\mathcal{P}_{1-2}$-problem, it is clear that the equations determining the interface values, namely, the influence and the Schur complement systems, must be the same. Nevertheless, it is instructive to make this identity more precise.

It is easy to see that the vector form of the discrete $\tilde{\mathcal{P}}$-problem (9.54)-(9.56) is

$$\mathcal{A}_{mm}\tilde{U}_m = F_m \quad \text{in } \Omega_m, \quad m = 1, 2, \qquad (9.66)$$

where $\tilde{U}_m$ is the vector made with the inner grid values of $\tilde{u}_{mN}$. With analogous notations, the discrete $\tilde{\mathcal{P}}_{m,l}$-problems are written

$$\mathcal{A}_{mm}\overline{U}_{m,l} = -\mathcal{A}_{m3}J_l \quad \text{in } \Omega_m, \quad m = 1, 2, \quad l = 1, \ldots, N_y - 1, \qquad (9.67)$$

where J_l is the $(N_y - 1)$-component vector such as

$$J_l = [\delta_{l,j}], \quad l = 1, \ldots, N_y - 1, \quad j = 1, \ldots, N_y - 1.$$

Now Eq.(9.58) can be written as

$$\mathcal{A}_{31}\left(\tilde{U}_1 + \sum_{l=1}^{N_y-1}\xi_l\overline{U}_{1,l}\right) + \mathcal{A}_{32}\left(\tilde{U}_2 + \sum_{l=1}^{N_y-1}\xi_l\overline{U}_{2,l}\right) + \mathcal{A}_{33}\Xi = 0. \quad (9.68)$$

From Eqs.(9.66) and (9.67), respectively, we have

$$\tilde{U}_m = \mathcal{A}_{mm}^{-1}F_m, \quad \overline{U}_{m,l} = -\mathcal{A}_{mm}^{-1}\mathcal{A}_{m3}J_l.$$

Then, bringing these expressions into Eq.(9.68) and observing that

$$\sum_{l=1}^{N_y-1}\xi_l J_l = \Xi,$$

it is easy to see that the resulting equation is identical to Eq.(9.64). This proves the identity between the influence system and the Schur complement system.

9.4   Time-dependent equations

When dealing with time-dependent equations, the interface conditions may have an influence on the stability. By stability we mean the stability of the Chebyshev approximation (Section 4.2.3) as well as the stability of the time-discretization scheme (Sections 4.3.2 and 4.4.2). These points are discussed now.

9.4.1   Stability of the Chebyshev approximation

In Section 4.2.3 it has been stated that, when advection dominates, the solution of the differential equations resulting from the Chebyshev approximation to the advection-diffusion equation (4.5) may be not bounded in time if the spatial resolution is not sufficiently high. In other words, for $\delta = \nu/a \ll 1$ (with $a > 0$), there exists a critical polynomial degree N_c such that the solution is unbounded if $N < N_c$. Approximate laws for the critical number N_c, according to the type of boundary conditions, are given in Section 4.2.3.

It may be useful to know what happens in a multidomain method. The question has been examined (Mofid, 1992 ; Peyret, 1990) in the case of the two-domain decomposition patching method associated with the Chebyshev tau approximation. This was done by a numerical study of the eigenvalue problem analogous to (4.25). The case where the interface is located at the origin ($a_1 = 0$), and the degree of the polynomial approximation is the same in each subdomain ($N_1 = N_2$), is considered first. It is found that the total critical number $N_c^{2D} = N_{1c} + N_{2c} = 2\,N_{1c}$ (i.e., corresponding to the total number of degrees of freedom) follows the approximate law

$$N_c^{2D} \cong 1.35\,N_c, \qquad (9.69)$$

where N_c is the critical number found in Section 4.2.3 for the one-domain solution. Relation (9.69) is valid for Dirichlet-Dirichlet conditions as well as for Dirichlet-Neumann conditions.

| $\delta = \nu/a$ | 1-Domain | 2-Domain | | | | | |
| | N_c | $a_1 = 0$ | | $a_1 = 0.6$ | | $a_1 = 0.8$ | |
		N_{1c}	N_{2c}	N_{1c}	N_{2c}	N_{1c}	N_{2c}
2.5×10^{-3}	35	7	24	9	14	12	10
1.0×10^{-3}	61	13	40	23	24	22	15
6.0×10^{-4}	81	18	54	32	32	38	22
5.0×10^{-4}	101	23	68	42	40	50	27

Table 9.2. Critical number for the advection-diffusion equation with Dirichlet conditions (tau method).

Then it is found that the total critical number $N_c^{2D} = N_{1c} + N_{2c}$ can be made smaller than N_c if the condition $N_1 = N_2$ is relaxed and/or if $a_1 > 0$. This is in agreement with the fact that the instability is due to the effect of the boundary condition at $x = 1$ (since $a > 0$) and, hence, the size of Ω_1 may be enlarged. Table 9.2 gives some typical results for the Dirichlet-Dirichlet case.

9.4.2 Stability of time-discretization schemes

In order to have an idea about the influence of the interface conditions on the stability of time-discretization schemes, we consider the diffusion equation with Dirichlet conditions

$$\partial_t u - \nu\,\partial_{xx} u = 0 \ \text{ in } \Omega = (-1, 1)\,, \ \ t > 0 \tag{9.70}$$

$$u(-1, t) = 0\,, \quad u(1, t) = 0 \tag{9.71}$$

$$u(x, 0) = u_0(x)\,, \tag{9.72}$$

solved by using the two-domain decomposition $\Omega_1 = (-1, 0)$ and $\Omega_2 = (0, 1)$, and prescribing the continuity condition

$$u_1(0, t) = u_2(0, t)\,, \quad \partial_x u_1(0, t) = \partial_x u_2(0, t)\,, \tag{9.73}$$

where u_1 and u_2 are the restrictions of u to Ω_1 and Ω_2, respectively.

Now let us consider the solution u_1 in Ω_1 and introduce the transformation $x \to -x$ (Gottlieb and Lutsman, 1983) so that the diffusion equation remains unchanged but Ω_1 is transformed into $\Omega_1' = (0, 1) = \Omega_2$. Let us set $u_1'(x, t) = u_1(-x, t)$, then it is easy to verify that $S = u_1' + u_2$ and $D = u_1' - u_2$ are, respectively, solutions to

$$\partial_t S - \nu\,\partial_{xx} S = 0\,, \quad 0 < x < 1$$

$$\partial_x S(0, t) = 0\,, \quad S(1, t) = 0 \tag{9.74}$$

$$S(x, 0) = S_0(x)\,,$$

and

$$\partial_t D - \nu\,\partial_{xx} D = 0\,, \quad 0 < x < 1$$

$$D(0, t) = 0\,, \quad D(1, t) = 0 \tag{9.75}$$

$$D(x, 0) = D_0(x)\,,$$

where S_0 and D_0 are deduced from Eq.(9.72).

Therefore, we are led to a mixed Neumann-Dirichlet problem for the sum S and to a pure Dirichlet problem for the difference D. The stability of a time-discretization scheme associated with the two-domain problem (assuming the same degree of approximation in each subdomain, i.e., $N_1 =$

$N_2 = N/2$) can be deduced from the stability of the same scheme associated with the one-domain problems (9.74) and (9.75).

For example, let us assume the diffusion equation (9.70) is discretized in time by means of the simple explicit scheme

$$\frac{u^{n+1} - u^n}{\Delta t} - \nu \partial_{xx} u^n = 0. \tag{9.76}$$

The stability of such a scheme associated with the Chebyshev (tau or collocation) method has been discussed in Section 4.3.2 for various types of boundary conditions. Therefore, we may apply these results to the problems (9.74) and (9.75), after having introduced the coordinate transform mapping the interval $[0, 1]$ onto $[-1, 1]$.

First, let us consider the case of the collocation method. The stability criterion for both problems (Neumann-Dirichlet and Dirichlet-Dirichlet) is

$$\Delta t \leq \frac{170.20}{\nu N^4} \tag{9.77}$$

which has to be compared to the criterion $\Delta t \leq 42.55 / (\nu N^4)$ associated with the one-domain problem (9.70)-(9.72). Therefore, the critical time-step for the two-domain solution is four times the critical time-step associated with the one-domain solution. This result is in agreement with the heuristic argument based on the variation of the time-step according to the size of the subdomains. However, it does not apply to the case where the original problem has Neumann conditions, namely, when conditions (9.71) are replaced by

$$\partial_x u (-1, t) = 0, \quad \partial_x u (1, t) = 0. \tag{9.78}$$

The same transformation $x \to -x$ as above leads to the same equations for S and D but with the boundary conditions

$$\partial_x S (0, t) = 0, \quad \partial_x S (1, t) = 0 \tag{9.79}$$

and

$$D (0, t) = 0, \quad \partial_x D (1, t) = 0. \tag{9.80}$$

Thus, the sum S is a solution of a pure Neumann problem and the associated stability criterion is $\Delta t \leq 571.40 / (\nu N^4)$. The difference D is the solution of a mixed Dirichlet-Neumann problem whose stability criterion is given by Eq.(9.77). The stability of the two-domain problem is dictated by the more stringent of these two criteria, that is, Eq.(9.77), while the critical time-step for the original Neumann problem is $\Delta t \leq 142.85 / (\nu N^4)$. Thus, the critical time-step for the two-domain problem is larger than the one associated with the one-domain problem, but it does not follow the heuristic argument based on the size of the subdomains : it is only 1.19 times larger.

Now, in the case of the Chebyshev tau method, the conclusion for the pure Dirichlet problem (9.70)-(9.72) is the same as that for the collocation method, namely, the two-domain problem admits a time-step four times larger than the one-domain problem does. On the other hand, if the original problem is of pure Neumann type, the two-domain critical time-step is dictated by the Dirichlet-Neumann criterion, namely,

$$\Delta t \le \frac{26.40}{\nu\,N^4}\,, \tag{9.81}$$

which is smaller than the one-domain Neumann-Neumann criterion $\Delta t \le 42.55/\left(\nu\,N^4\right)$.

In summary, the heuristic rule, which would consist of estimating the critical time-step in a multi-domain method by adapting the one-domain criterion to the size of the subdomains, is not of general application. The continuity conditions at the interface change the mathematical nature of the problem and, consequently, the associated stability criterion.

9.5   Navier-Stokes equations in vorticity-streamfunction variables

As discussed in Chapter 6, the solution of the time-dependent Navier-Stokes equations, using a semi-implicit scheme, reduces to the solution of a Stokes problem at each time-cycle. The influence matrix method constitutes an efficient algorithm for the solution of the Stokes problem, allowing us to accurately prescribe the boundary conditions on the streamfunction. In the domain decomposition method, the interface conditions are also efficiently handled by means of an influence matrix (Schur complement). With the Fourier-Chebyshev approximation, it is possible to combine together boundary conditions and interface conditions in the same influence matrix as was done in Section 9.2.2.a. On the other hand, for the two-dimensional Chebyshev-Chebyshev method, it is simpler to consider an interface influence matrix system to determine the interface values and to solve the associated Stokes problem in each subdomain using the boundary influence matrix.

The general multidomain solution method for the Stokes problem is based on the domain decomposition method developed above in the present chapter, and on the one-domain solution method given in Chapter 6. Consequently, in the following, we restrict ourselves to presenting the broad lines of the method by considering successively the case of Fourier-Chebyshev and Chebyshev-Chebyshev approximations.

9.5.1   Fourier-Chebyshev method

As seen in Section 6.3, the Fourier-Chebyshev method applied to the Stokes problem reduces to the solution of the following one-dimensional problem, for each Fourier mode k :

$$\omega'' - \sigma \omega = f, \quad -1 < y < 1 \tag{9.82}$$

$$\psi'' - \kappa \psi + \omega = 0, \quad -1 < y < 1 \tag{9.83}$$

$$\psi(-1) = g_-, \quad \psi(1) = g_+ \tag{9.84}$$

$$\psi'(-1) = h_-, \quad \psi'(1) = h_+. \tag{9.85}$$

where $\kappa = k^2$. The multidomain solution of this problem (Pulicani, 1988 ; Lacroix *et al.*, 1988) is similar to the one described in Section 9.2.2. The domain $\Omega = (-1,1)$ is decomposed in M subdomains $\Omega_m = (a_{m-1}, a_m)$, $m = 1, \ldots, M$, with $a_0 = -1$ and $a_M = 1$. We denote by ω_m and ψ_m, respectively, the restriction of ω and ψ to Ω_m. Therefore, (ω_m, ψ_m) satisfies Eqs.(9.82) and (9.83) in Ω_m, $m = 1, \ldots, M$, with the boundary conditions (9.84) and (9.85), that is,

$$\psi_1(-1) = g_-, \quad \psi_1'(-1) = h_-, \tag{9.86}$$

$$\psi_M(1) = g_+, \quad \psi_M'(1) = h_+, \tag{9.87}$$

and the interface conditions, for $m = 1, \ldots, M-1$:

$$\phi_m(a_m) = \phi_{m+1}(a_m), \tag{9.88}$$

$$\phi_m'(a_m) = \phi_{m+1}'(a_m), \tag{9.89}$$

where $\phi_m = \omega_m, \psi_m$.

As usual, we introduce the decomposition

$$\omega_m = \tilde{\omega}_m + \overline{\omega}_m, \quad \psi_m = \tilde{\psi}_m + \overline{\psi}_m, \quad m = 1, \ldots, M-1, \tag{9.90}$$

where the pair $\left(\tilde{\omega}_m, \tilde{\psi}_m \right)$, $m = 1, \ldots, M$, is the solution of the $\tilde{\mathcal{P}}_m$-*Problem* defined by

$$\tilde{\omega}_m'' - \sigma \tilde{\omega}_m = f \quad \text{in } \Omega_m \tag{9.91}$$

$$\tilde{\omega}_m(a_{m-1}) = 0, \quad \tilde{\omega}_m(a_m) = 0, \tag{9.92}$$

and

$$\tilde{\psi}_m'' - \kappa \tilde{\psi}_m = -\tilde{\omega}_m \quad \text{in } \Omega_m \tag{9.93}$$

$$\tilde{\psi}_m(a_{m-1}) = g_{m-1}, \quad \tilde{\psi}_m(a_m) = g_m, \tag{9.94}$$

with $g_0 = g_-$, $g_M = g_+$ and $g_m = 0$ for $m = 1, \ldots, M-1$.

The pair $\left(\overline{\omega}_m, \overline{\psi}_m \right)$, $m = 1, \ldots, M$, is the solution of the $\overline{\mathcal{P}}_m$-*Problem* :

$$\overline{\omega}_m'' - \sigma \overline{\omega}_m = 0 \quad \text{in } \Omega_m, \tag{9.95}$$

$$\overline{\psi}_m'' - \kappa \overline{\psi}_m + \overline{\omega}_m = 0 \quad \text{in } \Omega_m \tag{9.96}$$

$$\overline{\omega}_m(a_{m-1}) = \xi_{m-1}, \quad \overline{\omega}_m(a_m) = \xi_m \tag{9.97}$$

$$\overline{\psi}_m(a_{m-1}) = \eta_{m-1}, \quad \overline{\psi}_m(a_m) = \eta_m, \tag{9.98}$$

with $\eta_0 = \eta_M = 0$. The unknown quantities ξ_0 and ξ_M are the boundary values of ω at $x = -1$ and $x = 1$, respectively, and (ξ_m, η_m), $m = 1, \ldots, M - 1$, are the interface values of (ω, ψ). Now the solution $\overline{\phi}_m = \overline{\omega}_m, \overline{\psi}_m$ is sought in the form

$$\overline{\phi}_m = \xi_{m-1} \overline{\phi}_{m,1}^{(1)} + \xi_m \overline{\phi}_{m,2}^{(1)} + \eta_{m-1} \overline{\phi}_{m,1}^{(2)} + \eta_m \overline{\phi}_{m,2}^{(2)}. \tag{9.99}$$

Every $\overline{\phi}_{m,l}^{(n)} = \overline{\omega}_{m,l}^{(n)}, \overline{\psi}_{m,l}^{(n)}$, $m = 1, \ldots, M$, $l = 1, 2$, $n = 1, 2$, satisfies the homogeneous equations (9.95) and (9.96) in Ω_m with the boundary conditions

$$\overline{\omega}_{m,l}^{(1)}(a_{m-1}) = \delta_{1,l}, \quad \overline{\omega}_{m,l}^{(1)}(a_m) = \delta_{2,l}, \tag{9.100}$$

$$\overline{\omega}_{m,l}^{(2)}(a_{m-1}) = 0, \quad \overline{\omega}_{m,l}^{(2)}(a_m) = 0, \tag{9.101}$$

and

$$\overline{\psi}_{m,l}^{(1)}(a_{m-1}) = 0, \quad \overline{\psi}_{m,l}^{(1)}(a_m) = 0, \tag{9.102}$$

$$\overline{\psi}_{m,l}^{(2)}(a_{m-1}) = \delta_{1,l}, \quad \overline{\psi}_{m,l}^{(2)}(a_m) = \delta_{2,l}, \tag{9.103}$$

where $\delta_{j,l}$ is the Kronecker delta. Note that the boundary conditions (9.101) imply $\overline{\omega}_{m,1}^{(2)} = \overline{\omega}_{m,2}^{(2)} = 0$.

Finally, the boundary values ξ_0, ξ_M and the interface values (ξ_m, η_m), $m = 1, \ldots, M - 1$, are determined by requiring the boundary conditions (9.87) and the interface conditions (9.89) to be satisfied. We obtain an algebraic system which can be put in the form of a block-tridiagonal system by eliminating ξ_0 and ξ_M, each 2×2 block corresponding to an interface. The associated matrix is nonsingular.

The efficiency of the method lies essentially in the two facts : (1) the overall problem reduces to Dirichlet problems for the Helmholtz equation to be solved in each subdomain, and (2) the major part of these Dirichlet problems is time-independent and, consequently, can be solved in a preprocessing stage performed before the start of the time-integration.

This method has been applied to various double-diffusive convection problems by Lacroix et al. (1988), Peyret (1990), and Pulicani (1994). An application to the time-evolution of a wake submitted to a temperature field is described in Section 9.5.3.

9.5.2   Chebyshev-Chebyshev method

In this section, we are interested in the two-domain solution of the two-dimensional Stokes problem (6.96)-(6.99), that is,

$$\nabla^2 \omega - \sigma\,\omega = f \quad \text{in } \Omega \tag{9.104}$$

$$\nabla^2 \psi + \omega = 0 \quad \text{in } \Omega \tag{9.105}$$

$$\psi = g \quad \text{on } \Gamma \tag{9.106}$$

$$\partial_n \psi = h \quad \text{on } \Gamma, \tag{9.107}$$

where Ω is the square $(-1,1) \times (-1,1)$. The domain Ω is decomposed into Ω_1 and Ω_2 with the interface γ as shown in Fig.9.3.a. We denote by Γ_m the part of Γ associated with Ω_m. Let ω_m and ψ_m be the respective restriction of ω_m and ψ_m to Ω_m, $m = 1, 2$. In each Ω_m the pair (ω_m, ψ_m) is the solution of Eqs.(9.105) and (9.106) with the boundary conditions

$$\psi_m = g, \quad \partial_n \psi_m = h \quad \text{on } \Gamma_m, \tag{9.108}$$

and the interface conditions

$$\phi_1 = \phi_2 \quad \text{on } \gamma \tag{9.109}$$

$$\partial_x \phi_1 = \partial_x \phi_2 \quad \text{on } \gamma, \tag{9.110}$$

where $\phi_m = \omega_m, \psi_m$.

The global boundary-interface influence matrix method described in the previous section for the one-dimensional Stokes problem cannot be applied here because the influence matrix would be too large. Therefore, the enforcement of interface conditions is separated from one of the boundary conditions (Raspo *et al.*, 1996). The solution method using the Chebyshev collocation approximation is now briefly outlined.

As usual, each subdomain Ω_m is mapped onto the square domain $(-1,1) \times (-1,1)$ by means of the x-coordinate transform $\hat{x}^m = \Phi_m(x)$. The solution $\phi_m = \omega_m, \psi_m$ is approximated with the polynomial $\phi_{mN}(\hat{x}^m, y)$. Therefore, the unknowns are the grid values $\phi_{mN}(\hat{x}_i^m, y_j)$ at the collocation points defined by Eq.(9.52). We refer to Section 9.3.1 for definitions and notations.

The polynomial solution $\phi_{mN} = \omega_{mN}, \psi_{mN}$ is decomposed according to

$$\phi_{mN} = \tilde{\phi}_{mN} + \overline{\phi}_{mN}. \tag{9.111}$$

The grid values of $\tilde{\phi}_{mN} = \tilde{\omega}_{mN}, \tilde{\psi}_{mN}$ are determined by the discrete $\tilde{\mathcal{P}}_m$-*Problem* obtained from Eqs.(9.104) and (9.105) enforced at the collocation points belonging to Ω_{mN}, with the boundary conditions (9.108), namely,

$$\tilde{\psi}_{mN} = g, \quad \partial_n \tilde{\psi}_{mN} = h \quad \text{on } \Gamma_{mN}^I, \tag{9.112}$$

and the homogeneous boundary conditions

$$\tilde{\omega}_{mN} = 0, \quad \tilde{\psi}_{mN} = 0 \quad \text{on } \gamma_N^I. \tag{9.113}$$

This constitutes a discrete Stokes problem which is solved by means of the method given in Section 6.4.2. Note that the Neumann condition on the streamfunction is prescribed on part of the boundary only (i.e., not on γ_N^I), so that two collocation points on the boundary Γ_{mN}^I must be excluded when the boundary influence matrix is constructed. A convenient choice is constituted by the first two points near the corners on the horizontal part of Γ_{mN}^I.

Then the grid values of $\overline{\phi}_{mN} = \overline{\omega}_{mN}, \overline{\psi}_{mN}$ are determined by the discrete $\overline{\mathcal{P}}_m$-Problem constituted by the homogeneous equations derived from Eqs.(9.104) and (9.105) by setting $f = 0$, associated with the boundary conditions

$$\overline{\psi}_{mN} = 0, \quad \partial_n \overline{\psi}_{mN} = 0 \quad \text{on } \Gamma_{mN}^I \tag{9.114}$$

$$\overline{\omega}_{mN|y_j} = \xi_j, \quad \overline{\psi}_{mN|y_j} = \eta_j \quad \text{for } y_j \in \gamma_N^I. \tag{9.115}$$

The interface values $\xi_j, \eta_j, j = 1, \ldots, N_y - 1$, are determined by requiring the continuity condition (9.110) to be satisfied on γ_N^I. This is done thanks to the influence matrix technique.

The polynomial $\overline{\phi}_{mN} = \overline{\omega}_{mN}, \overline{\psi}_{mN}$ is sought in the form

$$\overline{\phi}_{mN} = \sum_{l=1}^{N_y - 1} \left(\xi_l \overline{\phi}_{m,l}^{(1)} + \eta_l \overline{\phi}_{m,l}^{(2)} \right). \tag{9.116}$$

The grid values of $\left(\overline{\omega}_{m,l}^{(n)}, \psi_{m,l}^{(n)} \right), n = 1, 2, l = 1, \ldots, N_y - 1$, are determined by the discrete $\overline{\mathcal{P}}_{m,l}$-Problem constituted by the same homogeneous equations as in the $\overline{\mathcal{P}}_m$-problem, with the homogeneous boundary conditions (9.114) and

$$\overline{\omega}_{m,l|y_j}^{(1)} = \delta_{l,j}, \quad \overline{\psi}_{m,l|y_j}^{(1)} = 0 \quad \text{for } y_j \in \gamma_N^I \tag{9.117}$$

$$\overline{\omega}_{m,l|y_j}^{(2)} = 0, \quad \overline{\psi}_{m,l|y_j}^{(2)} = \delta_{l,j} \quad \text{for } y_j \in \gamma_N^I, \tag{9.118}$$

with $l = 1, 2$ and $j = 1, \ldots, N_y - 1$. Therefore, each elementary solution is determined by a Stokes problem solved by the one-domain influence matrix method.

Finally, by requiring the solution $\phi_{mN} = \tilde{\phi}_{mN} + \overline{\phi}_{mN}$ to satisfy the continuity conditions (9.110) on γ_N^I, we get the system

$$\mathcal{M} \Psi = \tilde{E}, \tag{9.119}$$

where

$$\Psi = \begin{pmatrix} \Xi \\ H \end{pmatrix}$$

with

$$\Xi = \left(\xi_1, \ldots, \xi_{N_y-1}\right)^T, \quad H = \left(\eta_1, \ldots, \eta_{N_y-1}\right)^T.$$

It must be noticed that the Neumann condition $\partial_x \psi = g$ is not prescribed at the points A and B (see Fig.9.3.a). In fact, it can be shown that this Neumann condition is automatically satisfied. The reasoning is similar to that developed in Section 6.4.2 to show that the Neumann condition on ψ was satisfied at the points excluded from the construction of the influence matrix.

A constant and important question that arises when dealing with the Chebyshev polynomial approximation is the conditioning of the associated matrices, which may deteriorate when the resolution increases. As already discussed in Section 6.4.2 the boundary influence matrix is well-conditioned. On the other hand, the interface influence matrix $\mathcal{M}$ in Eq. (9.119) may be ill-conditioned. For example, the condition number of $\mathcal{M}$ is of order 10^5 for $a_1 = 0$, $\sigma = 3 \times 10^4$ and $N_{x1} = N_{x2} = 28$, $N_y = 60$. This is not catastrophic and the direct inversion of $\mathcal{M}$ (using elaborate inversion subroutines) produces accurate results (Raspo et al., 1996). However, in other circumstances (e.g. the four-domain decomposition of Fig.9.3.c) the conditioning may be worse, so that preconditioning becomes absolutely necessary. The ill-conditioning of $\mathcal{M}$ is due to the lack of balance between its elements. The form of matrix $\mathcal{M}$ is

$$\mathcal{M} = \begin{pmatrix} \mathcal{M}_{11} & \mathcal{M}_{12} \\ \mathcal{M}_{21} & \mathcal{M}_{22} \end{pmatrix}.$$

The elements of the two submatrices $\mathcal{M}_{11}$ and $\mathcal{M}_{22}$ depending, respectively, on the jumps of the derivatives $\partial_x \overline{\omega}_{m,l}^{(1)}$ and $\partial_x \overline{\psi}_{m,l}^{(2)}$ through γ, are of the same order of magnitude. But, relative to this scale, the elements of the submatrix $\mathcal{M}_{12}$, depending on the jumps of $\partial_x \overline{\omega}_{m,l}^{(2)}$, are too large, while those of $\mathcal{M}_{21}$, depending on the jumps of $\partial_x \overline{\psi}_{m,l}^{(1)}$, are too small. This defect of balance can be removed by preconditioning matrix $\mathcal{M}$ by means of a diagonal matrix $\mathcal{H}$ such that

$$\mathcal{H} = \begin{pmatrix} \mathcal{H}_{11} & \mathcal{O} \\ \mathcal{O} & \mathcal{H}_{22} \end{pmatrix},$$

where $\mathcal{H}_{11}$ and $\mathcal{H}_{22}$ are two diagonal matrices. A possible choice (Raspo et al., 1996) for the elements $(h_{11})_{i,i}$ and $(h_{22})_{i,i}$ of $\mathcal{H}_{11}$ and $\mathcal{H}_{22}$, respectively, is

$$(h_{11})_{i,i} = \frac{10\,N_y}{\displaystyle\sum_{j=1}^{J} \left|(M_{11})_{i,j}\right|}, \quad (h_{22})_{i,i} = \frac{\displaystyle\sum_{j=1}^{J} \left|(M_{22})_{i,j}\right|}{10\,N_y}, \quad i = 1, \ldots, J,$$

where $J = N_y - 1$ and $(M_{11})_{i,j}$ and $(M_{22})_{i,j}$ are the entries of $\mathcal{M}_{11}$ and $\mathcal{M}_{22}$, respectively.

Another possibility is to consider an ILU preconditioner rather than the matrix $\mathcal{H}$ defined above. Such a preconditioner, calculated, for example, by the subroutine F11DAE of the NAG library, is of broader utilization.

A third way to equilibrate the matrix $\mathcal{M}$ is to use a row and column scaling procedure defined by

$$(\mathcal{R} \mathcal{M} \mathcal{C}) \left(\mathcal{C}^{-1} \Psi \right) = \mathcal{R} \tilde{E}, \tag{9.120}$$

where $\mathcal{R}$ and $\mathcal{C}$ are the diagonal matrices containing, respectively, the row scale factors and the column scale factors calculated, for example, by the subroutine SGEEQU from the LAPACK library. The matrix $\mathcal{R} \mathcal{M} \mathcal{C}$ is well-conditioned and is accurately inverted, so that Eq.(9.120) gives the solution

$$\Psi = \mathcal{C} \left(\mathcal{R} \mathcal{M} \mathcal{C} \right)^{-1} \mathcal{R} \tilde{E}.$$

This last procedure has been found (Raspo, 2002) to be the most efficient in the case of the four-domain decomposition of Fig.9.3.c (see Section 9.5.3.b).

In summary, the general two-domain algorithm for the time-dependent solution of the Navier-Stokes equations is the following.

A. Preprocessing stage

A.1. Construct the boundary influence matrices associated with the Stokes problems in Ω_m, $m = 1, 2$, invert them, and store the inverses.

A.2. Construct the interface influence matrix $\mathcal{M}$, invert it, and store the inverse.

B. At each time-cycle

B.1. Calculate the right-hand side f of Eq.(9.104).

B.2. Calculate $\left(\tilde{\omega}_{mN}, \tilde{\psi}_{mN} \right)$, $m = 1, 2$, and the vector $\tilde{E}$.

B.3. Calculate the interface values Ψ.

B.4. Calculate the final solution (ω_{mN}, ψ_{mN}) by solving a $\tilde{\mathcal{P}}_m$-problem (9.91)-(9.94) where the homogeneous boundary conditions (9.92) are replaced by

$$\omega_{mN}|_{y_j} = \xi_j, \quad \psi_{mN}|_{y_j} = \eta_j \quad \text{for } y_j \in \gamma_N^I. \tag{9.121}$$

9.5.3 Examples of application

In this section we present two applications of the multidomain method developed in the previous sections. The first example concerns the evolution of a two-dimensional wake subjected to a temperature field and the second one deals with the axisymmetric flow in a rotating channel with a cavity.

(a) Wake subjected to a temperature field

This problem concerns the temporal evolution of the wake characterized
by the velocity profile (in dimensionless variables) :

$$u = 1 - \frac{1}{ch^2 y}, \quad v = 0, \tag{9.122}$$

subjected to the nonuniform temperature distribution

$$\theta = \theta_0 + \exp\left[-\beta\left(\frac{b-y}{b}\right)^2\right]. \tag{9.123}$$

By keeping the same notation for the dimensionless variables, the equa-
tions of motion are Eqs.(5.15), (5.32), and (5.41) with $\nu = 1/Re$, $F = Ri\,\partial_x T$, $T = \theta - \theta_0$ and $\kappa_T = 1/(Re\,Pr)$ where Re, Ri, and Pr are, respectively, the Reynolds, Richardson and Prandtl numbers. These equations
are solved in the domain $0 \le x \le A$, $-b \le y \le b$, assuming periodicity
(period A) in the x-direction and boundary conditions

$$\omega = 0, \quad \psi = 0, \quad \partial_y T = 0 \qquad \text{at } y = -b,$$
$$\omega = 0, \quad \psi = 2\,(b - \tanh b)\,, \quad \partial_y T = 0 \quad \text{at } y = b.$$

The initial conditions are obtained from Eqs.(9.122) and (9.123) with a
superimposed random disturbance of 10^{-10} amplitude on u and v.

The time-discretization is based upon the semi-implicit second-order
AB/BDI2 scheme (4.52). In the x-direction, the solution is expanded in
truncated Fourier series (Section 6.2) with $A = 20$ and $K = 200$. A three-
domain decomposition is considered in the y-direction, and the solution
method developed in Section 9.5.1 is applied in association with the Chebyshev tau method. With $b = 20$, the interfaces are $a_1 = -5$ and $a_2 = 5$. The
consideration of a multidomain method allows us to have a higher resolution ($N_2 = 60$) in the middle domain Ω_2 in which the initial gradients
of velocity and vorticity are high. In the other domains, the resolution is
$N_1 = N_3 = 48$. The time-step is $\Delta t = 10^{-2}$.

Figure 9.4 displays the instantaneous iso-vorticity lines at various times
in the case where $\beta = 40$, $Pr = 7$, $Re = 250$, and $Ri = 10$. We observe
the existence of elongated secondary vortices due to internal waves in the
region near the upper wall where the density stratification is significant.

(b) Flow in a rotating channel

This example (Raspo, 2002) concerns the axisymmetric flow in a semi-
infinite rotating channel with a cavity on its outer wall Γ_0 (Fig.9.5). This
boundary Γ_0 rotates with angular velocity Ω_0 and the inner cylinder Γ_i rotates with the angular velocity $\Omega_i = \alpha\,\Omega_0$. An axial flow (parabolic profile)
is imposed at the inlet of the channel whereas the flow is assumed to be
fully developed at the downstream infinity.

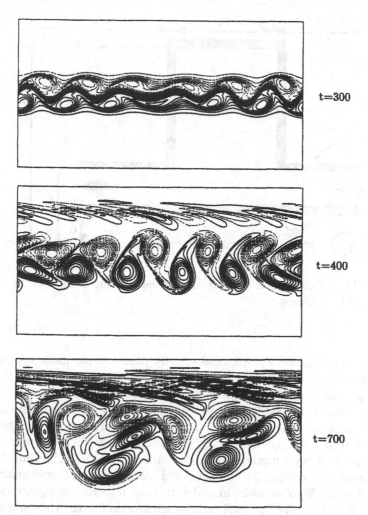

FIGURE 9.4. Instantaneous iso-vorticity lines of the wake flow subject to a nonuniform temperature field in the case $Pr = 7$, $Re = 250$, and $Ri = 10$.

The azimuthal velocity v, the azimuthal vorticity ω, and the stream-function ψ are determined by the equations given in Section 5.2.1. They are made dimensionless using the following characteristic quantities : b for length, Ω_0^{-1} for time, and $\Omega_0 b$ for velocity. The geometrical parameters are $G_1 = H/b$, $G_2 = s/b$, $g_1 = r_s/b$ and $g_2 = a/b$. The flow parameters are the Ekman number $E = \nu / \left(\Omega_0 b^2 \right)$ and the Reynolds number $Re = 2 \left(a - r_s \right) V_0 / \nu$, where V_0 is the maximal axial velocity at the inlet. Moreover, the equations are considered in a reference frame rotating with angular velocity Ω_0. By keeping the same notation for the dimensionless quantities, the dimensionless equations have the same form as Eqs.(5.33), (5.36), and (5.39) with $f_v = -2\,u$ and $F_\omega = 2\,\partial_z v$. No-slip conditions are

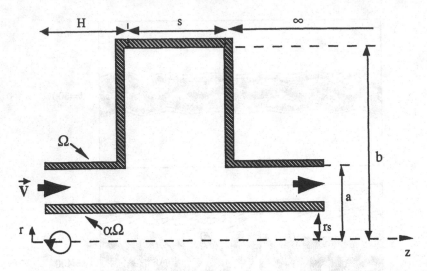

FIGURE 9.5. Geometrical configuration of the rotating channel-cavity system with axial throughflow.

prescribed on Γ_0 and Γ_i. An axial velocity profile of parabolic type is prescribed at the inlet $z = 0$, namely, $v = 0$, $\psi = \psi_e(r)$, and $\partial_z \psi = 0$. Fully developed flow conditions $\partial_z v = 0$, $\partial_z \omega = 0$, $\partial_z \psi = 0$ are prescribed for $z \to \infty$. The initial conditions are simply $u = v = w = 0$, therefore $\omega = 0$.

The equations are discretized in time by means of the second-order semi-implicit AB/BDI2 scheme (4.52), associated with the Chebyshev collocation multidomain method described in Section 9.5.2. The four-domain decomposition of Fig.9.3.c is considered. A coordinate transformation is applied in each subdomain in order to map it onto the square domain $(-1, 1) \times (-1, 1)$. For the semi-infinite subdomain Ω_4, the chosen mapping in the z-direction is the hyperbolic tangent mapping similar to Eq.(8.21).

Now we display some typical results obtained in the case where $G_1 = 0.24$, $g_1 = 0.08$, $g_2 = 0.32$, $E = 5 \times 10^{-4}$, and $Re = 196$, the time-step being $\Delta t = 10^{-4}$. The solution is calculated with $N_z = 90$ in each subdomain, $N_r = 40$ in Ω_1, and $N_r = 54$ in the other subdomains. Figure 9.6 shows the vorticity profiles at steady state, obtained with $G_2 = 1.36$ and $\alpha = 8$, in the vicinity of the interface γ_{13} separating the subdomain Ω_1 from Ω_3. Figure 9.6.a shows the vorticity profile on the first line of collocation points under γ_{13} in Ω_3, whereas Fig.9.6.b shows the same profile along the second line of points. In both cases, the profiles are calculated from an interpolation based on 401 equispaced points. Oscillations in Fig.9.6.a are a consequence of the vorticity singularity at the corners. More accurate results could be obtained by means of the subtraction method given in Section 8.4.

Figure 9.7 displays the instantaneous iso-streamfunction lines at two times during the transient stage of the flow corresponding to $Gr = 0.53$ and

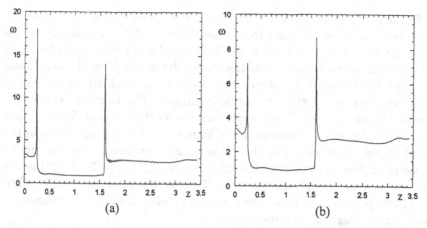

FIGURE 9.6. Vorticity profiles in the vicinity of the interface γ_{13} for $Re = 196$, $E = 5 \times 10^{-4}$, and $\alpha = 8$: (a) on the first line of collocation points in Ω_3, (b) on the second line of collocation points in Ω_3.

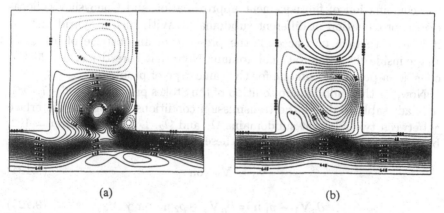

FIGURE 9.7. Instantaneous iso-streamfunction lines during the transient stage of the flow for $Re = 196$, $E = 5 \times 10^{-4}$, and $\alpha = 14$.

$\alpha = 14$. The figure shows the efficiency of the interface conditions allowing the eddies to move freely from the channel to the cavity and conversely.

9.6 Navier-Stokes equation in velocity-pressure variables

In chapter 7 it has been shown that, according to the time-discretization scheme, the Navier-Stokes problem reduces either to a Stokes problem or to the combination of Helmholtz and Darcy problems, to be solved at each time-cycle.

In the latter case, when the $I\!P_N - I\!P_N$ projection method is considered, the Darcy problem amounts to the solution of a Poisson equation for the pressure with Neumann boundary conditions. Pinelli *et al.* (1997) propose to solve this problem, as well as the Helmholtz problem for the provisional velocity, by means of the "Projection Decomposition Method" developed by Gervasio *et al.* (1997) for elliptic equations. The Legendre collocation method is based on a Galerkin approximation of the Poincaré-Steklov equation determining the values of the unknowns at the interface between two subdomains. With such a method, the pressure gradient is not continuous through an interface, so that the final velocity given by Eq.(7.246) is also not continuous through the interface. To avoid such a drawback, Pinelli *et al.* (1997) propose to enforce the equation (7.246) using a weak collocation technique rather than a strong one.

A multidomain version of the Chebyshev collocation $I\!P_N - I\!P_N$ projection method developed by Kräutle and Wielage (2001) is based on the "Conjugate Gradient Boundary Iteration" method mentioned at the end of Section 9.3.2. An interesting feature of the method proposed by these authors is the combination of finite-element approximation and Chebyshev collocation approximation in different subdomains. With such a hybrid method it is easy to take into account the presence of an obstacle of arbitrary shape inside the computational domain. Note that Forestier *et al.* (2000b) consider a penalty technique for the same type of problem.

Now, for the multidomain solution of the Stokes problem (7.196)-(7.198), it is advisable to define the transmission conditions through the interface γ between two adjacent subdomains Ω_1 and Ω_2. Let $(\mathbf{V}_m, p_m)$, $m = 1, 2$, be the restriction of $(\mathbf{V}, p)$ to Ω_m, these conditions are

$$\mathbf{V}_1 = \mathbf{V}_2 \quad \text{on } \gamma \tag{9.124}$$

$$\partial_n \mathbf{V}_1 - p_1 \mathbf{n} = \partial_n \mathbf{V}_2 - p_2 \mathbf{n} \quad \text{on } \gamma, \tag{9.125}$$

where $\mathbf{n}$ is the unit vector normal to γ, oriented from Ω_1 toward Ω_2. Resorting to a weak formulation of the problem, it can be shown (Marini and Quarteroni, 1989) that the two-domain formulation of the Stokes problem with the continuity conditions (9.124)-(9.125) is equivalent to the original one-domain Stokes problem.

The two-domain Stokes problem is considered in detail in the following two sections and we indicate the extent to an arbitrary number of subdomains. Finally, an example of the application to double-diffusive convection will be given for illustration.

9.6.1 *Stokes problem : Fourier-Chebyshev method*

In the case where the solution is periodic in the x-direction, the Stokes problem satisfied by each Fourier mode $k = 0, \ldots, K$ is defined by Eqs.(7.36)-

(7.40), namely,

$$u'' - \sigma u + k p = f_u, \quad -1 < y < 1 \tag{9.126}$$

$$v'' - \sigma v - p' = f_v, \quad -1 < y < 1 \tag{9.127}$$

$$k u + v' = 0, \quad -1 < y < 1 \tag{9.128}$$

$$u(\pm 1) = \gamma_\pm, \quad v(\pm 1) = g_\pm. \tag{9.129}$$

Let us consider the two-domain method (see Section 9.2.1 for definitions and notations). Let $\phi_m = u_m, v_m, p_m$ be the restriction of $\phi = u, v, p$ to Ω_m, $m = 1, 2$, this satisfies Eqs.(9.126)-(9.128) with the boundary conditions deduced from Eq.(9.129) :

$$u_1(-1) = \gamma_-, \quad v_1(-1) = g_- \tag{9.130}$$

$$u_2(1) = \gamma_+, \quad v_2(1) = g_+, \tag{9.131}$$

and with the continuity conditions at the interface $y = a_1$:

$$u_1 = u_2, \quad v_1 = v_2 \tag{9.132}$$

$$v_1' - p_1 = v_2' - p_2 \tag{9.133}$$

$$u_1' = u_2', \tag{9.134}$$

deduced from Eqs.(9.124)-(9.125).

For the Fourier mode $k = 0$, the problem simplifies and is easily solved, as is done in the one-domain case (Section 7.3.2.b). Therefore, it is assumed henceforth that $k \neq 0$.

The solution of the above two-domain problem is efficiently obtained owing to the influence matrix method developed in previous sections (especially Section 9.5.1). We begin by introducing the decomposition

$$\phi_m = \tilde{\phi}_m + \bar{\phi}_m, \quad m = 1, 2. \tag{9.135}$$

The part $\tilde{\phi}_m = \tilde{u}_m, \tilde{v}_m, \tilde{p}_m$ satisfies the $\tilde{\mathcal{P}}_m$-problem constituted by Eqs. (9.126)-(9.128), the boundary conditions at $y = \pm 1$ given by Eqs.(9.130)-(9.131), and the homogeneous boundary conditions at the interface

$$\tilde{u}_m(a_1) = 0, \quad \tilde{v}_m(a_1) = 0. \tag{9.136}$$

Therefore, $\tilde{\phi}_m$ is a solution of a Stokes problem solved by means of the influence matrix method (Section 7.3.2.c2) ensuring that the approximate velocity field is exactly solenoidal.

The part $\bar{\phi}_m = \bar{u}_m, \bar{v}_m, \bar{p}_m$ satisfies the $\bar{\mathcal{P}}_m$-problem constituted by Eqs.(9.126)-(9.128) with $f_u = f_v = 0$, associated with the homogeneous boundary conditions deduced from Eqs.(9.130)-(9.131), and with the boundary conditions at the interface

$$\bar{u}_m(a_1) = \xi, \quad \bar{v}_m(a_1) = \eta, \tag{9.137}$$

where ξ and η are such that the continuity conditions (9.133)-(9.134) are satisfied. Now the solution $\overline{\phi}_m$ is decomposed according to

$$\overline{\phi}_m = \xi \overline{\phi}_{m,1} + \eta \overline{\phi}_{m,2} \tag{9.138}$$

so that each elementary solution $\overline{\phi}_{m,l}$, $l = 1, 2$, satisfies the homogeneous Stokes equations, with homogeneous boundary conditions at $y = \pm 1$ deduced from Eqs.(9.130)-(9.131), and with the following boundary conditions at $y = a_1$:

$$\overline{u}_{m,l}(a_1) = \delta_{1,l}, \quad \overline{v}_{m,l} = \delta_{2,l}, \quad l = 1, 2. \tag{9.139}$$

The resulting Stokes problems are again solved by using the influence matrix technique. Therefore, the polynomial velocity fields approximating $(\overline{u}_{m,l}, \overline{v}_{m,l})$ are solenoidal. It results that the polynomial velocity field (u_{mN}, v_{mN}) approximating (u_m, v_m) is also solenoidal and, especially, at the interface $y = a_1$. Thus, from Eq.(9.128) written at $y = a_1$, we get

$$v_1'(a_1) = v_2'(a_1)$$

since it is assumed that $u_1(a_1) = u_2(a_1)$. The condition (9.133) reduces to

$$p_1(a_1) = p_2(a_1), \tag{9.140}$$

namely, the pressure must be continuous at the interface. Note that for $k \neq 0$, the pressure of the Stokes problem is determined in an unique way.

Finally, the constants ξ and η are determined by prescribing the continuity conditions (9.134) and (9.140). We get the 2×2 algebraic system

$$\mathcal{M}\Psi = \tilde{E}, \tag{9.141}$$

where $\Psi = (\xi, \eta)^T$ and $\mathcal{M}$ is the influence matrix whose elements depend on the jumps at $y = a_1$ of the polynomial approximation to $\overline{p}_{m,l}$ and $\overline{u}'_{m,l}$. We refer to previous sections for the details concerning the implementation of the methods. The extension to the multidomain case is straightforward. The only difference is the necessity to consider two supplementary elementary solutions in Eq.(9.138) when the subdomain Ω_m is bounded by two interfaces $y = a_{m-1}$ and $y = a_m$ (see Section 9.5.1).

9.6.2 Stokes problem: Chebyshev-Chebyshev method

The two-dimensional Stokes problem is defined by Eqs.(7.196)-(7.198), that is,

$$\nabla^2 \mathbf{V} - \sigma \mathbf{V} - \nabla p = \mathbf{F} \quad \text{in } \Omega \tag{9.142}$$

$$\nabla \cdot \mathbf{V} = 0 \quad \text{in } \Omega \tag{9.143}$$

$$\mathbf{V} = \mathbf{V}_\Gamma \quad \text{on } \Gamma, \tag{9.144}$$

where $\mathbf{V}_\Gamma = (u_\Gamma, v_\Gamma)$ satisfies the total flow rate condition

$$\int_\Gamma \mathbf{V}_\Gamma \cdot \mathbf{n} \, d\Gamma = 0 \qquad (9.145)$$

with $\mathbf{n}$ the outer unit vector normal to Γ.

Let us consider the two-domain decomposition of Fig.9.3.a. We denote by $(\mathbf{V}_m, p_m)$ the restriction of $(\mathbf{V}, p)$ to the subdomain Ω_m, $m = 1, 2$, and by Γ_m the part of Γ associated with Ω_m. The two-domain formulation of the above Stokes problem is

$$\nabla^2 \mathbf{V}_m - \sigma \mathbf{V}_m - \nabla p_m = \mathbf{F} \quad \text{in } \Omega_m \qquad (9.146)$$

$$\nabla \cdot \mathbf{V}_m = 0 \quad \text{in } \Omega_m \qquad (9.147)$$

$$\mathbf{V}_m = \mathbf{V}_\Gamma \quad \text{on } \Gamma_m, \qquad (9.148)$$

and the transmission conditions

$$\mathbf{V}_1 = \mathbf{V}_2 \quad \text{on } \gamma \qquad (9.149)$$

$$\partial_x u_1 - p_1 = \partial_x u_2 - p_2 \quad \text{on } \gamma \qquad (9.150)$$

$$\partial_x v_1 = \partial_x v_2 \quad \text{on } \gamma, \qquad (9.151)$$

deduced from Eqs.(9.124)-(9.125).

The solution of these problems by the Chebyshev method requires that each subdomain Ω_m is transformed into the square domain $(-1, 1) \times (-1, 1)$ by the coordinate transformation

$$\hat{x}^m = \frac{2}{l_m}(x - a_1) - 1 \qquad (9.152)$$

with $l_1 = 1 + a_1$, $l_2 = 1 - a_1$. We denote by $\mathbf{V}_{mN}(\hat{x}^m, y)$ and $p_{mN}(\hat{x}^m, y)$ the polynomial approximations, respectively, to $\mathbf{V}_m(x, y)$ and $p_m(x, y)$.

The solution of the problem (9.146)-(9.151) may be obtained by a direct method (influence matrix method) or by an iterative one (Dirichlet/Neumann method) as discussed in Section 9.2.1.

In the iterative method (Quarteroni, 1989, 1991 ; Marini and Quarteroni, 1989) a "Dirichlet" problem is solved in Ω_1 with $\mathbf{V}_1^n = (u_1^n, v_1^n)$ prescribed on the interface γ. Then the "Neumann" problem in Ω_2 is characterized by the boundary conditions

$$\partial_x u_2^n - p_2^n = \partial_x u_1^n - p_1^n$$
$$\partial_x v_2^n = \partial_x v_1^n,$$

prescribed on γ. We refer to works quoted above for the mathematical analysis of the iterative procedure.

In the following we present the two-domain version of the direct method developed by Sabbah and Pasquetti (1998) for the multidomain decomposition of Fig.9.3.c. The basic principle is to find the common value $\mathbf{V}_\gamma$ of $\mathbf{V}_1 = \mathbf{V}_2$ at the interface γ such that the continuity conditions (9.150) and (9.151) are satisfied. With such a method, the global problem (9.146)-(9.151) is replaced by separated Stokes problems which can be solved independently in each subdomain.

According to the usual technique, we set

$$\phi_m = \tilde{\phi}_m + \overline{\phi}_m\,, \quad m = 1, 2\,, \tag{9.153}$$

where $\phi_m = \mathbf{V}_m, p_m$. The set $\tilde{\phi}_m = \tilde{\mathbf{V}}_m, \tilde{p}_m$ satisfies the $\tilde{\mathcal{P}}_m$-Problem defined by

$$\nabla^2 \tilde{\mathbf{V}}_m - \sigma \tilde{\mathbf{V}}_m - \nabla \tilde{p}_m = \mathbf{F} \quad \text{in } \Omega_m \tag{9.154}$$

$$\nabla \cdot \tilde{\mathbf{V}}_m = 0 \quad \text{in } \Omega_m \tag{9.155}$$

$$\tilde{\mathbf{V}}_m = \mathbf{V}_\Gamma \quad \text{on } \Gamma_m \tag{9.156}$$

$$\tilde{\mathbf{V}}_m = \tilde{\mathbf{V}}_\gamma \quad \text{on } \gamma. \tag{9.157}$$

The interface velocity $\tilde{\mathbf{V}}_\gamma(y) = (\tilde{u}_\gamma(y), \tilde{v}_\gamma(y))$ is arbitrary, but must satisfy some compatibility conditions at points A and B and ensure the conservation of the flow rate through γ. These conditions are :

(1) Continuity of the velocity at A and B :

$$\tilde{u}_\gamma(-1) = u_{\Gamma|A}\,, \quad \tilde{v}_\gamma(-1) = v_{\Gamma|A}\,, \tag{9.158}$$

$$\tilde{u}_\gamma(1) = u_{\Gamma|B}\,, \quad \tilde{v}_\gamma(1) = v_{\Gamma|B}\,. \tag{9.159}$$

(2) Incompressibility condition at A and B :

$$\tilde{v}'_\gamma(-1) = -\partial_x u_{\Gamma|A}\,, \quad \tilde{v}'_\gamma(1) = -\partial_x u_{\Gamma|B}\,. \tag{9.160}$$

(3) Conservation of the flow rate through γ :

$$\int_{-1}^{1} \tilde{u}_\gamma(y)\, dy = - \int_{\Gamma_1} \mathbf{V}_\Gamma \cdot \mathbf{n}\, d\Gamma \left(= \int_{\Gamma_2} \mathbf{V}_\Gamma \cdot \mathbf{n}\, d\Gamma \right). \tag{9.161}$$

Therefore, in order to satisfy these constraints, $\tilde{u}_\gamma(y)$ and $\tilde{v}_\gamma(y)$ are chosen as polynomials of degree 2 and 3, respectively.

The $\tilde{\mathcal{P}}_m$-problem (9.154)-(9.157) is a Stokes problem whose solution is obtained by means of the Chebyshev collocation influence matrix method given in Section 7.4.1.b. This method ensures that the polynomial approximation $\tilde{\mathbf{V}}_{mN}$ to $\tilde{\mathbf{V}}_m$ is solenoidal, including at the interface γ.

The part $\overline{\phi}_m = \overline{\mathbf{V}}_m, \overline{p}_m$ satisfies the $\overline{\mathcal{P}}_m$-Problem defined by

$$\nabla^2 \overline{\mathbf{V}}_m - \sigma \overline{\mathbf{V}}_m - \nabla \overline{p}_m = \mathbf{0} \quad \text{in } \Omega_m \tag{9.162}$$

$$\nabla \cdot \overline{\mathbf{V}}_m = 0 \quad \text{in } \Omega_m \tag{9.163}$$

$$\overline{\mathbf{V}}_m = 0 \quad \text{on } \Gamma_m \tag{9.164}$$

$$\overline{\mathbf{V}}_m = \mathbf{V}_\gamma - \tilde{\mathbf{V}}_\gamma \quad \text{on } \gamma, \tag{9.165}$$

where the unknown vector $\mathbf{V}_\gamma(y) = (u_\gamma(y), v_\gamma(y))$ is the interface values $\mathbf{V}_1(a_1, y) = \mathbf{V}_2(a_1, y)$ and must be found such that the continuity conditions (9.150) and (9.151) are satisfied. The interface values $\overline{\mathbf{V}}_1(a_1, y) = \overline{\mathbf{V}}_2(a_1, y) = \overline{\mathbf{V}}_\gamma(y) = (\overline{u}_\gamma(y), \overline{v}_\gamma(y))$ may be sought in the form

$$\overline{u}_\gamma(y) = \sum_{l=1}^{L_u} \xi_l U_l(y), \quad \overline{v}_\gamma(y) = \sum_{l=1}^{L_v} \eta_l V_l(y). \tag{9.166}$$

If we followed the basic lines of the influence matrix technique considered in previous sections, the bases U_l and V_l would be canonical bases, so that $U_l(y_j) = V_l(y_j) = \delta_{l,j}$ and, consequently, (ξ_l, η_l) would simply be the values of $(u_\gamma - \tilde{u}_\gamma, v_\gamma - \tilde{v}_\gamma)$ at the collocation points $y_j = \cos(\pi j/N_y)$, $j = 0, \ldots, N_y$, on the interface γ. However, in the present problem such bases cannot be considered because $\overline{u}_\gamma$ and $\overline{v}_\gamma$ must satisfy some constraints which are now detailed :

(1) Continuity of the velocity at A and B :

$$\overline{u}_\gamma(\pm 1) = 0, \quad \overline{v}_\gamma(\pm 1) = 0. \tag{9.167}$$

(2) Incompressibility condition at A and B :

$$\partial_y \overline{v}(\pm 1) = 0. \tag{9.168}$$

(3) Conservation of the flow rate through γ :

$$\int_{-1}^{1} \overline{u}_\gamma(y)\, dy = 0. \tag{9.169}$$

(4) The incompressibility equation on γ implies that $\overline{u}_\gamma(y)$ is a polynomial of degree $N_y - 1$ because the approximation of $\partial_x u$ is a polynomial of degree $N_y - 1$ in the y-direction, since $\partial_x u = -\partial_y v$.

Each component $\overline{u}_\gamma$ or $\overline{v}_\gamma$ must satisfy four constraints. Therefore, the number of basis functions $L_u = L_v$ in Eq.(9.166) is equal to the number $N_y + 1$ of degrees of freedom diminished by the number of constraints, namely, $L_u = L_v = N_y - 3$.

A possible basis $U_l(y), l = 1, \ldots, N_y - 3$, satisfying the four constraints is defined by

$$U_l(y) = T_l(y) - T_1(y) \quad \text{for } l \text{ odd and } l \geq 1,$$

and

$$U_l(y) = T_{l+2}(y) - \alpha T_2(y) - \beta \quad \text{for } l \text{ even and } l \geq 2,$$

where

$$\alpha = \frac{3 (l+2)^2}{4 (l+1) (l+3)}, \quad \beta = \frac{l (l+4)}{4 (l+1) (l+3)}.$$

A possible basis $V_l(y)$, $l = 1, \ldots, N_y - 1$, is defined by

$$V_l(y) = T_{l+3}(y) - \alpha T_2(y) - \beta \quad \text{for} \quad l \text{ odd and } l \geq 1,$$

with

$$\alpha = \frac{1}{4}(l+3)^2, \quad \beta = -\frac{1}{4}(l+1)(l+5),$$

and

$$V_l(y) = T_{l+3}(y) - \alpha T_3(y) - \beta \quad \text{for} \quad l \text{ even and } l \geq 2,$$

with

$$\alpha = \frac{1}{8}(l+2)(l+4), \quad \beta = -\frac{1}{8}l(l+6).$$

Another possible choice is constituted by the pseudocanonical basis $V_l(y)$, $l = 2, \ldots, N_y - 2$, such that $V_l(y_j) = \delta_{l,j}$ for $j = 2, \ldots, N_y - 2$, $V_l(-1) = V_l(1) = 0$ and the values $V_l(y_1)$ and $V_l(y_{N_y-1})$ determined from Eq.(9.168), using the differentiation formulas (3.45)-(3.46). Note that a similar basis is considered by Forestier *et al.* (2000a) to deal with the case of a stress-free boundary.

Now we may express the solution $\overline{\phi}_m = \overline{\mathbf{V}}_m, \overline{p}_m$ as

$$\overline{\phi}_m = \sum_{l=1}^{N_y-3} \left(\xi_l \overline{\phi}_{m,l}^{(1)} + \eta_l \overline{\phi}_{m,l}^{(2)} \right), \quad m = 1, 2, \tag{9.170}$$

where the elementary solutions $\overline{\phi}_{m,l}^{(n)}$, $n = 1, 2$, satisfy the $\overline{\mathcal{P}}_{m,l}$-*Problem* constituted by the homogeneous equations (9.162)-(9.163), the homogeneous boundary conditions (9.164) on Γ_m, and by the following boundary conditions

$$\overline{\mathbf{V}}_{m,l}^{(1)} = (U_l, 0), \quad \overline{\mathbf{V}}_{m,l}^{(2)} = (0, V_l) \quad \text{on } \gamma. \tag{9.171}$$

Therefore, each $\overline{\phi}_{m,l}^{(n)}$ satisfies a Stokes problem solved, as above, by means of the influence matrix method of Section 7.4.1.b, producing a solenoidal polynomial velocity field. Equation (9.171) implies that the polynomial velocity field $\overline{\mathbf{V}}_{mN}$ approximating $\overline{\mathbf{V}}_m$ is continuous through γ (but this is not the case for $\overline{p}_{mN}$). Since the final approximate velocity field $\mathbf{V}_{mN} = \tilde{\mathbf{V}}_{mN} + \overline{\mathbf{V}}_{mN}$ is such that $\nabla \cdot \mathbf{V}_{mN}$ is the null polynomial, it follows that the x-derivative of the component u_{mN} is continuous through γ and, consequently, the condition (9.150) reduces to

$$p_{1N} = p_{2N} \quad \text{on } \gamma. \tag{9.172}$$

By the way, it is interesting to observe that the second-order x-derivative of u_{mN} and the first-order derivative of the vorticity are also continuous through γ.

The coefficients ξ_l and η_l, $l = 1, \ldots, N_y - 3$, are determined from the continuity conditions (9.151) and (9.172). There is no difficulty in enforcing the condition (9.151), on the other hand, the presence of spurious modes in the pressure solution prevents the strict enforcement of the condition (9.172).

The approximate pressure $p_{mN} = \tilde{p}_{mN} + \overline{p}_{mN}$ on γ is a polynomial in y of degree N_y affected by the "spurious" modes (see Remark 1 of Section 7.4.1.b) :

(1) the constant mode $T_0(y) = 1$ (this mode is not really spurious since it has a physical meaning),

(2) the mode $T_{N_y}(y)$, and

(3) the "corner" modes constituted by two polynomials vanishing at inner collocation points on γ and equal to 1 at points A and B, that is, at $y = \pm 1$.

The remedy proposed by Sabbah and Pasquetti (1998) to cope with the presence of these spurious modes (including the constant mode) is to enforce the continuity condition on the pressure in a weak sense, using, as test functions, a set of polynomials orthogonal to the spurious modes. Therefore, the continuity condition (9.172) is replaced by the set of $N_y - 3$ integral relations

$$\int_{-1}^{1} (p_{1N} - p_{2N})_\gamma \, [T_m(y) - T_q(y)] \, w(y) \, dy = 0, \quad m = 3, \ldots, N_y - 1,$$

$$(9.173)$$

where $q = 1$ if m is odd and $q = 2$ if m is even. Owing to the orthogonality property of the Chebyshev polynomials, the spurious part of the jump $(p_{1N} - p_{2N})$ associated with the polynomials T_{N_y} and T_0 is cancelled. Moreover, the test functions $T_m(y) - T_q(y)$ are also orthogonal to the corner modes since they vanish at $y = \pm 1$. The integrals (9.173) are approximated by means of the Gauss-Lobatto quadrature formula, and yield the following weak continuity conditions

$$\sum_{j=1}^{N_y-1} (p_{1N} - p_{2N})_{y_j} \, [T_m(y_j) - T_q(y_j)] = 0, \quad m = 3, \ldots, N_y - 1. \quad (9.174)$$

Then the enforcement of condition (9.151) at collocation points y_j, $j = 2, \ldots, N_y - 2$, on the interface γ gives the set of equations

$$[(1 - a_1) \partial_{\hat{x}^1} v_{1N} - (1 + a_1) \partial_{\hat{x}^2} v_{2N}]_{y_j} = 0, \quad j = 2, \ldots, N_y - 2. \quad (9.175)$$

The sets of equations (9.174) and (9.175), with p_{mN} and v_{mN} defined by Eqs.(9.153) and (9.170), constitute an algebraic system of $2(N_y - 3)$ equations determining the coefficients ξ_l and η_l, $l = 1, \ldots, N_y - 3$. The associated matrix (interface influence matrix) is found to be invertible and relatively well-conditioned. We refer to Section 9.5.2 for the question of conditioning.

We observe that the equation (9.175) is not enforced at points y_1 and y_{N_y-1}. As a matter of fact, if the polynomial $\partial_{\hat{x}^m} v_{mN}$ is known at all collocation points on γ except at y_1 and y_{N_y-1}, its values at these points are determined in a unique way from the compatibility conditions

$$\partial_{yx} v_{|A} = -\partial_{xx} u_{\Gamma|A}, \quad \partial_{yx} v_{|B} = -\partial_{xx} u_{\Gamma|B},$$

resulting from the incompressibility equation. Therefore, the continuity condition (9.175) is automatically satisfied at y_1 and y_{N_y-1} since it is at all other points.

When the coefficients ξ_l and η_l have been determined, the final solution $(\mathbf{V}_{mN}, p_{mN})$ can be reconstituted thanks to Eqs. (9.153) and (9.170). However, when included in a time-dependent solution, such a procedure is too costly in terms of storage and CPU time. Therefore, it is more convenient to determine the interface values of $\mathbf{V}_{mN} = (u_{mN}, v_{mN})$, namely,

$$u_{\gamma N}\left(y_j\right) = \tilde{u}_\gamma\left(y_j\right) + \sum_{l=1}^{N_y-1} \xi_l U_l\left(y_j\right), \quad j = 1, \ldots, N_y - 1, \qquad (9.176)$$

$$v_{\gamma N}\left(y_j\right) = \tilde{v}_\gamma\left(y_j\right) + \sum_{l=1}^{N_y-1} \eta_l V_l\left(y_j\right), \quad j = 1, \ldots, N_y - 1, \qquad (9.177)$$

and, then, to obtain the final solution $(\mathbf{V}_{mN}, p_{mN})$ by solving a Stokes problem in each Ω_m with the boundary conditions (9.176) and (9.177) on γ.

In summary, the general two-domain algorithm for the time-dependent solution of the Navier-Stokes equations is the following :

A. Preprocessing stage

A.1. Construct the boundary influence matrices associated with the Stokes problems in Ω_m, $m = 1, 2$. Invert these matrices and store the inverses.

A.2. Construct the interface influence matrix, invert it, and store the inverse.

B. At each time-cycle

B.1. Calculate the right-hand side $\mathbf{F}$ of Eq.(9.142).

B.2. Calculate $\left(\tilde{\mathbf{V}}_{mN}, \tilde{p}_{mN}\right)$, $m = 1, 2$.

B.3. Calculate the coefficients (ξ_l, η_l), $l = 1, \ldots, N_y - 1$.

B.4. Calculate the final solution $(\mathbf{V}_{mN}, p_{mN})$ by solving a $\tilde{\mathcal{P}}_m$-problem where the boundary conditions (9.157) are replaced by (9.176) and (9.177).

In the case of the multidomain decomposition of Fig.9.3.b, the method is similar, except that supplementary elementary solutions have to be considered in Eq.(9.170) for the inner subdomains Ω_m delimited by the two interfaces γ_{m-1} and γ_m. More precisely, for such a subdomain,

$$\overline{\phi}_m = \sum_{l=1}^{N_y-1} \left(\xi_l \, \overline{\phi}_{m,l}^{(1)} + \eta_l \, \overline{\phi}_{m,l}^{(2)} + \lambda_l \, \overline{\phi}_{m,l}^{(3)} + \mu_l \, \overline{\phi}_{m,l}^{(4)} \right) . \qquad (9.178)$$

The coefficients (ξ_l, η_l) correspond to the interface quantities $(\overline{u}_{\gamma_m}, \overline{v}_{\gamma_m})$ and are expressed by Eq.(9.166). The coefficients (λ_l, μ_l) have a similar definition but they are associated with the interface quantities $(\overline{u}_{\gamma_{m-1}}, \overline{v}_{\gamma_{m-1}})$.

The elementary solutions $\overline{\phi}_{m,l}^{(n)}$, $n = 1, \ldots, 4$, are determined by the homogeneous equations (9.162)-(9.163), the homogeneous boundary conditions (9.164), and the following boundary conditions on the interfaces

$$\overline{u}_{m,l}^{(n)} = \delta_{3,m} , \quad \overline{v}_{m,l}^{(n)} = \delta_{4,m} \quad \text{on } \gamma_{m-1} ,$$

$$\overline{u}_{m,l}^{(n)} = \delta_{1,m} , \quad \overline{v}_{m,l}^{(n)} = \delta_{2,m} \quad \text{on } \gamma_m .$$

Now, by expressing the continuity conditions at the interfaces γ_m, $m = 1, \ldots, M-1$, we get an algebraic system for the coefficients $(\xi_l, \eta_l, \lambda_l, \mu_l)$, $l = 1, \ldots, N_y-1$. The matrix is block-tridiagonal, each block corresponding to an interface. The inversion and storage of the inverse matrix cannot be envisaged. Sabbah and Pasquetti (1998) solve the algebraic system, at each time-cycle, by the block-tridiagonal version of the LU decomposition algorithm (see, e.g., Isaacson and Keller, 1966).

9.6.3 Example of application

In this section we present an application of the multidomain method described in the previous section to the calculation of the thermohaline convection in an enclosure whose a vertical wall is heated (Sabbah et al., 2001). Such a physical problem is a typical example for which the choice of a Chebyshev method is to be recommended. First, the rectangular shape of the computational domain is naturally well-suited to the Chebyshev polynomial approximation. Then the high accuracy of the approximation allows a fine description of the physical phenomena whose complexity might be hidden by the numerical errors associated with a low-order method. Lastly, the necessity to have a very high resolution in the vertical direction, for accurately representing the thin physical interfaces separating the convective cells, requires the use of a multidomain method. This leads to an efficient algorithm well-adapted to parallel computations.

The rectangular enclosure $0 \leq x \leq L$, $0 \leq y \leq H$ is filled with a fluid initially at rest, at uniform temperature T_0, and stably stratified by a linear negative concentration gradient $-\Delta C/H$. At $t = 0$, the temperature field is weakly randomly disturbed and, at subsequent times, the wall $x = 0$ is heated by a constant heat flux φ_T on $y_- = H/5 \leq y \leq y_+ = 4H/5$, and is null elsewhere on the boundary. A smooth transition is applied at $y = y_\pm$ through a hyperbolic tangent function. Moreover, a zero concentration flux is imposed on the boundary and no-slip conditions are prescribed for the velocity.

The motion is governed by the Navier-Stokes equations (5.1)-(5.2) with the temperature equation (5.41) and the concentration equation (5.46). The forcing term $\mathbf{f}$ is given by Eq.(5.40) with the density ρ defined by Eq.(5.45). Using the same notations for the dimensionless variables, the equations to be solved in the domain $0 \leq x \leq 1$, $0 \leq y \leq A = H/L$, are

$$\partial_t T + \mathbf{V} \cdot \nabla T - \frac{1}{\sqrt{Ra}} \nabla^2 T = 0 \qquad (9.179)$$

$$\partial_t C + \mathbf{V} \cdot \nabla C - \frac{1}{Le\sqrt{Ra}} \nabla^2 C = 0 \qquad (9.180)$$

$$\partial_t \mathbf{V} + (\mathbf{V} \cdot \nabla) \mathbf{V} + \nabla p - \frac{Pr}{\sqrt{Ra}} \nabla^2 \mathbf{V} = Pr\,(T - R_\rho C)\,\mathbf{j} \qquad (9.181)$$

$$\nabla \cdot \mathbf{V} = 0, \qquad (9.182)$$

where $\mathbf{j}$ is the unit vector in the vertical direction. The Rayleigh number Ra, the buoyancy ratio R_ρ, the Lewis number Le, and the Prandtl number Pr are, respectively, defined by

$$Ra = \frac{\alpha_T \, \varphi_T \, g \, L^4}{k_T \, \kappa_T \, \nu}, \quad R_\rho = \frac{k_T \, \alpha_C \, \Delta C}{\alpha_T \, \varphi_T \, L}, \quad Le = \frac{\kappa_T}{\kappa_C}, \quad Pr = \frac{\nu}{\kappa_T}.$$

We refer to Section 5.3 for the definitions of the various physical quantities. Without entering into the details, we mention that the length, time, and velocity have been made dimensionless by using, as characteristic quantities, L, $L^2 / \left(\kappa_T \sqrt{Ra} \right)$, and $\kappa_T / \left(L \sqrt{Ra} \right)$, respectively.

For the numerical solution, the mapping (8.15) is considered in order to cluster the collocation points in the region ($x \approx 0.20$) where the complexity of the flow is higher. The discretization in time makes use of the second-order AB/BDI2 scheme (4.52), except for the concentration equation (9.181) which is treated more implicitly. For this equation, the large value of the Lewis number would lead to very small time-steps if the term $\mathbf{V} \cdot \nabla C$ was treated in an explicit way. Therefore, this term is approximated according to

$$(\mathbf{V} \cdot \nabla C)^{n+1} \cong \left(2\mathbf{V}^n - \mathbf{V}^{n-1} \right) \cdot \nabla C^{n+1},$$

which leads to the solution, at each time-cycle, of a steady advection-diffusion equation with variable velocity. This solution is obtained through

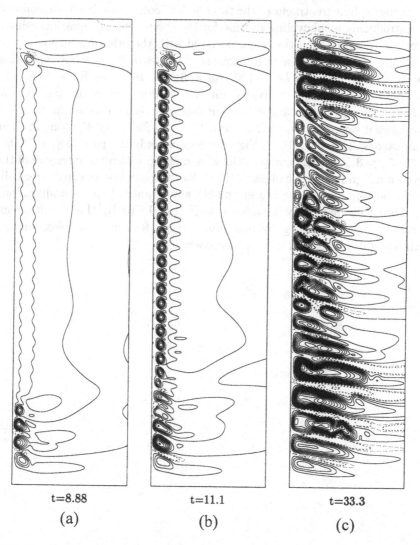

t=8.88   t=11.1   t=33.3

(a)   (b)   (c)

FIGURE 9.8. Instantaneous streamlines of thermohaline convection in the case $Pr = 7$, $Le = 100$, $Ra = 7.86 \times 10^8$, and $R_\rho = 1.66$.

a preconditioned iterative procedure, described in Section 3.8, and using as preconditioner, the collocation approximation to the Helmholtz operator. Amongst the three PR, PMR, and PCR methods, the latter was found to be the more efficient (less than 10 iterations per time-cycle) and is used in the calculations reported below. Note that the stability problem resulting from the explicit treatment of the term $\mathbf{V} \cdot \nabla C$ could have been surmounted by introducing a subcycling in time for the concentration equation using a time-step smaller than the one associated with the other equations.

Figure 9.8 shows the instantaneous streamlines of the flow corresponding to $A = 25/6$, $Pr = 7$, $Le = 100$, $Ra = 3.86 \times 10^8$, and $R_\rho = 1.66$. The computational domain is divided into 34 subdomains in the y-direction, the interface of the extreme two subdomains being located at $y_\pm$. The polynomial degree is $N_x = 250$ and $N_y = 70$ for every subdomain. The time-step is $\Delta t = 3.7 \times 10^{-3}$. Figure 9.8 shows only the part ($0 \leq x \leq 0.68$, $0.75 \leq y \leq 3.50$) of the computational domain in which the intensity of the motion is significant. We observe (Fig.9.8.a) the creation of convective cells in the bottom part of the heater and the appearance of an instability along it. Then this instability manifests itself (Fig.9.8.b) by the simultaneous creation of cells which grow and evolve (Fig.9.8.c) in a complex manner exhibiting doubling and merging processes.

Appendix A
Formulas on Chebyshev polynomials

A.1 Definition and general properties

Definition
The Chebyshev polynomial of first kind $T_k(x)$ is defined by

$$T_k(x) = \cos\left(k \cos^{-1} x\right), \quad -1 \le x \le 1, \tag{A.1}$$

where $k = 0, 1, 2, \ldots$, therefore, $-1 \le T_k(x) \le 1$.

$$T_k(x) = \cos k z \quad \text{with} \quad x = \cos z \tag{A.2}$$

$$T_k(-x) = (-1)^k T_k(x) \tag{A.3}$$

$$T_k(T_l(x)) = T_{kl}(x). \tag{A.4}$$

The polynomial $T_k(x)$ can be expressed as

$$T_k(x) = \frac{k}{2} \sum_{m=0}^{K} (-1)^m \frac{(k-m-1)!}{m!\,(k-2m)!} (2x)^{k-2m}, \tag{A.5}$$

where K is the integer part of $k/2$.
First Chebyshev polynomials :

$$T_0 = 0, \quad T_1 = x, \quad T_2 = 2x^2 - 1, \quad T_3 = 4x^3 - 3x, \quad T_4 = 8x^4 - 8x^2 + 1. \tag{A.6}$$

Differential equation

The polynomial $T_k(x)$ is a solution of the differential equation

$$\left[\sqrt{1-x^2}T_k'(x)\right]' + \frac{k^2}{\sqrt{1-x^2}}T_k(x) = 0. \tag{A.7}$$

Orthogonality property

$$\int_{-1}^{1} T_k(x)\, T_l(x)\, w(x)\, dx = \frac{\pi}{2}\, c_k\, \delta_{k,l}, \tag{A.8}$$

where $w(x) = \left(1-x^2\right)^{-1/2}$ is the Chebyshev weight, $\delta_{k,l}$ is the Kronecker delta, and

$$c_k = \begin{cases} 2 & \text{if } k = 0 \\ 1 & \text{if } k \geq 1. \end{cases} \tag{A.9}$$

Recurrence formulas

$$T_{k+1}(x) - 2\,x\,T_k(x) + T_{k-1}(x) = 0, \quad k \geq 1 \tag{A.10}$$

$$T_{k+2}(x) - 2\left(2\,x^2 - 1\right)T_k(x) + T_{k-2}(x) = 0, \quad k \geq 2 \tag{A.11}$$

and, more generally

$$T_{k+l}(x) - 2\,T_l(x)\,T_k(x) + T_{k-l}(x) = 0, \quad k \geq l. \tag{A.12}$$

Special values

$$T_k(0) = \begin{cases} 0 & \text{if } k \text{ odd} \\ 1 & \text{if } k = 4q, \ q \geq 0 \\ -1 & \text{if } k = 4q + 2, \ q \geq 0 \end{cases} \tag{A.13}$$

$$T_k(\pm 1) = (\pm 1)^k. \tag{A.14}$$

A.2   Differentiation

First- and second-order derivatives

$$T_k'(x) = 2\,k \sum_{n=0}^{K'} \frac{1}{c_{k-1-2n}} T_{k-1-2n}(x), \tag{A.15}$$

where K' is the integer part of $(k-1)/2$.

$$T_k''(x) = 4\,k \sum_{n=0}^{K''} \frac{1}{c_{k-2(n+1)}}(n+1)\left[k - (n+1)\right] T_{k-2(n+1)}(x), \tag{A.16}$$

where K'' is the integer part of $(k-2)/2$.

Recurrence formulas

$$\frac{T'_{k+1}(x)}{k+1} - \frac{T'_{k-1}(x)}{k-1} = 2T_k(x), \quad k > 1 \tag{A.17}$$

and

$$k = 0, \quad T'_1 = T_0, \qquad k = 1, \quad T'_2 = 4T_1.$$

More generally,

$$\frac{T^{(p)}_{k+1}(x)}{k+1} - \frac{T^{(p)}_{k-1}(x)}{k-1} = 2T^{(p-1)}_k(x), \quad k > 1, \quad p \geq 1, \tag{A.18}$$

and

$$k = 0, \quad T^{(p)}_1 = 0 \quad \text{for} \quad p \geq 2,$$

$$k = 1, \quad T'_2 = 4T_1, \quad T''_2 = 4T_0, \quad T^{(p)}_2 = 0 \quad \text{for} \quad p \geq 3.$$

Special values

$$T'_k(0) = \begin{cases} 0 & \text{if } k = 2q, \ q \geq 0, \\ (-1)^q k & \text{if } k = 2q+1, \ q \geq 0, \end{cases} \tag{A.19}$$

$$T'_k(\pm 1) = (\pm 1)^{k+1} k^2, \tag{A.20}$$

$$T''_k(0) = -k^2 T_k(0), \tag{A.21}$$

$$T''_k(\pm 1) = (\pm 1)^k \frac{k^2}{3}(k^2 - 1), \tag{A.22}$$

$$T^{(p)}_k(\pm 1) = (\pm 1)^{k+p} \prod_{n=0}^{p-1} \frac{k^2 - n^2}{2n+1}. \tag{A.23}$$

A.3   Collocation points

Gauss-Lobatto [zeros of $(1 - x^2) T'_N(x)$] :

$$x_i = \cos \frac{\pi i}{N}, \quad i = 0, \ldots, N. \tag{A.24}$$

Gauss [zeros of $T_{N+1}(x)$] :

$$x_i = \cos \frac{\pi(2i+1)}{2(N+1)}, \quad i = 0, \ldots, N. \tag{A.25}$$

Gauss-Radau :

$$x_i = \cos \frac{2\pi i}{2N+1}, \quad i = 0, \ldots, N.$$ (A.26)

Special values at Gauss-Lobatto points :

$$T_N(x_i) = (-1)^i, \quad i = 0, \ldots, N,$$ (A.27)

$$T_N'(x_i) = 0, \quad i = 1, \ldots, N-1,$$ (A.28)

$$T_N''(x_i) = \frac{(-1)^{i+1}}{(1-x_i^2)} N^2, \quad i = 1, \ldots, N-1.$$ (A.29)

A.4 Truncated series expansion

For $u(x) \in [-1, 1]$, the approximation $u_N(x)$ is

$$u_N(x) = \sum_{k=0}^{N} \hat{u}_k T_k(x)$$ (A.30)

with

$$\hat{u}_k = \frac{2}{\pi c_k} \int_{-1}^{1} u(x) T_k(x) w(x) \, dx.$$ (A.31)

Discrete series expansion using Gauss-Lobatto points :

$$u_N(x_i) = \sum_{k=0}^{N} \hat{u}_k T_k(x_i), \quad i = 0, \ldots, N,$$ (A.32)

with

$$\hat{u}_k = \frac{2}{\bar{c}_k N} \sum_{i=0}^{N} \frac{1}{\bar{c}_i} u(x_i) T_k(x_i), \quad i = 0, \ldots, N,$$ (A.33)

and $T_k(x_i) = \cos(k\pi i/N)$.

Derivatives

$$u_N^{(p)}(x) = \sum_{j=0}^{N} \hat{u}_k^{(p)} T_k(x)$$ (A.34)

with

$$\hat{u}_k^{(1)} = \frac{2}{c_k} \sum_{\substack{p=k+1 \\ (p+k)\ \text{odd}}}^{N} p\, \hat{u}_p, \quad k = 0, \ldots, N-1$$ (A.35)

$$\hat{u}_N^{(1)} = 0,$$

and

$$\hat{u}_k^{(2)} = \frac{1}{c_k} \sum_{\substack{p=k+2 \\ (p+k) \text{ even}}}^{N} p\left(p^2 - k^2\right) \hat{u}_p, \quad k = 0, \ldots, N-2$$

(A.36)

$$\hat{u}_{N-1}^{(2)} = \hat{u}_N^{(2)} = 0.$$

A.5   Lagrange interpolation polynomial

For $u(x) \in [-1, 1]$, whose values $u(x_i)$ are known at Gauss-Lobatto points, the approximation $u_N(x)$ is

$$u_N(x) = \sum_{j=0}^{N} h_j(x)\, u(x_j)$$

(A.37)

with

$$h_j(x) = \frac{(-1)^{j+1}\left(1 - x^2\right) T_N'(x)}{\bar{c}_j N^2 (x - x_j)},$$

(A.38)

where

$$\bar{c}_k = \begin{cases} 2 & \text{if } k = 0, \\ 1 & \text{if } 1 \leq k \leq N-1, \\ 2 & \text{if } k = N. \end{cases}$$

(A.39)

A.6   Derivatives at Gauss-Lobatto points

$$u_N^{(p)}(x_i) = \sum_{j=0}^{N} d_{i,j}^{(p)}\, u_N(x_j)$$

(A.40)

with

$$d_{i,j}^{(1)} = \frac{\bar{c}_i}{\bar{c}_j} \frac{(-1)^{i+j}}{(x_i - x_j)}, \qquad \begin{array}{l} 0 \leq i, j \leq N, \\[4pt] i \neq j, \end{array}$$

$$d_{i,i}^{(1)} = -\frac{x_i}{2(1 - x_i^2)}, \qquad 1 \leq i \leq N-1,$$

(A.41)

$$d_{0,0}^{(1)} = -d_{N,N}^{(1)} = \frac{2N^2 + 1}{6},$$

and

$$d_{i,j}^{(2)} = \frac{(-1)^{i+j}}{\overline{c}_j} \frac{x_i^2 + x_i x_j - 2}{(1 - x_i^2)(x_i - x_j)^2}, \qquad \begin{array}{l} 1 \le i \le N - 1, \\ 0 \le j \le N, \ i \ne j, \end{array}$$

$$d_{i,i}^{(2)} = -\frac{(N^2 - 1)(1 - x_i^2) + 3}{3(1 - x_i^2)^2}, \qquad 1 \le i \le N - 1,$$

$$d_{0,j}^{(2)} = \frac{2}{3} \frac{(-1)^j}{\overline{c}_j} \frac{(2N^2 + 1)(1 - x_j) - 6}{(1 - x_j)^2}, \qquad 1 \le j \le N,$$

$$d_{N,j}^{(2)} = \frac{2}{3} \frac{(-1)^{j+N}}{\overline{c}_j} \frac{(2N^2 + 1)(1 + x_j) - 6}{(1 + x_j)^2}, \quad 0 \le j \le N - 1,$$

$$d_{0,0}^{(2)} = d_{N,N}^{(2)} = \frac{N^4 - 1}{15}.$$

$$\text{(A.42)}$$

A.7   Integration

$$\int T_k(x) \, dx = \frac{1}{2} \left[\frac{T_{k+1}(x)}{k+1} - \frac{T_{k-1}(x)}{k-1} \right], \quad k > 1 \qquad \text{(A.43)}$$

and

$$\int T_0 \, dx = T_1(x), \qquad \int T_1(x) \, dx = \frac{1}{4} [T_0 + T_2(x)],$$

$$\int_{-1}^{1} T_k(x) \, dx = \begin{cases} 0 & \text{if } k \text{ odd}, \\ \dfrac{2}{1 - k^2} & \text{if } k \text{ even}. \end{cases} \qquad \text{(A.44)}$$

A.8   Numerical integration based on Gauss-Lobatto points

Gauss-Lobatto formula

$$\int_{-1}^{1} u(x) \, w(x) \, dx \cong \frac{\pi}{N} \sum_{i=0}^{N} \frac{u(x_i)}{\overline{c}_i} \qquad \text{(A.45)}$$

which is exact if $u(x)$ is a polynomial of degree at most $2N - 1$.

$$\int_{-1}^{1} u(x) \, dx \cong \frac{\pi}{N} \sum_{i=0}^{N} \frac{u(x_i)}{\overline{c}_i \, w(x_i)}. \qquad \text{(A.46)}$$

Clenshaw-Curtis formula

$$\int_{-1}^{1} u\left(x\right) dx \cong \sum_{i=0}^{N} \omega_i\, u\left(x_i\right), \qquad (A.47)$$

where the weights ω_i are :

  For N even

$$\omega_i = \begin{cases} \dfrac{1}{N^2 - 1} & \text{for } i = 0 \text{ and } i = N, \\[2mm] \dfrac{4}{N} \displaystyle\sum_{k=0}^{N/2} \dfrac{1}{\bar{c}_k} \dfrac{\cos\left(2\,\pi\,i\,k/N\right)}{1 - 4k^2} & \text{for } i = 1, \ldots, N-1, \end{cases} \qquad (A.48)$$

with $\bar{c}_0 = \bar{c}_{N/2} = 2$ and $\bar{c}_k = 1$ for $1 \le k \le N/2 - 1$.

  For N odd

$$\omega_i = \begin{cases} \dfrac{1}{N^2} & \text{for } i = 0 \text{ and } i = N, \\[2mm] \dfrac{4}{N} \displaystyle\sum_{k=0}^{(N-1)/2} \dfrac{1}{\bar{c}_k} \dfrac{\cos\left(2\,\pi\,i\,k/N\right)}{1 - 4\,k^2} + (-1)^i \dfrac{2\cos\left(\pi\,i/N\right)}{N^2(2 - N)} \\[2mm] \qquad \text{for } i = 1, \ldots, N-1, \end{cases} \qquad (A.49)$$

with $\bar{c}_0 = \bar{c}_{(N-1)/2} = 2$ and $\bar{c}_k = 1$ for $1 \le k \le (N - 1)/2 - 1$. Formula (A.47) is exact if $u\left(x\right)$ is a polynomial of degree at most N.

Appendix B
Solution of a quasi-tridiagonal system

Let us consider the quasi-tridiagonal system

$$p_i\, w_{i-1} + q_i\, w_i + r_i\, w_{i+1} = f_i, \quad i = 1, \ldots, I-1, \tag{B.1}$$

$$p_I\, w_{I-1} + q_I\, w_I = f_I, \tag{B.2}$$

$$c_0\, w_0 + c_1\, w_1 + \ldots + c_I\, w_I = g. \tag{B.3}$$

The solution (Thual, 1986) is based on the recurrence formula :

$$w_{i+1} = X_i\, w_i + Y_i, \quad i = 0, \ldots, I-1. \tag{B.4}$$

The coefficients X_i and Y_i are also determined by recurrence relations obtained in the following manner : (1) w_{i+1} is eliminated from (B.2) using (B.4) ; (2) the resulting equation is identified with (B.4) written for $i - 1$:

$$w_i = X_{i-1}\, w_{i-1} + Y_{i-1}. \tag{B.5}$$

The identification yields

$$X_{i-1} = \frac{-p_i}{q_i + r_i\, X_i}, \quad Y_{i-1} = \frac{f_i - r_i\, Y_i}{q_i + r_i\, X_i}, \quad i = I-1, \ldots, 1. \tag{B.6}$$

The coefficients X_{I-1} and Y_{I-1} must be known to initiate the recurrence. They are determined from (B.3) and (B.4) with $i = I - 1$, that is,

$$X_{I-1} = -\frac{p_I}{q_I}, \quad Y_{I-1} = \frac{f_I}{q_I}. \tag{B.7}$$

Therefore, with the coefficients X_i, Y_i, $i = 0, \ldots, I - 1$, formula (B.4) permits the calculation of the solution $w_1, \ldots, w_I$ provided w_0 is known.

The quantity w_0 is calculated from (B.3), in which $w_1, \ldots, w_I$ are expressed in terms of w_0 by

$$w_i = \theta_i \, w_0 + \lambda_i \, , \quad i = 1, \ldots, I \, . \tag{B.8}$$

By setting $i = 0$ in this expression we get

$$\theta_0 = 1 \, , \quad \lambda_0 = 0 \, . \tag{B.9}$$

Recurrence relations for θ_i and λ_i are obtained by considering (B.5) in which w_{i-1} is replaced by its expression deduced from (B.8). Then, by identification of the resulting equation with (B.8), we get

$$\theta_i = X_{i-1} \theta_{i-1} \, , \quad \lambda_i = X_{i-1} \lambda_{i-1} + Y_{i-1} \, , \quad i = 1, \ldots, I \, . \tag{B.10}$$

Finally, substituting (B.8) into (B.3), we get

$$w_0 = \frac{g - \Lambda}{\Theta} \tag{B.11}$$

with

$$\Theta = \sum_{i=0}^{I} c_i \, \theta_i \, , \quad \Lambda = \sum_{i=1}^{I} c_i \, \lambda_i \, . \tag{B.12}$$

In summary, the algorithm is :

1. Calculate X_i, Y_i, $i = 0, \ldots, I - 1$, from (B.6) and (B.7).
2. Calculate θ_i, λ_i, $i = 1, \ldots, I$, from (B.10) and (B.9).
3. Calculate w_0 from (B.11) and (B.12).
4. Calculate w_i, $i = 1, \ldots, I$, from (B.4).

The algorithm is stable, that is, the solution remains nicely bounded, if $|X_i| < 1$, $i = 0, \ldots, I - 1$, and if the coefficients Y_i, $i = 0, \ldots, I - 1$, are bounded. Following the proof given by Isaacson and Keller (1966) for the classical tridiagonal case, it can easily be shown that sufficient conditions for the above properties to be satisfied are

$$|q_i| \geq |p_i| + |r_i| \, , \quad r_i \neq 0 \, , \quad i = 2, \ldots, I - 1 \, , \tag{B.13}$$

$$|q_I| > |p_I| \, . \tag{B.14}$$

Note that the strict inequality must be taken in (B.13) if $r_i = 0$.

Appendix C
Theorems on the zeros of a polynomial

The theorems proved by Miller (1971) concern the location of the zeros z of a given polynomial relative to the unit circle in the complex plane. Two types of polynomials are considered :

(1) Von Neumann polynomials, the zeros of which are in the closed unit disk ($|z| \leq 1$).

(2) Schur polynomials, the zeros of which are in the open unit disk ($|z| < 1$).

Other types of polynomials are also considered by Miller (1971).

Let us consider the polynomial of degree q :

$$f(z) = a_0 + a_1 z + \ldots + a_q z^q = \sum_{j=0}^{q} a_j z^j, \tag{C.1}$$

where the coefficients can be complex numbers. It is assumed that $a_0 \neq 0$ and $a_q \neq 0$.

First, we define the polynomial $\tilde{f}(z)$ constructed from (C.1) according to

$$\tilde{f}(z) = \bar{a}_0 z^q + \bar{a}_1 z^{q-1} + \ldots + \bar{a}_q = \sum_{j=0}^{q} \bar{a}_{q-j} z^j, \tag{C.2}$$

where $\bar{a}_i$ is the complex conjugate of a_i. Then we consider the polynomial $f_1(z)$ defined by

$$f_1(z) = \frac{1}{z} \left[\tilde{f}(0) f(z) - f(0) \tilde{f}(z) \right] \tag{C.3}$$

whose degree is lower than q (at most equal to $q-1$).

Theorem 1. *The polynomial f is a Von Neumann polynomial iff either :*

(i) $|\tilde{f}(0)| > |f(0)|$ *and f_1 is a Von Neumann polynomial,*

or

(ii) $f_1 \equiv 0$ *and f' is a Von Neumann polynomial.*

Theorem 2. *The polynomial f is a Schur polynomial iff $|\tilde{f}(0)| > |f(0)|$ and f_1 is a Schur polynomial.*

By using these theorems, the study of the zeros of the polynomial f is reduced to the study of a polynomial f_1 with lower degree. This can be repeated until this polynomial is sufficiently simple. Obviously, if the degree of f is large, the successive conditions obtained by a repeated application of the theorem could be too complicated to be useful and numerical calculations are then needed.

References

Augenbaum, J.M. (1989) An adaptive pseudospectral method for discontinuous problems. Appl. Numer. Math. **5**, 459-480.

Auteri, F., Quartapelle, L. (1999) Galerkin spectral method for the vorticity and stream function equations. J. Comput. Phys. **149**, 306-332.

Azaïez, M., Bernardi, C., Grundmann, M. (1994a) Spectral methods applied to porous media equations. East-West J. Numer. Math. **2**, 91-105.

Azaïez, M., Fikri, A., Labrosse, G. (1994b) A unique grid spectral solver of the nD Cartesian unsteady Stokes system. Illustrative numerical results. Finite Element Anal. Design **16**, 247-260.

Babuška, I., Oh, H.-S. (1990) The p-version of the finite element method for domains with corners and for infinite domains. Numer. Methods Partial Diff. Eqs. **6**, 371-392.

Babuška, I., Suri, M. (1987) The optimal convergence rate of the p-version of the finite element method. SIAM J. Numer. Anal. **24**, 750-776.

Babuška, I., Suri, M. (1994) The p and h-p versions of the finite element method, basic principles and properties. SIAM Review **36**, 578-632.

Balachandar, S., Madabhushi, R.K. (1994) Spurious modes in spectral methods with two non-periodic directions. J. Comput. Phys. **113**, 151-153.

Barrett, R., Berry, M., Chan, T., Demmel, J., Donato, J., Dongarra, J., Eijkhout, V., Pozo, R., Romine, C., Van der Vorst, H. (1994) Templates for the Solution of Linear Systems Building Blocks for Iterative Methods. SIAM, Philadelphia.

Barthels, R.H., Stewart, G.W. (1972) Solution of the matrix equation $AX + XB = C$. Comm. ACM **15**, 820-826.

Basdevant, C. (1982) Le modèle de simulation numérique de turbulence bidimensionnelle du L.M.D. Note interne LMD No.114, Laboratoire de Météorologie Dynamique du CNRS, Paris.

Basdevant, C., Deville, M.O., Haldenwang, P., Lacroix, J.-M., Ouazzani, J., Peyret, R., Orlandi, P., Patera, A.T. (1986) Spectral and finite difference solutions of the Burgers equation. Computers and Fluids **14**, 23-41.

Batoul, A., Khallouf, H., Labrosse, G. (1994) Une méthode de résolution directe (pseudo-spectrale) du problème de Stokes instationnaire. Application à la cavité entraînée carrée. C. R. Acad. Sci. Paris **319(II)**,1455-1461.

Bayliss, A. (1978) On the use of co-ordinate stretching in the numerical computation of high frequency scattering. J. Sound Vibr. **60**, 543-553.

Bayliss, A., Belytschko, T., Hansen, D., Turkel, E. (1992) Adaptive Multi-Domain Spectral Methods. In: Keyes, D.E., Chan, T.F., Meurant, G., Scroggs, J.S., Voigt, R.G. (eds.): Fifth SIAM Conference on Domain Decomposition Methods for Partial Differential Equations. SIAM, Philadelphia, pp.195-203.

Bayliss, A., Class, A., Matkowsky, B.J. (1994) Roundoff error in computing derivatives using the Chebyshev differentiation matrix. J. Comput. Phys. **116**, 380-383.

Bayliss, A., Class, A., Matkowsky, B.J. (1995a) Adaptive approximation of solutions to problems with multiple layers by Chebyshev pseudo-spectral methods. J. Comput. Phys. **116**, 160-172.

Bayliss, A., Garbey, M., Matkowsky, B.J. (1995b) Adaptive pseudo-spectral domain decomposition and the approximation of multiple layers. J. Comput. Phys. **119**, 132-141.

Bayliss, A., Gottlieb, D., Matkowsky, B.J. (1989) An adaptive pseudo-spectral method for reaction diffusion problems. J. Comput. Phys. **81**, 421-443.

Bayliss, A., Matkowsky, B.J. (1987) Fronts, relaxation oscillations and period doubling in solid fuel combustion. J. Comput. Phys. **71**, 147-168.

Bayliss, A., Turkel, E. (1992) Mappings and accuracy for Chebyshev pseudo-spectral approximations. J. Comput. Phys. **101**, 349-359.

Bernardi, C., Canuto, C., Maday, Y. (1988) Generalized inf-sup conditions for Chebyshev spectral approximation of the Stokes problem. SIAM J. Numer. Anal. **28**, 1237-1271.

Bernardi, C., Canuto, C., Maday, Y., Metivet, B. (1990) Single-grid spectral collocation for the Navier-Stokes equations. IMA J. Numer. Anal. **10**, 253-297.

Bernardi, C., Maday, Y. (1988) A collocation method over staggered grids for the Stokes problem. Int. J. Numer. Methods Fluids **8**, 537-557.

Bernardi, C., Maday, Y. (1991) Polynomial approximation of some singular functions. Applicable Anal. **42**, 1-32.

Bernardi, C., Maday, Y. (1992) Approximations Spectrales de Problèmes aux Limites Elliptiques. Springer-Verlag, Paris.

Bjørstadt, P., Widlund, O.B. (1986) Iterative methods for the solution of elliptic problems on regions partitioned into substructures. SIAM J. Numer. Anal. **23**, 1097-1120.

Blaisdell, G.A., Spyropoulos, E.T., Qin, J.H. (1996) The effect of the formulation of the nonlinear terms on aliasing errors in spectral methods. Appl. Numer. Math. **21**, 207-219.

Blum, E.K. (1962) A modification of the Runge-Kutta fourth-order method. Math. Comput. **16**, 176-187.

Blumen, W. (1970) Shear layer instability in an inviscid compressible fluid. J. Fluid Mech. **40**, 769-781.

Borchers, W., Forestier, M.Y., Kräutle, S., Pasquetti, R., Peyret, R., Rautmann, R., Ross, N., Sabbah, C. (1998) A Parallel Hybrid Highly Accurate Elliptic Solver for Viscous Flow Problems. In: Hirschel, E.H. (ed): Numerical Flow Simulation I. Vieweg, Braunschweig, pp.3-24.

Botella, O. (1997) On the solution of the Navier-Stokes equations using projection schemes with third-order accuracy in time. Computers and Fluids **26**, 107-116.

Botella, O. (1998) Résolution numérique de problèmes de Navier-Stokes singuliers par une méthode de projection Tchebychev. Thèse de Doctorat, Université de Nice-Sophia Antipolis.

Botella, O., Forestier, M.Y., Pasquetti, R., Peyret, R., Sabbah, C. (2001) Chebyshev methods for the Navier-Stokes equations : Algorithms and applications. Nonlinear Anal. **47**, 4157-4168.

Botella, O., Peyret, R. (1998) Benchmark spectral results on the lid-driven cavity flow. Computers and Fluids **27**, 421-433.

Botella, O., Peyret, R. (2001) Computing singular solutions of the Navier-Stokes equations with the Chebyshev-collocation method. Int. J. Numer. Methods Fluids **36**, 125-163.

Boyd, J.P. (1982) The optimization of convergence for Chebyshev polynomial methods in an unbounded domain. J. Comput. Phys. **45**, 43-79.

Boyd, J.P. (1986) An analytical and numerical study of the two-dimensional Bratu equation. J. Scient. Comput. **1**, 183-206.

Boyd, J.P. (1987a) Spectral methods using rational basis functions on an infinite interval. J. Comput. Phys. **69**, 112-142.

Boyd, J.P. (1987b) Orthogonal rational functions on a semi-infinite interval. J. Comput. Phys. **70**, 63-88.

Boyd, J.P. (1989) Chebyshev and Fourier Spectral Methods. Springer, Berlin.

Boyd, J.P. (1992) The arctan/tan and Kepler-Burgers mappings for periodic solutions with a shock, front or internal boundary layer. J. Comput. Phys. **98**, 181-193.

Boyd, J.P. (1996) The Erfc-Log Filter and the Asymptotics of the Euler and Vandeven Sequence Acceleration. In: Ilin, A.V., Scott, L.R. (eds.): Third ICOSAHOM. J. of Mathematics, Houston, pp.267-275.

Boyd, J.P. (1998) Two comments on filtering (artificial viscosity) for Chebyshev and Legendre spectral and spectral element methods : preserving boundary conditions and interpretation of the filter as a diffusion. J. Comput. Phys. **143**, 283-288.

Breuer, K.S., Everson, R.M. (1992) On the errors incurred calculating derivatives using Chebyshev polynomials. J. Comput. Phys. **99**, 56-67.

Brigham, E.O. (1988) The Fast Fourier Transform and its Applications. Prentice-Hall, Englewood Cliffs, NJ.

Bwemba, R. (1994) Résolution numérique des formulations $\Omega - \Psi$ des équations de Stokes et de Navier-Stokes par méthode spectrale. Thèse de doctorat, Université de Nice-Sophia Antipolis.

Bwemba, R., Pasquetti, R. (1995) On the influence matrix used in the spectral solution of the 2D Stokes problem (vorticity-stream function formulation). Appl. Numer. Math. **16**, 299-315.

Cahouet, J., Chabard, J.P. (1988) Some fast 3-D finite element solvers for the generalized Stokes problem. Int. J. Numer. Methods Fluids **8**, 869-895.

Canuto, C., Hussaini, M.Y., Quarteroni, A., Zang, T.A. (1988) Spectral Methods in Fluid Dynamics. Springer, New York.

Canuto, C., Quarteroni, A. (1985) Preconditioned minimal residual methods for Chebyshev spectral calculations. J. Comput. Phys. **60**, 315-337.

Canuto, C., Sacchi-Landriani, G. (1986) Analysis of the Kleiser-Schumann method. Numer. Math. **50**, 217-243.

Carlenzoli, C., Quarteroni, A. (1995) Adaptive Domain-Decomposition Methods for Advection-Diffusion Problem. In: Babuška, I. *et al.* (eds): Modeling, Mesh Generation and Adaptive Numerical Methods for Partial Differential Equations. IMA Volumes in Mathematics and its Applications, Vol.75. Springer, New York, pp.165-199.

Carpenter, M.H., Gottlieb, D., Abarbanel, S. (1994) Time-stable boundary conditions for finite-difference schemes solving hyperbolic systems : methodology and application to high-order compact schemes. J. Comput. Phys. **111**, 220-236.

Carpenter, M.H., Kennedy, C.A. (1994) A fourth-order 2N-storage Runge-Kutta scheme. NASA TM 109112.

Chan, T.F., Gallopoulos, E., Simoncini, V., Szeto, T., Tong, C.H. (1994) A quasi-minimal variant of the Bi-CGSTAB algorithm for nonsymmetric systems. SIAM J. Sci. Comput. **15**, 338-347.

Chan, T.F., Goovaerts, D. (1989/90) Schur complement domain decomposition algorithms for spectral methods. Appl. Numer. Math. **6**, 53-64.

Chan, T.F., Mathew, T.R. (1994) Domain Decomposition Algorithms. In: Acta Numerica, Cambridge University Press, London, pp. 61-143.

Chaouche, A. (1990) Une méthode de collocation-Chebyshev pour la simulation des écoulements axisymétriques dans un domaine annulaire en rotation. Rech. Aérosp. 1990-5, 1-13.

Chaouche, A., Randriamampianina, A., Bontoux, P. (1990) A collocation method based on the influence matrix technique for Navier-Stokes problems in annular domains. Comput. Methods Appl. Mech. Engrg. **80**, 237-244.

Chorin, J.A. (1968) Numerical solution of the Navier-Stokes equations. Math. Comput. **22**, 745-762.

Clercx, H.J.H. (1997) A spectral solver for the Navier-Stokes equations in the velocity-vorticity formulation for flows with two nonperiodic directions. J. Comput. Phys. **137**, 186-211.

Collatz, L. (1966) The Numerical Treatment of Differential Equations. Springer, Berlin.

Cooley, J.W., Tukey, J.W. (1965) An algorithm for the machine calculation of complex Fourier series. Math. Comput. **19**, 297-301.

Couzy, W., Deville, M.O. (1995) A fast Schur complement method for the spectral element discretization of the incompressible Navier-Stokes equations. J. Comput. Phys. **116**, 135-142.

Crespo del Arco, E., Maubert, P., Randriamampianina, A., Bontoux, P. (1996) Spatio-temporal behaviour in a rotating annulus with a source-sink flow. J. Fluid Mech. **328**, 271-296.

Crouzeix, M. (1980) Une méthode multipas implicite-explicite pour l'approximation des équations d'évolution paraboliques. Numer. Math. **35**, 257-276.

Dahlquist, G. (1963) A special stability problem for linear multistep methods. BIT **3**, 27-43.

Davis, P.J., Rabinowitz, P. (1984) Methods of Numerical Integration. Academic Press, New York.

Dean, E.J., Glowinski, R., Pironneau, O. (1991) Iterative solution of the stream function-vorticity formulation of the Stokes problem. Applications to the numerical simulation of incompressible viscous flow. Comput. Methods Appl. Mech. Engrg. **87**, 117-155.

Dennis, S.C.R., Quartapelle, L. (1983) Direct solution of the vorticity-stream function ordinary differential equations by a Chebyshev approximation. J. Comput. Phys. **52**, 448-463.

Dennis, S.C.R., Quartapelle, L. (1985) Spectral algorithms for vector elliptic equations in a spherical gap. J. Comput. Phys. **61**, 218-241.

Dennis, J.E. Jr, Schnabel, R.B. (1983) Numerical Methods for Unconstrained Optimization and Nonlinear Equations. Prentice-Hall, Englewood Cliffs, NJ.

Deville, M.O., Mund, E. (1985) Chebyshev pseudospectral solution of second-order elliptic equations with finite element preconditioning. J. Comput. Phys. **60**, 517-533.

Deville, M.O., Mund, E. (1990) Finite element preconditioning for pseudospectral solutions of elliptic problems. SIAM J. Sci. Stat. Comput. **11**, 311-342.

Dimitropoulos, C.D., Beris, A.N. (1997) An efficient and robust spectral solver for nonseparable elliptic equations. J. Comput. Phys. **133**, 186-191.

Don, W.S., Gottlieb, D. (1990) Spectral simulation of an unsteady compressible flow past a circular cylinder. Comput. Methods Appl. Mech. Engrg. **80**, 39-58.

Dubiner, M. (1991) Spectral methods on triangle and others domains. J. Sci. Comput. **6**, 345-390.

Ehrenstein, U. (1986) Méthodes spectrales de résolution des équations de Stokes et de Navier-Stokes. Application à des écoulements de convection double-diffusive. Thèse de Doctorat, Université de Nice-Sophia Antipolis.

Ehrenstein, U. (1990) The spectrum of a Chebyshev-Fourier approximation for the Stokes equations. J. Sci. Comput. **5**, 55-84.

Ehrenstein, U., Peyret, R. (1986) A collocation Chebyshev method for solving Stokes-type equations. In: Bristeau, M.O., Glowinski, R., Hauguel, A., Périaux, J. (eds.): Sixth Int. Symp. Finite Elements in Flow Problems. INRIA, pp. 213-218.

Ehrenstein, U., Peyret, R. (1989) A Chebyshev collocation method for the Navier-Stokes equations with application to double-diffusive convection. Int. J. Numer. Methods Fluids **9**, 427-452.

Eisenstat, S.C., Elman, A.C., Schultz, M.H. (1983) Variational iterative methods for nonsymmetric systems of linear equations. SIAM J. Numer. Anal. **20**, 345-357.

Elghaoui, M., Pasquetti, R. (1999) Mixed spectral-boundary element embedding algorithms for the Navier-Stokes equations in the vorticity-stream-function formulation. J. Comput. Phys. **153**, 82-100.

Elliott, D. (1961) A method for the numerical integration of the one-dimensional heat equation using Chebyshev series. Proc. Cambridge Philos. Soc. **57**, 823-832.

408   References

Erlebacher, G., Hussaini, M.Y., Kreiss, K.O., Sarkar, S. (1990) The analysis and simulation of compressible turbulence. Theoret. Comput. Fluid Dynamics 2, 73-95.

Erlebacher, G., Hussaini, M.Y., Shu, C.-W. (1997) Interaction of the shock with a longitudinal vortex. J. Fluid Mech. 337, 129-154.

Farhat, C., Roux, F.-X. (1994) Implicit parallel processing in structural mechanics. Comput. Mech. Adv. 2, 1-124.

Ferziger, J.H. (1981) Numerical Methods for Engineering Application. J Wiley and Sons, New York.

Finlayson, B.A. (1972) The Method of Weighted Residuals and Variational Principles. Academic Press, New York.

Forestier, M.Y. (2000) Etude par méthode spectrale de sillages tridimensionnels en fluides stratifiés. Thèse de Doctorat, Université de Nice-Sophia Antipolis.

Forestier, M.Y., Pasquetti, R., Peyret, R., Sabbah, C. (2000a) Spatial development of wakes using a spectral multi-domain method. Appl. Numer. Math. 33, 207-216.

Forestier, M.Y., Pasquetti, R., Peyret, R. (2000b) Calculations of 3D wakes in stratified fluids. In : European Congress on Computational Methods in Applied Sciences and Engineering, ECCOMAS 2000, Barcelona, 11-14 September (CD-Rom).

Fortin, M., Peyret, R., Temam, R. (1971) Résolution numérique des équations de Navier-Stokes pour un fluide incompressible. J. Méca. 10, 357-390.

Fox, L., Parker, I.B. (1968) Chebyshev Polynomials in Numerical Analysis. Oxford University Press, London.

Fröhlich, J., Gauthier, S. (1993) Numerical investigations from compressible to isobaric Rayleigh-Bénard convection in two dimensions. Europ. J. Mech., B/Fluids 12, 141-159.

Fröhlich, J., Gerhold, T., Lacroix, J.M., Peyret, R. (1991) Fully Implicit Spectral Methods for Convection. In: Durand, M., El Dabaghi, F. (eds.): High Performance Computing II. North-Holland, Amsterdam, pp.585-596.

Fröhlich, J., Laure, P., Peyret, R. (1992) Large departures from Boussinesq approximation in the Rayleigh-Bénard problem. Physics of Fluids A4, 1355-1372.

Fröhlich, J., Peyret, R. (1990) Calculations of non-Boussinesq convection by a pseudospectral method. Comput. Methods Appl. Mech. Engrg. **80**, 425-433.

Fröhlich, J., Peyret, R. (1992) Direct spectral methods for the low Mach number equations. Int. J. Numer. Methods Heat Fluid Flow **2**, 195-213.

Fulton, S.R., Taylor, G.D. (1984) On the Gottlied-Turkel time filter for Chebyshev spectral methods. J. Comput. Phys. **55**, 302-312.

Funaro, D. (1988) Domain decomposition methods for pseudospectral approximation, I : second order equations in one dimension. Numer. Math. **52**, 329-344.

Funaro, D. (1992) Polynomial Approximation of Differential Equations. Springer, Berlin.

Funaro, D. (1997) Spectral Elements for Transport-Dominated Equations. Springer, Berlin.

Funaro, D., Gottlieb, D. (1988) A new method of imposing boundary conditions in pseudospectral approximations of hyperbolic equations. Math. Comput. **51**, 599-613.

Funaro, D., Quarteroni, A., Zanolli, P. (1988) An iterative procedure with interface relaxation for domain decomposition methods. SIAM J. Numer. Anal. **25**, 1213-1236.

Fyfe, D.J. (1966) Economical evaluation of Runge-Kutta formulae. Math. Comput. **20**, 392-398.

Gallopoulos, E., Saad, Y. (1989) Some fast elliptic solvers on parallel architectures and their complexities. Int. J. High Speed Comput. **1**, 113-141.

Garbey, M. (1994) Domain decomposition to solve transition layers and asymptotics. SIAM J. Sci. Comput. **15**, 866-891.

Gauthier, S. (1988) A spectral collocation method for two-dimensional compressible convection. J. Comput. Phys. **75**, 217-235.

Gauthier, S. (1991) A semi-implicit collocation method : Application to thermal convection in 2D compressible fluids. Int. J. Numer. Methods Fluids **12**, 985-1002.

Gauthier, S., Guillard, H., Lumpp, T., Malé. J.-M., Peyret, R., Renaud, F. (1996) A Spectral Domain Decomposition Method with Moving Interfaces for Viscous Compressible Flows. In: Désidéri, J.A.*et al.* (eds.): Computational Fluid Dynamics'96. J. Wiley and Sons, Chichester, pp.839-844.

Gavrilakis, S., Machiels, L., Deville, M.O. (1996) A Spectral Element Method for Direct Simulation of Turbulent Flows with Two Inhomogeneous Directions. In: Désidéri, J.A. *et al.* (eds.): Computational Fluid Dynamics'96. J. Wiley and Sons, Chichester, pp.1080-1084.

Gear, C.W. (1971) Numerical Initial Value Problems in Ordinary Differential Equations. Prentice-Hall, Englewood Cliffs, NJ.

Gervasio, P., Ovtchinnikov, E., Quarteroni, A. (1997) The spectral projection decomposition method for elliptic equations in two dimensions. SIAM J. Numer. Anal. **34**, 1616-1639.

Glowinski, R., Pironneau, O. (1979) Numerical methods for the first biharmonic equation and for the two-dimensional Stokes problem. SIAM Review **21**, 167-212.

Goda, K. (1979) A multistep technique with implicit difference schemes for calculating two- or three-dimensional cavity flows. J. Comput. Phys. **30**, 76-95.

Gottlieb, D., Hussaini, M.Y., Orszag, S.A. (1984) Theory and Applications of Spectral Methods. In: Voigt, R.G., Gottlieb, D., Hussaini, M.Y. (eds.): Spectral Methods for Partial Differential Equations. SIAM, Philadelphia, pp.1-54 .

Gottlieb, D., Lutsman, L. (1983) The spectrum of the Chebyshev collocation operator for the heat equation. SIAM J. Numer. Anal. **20**, 909-921.

Gottlieb, D., Orszag, S.A. (1977) Numerical Analysis of Spectral Methods: Theory and Applications. Regional Conference Series in Applied Mathematics, Vol.28. SIAM, Philadelphia.

Gottlieb, D., Shu, C.-W. (1997) On the Gibbs phenomenon and its resolution. SIAM Review **39**, 644-668.

Grasso, F., Meola, C. (1996) Euler and Navier-Stokes Equations for Compressible Flows : Finite-Volume Methods. In: R. Peyret (ed): "Handbook of Computational Fluid Mechanics". Academic Press, London, pp.159-282.

Grisvard, P. (1985) Elliptic Problems in Nonsmooth Domains. Pitman, London.

Grosch, C.E., Orszag, S.A. (1977) Numerical solution of problems in unbounded regions : coordinate transforms. J. Comput. Phys. **25**, 273-296.

Guermond, J.-L. (1996) Some implementation of projection methods for Navier-Stokes equations. Math. Model. Numer. Anal. (M2AN) **30**, 637-667.

Guermond, J.-L. (1999) Un résultat de convergence d'ordre deux en temps pour l'approximation des équations de Navier-Stokes par une technique de projection incrémentale. Math. Model. Numer. Anal. (M2AN) **33**, 169-189.

Guermond, J.-L., Quartapelle, L. (1994) Equivalence of u-p and ζ-ψ formulations of the time-dependent Navier-Stokes equations. Int. J. Numer. Methods Fluids **18**, 471-487.

Guermond, J.-L., Quartapelle, L. (1998) On the approximation of the unsteady Navier-Stokes equations by finite element projection methods. Numer. Math. **80**, 207-238.

Guillard, H., Désidéri, J.A. (1990) Iterative methods with spectral preconditioning for elliptic equations. Comput. Methods Appl. Mech. Engrg. **80**, 305-312.

Guillard, H., Malé, J.-M., Peyret, R. (1992) Adaptative spectral methods with application to mixing layer computations. J. Comput. Phys. **102**, 114-127.

Guillard, H., Peyret, R. (1988) On the use of the spectral methods for the numerical solution of stiff problems. Comput. Methods Appl. Mech. Engrg. **66**, 17-43.

Gupta, M.M., Manohar, R.P., Noble, B. (1981) Nature of viscous flows near sharp corners. Computers and Fluids **9**, 379-388.

Haidvogel, D.B. (1977) Quasigeostrophic Regional and General Circulation Modelling : an Efficient Pseudospectral Approximation Technique. In: Shaw, R.P. (ed.): Computing Methods in Geophysical Mechanics Vol.25. ASME, New York, pp.131-153.

Haidvogel, D.B. (1979) Resolution of downstream boundary layers in the Chebyshev approximation to viscous flow problems. J. Comput. Phys. **33**, 313-324.

Haidvogel, D.B., Zang, T.A. (1979) The accurate solution of Poisson's equation by expansion in Chebyshev polynomials. J. Comput. Phys. **30**, 167-180.

Hairer, E., Nørsett, S.P., Wanner, G. (1987) Solving Ordinary Differential Equations I. Nonstiff Problems. Springer,Berlin.

Hairer, E. and Wanner, G. (1991) Solving Ordinar Differential Equations II. Stiff and Differential-Algebraic Problems. Springer, Berlin.

Haldenwang, P., Labrosse, G., Abboudi, S., Deville, M.O. (1984) Chebyshev 3-D spectral and 2-D pseudospectral solvers for the Helmholtz equation. J. Comput. Phys. **55**, 115-128.

Halmos, P. (1958) Finite Dimensional Vector Spaces. Van Nostrand, Princeton, NJ.

Hancock, C., Lewis, E., Moffatt, H.K. (1981) Effect of inertia in forced corner flows. J. Fluid Mech. **112**, 315-327.

Hansen, E.B., Kelmanson, M.A. (1994) An integral equation justification of the boundary conditions of the driven cavity problems. Computers and Fluids **26**, 225-240.

Heinrichs, W. (1993) Splitting techniques for the pseudospectral aproximation of the unsteady Stokes equations. SIAM J. Numer. Anal. **30**, 19-39.

Heinrichs, W. (1998) Splitting techniques for the unsteady Stokes equations. SIAM J. Numer. Anal. **35**, 1646-1662.

Hesthaven, J.S., Gottlieb, D. (1996) A stable penalty method for the compressible Navier-Stokes equations. I. Open boundary conditions. SIAM J. Sci. Comput. **17**, 579-612.

Heywood, J.C., Rannacher, R. (1982) Finite element approximation of the nonstationary Navier-Stokes problem. I : Regularity of the solutions and second-order estimates for spatial discretization. SIAM J. Numer. Anal. **19**, 275-311.

Hirsch, C. (1988) Numerical Computation of Internal and External Flow, Vol.I : Fundamental of Numerical Discretization. J. Wiley and Sons, Chichester, UK.

Hugues, S., Randriamampianina, A. (1998) An improved projection scheme applied to pseudospectral methods for the incompressible Navier-Stokes equations. Int. J. Numer. Methods Fluids **28**, 501-521.

Hussaini, M.Y., Zang, T.A. (1982) Iterative spectral methods and spectral solutions to compressible flows. ICASE Report 82-40, NASA Langley Research Center, Hampton, VA.

Isaacson, E., Keller, H.B. (1966) Analysis of Numerical Methods. J. Wiley and Sons, New York.

Joseph, D.D. (1976) Stability of Fluid Motions II. Springer, Berlin.

Joslin, R.D., Streett, C.L., Chang, C.-L. (1992) Validation of a Three-Dimensional Incompressible Spatial Direct Numerical Simulation Code - a Comparison with Linear Stability and Parabolic Stability Theories for Boundary-Layer Transition on a Flat Plate. NASA TP-3205.

Karniadakis, G.E., Henderson, R.D. (1998) Spectral Element Methods for Incompressible Flows. In: Johnson, R.W. (ed.): Handbook of Fluid Dynamics. CRC Press, Boca Raton, FL, pp.29.1-29.41.

Karniadakis, G.E., Israeli, M., Orszag, S.A. (1991) High-order splitting methods for the incompressible Navier-Stokes equations. J. Comput. Phys. **97**, 414-443.

Karniadakis, G.E., Sherwin, S.J. (1999) Spectral/hp Element Methods for CFD. Oxford University Press, New York.

Kim, J., Moin, P. (1985) Application of a fractional step method to incompressible Navier-Stokes equations. J. Comput. Phys. **59**, 308-323.

Kleiser, L., Schumann, U. (1980) Treatment of Incompressibility and Boundary Conditions in 3-D Numerical Spectral Simulations of Plane Channel Flows. In: Hirschel, E.H. (ed.): Third GAMM Conference Numerical Methods in Fluid Mechanics. Vieweg, Braunschweig, pp.165-173.

Kopriva, D.A. (1987) A practical assessment of spectral accuracy for hyperbolic problems with discontinuities. J. Sci. Comput. **2**, 249-262.

Kopriva, D.A. (1994) Multidomain spectral solution of compressible viscous flows. J. Comput. Phys. **115**, 184-199.

Korczak, K.Z., Patera, A.T. (1986) Isoparametric spectral element method for solution of the Navier-Stokes equations in complex geometry. J. Comput. Phys. **62**, 361-382.

Körner, T.W. (1988) Fourier Analysis. Cambridge University Press, Cambridge, UK.

Kosloff, D., Tal-Ezer, H. (1993) A modified Chebyshev pseudospectral method with $O(N^{-1})$ time step restriction. J. Comput. Phys. **104**, 457-469.

Kräutle, S., Wielage, K. (2001) The CGBI method for viscous channel flows and its preconditioning. Nonlinear Anal. **47**, 4191-4203.

Kravchenko, A.G., Moin, P. (1997) On the effect of numerical errors in large eddy simulations of turbulent flows. J. Comput. Phys. **131**, 310-322.

Lacroix, J.-M., Peyret, R., Pulicani, J.-P. (1988) A Pseudospectral Multi-Domain Method for the Navier-Stokes Equations with Application to Double-Diffusive Convection. In: Deville, M.O. (ed.): Seventh GAMM Conference on Numerical Methods in Fluid Mechanics. Vieweg, Braunschweig, pp.167-174.

Lanczos, C. (1956) Applied Analysis. Prentice-Hall, Englewood Cliffs, NJ.

Larroudé, P., Raspo, I., Ouazzani, J., Peyret, R. (1994) Application of the Spectral Multi-Domain Method to Flow in Crystal Growth System. In: Wagner, S. *et al* (eds.): Computational Fluid Dynamics'94. J. Wiley and Sons, Chichester, pp. 964-970.

Le Marec, C., Guérin, R., Haldenwang, P. (1996) Collocation method for convective flow induced by directional solidification in a cylinder. Int. J. Numer. Methods Fluids **22**, 293-409.

Le Quéré, P. (1987) Etude de la transition à l'instationnarité des écoulements de convection naturelle en cavité verticale différentiellement chauffée par méthodes spectrales Chebyshev. Thèse de Doctorat, Université de Poitiers.

Le Quéré, P. (1989) Mono and Multidomain Chebyshev Algorithms on Staggered Grid. In: Chung, T.J., Karr, G.R. (eds.): Finite Element Analysis in Fluids. University of Alabama in Huntsville Press, pp.1434-1439.

Le Quéré, P. (1990) Transition to unsteady natural convection in a tall water-filled cavity. Phys. Fluids **A2**, 503-515.

Le Quéré, P., Alziary de Roquefort, T. (1985) Computation of natural convection in two-dimensional cavities with Chebyshev polynomials. J. Comput. Phys. **57**, 210-228.

Le Quéré, P., Masson, R., Perrot, P. (1992) A Chebyshev collocation algorithm for 2D non-Boussinesq convection. J. Comput. Phys. **103**, 320-335.

Lynch, R.E., Rice, J.R., Thomas, D.H. (1964) Direct solution of partial differential equations by tensor product method. Numer. Math. **6**, 185-199.

Macaraeg, M.G., Streett, C.L. (1986) Improvements in spectral collocation through a multiple domain technique. Appl. Numer. Math. **2**, 95-108.

Madabhushi, R.K., Balachandar, S., Vanka, S.P. (1993) A divergence-free Chebyshev collocation procedure for incompressible flows with two non-periodic directions. J. Comput. Phys. **105**, 199-206.

Maday, Y., Patera, A.T. (1989) Spectral Element Methods for the Incompressible Navier-Stokes Equations. In: Noor, A.K., Oden, J.T. (eds.): State-Of-The-Art Surveys, ASME, pp. 71-143.

Maday, Y., Patera, A.T., Ronquist, E.M. (1990) An operator-integration-factor splitting method for time-dependent problems : Application to incompressible fluid flow. J. Sci. Comput. **5**, 263-292.

Majda, A., McDonough, J., Osher, S. (1978) The Fourier method for non-smooth initial data. Math. Comput. **32**, 1041-1081.

Malé, J.-M. (1992) Calcul d'écoulements visqueux compressibles par méthodes spectrales auto-adaptatives. Application aux couches de mélange. Thèse de Doctorat, Université de Nice-Sophia Antipolis.

Malik, M.R., Zang, T.A., Hussaini, M.Y. (1985) A spectral method for the Navier-Stokes equations. J. Comput. Phys. **61**, 64-88.

Manna, M., Vacca, A. (1999) An efficient method for the solution of the incompressible Navier-Stokes equations in cylindrical geometries. J. Comput. Phys. **151**, 563-584.

Marini, L.D., Quarteroni, A. (1989) A relaxation procedure for domain decomposition methods using finite elements. Numer. Math. **55**, 575-598.

Marion, Y., Gay, B. (1986) Résolution des équations de Navier-Stokes par méthode pseudo-spectrale via une technique de coordination. In: Bristeau, M.-O., Glowinski, R., Hauguel, A., Periaux, J. (eds.): Sixth Int. Symp. Finite Elements Methods in Flow Problems, pp.239-243.

Marion, M., Temam, R. (1998) Navier-Stokes Equations , Theory and approximation. In: Ciarlet, P.G., Lions, J.-L. (eds.): Handbook of Numerical Analysis. North-Holland, Amsterdam, pp.503-688.

Marquillie, M., Ehrenstein, U. (2002) Numerical simulation of separating boundary-layer flow. Computers and Fluids, to appear.

Mercier, B. (1989) An Introduction to the Numerical Analysis of Spectral Methods. Springer, Berlin.

Mihaljan, J.M. (1962) A rigorous exposition of the Boussineq approximation to a thin layer of fluid. Astrophys. J. **136**, 1126-1133.

Miller, J.J.H. (1971) On the location of zeros of certain classes of polynomials with applications to numerical analysis. J. Inst. Math. Applcs. **8**, 379-406.

Mitchell, A.R., Griffiths, D.F. (1980) The Finite Difference Method in Partial Differential Equations. John Wiley and Sons, Chichester, UK.

Moffatt, H.K. (1964) Viscous and resistives eddies near a sharp corner. J. Fluid. Mech. **18**, 1-18.

Moffatt, H.K., Duffy, B.R. (1980) Local similarity solutions and their limitations. J. Fluid. Mech. **96**, 299-313.

Mofid, A. (1992) Application des méthodes spectrales à l'étude de l'effet d'une source de chaleur dans un fluide stratifié. Thèse de Doctorat, Université de Nice-Sophia Antipolis.

Mofid, A., Peyret, R. (1993) Stability of the Chebyshev collocation approximation to the advection-diffusion equation. Computers and Fluids **22**, 453-465.

Morchoisne, Y. (1979) Résolution des équations de Navier-Stokes par une méthode pseudo-spectrale en espace-temps. Rech. Aérosp. **1979-5**, 293-306.

Morchoisne, Y. (1981) Pseudo-spectral space-time calculations of incompressible viscous flows. AIAA paper No.81-0109.

Morchoisne, Y. (1983) Résolution des équations de Navier-Stokes par une méthode spectrale de sous-domaines. In: Third Int. Conf. Numerical Methods Sci. Engrg. GAMNI, Paris.

Nana Kouamen, C. (1992) Analyse du spectre et préconditionnement de l'opérateur d'advection-diffusion pseudospectral Tchebychev et Legendre: Applications à la stabilité temporelle et à la convergence des méthodes spectrales itératives à haut nombre de Reynolds. Thèse de Doctorat, Université de Paris-Sud.

Orlanski, I. (1976) A simple boundary condition for unbounded hyperbolic flows. J. Comput. Phys. **21**, 251-269.

Orszag, S.A. (1969) Numerical methods for the simulation of turbulence. Phys. of Fluids **12**(Suppl.II), 250-257.

Orszag, S.A. (1971) Numerical simulation of incompressible flows within simple boundaries I : Galerkin (spectral) representations. Stud. Appl. Math. **50**, 293-327.

Orszag, S.A. (1972) Comparison of pseudospectral and spectral approximations. Stud. Appl. Math. **51**, 253-259.

Orszag, S.A. (1980) Spectral methods for problems in complex geometries. J. Comput. Phys. **37**, 70-90.

Orszag, S.A., Israeli, M., Deville, M.O. (1986) Boundary conditions for incompressible flows. J. Sci. Comput. **1**, 75-111.

Orszag, S.A., Patera, A.T. (1983) Secondary instability of wall-bounded shear flow. J. Fluid Mech. **128**, 347-385.

Orszag, S.A., Patterson, J.S. Jr (1972) Numerical simulation of three-dimensional homogeneous isotropic turbulence. Phys. Rev. Letters **28**, 76-79.

Ouazzani, J., Peyret, R. (1984) A Pseudo-Spectral Solution of Binary Gas Mixture Flows. In: Pandolfi, M., Piva, R. (eds.): Fifth GAMM Conference on Numerical Methods in Fluid Mechanics. Vieweg, Braunschweig, pp.275-282.

Ouazzani, J., Peyret, R., Zakaria, A. (1986) Stability of Collocation-Chebyshev Schemes with Application to the Navier-Stokes Equations. In: Rues, D., Kordulla, W. (eds.): Sixth GAMM Conference on Numerical Methods in Fluid Mechanics. Vieweg, Braunschweig, pp.287-294.

Pagès, T., Pasquetti, R., Peyret, R. (1995) Interaction of vortices with free surfaces. Z. Angew. Math. Mech. **76**, 305-311.

Pasquetti, R., Bwemba, R. (1994) A spectral algorithm of the Stokes problem in vorticity vector-potential formulation and cylindrical geometry. Comput. Methods Appl. Mech. Engrg. **117**, 71-90.

Passot, T., Pouquet, A. (1987) Numerical simulation of compressible homogeneous flows in the turbulent regime. J. Fluid Mech. **181**, 441-466.

Patera, A.T. (1984) A spectral element method for fluid dynamics : laminar flow in a channel expansion. J. Comput. Phys. **54**, 468-488.

Patera, A.T. (1986) Fast direct Poisson solvers for high-order finite element discretizations in rectangularly decomposable domains. J. Comput. Phys. **65**, 474-480.

Pathria, D., Karniadakis, G.E. (1995) Spectral element methods for elliptic problems in nonsmooth domains. J. Comput. Phys. **122**, 83-95.

Peyret, R. (1990) The Chebyshev multidomain approach to stiff problems in fluid mechanics. Comput. Methods Appl. Mech. Engrg. **80**, 129-145.

Peyret, R. (2000) Introduction to High-Order Approximation Methods for Computational Fluid Dynamics. In: Peyret, R., Krause, E. (eds.): Advanced Turbulent Flow Computations. Springer, Wien-New York, pp.1-79.

Peyret, R., Taylor, T.D. (1983) Computational Methods for Fluid Flow. Springer, New York.

Phillips, T.N., Roberts, G.W. (1993) The treatment of spurious pressure modes in spectral incompressible flow calculations. J. Comput. Phys. **105**, 150-164.

Pinelli, A., Vacca, A., Quarteroni, A. (1997) A spectral multidomain method for the numerical simulation of turbulent flows. J. Comput. Phys. **136**, 546-548.

Polak, E. (1971) Computational Methods in Optimization. Academic Press, New York.

Press, W.H., Teukolsky, S.A., Vetterling, W.T., Flannery, B.P. (1992) Numerical Recipes in Fortran, 2nd ed. Cambridge University Press, Cambridge, UK.

Priymak, V.G. (1995) Pseudospectral algorithms for Navier-Stokes simulation of turbulent flows in cylindrical geometry with coordinate singularities. J. Comput. Phys. **118**, 366-379.

Pulicani, J.-P. (1988) A spectral multi-domain method for the solution of 1-D Helmholtz and Stokes-type equations. Computers and Fluids **16**, 207-215.

Pulicani, J.-P. (1994) Modélisation isotherme d'une interface diffusive en convection de double-diffusion. Int. J. Heat Mass Transfer **37**, 2835-2858.

Pulicani, J.-P., Ouazzani, J. (1991) A Fourier-Chebyshev pseudospectral method for solving steady 3-D Navier-Stokes equations in cylindrical cavities. Computers and Fluids **20**, 93-109.

Quarteroni, A. (1987) Domain decomposition techniques using spectral methods. Calcolo **24**, 141-177.

Quarteroni, A. (1989) Domain Decomposition Algorithms for the Stokes Equations. In: Chan, T.F., Glowinski, R., Périaux, J., Widlund, O.B. (eds.): Domain Decomposition Methods for Partial Differential Equations, Vol.2. SIAM, Philadelphia, pp.431-442.

Quarteroni, A. (1991) Domain decomposition and parallel processing for the numerical solution of partial differential equations. Survey Math. Ind. **1**, 75-118.

Quarteroni, A., Sacchi-Landriani, G. (1988) Domain decomposition pre-conditioners for the spectral collocation method. J. Scientific Comput. **3**, 45-75.

Quarteroni, A., Valli, A. (1994) Numerical Approximation of Partial Differential Equations. Springer, Berlin.

Quarteroni, A., Valli, A. (1999) Domain Decomposition Methods for Partial Differential Equations. Clarendon Press, Oxford, UK.

Quarteroni, A., Zampieri, E. (1992) Finite element preconditioning for Legendre spectral collocation approximations to elliptic equations and systems. SIAM J. Numer. Anal. **29**, 917-936.

Raspo, I. (2002) A direct spectral domain decomposition method for the computation of rotating flows in a T-shape geometry. Computers and Fluids, to appear.

Raspo, I., Hugues, S., Serre, E., Randriamampianina, A., Bontoux, P. (2002) Spectral projection methods for the simulation of complex three-dimensional rotating flows. Computers and Fluids, to appear.

Raspo, I., Ouazzani, J., Peyret, R. (1996) A spectral multidomain technique for the computation of the Czochralski melt configuration. Int. J. Numer. Methods Heat Fluid Flow **6**, 31-58.

Reddy, S.C., Trefethen, L.N. (1994) Pseudospectra of the convection-diffusion operator, SIAM J. Appl. Math. **6**, 1634-1649.

Renaud, F., Gauthier, S. (1997) A dynamical pseudo-spectral domain decomposition technique : Application to viscous compressible flows. J. Comput. Phys. **131**, 89-108.

Renaut, R., Fröhlich, J. (1996) A pseudospectral Chebyshev method for the 2D wave equation with domain stetching and absorbing boundary conditions. J. Comput. Phys. **124**, 324-336.

Rigal, A. (1988) Analyse de stabilité de problèmes mixtes de diffusion-convection. C. R. Acad. Sci. Paris **306** (Série I),803-808.

Rivlin, T.J. (1974) The Chebyshev Polynomials. John Wiley and Sons, New York.

Rogallo, R.S. (1977) An ILLIAC Program for the Numerical Simulation of Homogeneous Incompressible Turbulence. NASA TM 73-203.

Rothman, E.E. (1991) Reducing round-off error in Chebyshev pseudospectral computations. In: Durand, M., El Dabaghi, F. (eds.): High Performance Computing II. North-Holland, Amsterdam, pp.423-439.

Roy, P. (1982) Numerical Simulation of Homogeneous Anisotropic Turbulence. In: Krause, E. (ed): 8th Int Conf Numerical Methods in Fluid Dynamics. Springer, Berlin, pp.440-447.

Roy, P. (1986) Simulation numérique d'un champ turbulent homogène incompressible soumis à des gradients de vitesse moyenne. ONERA, Note Technique 1986-1.

Rubin, S.G., Khosla, P.K. (1977) Turbulent boundary layers with and without mass injection. Computers and Fluids **5**, 241-259.

Saad, Y. (1998) Enhanced Acceleration and Preconditioning Techniques. In: Hafez, M., Oshima, K. (eds.): Computational Fluid Dynamics Review 1998. World Scientific, Singapore, pp.478-487.

Sabbah, C. (2000) Etude par méthode spectrale multi-domaine et calcul parallèle d'écoulements de convection thermosolutale en cavité chauffée latéralement. Thèse de Doctorat, Université de Nice-Sophia Antipolis.

Sabbah, C., Pasquetti, R. (1998) A divergence-free multidomain spectral solver of the Navier-Stokes equations in geometries of high aspect ratio. J. Comput. Phys. **139**, 359-379.

Sabbah, C., Pasquetti, R., Peyret, R., Levitsky, V., Chaschechkin, Y.D. (2001) Numerical and laboratory experiments of sidewall heating thermoline convection. Int. J. Heat Mass Transfer **44**, 2681-2697.

Sacchi-Landriani, G. (1986) Convergence of the Kleiser Schuman method for the Navier-Stokes equations. Calcolo **23**, 1-24.

Schultz, W.W., Lee, N.Y., Boyd, J.P. (1989) Chebyshev pseudospectral method of viscous flows with corner singularities. J. Sci. Comput. **4**, 1-24.

Schumack, M.R., Schultz, W.W., Boyd, J.P. (1991) Spectral solution of the Stokes equations on nonstaggered grids. J. Comput. Phys. **94**, 30-58.

Serre, E., Hugues, S., Crespo del Arco, E., Randriamampianina, A., Bontoux, P. (2001) Axisymmetric and three-dimensional instabilities in an Ekman boundary layer flow. Int. J. Heat Fluid Flow **22**, 89-93.

Serre, E., Pulicani, J.-P. (2001) A 3D pseudospectral method for rotating flows in a cylinder. Computers and Fluids, **30**, 491-519.

Shen, J. (1994) Efficient spectral-Galerkin method, I. Direct solvers of second- and fourth-order equations using Legendre polynomials. SIAM J. Sci. Comput. **15**, 1489-1505.

Sherwin, S.J., Karniadakis, G.E. (1995) A triangular spectral element method; applications to the incompressible Navier-Stokes equations. Comput. Methods Appl. Mech. Engrg **123**, 189-229.

Shizgal, B. (2002) Spectral methods based on nonclassical basis functions: the advection-diffusion equation. Computers and Fluids, to appear.

Shu, C.-W. (1998) Essentially Non-Oscillatory and Weighted Essentially Non-Oscillatory Schemes for Hyperbolic Conservation Laws. In: Cockburn, B., Johnson, C., Shu, C.-W., Tadmor, E., Quarteroni, A. (eds.): Advanced Numerical Approximation of Nonlinear Hyperbolic Equations. Lecture Notes in Mathematics, Vol.1697, Springer, Berlin, pp.325-432.

Shu, C.-W., Osher, S. (1988) Efficient implementation of essentially non-oscillatory shock-capturing schemes. J. Comput. Phys. **77**, 439-471.

Sleijpen, G.L.G., Fokkema, D.R. (1993) Bi-CGSTAB(l) for linear equations involving unsymmetric matrices with complex spectrum. Electron. Trans. Numer. Anal. **1**, 11-32.

Solomonoff, A., Turkel, E. (1989) Global properties of pseudospectral methods. J. Comput. Phys. **81**, 239-276.

Streett, C.L., Hussaini, M.Y. (1986) Finite Length Effects in Taylor-Couette Flow. ICASE Report 86-59, NASA Langley Research Center, Hampton, VA.

Streett, C.L., Macaraeg, M.G. (1989/90) Spectral multi-domain for large-scale fluid dynamic simulations. Appl. Numer. Math. **6**, 123-139.

Taylor, G.I. (1962) On scraping viscous fluid from a plane surface. In: Schäfer, M. (ed.): Miszellanen der Angerwandten Mechanik. Akademic-Verlag, Berlin, pp.313-315.

Temam, R. (1969) Sur l'approximation de la solution des équations de Navier-Stokes par la méthode des pas fractionnaires II. Archiv. Rat. Mech. Anal. **32**, 377-385.

Temam, R. (1982) Behaviour at time $t = 0$ of the solutions of semi-linear evolution equations. J. Diff. Eqs. **43**, 73-92.

Temam, R. (1991) Remark on the pressure boundary condition for the projection method. Theoret. Comput. Fluid Dynamics **3**, 181-184.

Temperton, C. (1983) Self-sorting mixed-radix fast Fourier transforms. J. Comput. Phys. **52**, 1-23.

Thual, O. (1986) Transition vers la turbulence dans des systèmes dynamiques apparentés à la convection. Thèse de Doctorat, Université de Nice-Sophia Antipolis.

Timan, A.F. (1963) Theory of Approximation of Functions of a Real Variable. Pergamon Press, Oxford, UK.

Trujillo, J., Karniadakis, G.E. (1999) A penalty method for the vorticity-velocity formulation. J. Comput. Phys. **149**, 32-58.

Tuckerman, L. (1983) Formation of Taylor Vortices in Spherical Couette Flow. PhD thesis, MIT, Cambridge, MA.

Tuckerman, L. (1989) Divergence-free velocity fields in nonperiodic geometries. J. Comput. Phys. **80**, 403-441.

Vanderstraeten, D., Keunings, R. (1998) A parallel solver based on the dual Schur decomposition of general finite element matrices. Int. J. Numer. Methods Fluids **28**, 23-46.

Van der Vorst, H.A. (1992) Bi-CGSTAB : a fast and smoothly converging variant of BICG for the solution of nonsymmetric linear systems. SIAM J. Sci. Stat. Comput. **13**, 631-644.

Vandeven, H. (1991) Family of spectral filters for discontinuous problems. J. Sci. Comput. **6**, 159-192.

Vanel, J.-M., Peyret, R., Bontoux, P. (1986) A Pseudo-Spectral Solution of Vorticity-Stream Function Equations Using the Influence Matrix Technique. In: Morton, K.W., Baines, M.J. (eds.): Numerical Methods for Fluid Dynamics II. Clarendon Press, Oxford, pp.463-475.

Van Kan, J. (1986) A second-order accurate pressure-correction scheme for viscous incompressible flow. SIAM J. Sci. Stat. Comput. **7**, 870-891.

Van Loan, C. (1992) Computational Frameworks for the Fast Fourier Transform. SIAM, Philadelphia.

Vincent, A., Meneguzzi, M. (1991) The spatial structure and statistical properties of homogeneous turbulence. J. Fluid Mech. **225**, 1-20.

Weideman, J.A.C., Trefethen, L.N. (1988) The eigenvalues of second-order spectral differentiation matrices. SIAM J. Numer. Anal. **25**, 1279-1298.

Wengle, H. (1979) Numerical Solution of Advection-Diffusion Problems by Collocation Methods. In: Hirschel, E. (ed.): Third GAMM Conference on Numerical Methods in Fluid Mechanics. Vieweg, Braunschweig, pp.295-304.

Wengle, H., Seinfeld, J.H. (1978) Pseudospectral solution of atmospheric diffusion problems. J. Comput. Phys. **26**, 87-106.

Widlund, O.B. (1967) A note on unconditionally stable linear multistep methods. BIT **7**, 65-70.

Wigley, N.M. (1988) An efficient method for subtracting off singularities at corners for Laplace's equation. J. Comput. Phys. **78**, 369-377.

Wilhelm, D., Kleiser, L. (2000) Stable and unstable formulations of the convection operator in spectral element simulations. Appl. Numer. Math. **33**, 275-280.

Wilkinson, J.H. (1965) The Algebraic Eigenvalue Problem. Clarendon Press, Oxford, UK.

Williamson, J.H. (1980) Low-storage Runge-Kutta schemes. J. Comput. Phys. **35**, 48-56.

Wilson, R.V., Demuren, A.O., Carpenter, M. (1998) Higher-Order Compact Schemes for Numerical Simulation of Incompressible Flows. ICASE Report 98-13, NASA Langley Research Center, Hampton, VA.

Xu, C., Pasquetti, R. (2001) On the efficiency of semi-implicit and semi-Lagrangian spectral methods for the calculation of incompressible flows. Int. J. Numer. Methods Fluids **35**, 319-340.

Zang, T.A. (1991) On the rotation and skew-symmetric forms for incompressible flow simulations. Appl. Numer. Math. **7**, 27-40.

Zang, T.A., Hussaini, M.Y. (1985) Numerical Experiments on Subcritical Transition Mechanisms. AIAA paper 85-0296.

Index

Applied Mathematical Sciences

(continued from page ii)

(continued on next page)

Applied Mathematical Sciences